T0187983

QUATERNARY VEGETATION DYNAMICS –
THE AFRICAN POLLEN DATABASE

Palaeoecology of Africa

International Yearbook of Landscape Evolution and Palaeoenvironments

ISSN 2372-5907

Volume 35

Editor in Chief

J. Runge, Frankfurt, Germany

Editorial board

Quaternary Vegetation Dynamics – The African Pollen Database

Series Editor

Jürgen Runge

Goethe Universität, FB 11: Institut für Physische Geographie & Zentrum für interdisziplinäre Afrikaforschung (ZIAF), Frankfurt am Main, Germany

Guest Editors

William D. Gosling

Ecosystem & Landscape Dynamics, Institute for Biodiversity & Ecosystem Dynamics (IBED), Faculty of Science, University of Amsterdam, Amsterdam, The Netherlands

Anne-Marie Lézine

Laboratoire d'Océanographie et du Climat: expérimentations et modélisations numériques, Sorbonne Université, Paris, France

Louis Scott

Faculty of Natural and Agricultural Sciences, University of the Free State, Bloemfontein, Republic of South Africa

CRC Press
Taylor & Francis Group
Boca Raton London New York Leiden

CRC Press is an imprint of the
Taylor & Francis Group, an **informa** business

A BALKEMA BOOK

Published by:
CRC Press/Balkema
Schipholweg 107C, 2316 XC Leiden, The Netherlands

ISBN: 978-0-367-75508-9 (hbk)
ISBN: 978-1-003-16276-6 (eBook)
ISBN: 978-0-367-75510-2 (pbk)

Typeset by MPS Limited, Chennai, India

Visit the Taylor & Francis Web site at
http://www.taylorandfrancis.com

and the CRC Press Web site at
http://www.crcpress.com

Library of Congress Cataloging-in-Publication Data
A catalog record for this book has been requested

DOI: 10.1201/9781003162766

Contents

Foreword

The "Quaternary Vegetation Dynamics", and the 35th volume of *Palaeoecology of Africa* (PoA) was born out of two workshops organised to relaunch the African Pollen Database initiative in Bondy near Paris (October 2019) and Amsterdam (January 2020). These workshops brought together researchers from around the world to showcase their work from across the African continent and to train them in data management. The training at the workshop was led by the late Eric Grimm who sadly passed away in November 2020. We hope that Eric's enthusiasm for collating, standardising, and disseminating palaeoecologicial data is captured in this volume.

In total this volume of PoA, that for the first time is available as an OPEN ACCESS publication, contains twenty-four papers and is organised regionally following the format of the first meeting: western Africa (chapters 3–6), eastern Africa and Arabia (chapters 7–11), central Africa (chapters 12–13), southern Africa (chapters 14–19), islands in the vicinity of Africa (chapter 20), and pan-African syntheses (chapters 21–24). With a view to stimulate primary research and mobilise open access publication of palaeoecological data two types of primary research papers are included: (i) short data papers containing descriptions of new palaeoecological data sets (chapters 9, 18, 19, 20), and (ii) longer research papers providing more in-depth analysis (chapters 3, 4, 5, 7, 8, 12, 13, 15, 16 and 17). To provide a synthetic overview of the state of research into Quaternary Vegetation Dynamics from in and around Africa we also include short perspective papers to provide insights into key regions or issues (chapters 1, 2, 14, and 24), and extensive review papers which bring together the growing body of research from around the continent to address particular scientific questions (chapters 6, 10, 11, 21, 22 and 23).

We are delighted to have been able to bring together work from all parts of Africa focused on the Quaternary. We believe that this reflects the enthusiasm and urgency for this type of palaeoecological research. As current climatic and land-use changes continue to exert pressure on the modern ecosystems, turning to the record of past environmental change to shed light on their origins, resilience and trajectories of change becomes increasingly pertinent. While the record of past change cannot be used to define how we should manage and develop ecosystems today, we hope that it can provide a frame of reference and inspiration for managers and policy makers to work with. We hope that the data, research and syntheses provided in this volume forms a solid basis for the next generation of palaeoecological researchers from Africa to provide scientifically and societally relevant information that will assist with the development of sustainable management practices.

Finally, we would like to thank all the participants in the meetings, the authors and the reviewers whose contributions have made this volume possible. We would like to also thank the French Agence Nationale pour la Recherche and the Belmont-Forum (VULPES: ANR-15-MASC-0003 and ACCEDE: 18 BELM 0001 05) who have provided financial support for the meetings and publication of this volume. We also take this opportunity to place on record our gratitude to Senior Publisher Janjaap Blom and his team from Routledge/Taylor & Francis/CRC Press for the continuous support to PoA.

Guest Editors: William D. Gosling (Amsterdam),
Anne-Marie Lézine (Paris),
and Louis Scott (Bloemfontein)
Series Editor: Jürgen Runge (Frankfurt)
February 2021

Contributors

Gaston Achoundong

National Herbarium, IRAD, Yaoundé, Cameroon, Email: gachoundong@yahoo.fr
ORCID: https://orcid.org/0000-0001-6298-5763

Marion. K. Bamford

The Evolutionary Studies Institute, University of the Witwatersrand, Johannesburg Private Bag 3. Wits, 2050, South Africa, Email: marion.bamford@wits.ac.za
ORCID: https://orcid.org/0000-0003-0667-130X

Asmae Baqloul

Faculté des Sciences, Ibn Zohr University, B.P. 8106 Agadir, Morocco,
Email: asmaebaqloul@gmail.com
ORCID: https://orcid.org/0000-0002-7700-9535

Anna K. Behrensmeyer

Department of Paleobiology, Smithsonian Institution, P.O Box 37012, Washington, DC 20013-7012, USA, Email: behrensa@si.edu
ORCID: https://orcid.org/0000-0001-6857-368X

Lucas Bittner

Heisenberg Chair of Physical Geography, Technische Universität, Dresden, Dresden, Germany, Email: lucas.bittner@tu-dresden.de
ORCID: https://orcid.org/0000-0003-2521-5596

Mick. N. T. Bönnen

Institute for Biodiversity & Ecosystems Dynamics, University of Amsterdam, Amsterdam, The Netherlands, Email: mickbonnen@hotmail.com
ORCID: https://orcid.org/0000-0002-0867-9949

Raymonde Bonnefille

Honorary Director of Research, CNRS, CEREGE, Aix Marseille University, CNRS, IRD, INRAE, Coll. France, Technopole Arbois-Méditerranée,
13545 Aix en Provence cedex 4, France, Email: rbonnefille@orange.fr
ORCID: https://orcid.org/0000-0002-7912-9186

Ilham Bouimetarhan

Faculté des Sciences appliquées, CUAM, Ibn Zohr University, Agadir, Morocco; and MARUM-Center for Marine Environmental Sciences, University of Bremen, Leobener Str. 8, 28359 Bremen, Germany, Email: bouimetarhan@uni-bremen.de
ORCID: https://orcid.org/0000-0003-3369-3811

Benjamin Bourel

CEREGE, Aix Marseille University, CNRS, IRD, INRAE, Coll. France, Technopole Arbois-Méditerranée, 13545 Aix en Provence cedex 4, France, Email: benjaminbourel1@gmail.com
ORCID: https://orcid.org/0000-0001-8312-5471

Jane Bunting

Department of Geography, Geology and Environment, Faculty of Science and Engineering, University of Hull, UK, Email: m.j.bunting@hull.ac.uk
ORCID: https://orcid.org/0000-0002-3152-5745

Brian M. Chase

Institut des Sciences de l'Evolution-Montpellier (ISEM), University of Montpellier, Centre National de la Recherche Scientifique (CNRS), EPHE, IRD, 34095 Montpellier, France; and Department of Environmental and Geographical Science, University of Cape Town, Rondebosch, South Africa, Email: brian.chase@umontpellier.fr
ORCID: https://orcid.org/0000-0001-6987-1291

Manuel Chevalier

Institute of Earth Surface Dynamics, Geopolis, University of Lausanne, Switzerland; and Institute of Geosciences, Sect. Meteorology, Rheinische Friedrich-Wilhelms-Universität Bonn, Auf dem Hügel 20, 53121 Bonn, Germany, Email: chevalier.manuel@gmail.com
ORCID: https://orcid.org/0000-0002-8183-9881

Colin J. Courtney Mustaphi

Geoecology, Department of Environmental Sciences, University of Basel, Basel, Switzerland; Water Infrastructure and Sustainable Energy (WISE) Futures, Nelson Mandela African Institution of Science & Technology, Tengeru, Arusha, Tanzania, and York Institute for Tropical Ecosystems, Department of Environment and Geography, University of York, United Kingdom, Email: colin.courtney-mustaphi@york.ac.uk
ORCID: https://orcid.org/0000-0002-4439-2590

Frank Darius

Technische Universität Berlin, Institute for Ecology, Berlin, Germany, Email: frank.darius@fu-berlin.de

Michèle Dinies

Freie Universität Berlin, Institute of Geographical Sciences, Berlin, Germany; and German Archaeological Institute (DAI), Scientific Department of the Head Office, Berlin, Germany, Email: michele.dinies@fu-berlin.de
ORCID: https://orcid.org/0000-0003-2196-4418

Nadia du Plessis

Department of Environmental and Geographical Science, University of Cape Town, Rondebosch, South Africa, Email: dup.nadia@gmail.com
ORCID: https://orcid.org/0000-0003-0290-4866

Lydie Dupont

MARUM – Center for Marine Environmental Sciences, University of Bremen, Leobener Str. 8, DE 28359 Bremen, Germany, Email: ldupont@marum.de
ORCID: https://orcid.org/0000-0001-9531-6793

Tristan J. Duthie

Discipline of Geography, School of Agriculture, Earth and Environmental Science, University of KwaZulu-Natal, Pietermaritzburg, South Africa; and University of Suwon, South Korea, Email: tjduthie@gmail.com

Mekbib Fekadu

Department of Ecology, Philipps-Marburg University, Marburg, Germany; and Department of Biology, Addis Ababa University, Addis Ababa, Ethiopia, Email: mekbib_fekadu@yahoo.com

Jemma Finch

Discipline of Geography, School of Agricultural, Earth and Environmental Sciences, University of KwaZulu-Natal, Pietermaritzburg, South Africa, Email: finchj@ukzn.ac.za
ORCID: https://orcid.org/0000-0002-6678-6910

Marie-José Gaillard

Department of Biology and Environmental Science, Linnaeus University, Kalmar, Sweden, Email: marie-jose.gaillard-lemdahl@lnu.se
ORCID: https://orcid.org/0000-0002-2025-410X

Graciela Gil-Romera

Department of Ecology, Philipps-Marburg University, Marburg, Germany; and IPE-CSIC, Avda. Montañana, 1005, 50059, Zaragoza, Spain, Email: graciela.gil@ipe.csic.es
ORCID: https://orcid.org/0000-0001-5726-2536

Lindsey Gillson

Plant Conservation Unit, Department of Biological Sciences, University of Cape Town, South Africa, Email: lindsey.gillson@uct.ac.za
ORCID: https://orcid.org/0000-0001-9607-6760

Esther N. Githumbi

Department of Physical Geography and Ecosystem Science, Lund University, Lund, Sweden; and Department of Biology and Environmental Science, Linnaeus University, Kalmar, Sweden, Email: esther.githumbi@lnu.se
ORCID: https://orcid.org/0000-0002-6470-8986

William D. Gosling

Institute for Biodiversity and Ecosystems Dynamics, University of Amsterdam, Amsterdam, The Netherlands, Email: W.D.Gosling@uva.nl
ORCID: https://orcid.org/0000-0001-9903-8401

David Grady

Department of Geography and Earth Sciences, Aberystwyth University, Aberystwyth, United Kingdom, Email: gradydai524@gmail.com
ORCID: https://orcid.org/0000-0002-8286-5010

Zoë Groenewoud

Institute for Biodiversity and Ecosystems Dynamics, University of Amsterdam, Amsterdam, The Netherlands, Email: groenewoud.zoe@gmail.com

Torsten Haberzettl

Physical Geography, Institute of Geography and Geology, University of Greifswald, Germany, Email: torsten.haberzettl@uni-greifswald.de
ORCID: https://orcid.org/0000-0002-6975-9774

Donna Hawthorne

School of Geography and Sustainable Development, Irvine Building, University of St Andrews, UK, Email: dj43@st-andrews.ac.uk
ORCID: https://orcid.org/0000-0002-0104-473X

Christelle Hély

Institut des Sciences de l'Evolution de Montpellier, Université de Montpellier and Ecole Pratique des Hautes Etudes, Université PSL, Paris, France, Email: christelle.hely@ephe.psl.eu
ORCID: https://orcid.org/0000-0002-7549-3239

Trevor Hill

Discipline of Geography, School of Agricultural, Earth and Environmental Sciences, University of KwaZulu-Natal, Pietermaritzburg, South Africa, Email: HillT@ukzn.ac.za
ORCID: https://orcid.org/0000-0001-7015-6906

Philipp Hoelzmann

Freie Universität Berlin, Institute of Geographical Sciences, Berlin, Germany,
Email: philipp.hoelzmann@fu-berlin.de
ORCID: https://orcid.org/0000-0001-8709-8474

Henry Hooghiemstra

Institute for Biodiversity and Ecosystems Dynamics, University of Amsterdam, Amsterdam, The Netherlands, Email: H.Hooghiemstra@uva.nl
ORCID: https://orcid.org/0000-0003-2502-1644

Sarah J. Ivory

Department of Geosciences, Pennsylvania State University, University Park, USA,
Email: ivorysj@gmail.com
ORCID: https://orcid.org/0000-0003-4709-4406

Adele C.M. Julier

Plant Conservation Unit, University of Cape Town, Cape Town, South Africa,
Email: adelejulier@gmail.com
ORCID: https://orcid.org/0000-0003-1106-6008

Rahab N. Kinyanjui

Department of Earth Sciences, National Museums of Kenya, P.O. Box 40658-00100 Nairobi, Kenya; Department of Environmental & Geographical Science, University of Cape Town, Rondebosch 7701, South Africa; School of Geographical Sciences, East China Normal University, Shanghai 200241, China; and College of Geography and Environmental Sciences, Zhejiang Normal University, Jinhua 321004, China, Email: rkinyanji@gmail.com
ORCID: https://orcid.org/0000-0003-2032-8321

Stefan Kröpelin

University of Cologne, Institute of Prehistoric Archaeology, Africa Research Unit, Cologne, Germany, Email: s.kroe@uni-koeln.de
ORCID: https://orcid.org/0000-0002-6377-3401

Henry F. Lamb

Department of Geography and Earth Sciences, Aberystwyth University, Aberystwyth, United Kingdom; and Department of Botany, Trinity College Dublin, Dublin, Ireland, Email: hfl@aber.ac.uk
ORCID: https://orcid.org/0000-0003-0025-0766

Judicaël Lebamba

Département de Biologie, Université des Sciences et Techniques de Masuku, Franceville, Gabon, Email: jlebamba@yahoo.fr
ORCID: https://orcid.org/0000-0002-4890-4215

Kévin Lemonnier

Laboratoire d'Océanographie et du Climat, Expérimentation et Approche Numérique/IPSL, Sorbonne Université, CNRS-IRD-MNHN, Paris, France, Email: kevin.lemonnier@locean.ipsl.fr

Anne-Marie Lézine

Laboratoire d'Océanographie et du Climat, Expérimentation et Approche Numérique/IPSL, Sorbonne Université, CNRS-IRD-MNHN, Paris, France, Email: anne-marie.lezine@locean.ipsl.fr
ORCID: https://orcid.org/0000-0002-3555-5124

Furong Li

School of Ecology, Sun Yat-sen University, Guangzhou, China, Email: lifr5@mail.sysu.edu.cn
ORCID: https://orcid.org/0000-0001-8717-502X

Saúl Manzano

Plant Conservation Unit, University of Cape Town, Cape Town, South Africa, Email: saul.manzano.rodriguez@gmail.com
ORCID: https://orcid.org/0000-0002-5720-2768

Robert Marchant

York Institute for Tropical Ecosystems, Department of Environment and Geography, University of York, York, United Kingdom, Email: robert.marchant@york.ac.uk
ORCID: https://orcid.org/0000-0001-5013-4056

Laurent Marquer

Institute of Botany, University of Innsbruck, Innsbruck, Austria, and Research Group for Terrestrial Palaeoclimates, Max Planck Institute for Chemistry, Mainz, Germany, Email: Laurent.Marquer@uibk.ac.at
ORCID: https://orcid.org/0000-0002-5772-3782

Florence Mazier

Environmental Geography Laboratory, GEODE UMR-CNRS 5602, Université de Toulouse Jean Jaurès, Toulouse, France, Email: florence.mazier@univ-tlse2.fr
ORCID: https://orcid.org/0000-0003-2643-0925

Michael E. Meadows

Department of Environmental and Geographical Science, University of Cape Town, Rondebosch, South Africa; School of Geographic Sciences, East China Normal University, Shanghai, China; and College of Geography and Environmental Sciences, Zhejiang Normal University, China, Email: michael.meadows@uct.ac.za
ORCID: https://orcid.org/0000-0001-8322-3055

Crystal N.H. McMichael

Institute for Biodiversity and Ecosystems Dynamics, University of Amsterdam, Amsterdam, The Netherlands, Email: C.N.H.McMichael@uva.nl
ORCID: https://orcid.org/0000-0002-1064-1499

Georg Miehe

Department of Geography, Philipps-Marburg University, Marburg, Germany, Email: miehe@staff.uni-marburg.de
ORCID: https://orcid.org/0000-0002-9813-4909

Charlotte S. Miller

Leeds Trinity University, Leeds, United Kingdom, Email: lottie.miller2@gmail.com
ORCID: https://orcid.org/0000-0003-1547-1641

Reinder Neef

German Archaeological Institute (DAI), Scientific Department of the Head Office, Berlin, Germany, Email: reinder.neef@dainst.de

Glory Oden

Department of Plant and Ecological Studies, University of Calabar, Calabar, Nigeria, Email: gloryoden@unical.edu.ng
ORCID: https://orcid.org/0000-0002-3124-7853

Lars Opgenoorth

Department of Ecology, Philipps-Marburg University, Marburg, Germany, Email: Opgenoorth@uni-marburg.de
ORCID: https://orcid.org/0000-0003-0737-047X

Richard Potts

Human Origins Program, P.O. Box 37012, Smithsonian Institution, Washington, DC 20013-7012 USA, Email: pottsr@si.edu
ORCID: https://orcid.org/0000-0001-6008-0100

Lynne J. Quick

African Centre for Coastal Palaeoscience, Nelson Mandela University, Port Elizabeth, South Africa, Email: lynne.j.quick@gmail.com
ORCID Number: https://orcid.org/0000-0002-6735-7106

Andriantsilavo H.I. Razafimanantsoa

Plant Conservation Unit, University of Cape Town, Cape Town, South Africa, Email: tsilamiezaka3@gmail.com
ORCID: https://orcid.org/0000-0002-9933-6991

Estelle Razanatsoa

Plant Conservation Unit, Department of Biological Sciences, University of Cape Town, South Africa, Email: estelle.razanatsoa@uct.ac.za
ORCID: https://orcid.org/0000-0002-7219-1411

Hanane Reddad

L'Ecole Supérieure de Technologie & Faculté des Lettres et des Sciences Humaines, Sultan Moulay Slimane University, Beni Mellal, Morocco, Email: h.reddad@usms.ma
ORCID: https://orcid.org/0000-0002-6238-165X

Keith Richards

KrA Stratigraphic, 116 Albert Drive, Conwy, North Wales, LL31 9YY, United Kingdom; and Department of Geography and Planning, University of Liverpool, L69 7ZT, United Kingdom, Email: kr@paly.co.uk
ORCID: https://orcid.org/0000-0002-5158-8209

Eleonora Roding

Institute for Biodiversity and Ecosystems Dynamics, University of Amsterdam, Amsterdam, The Netherlands, Email: eleonora.roding@gmail.com

Lena Schimmel

Freie Universität Berlin, Institute of Geographical Sciences, Berlin, Germany,
Email: lena.schimmel@fu-berlin.de
ORCID: https://orcid.org/0000-0002-8084-2915

Lisa Schüler

Department of Palynology and Climate Dynamics, Albrecht-von-Haller Institute for Plant Sciences, Göttingen University, Göttingen, Germany, Email: lschuel@gwdg.de
ORCID: https://orcid.org/0000-0002-4116-1105

Louis Scott

Department of Plant Sciences, University of the Free State, Bloemfontein, South Africa, Email: ScottL@ufs.ac.za
ORCID: https://orcid.org/0000-0002-4531-0497

Paul Strobel

Physical Geography, Institute of Geography, Friedrich Schiller University Jena, Germany, Email: paul.strobel@uni-jena.de
ORCID: https://orcid.org/0000-0002-7860-5398

Shinya Sugita

Institute of Ecology, Tallinn University, Estonia, Email: sugita@tlu.ee
ORCID: https://orcid.org/0000-0002-3634-7095

Monique Tossou

Department of Plant Biology, University of Abomey-Calavi, Bénin, Email: tossoumonique@gmail.com
ORCID: http://orcid.org/0000-0003-2397-728X

Malika Virah-Sawmy

Humboldt-Universitat zu Berlin, Geography, Germany, Email: malikavs@gmail.com
ORCID: https://orcid.org/0000-0003-3646-5646

Martyn P. Waller

Department of Geography and Geology, Kingston University London, Kingston upon Thames, Surrey, UK, Email: martyn.waller@kingston.ac.uk
ORCID: https://orcid.org/0000-0003-3876-2492

Stephan Woodborne

iThemba LABS, Johannesburg, South Africa, Email: swoodborne@ tlabs.ac.za
ORCID: https://orcid.org/0000-0001-8573-8626

Michael Zech

Heisenberg Chair of Physical Geography, Technische Universität Dresden, Dresden, Germany, Email: mzech@msx.tu-dresden.de
ORCID: https://orcid.org/0000-0002-9586-0390

CHAPTER 1

Rise of the Palaeoecology of Africa series

Louis Scott

Department of Plant Sciences, University of the Free State, Bloemfontein, South Africa

ABSTRACT: The *Palaeoecology of Africa* series was founded by E.M. van Zinderen Bakker, a biologist and palynologist who emigrated from Europe to South Africa. Ever since its origin during the nineteen-sixties, it gave a reflection of the history and development of palaeoscience in connection with the African continent and played a role in promoting international collaboration with multidisciplinary reports, conference proceedings and other papers in this field.

1.1 INTRODUCTION

Palaeoecology of Africa (PoA) of which this is the 35th volume, is a series traditionally focusing on multi-disciplinary studies on palaeoenvironments of Africa, especially on more recent parts of geological time like the Neogene and Quaternary. The early PoA volumes reveal the development history of these palaeoscience aspects in Africa. Thanks to the pioneering and visionary efforts of Eduard Meine van Zinderen Bakker and the publisher A. A. Balkema, the early volumes of PoA was a unique academic initiative out of Africa in the nineteen sixties. The series served as a mouth piece for an international group of specialists at a time when the important role palaeoenvironmental changes did not yet receive as much attention as today. Palaeoscience was then, a little-known discipline for the continent except for a rich heritage of Palaoeozoic fossils and a few important hominid cranial finds such as at such as Taung, Florisbad and a few other sites (e.g. Dart 1925; Dryer 1935).

1.2 ORIGIN

PoA had its humble origins in the late nineteen-fifties and early sixties when van Zinderen Bakker, a Dutch naturalist who immigrated to South Africa in 1947 (Meadows 2015; Neumann and Scott 2018) started lecturing at the University of the Orange Free State in Bloemfontein (now shortened to University of the Free State). Here, he formed the Palynological Research Unit that was sponsored by the South African Council for Scientific and Industrial Research, and by the 1950's he was already involved in some of the first palynogical research projects in Africa, including modern pollen surveys in Southern Africa (Coetzee and van Zinderen Bakker 1952) and analysis of the hominin-bearing spring deposits of Florisbad near Bloemfontein (Dreyer 1935; van Zinderen Bakker 1957).

PoA started as eight soft-cover reports on pollen analysis entitled 'Palynology of Africa' in which van Zinderen Bakker reported on research news and activities in palynology and related aspects of palaeosciences in Africa covering the period of 1950–1963 (referenced in Neumann and Scott 2018). They were later re-published by Balkema, Cape Town in 1966 in book form as Volume 1 of *Palaeoecology of Africa* with the sub-title *and the Surrounding Islands &*

DOI 10.1201/9781003162766-1

1

Antarctica (Van Zinderen Bakker 1966). The sub-title reflected van Zinderen Bakker's wide multi-disciplinary interests, which included biological and palynological research at the Marion and Prince Edward Islands, *c.* 2000 km south east of Cape Town in the Southern Ocean, to which he organized the first biological and geological research expedition in 1965 (van Zinderen Bakker 1976a; Van Zinderen Bakker *et al.* 1971). PoA Volume 1 (Van Zinderen Bakker 1966) included accounts of activities of several well-known international scientist at the time like D.A. Livingstone, L.S.B. Leakey, J.D. Clarke, R.J. Mason, C.K. Brain, W.W. Bishop, A.R.H. Martin, M. van Campo, E.P. Plumstead, H. Rakotoarivelo, H. Straka, O. Hedberg, R.E. Moreau, W.F. Libby, and others.

Volume 2, 4 and 6 followed basically the same format as Volume 1, but contained chapters under specific topics as short reports by various authors (Van Zinderen Bakker 1966, 1967b, 1972). The chapter topics included climatology and palaeoclimatology, geology and palaeontology, archaeology, biogeography, palynology in Africa, Antarctica and the Southern Ocean, isotope dates and morphology of microfossils, periglacial evidence, sea-levels, and oceanography.

The third volume of Palaeoecology of Africa contained the doctoral thesis of Joey Coetzee (Coetzee 1967), which is a landmark publication for Quaternary palynology in Africa (Meadows 2007). The early volumes also included additional lists with researcher's addresses and publications.

Maintaining the wide multi-disciplinary character from an international group of palaeo-scientists, subsequent volumes of PoA began to include separately authored papers, e.g., Volume 7 that included an extended study of 106 pages by A.C. Hamilton on palynology in Uganda (Hamilton 1972).

1.3 EDITORIAL CHANGES

Over the years the series eventually changed editors or were assisted by guest editors especially in the case of conference proceedings. Volume 18 was edited by Joey Coetzee and served as a dedication to the work of van Zinderen Bakker (Table S1).

Klaus Heine of Regensburg took over as editor for Volumes 19-27, of which Volume 27 was co-edited by Jürgen Runge (Frankfurt) who took over the duties for the rest of the volumes. After 1999, CRC Press (Taylor & Francis Group/'A Balkema Book') became the publisher and introduced the new current enlarged layout format and new series sub-title, *Landscape evolution and Palaeoenvironments* since 2007 (Volumes 28–34).

1.4 CONFERENCE PROCEEDINGS

With the appearance of Volume 22 in 1991, the following seven of eleven conference proceedings, were already published in PoA of which four represented meetings of the Southern African Society for Quaternary Research (SASQUA) (see Table S1 for the full list):

(1) Volume 5, Scientific Committee on Antarctic Research (SCAR) of the International Council of Scientific Unions (ICSU) Cambridge in 1968 (Van Zinderen Bakker 1969). This was later followed up by Volume 8 (Van Zinderen Bakker 1973), which did not cover the proceedings of a specific event but expanded the topic and inspired further interest in history of Antarctica.
(2) Volume 12, Sahara and Surrounding Seas held in Mainz in 1979 under the auspices of the *Akademie der Wissenschaften und der Literatur*, guest edited by M. Sarnthein, E. Siebold and P. Rognon, (Van Zinderen Bakker and Coetzee 1979).
(3) Volume 15, Southern African Society for Quaternary Research, SASQUA VI (1981) Pretoria, guest edited by J.C. Vogel, E.A. Voigt and T.C. Partridge (Coetzee and van Zinderen Bakker 1982).

(4) Volume 17, Southern African Society for Quaternary Research, SASQUA VII (1985) Stellenbosch, guest edited by H. Deacon (Van Zinderen Bakker *et al.* 1986).
(5) Volume 19, Southern African Society for Quaternary Research, SASQUA VIII (1987) Bloemfontein, guest edited by J.A. Coetzee (Heine 1988).
(6) Volume 21 Southern African Society for Quaternary Research, SASQUA IX (1989) Durban, guest edited by R.R. Maud (Heine 1990).
(7) Volume 22 (African Palynology) Rabat, guest edited by A. Ballouche and J. Maley (Heine 1991).

REFERENCES

Coetzee, J.A., 1967, Pollen analytical studies in east and Southern Africa. *Palaeoecology of Africa and the Surrounding Islands,* **3** (A.A. Balkema: Cape Town).

Coetzee, J.A., van Zinderen Bakker, E.M., 1952, Pollen spectrum of the Southern Middleveld of the Orange Free State. *South African Journal of Science*, **48** (9), pp. 275–281.

Coetzee, J.A., van Zinderen Bakker, E.M. 1982, *Palaeoecology of Africa and the Surrounding Islands*, **15**, (Balkema: Cape Town).

Dart. R.A. 1925, Australopithecus africanus: The Man-Ape of South Africa. *Nature*, **145** (2884), pp. 195–199, 10.1038/115195a0.

Dreyer, T.F., 1935, A Human Skull from Florisbad, Orange Free State, with a Note on the Endocranial Cast, by C.U. Ariens Kappers. *Verhandelingen der Koninklijke Nederlandse Akademie van Wetenschappen*, **38**, pp. 3–12.

Hamilton, A.C. 1972, The interpretation of pollen diagrams from highland Uganda. *Palaeoecology of Africa,* **7**, pp. 45–149 (A.A. Balkema: Cape Town).

Heine, K. 1988, *Palaeoecology of Africa and the Surrounding Islands*, **19** (A.A. Balkema: Cape Town).

Heine, K. 1990, *Palaeoecology of Africa and the Surrounding Islands*, **21** (A.A. Balkema: Cape Town).

Heine, K. 1991, *Palaeoecology of Africa and the Surrounding Islands*, **22** (A.A. Balkema: Cape Town).

Meadows, M.E., 2007, Classics revisited – Coetzee, J.A. 1967: pollen analytical studies in east and southern Africa. *Palaeoecology of Africa* **3**, 1–146. *Prog. Phys. Geogr.*, **31** (3), pp. 313–317, 10.1177/0309133307079056.

Meadows, M.E., 2015, Seven decades of Quaternary palynological studies in southern Africa: a historical perspective. *Trans. R. Soc. S. Afr.*, **70** (2), pp. 103–108, 10.1080/0035919X.2015.10 04139.

Neumann, F. and Scott, L., 2018, E.M. van Zinderen Bakker (1907–2002) and the study of African Quaternary palaeoenvironments, *Quaternary International*, **495**, 153–168, 10.1016/j.quaint.2018.04.017.

Van Zinderen Bakker, E.M., 1966, *Palaeoecology of Africa and the Surrounding Islands 1950–1963*, **1,** (Balkema: Cape Town).

Van Zinderen Bakker, E.M., 1967a, The South African biological-geological survey of the Marion and Prince Edward Islands and the meteorological expedition to Bouvet Island – introduction. *South African. Journal of Science* **63**(6), 217–218.

Van Zinderen Bakker, E.M., 1967b, *Palaeoecology of Africa and the Surrounding Islands 1964–1965*, **2** (A.A. Balkema: Cape Town).

Van Zinderen Bakker, E.M., 1967, *Palaeoecology of Africa and the Surrounding Islands 1966–1965*, **4** (A.A. Balkema: Cape Town).

Van Zinderen Bakker, E.M., 1969, *Palaeoecology of Africa and the Surrounding Islands*, **5** (A.A. Balkema: Cape Town).

Van Zinderen Bakker, E.M., 1972, *Palaeoecology of Africa and the Surrounding Islands 1969–1971*, **6**, (Balkema: Cape Town).

Van Zinderen Bakker, E.M., 1973, *Palaeoecology of Africa and the Surrounding Islands* **8** (A.A. Balkema: Cape Town).

Van Zinderen Bakker, E.M., and Coetzee, J.A., 1979, *Palaeoecology of Africa and the Surrounding Islands*, **12**, (A.A. Balkema: Cape Town).

Van Zinderen Bakker, E.M., Coetzee, J.A. and Scott, L., 1986, *Palaeoecology of Africa and the Surrounding Islands*, **17**, (A.A. Balkema: Cape Town).

Van Zinderen Bakker, E.M., Winterbottom, J.M., and Dyer, R.A. (Eds.), 1971, Marion and Prince Edward Islands. Report on the South African Biological and Geological Expedition, 1965–1966. (A.A. Balkema: Cape Town).

CHAPTER 2

The African Pollen Database (APD) and tracing environmental change: State of the Art

Anne-Marie Lézine

Laboratoire d'Océanographie et du Climat, Expérimentation et Approche numérique/IPSL, Sorbonne Université, CNRS-IRD-MNHN, Paris, France

Sarah J. Ivory

Department of Geosciences, Pennsylvania State University, University Park, USA

William D. Gosling

Institute for Biodiversity and Ecosystems Dynamics, University of Amsterdam, Amsterdam, The Netherlands

Louis Scott

Department of Plant Sciences, University of the Free State, Bloemfontein, South Africa

ABSTRACT: The African Pollen Database is a scientific network with the objective of providing the international scientific community with data and tools to develop palaeoenvironmental studies in sub-Saharan Africa and to provide the basis for understanding the vulnerability of ecosystems to climate change. This network was developed between 1996 and 2007. It promoted the collection, homogenization and validation of pollen data from modern (trap, soils, lake and river mud) and fossil materials (Quaternary sites) and developed a tool to determine pollen grains using digital photographs from international herbaria. Discontinued in 2007 due to a lack of funding, this network now resumes its activity in close collaboration with international databases: Neotoma, USA, Pangaea, DE, and the Institut Pierre Simon Laplace, FR.

2.1 INTRODUCTION

International cooperation in research and decision-making is critical for solving global environmental problems linked to climate and/or human impact on ecosystems, particularly at regional level. These environmental problems include forest degradation, accelerating loss of biodiversity and water resources, instability of transitional ecological domains, and change in coastal zones. Tropical ecosystems are especially at risk as future states are likely to be beyond the range of observations, yet their preservation is of crucial importance for the maintenance of the Earth's biosphere and climate. For example, the role of the equatorial and tropical forests

DOI 10.1201/9781003162766-2

in global exchanges, such as the global carbon cycle, is widely recognised (Detwiler and Hall 1988). Research developed within the general framework of international scientific programmes (e.g., the International Geosphere Biosphere Programme, IGBP – http://www.igbp.net/) have long highlighted the need to provide quantitative understanding of the Earth's past environment in order to define the envelope of natural environmental variability within which we can assess anthropogenic impact on the Earth's biosphere, geosphere and atmosphere. Following these recommendations, the scientific community has developed a set of analytical techniques to recover high-resolution environmental and ecological records from different natural archives such as tree-rings, and lake and ocean sediments, thus providing accurate scientific information for the development of predictive models of regional and global change. The use of these data to test Earth system model simulations requires the collection, assemblation, standardization and subsequently the access to the wider scientific and modelling community in the form of specific regional databases.

2.2 MAIN ACHIEVEMENTS

The African Pollen Database (APD) was first developed in 1996 in close cooperation with its European counterpart (EPD) (European Pollen Database; http://www.europeanpollendatabase.net) and the Global Pollen Database hosted at the National Oceanic and Atmospheric Administration (NOAA) Paleoclimatology Database (USA) (https://www.ncdc.noaa.gov/data-access/paleoclimatology-data). The initial workshop and subsequent work, funded by the French National Centre for Scientific Research (CNRS), the European Union (International Cooperation for development (INCO-DEV) and European network of research and innovation centres (ENRICH) programmes and the UNESCO International Geoscience Programme (PICG), established methods of collating pollen data, developed a standardized pollen nomenclature (Vincens *et al.* 2007), generated updated age models, composed images of pollen grains from internationally recognized herbaria, and created a searchable web interface. In the first stage of its development, the APD contained 288 fossil sites and 1985 modern samples (Figure 1). Among the numerous achievements of the APD one can highlight the following topics.

2.2.1 Pollen based biome reconstructions

Jolly *et al.* (1998) first demonstrated that the 'biomization' method for assigning pollen taxa to plant functional types and biomes was able to predict the potential natural vegetation of tropical Africa, despite uncertainty and variability of pollen production and dissemination (Ritchie 1995) of an extremely biodiverse flora (26,000 plant species; Lebrun and Stork 1991–1997). This allowed palaeoecologists working in Africa to participate in the 'BIOME 6000' project sponsored by IGBP (Prentice and Webb 1998). This project aimed to use palaeoecological data from the mid-Holocene as a benchmark to evaluate simulations with coupled climate-biosphere models and thus to assess the extent of biogeophysical (vegetation-atmosphere) feedbacks in the global climate system (Prentice *et al.* 2000). After the validation of the modern pollen dataset (Gajewski *et al.* 2002; Jolly *et al.* 1998), the biomization method was successfully applied to reconstruct modern (Lézine *et al.* 2009; Vincens *et al.* 2006) and past biomes for selected time periods, typically the Last Glacial Maximum (LGM) and the Holocene (Jolly *et al.* 1998; Elenga *et al.* 2000). Past biome reconstructions have also been performed for the Plio-Pleistocene (Bonnefille *et al.* 2004; Novello *et al.* 2015) and more recent time periods such as the last glacial-interglacial cycle and the Holocene (Amaral *et al.* 2013; Izumi and Lézine 2016; Lebamba *et al.* 2012; Lézine *et al.* 2019).

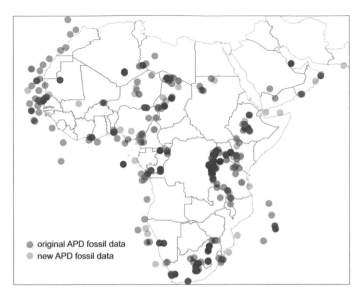

Figure 1. Late Quaternary African Pollen Database (APD) sites. In red: pollen data gathered during the first phase of the APD (1994–2007). In yellow: new pollen sites to be entered into the new version of APD, in construction).

2.2.2 Quantitative reconstructions of climate variables from pollen data

Modern-analogue, regression, and model-inversion techniques have been developed to reconstruct past climates from pollen assemblages or pollen-based reconstructed biomes worldwide. Using the APD modern pollen dataset, Peyron *et al.* (2007) provided the first quantitative pollen-based reconstruction of precipitation for all of Africa at 6000 yr BP based on the Modern Analogues Technique (MAT) and the Plant-Functional Types (PFT) climate relationships. Results were then compared with atmospheric general circulation model output and coupled ocean atmosphere-vegetation models developed in the frame of the Paleoclimate Model Intercomparison Project (PMIP) international project (Joussaume and Taylor 1995). More recently, Wu *et al.* (2007) then Izumi and Lézine (2016) developed an inverse modelling approach based on the BIOME model to quantitatively reconstruct past climates based on pollen biome scores. The advantage of this method was to provide quantitative climate reconstructions for periods when CO_2 concentrations were different from today. While reconstruction attempts of climate by means of pollen data in South Africa, were mostly qualitative (Scott *et al.* 2012), Chevalier and Chase (2015) applied a method that related the pollen to plant distribution data to obtain quantitative estimates.

2.2.3 Vegetation reconstructions from pollen data

Palaeoecological data from the APD led to a series of reconstructions of vegetation at a continental scale. Site-based global biome maps for Africa for the mid-Holocene and LGM were first completed within the framework of the 'BIOME 6000' project (Prentice *et al.* 2000). More recently, APD palaeoecological data was included in a global synthesis of changes in composition and structure of past vegetation since the LGM performed by Nolan *et al.* (2018). This study provided a baseline for evaluation of the magnitudes of ecosystem transformations under future emission scenarios.

At a regional scale, APD data were used to reconstruct the Green Sahara and evaluate plant migration rates during the African Humid Period (Hély *et al.* 2014; Watrin *et al.* 2009). In

East Africa, compilation of pollen and archaeological data was used to discuss the cumulative effects of climate and land-use on the environment (Marchant *et al.* 2018). Furthermore, with the increasing recognition that fossil data can improve information about fundamental climatic tolerances, modern and palaeoecological data from the APD have been included in estimates of climatic niches of at-risk plant taxa (Ivory *et al.* 2016). This information was then used to provide forecasts of future impacts to ranges under climate and land-use trajectories for the end of the 21st century (Ivory *et al.* 2019).

2.3 CHALLENGES AND FUTURE DEVELOPMENTS

All these realizations suffer from (1) an highly uneven geographic distribution of data. Pollen data are relatively abundant in Eastern and Southern Africa where palynological research has been ongoing since the early 1950s (Hedberg 1954; van Zinderen Bakker and Coetzee 1952). In the former region, the abundance of lakes and swamps also provides favourable conditions for pollen preservation and long-time series. In North and Central Africa, on the other hand, data are less numerous and often discontinuous in time. The drying out of Holocene lakes in the Sahara, the difficulties of access to the sites, and their rarity are all limitations to regional geographical reconstructions; (2) the scarcity of long time series beyond the LGM (Ivory *et al.* 2017; Lézine *et al.* 2019; Miller and Gosling 2014; Scott 2016), and therefore long-term vegetation changes are mostly derived from marine cores (e.g., Dupont and Kuhlmann 2017; Hessler *et al.* 2010).

Thanks to a recent funding from the Belmont Forum for Science-driven e-infrastructure innovation for the project 'Abrupt Change in Climate and Ecosystems: Where are the Tipping points?', the APD is now being relaunched and developing further collaborations with international databases ('NEOTOMA', 'PANGAEA') and the French Institute Pierre Simon Laplace (IPSL). One of the priorities is to gather and validate data published since 2007, the date of the closure of the French data centre Medias-France where APD was stored. Today, 67 new late Quaternary, 17 Plio-Pleistocene and 20 marine new pollen series have been collated. Strong links are being developed with NEOTOMA (https://www.neotomadb.org/) and PANGAEA (https://pangaea.de/) databases.

These new datasets benefit from improved dating techniques and age modelling methods. The result of which is that newly acquired pollen series have reduced temporal uncertainty and improved resolution, allowing to more precision in interpretation of local and regional ecosystem dynamics and climate-vegetation interactions. All this allows to envisage new scientific developments of which one can cite three examples here:

2.3.1 Better understanding of ecosystem-human interactions

The identification and quantification of human-induced alterations to the Earth's surface are critical to understand the role of land-use change on ecosystems and climate. The Land Cover 6k (6000 yrs BP) project of IGBP PAGES (http://www.pastglobalchanges.org/science/wg/landcover6k) described in Gaillard *et al.* (2018) is a unique opportunity to develop a methodological approach to carefully reconstruct land cover change from pollen data and evaluate anthropogenic land-cover change scenarios for palaeoclimate modelling. The major limitation of such a quantification in tropical Africa is that 'with very few exceptions, tropical trees have zoophilous pollinating systems and relatively low pollen productivity' (Ritchie 1995; p. 487). The relationship between pollen percentages in diagrams and actual vegetation cover is thus extremely difficult to assess. Within the equatorial forest for instance, many tree taxa are under-represented or even absent from the pollen assemblages.

The reliability of any landscape reconstruction requires the spatial scale represented by the pollen assemblage to be carefully taken into account. This requires the most accurate possible evaluation of pollen productivity of species and pollen-rain settling time. Following the work of Duffin and Bunting (2008) in South Africa, Gaillard and colleagues are currently applying the LOVE (Local Vegetation Estimates) and REVEALS (Regional Estimates of VEgetation Abundance from Large Sites) models developed by Sugita (2007a,b) in Central Africa. The development of such an approach as well as the compilation of pollen and archaeological data (Marchant *et al.* 2018) can greatly improve the involvement of the palynological community in the study of land-use as a climate forcing.

2.3.2 Improved constraint of climate variability over the last millennia

The recent publication of Nash *et al.* (2016) within the framework of the PAGES 2ka working group (http://www.pastglobalchanges.org/science/wg/2k-network) showed that Africa is one of the world's most poorly documented regions in terms of climate reconstructions over the last millennia. Historical archives are very rare, as are natural archives with adequate temporal resolution and age control. Time frames where historical records overlap the prehistorical data are important to obtain seamless reconstructions and validate reconstructions. Improving the spatial coverage and resolution of palaeoecological records is crucial for studying natural decadal (or multi-decadal) climate variability and associated mechanisms. It is also essential to analyze the complexity of spatial hydroclimate patterns, such as that suggested for the Little Ice Age (1250–1750 CE) in Africa, for which wet or dry regions have been identified.

2.3.3 Understanding vegetation responses to abrupt climate change

The evolution of northern Africa from a 'Green Sahara' state to one of the most arid deserts today (Kröpelin *et al.* 2008) or the collapse of the equatorial forests (e.g., Lézine *et al.* 2013) occurred at the end of the African Humid Period. These are among the most emblematic examples of the extreme changes that can affect the global environment with dramatic consequences for human populations. The tipping points and climatic drivers between extreme states remain to be studied, as do the early warning signals of these environmental crises, which may date back several millennia, and still need to be identified. Long-term, high-resolution and evenly distributed pollen records are critical to address these questions.

2.4 CONCLUSION

The African Pollen Database aims at providing the scientific community, students and teachers with scientific and educational tools (pollen grain determination tools, modern and fossil data duly validated and dated, publications) to address key issues for understanding the vulnerability of ecosystems facing climate change. It ensures interoperability with international databases for multi-proxy reconstructions of the past environment. Beyond these activities, the African Pollen Database is a scientific network to develop the sharing of data and knowledge between researchers in different countries.

REFERENCES

Amaral, P.G.C., Vincens, A., Guiot, J., Buchet, G., Deschamps, P., Doumnang, J.C. and Sylvestre, F., 2013, Palynological evidence for gradual vegetation and climate changes during the African Humid Period termination at 13° N from a Mega-Lake Chad sedimentary sequence. *Climate of the Past*, **9**(1), pp. 223–241, 10.5194/cp-9-223-2013.

Bonnefille, R., Potts, R., Chalié, F., Jolly, D. and Peyron, O., 2004, High-resolution vegetation and climate change associated with Pliocene Australopithecus afarensis. *Proceedings of the National Academy of Sciences*, **101**(33), pp. 12125–12129, 10.1073/pnas.0401709101.

Chevalier, M. and Chase, B.M., 2015, Southeast African records reveal a coherent shift from high- to low-latitude forcing mechanisms along the east African margin across last glacial-interglacial transition. *Quaternary Science Reviews*, **125**, pp.117–130, 10.1016/j.quascirev.2015.07.009.

Detwiler, R.P. and Hall, C.A., 1988, Tropical forests and the global carbon cycle. *Science*, **239**(4835), pp.42–47, 10.1126/science.239.4835.42.

Duffin, K.I. and Bunting, M.J., 2008, Relative pollen productivity and fall speed estimates for southern African savanna taxa. *Vegetation History and Archaeobotany*, **17**(5), pp. 507–525, 10.1007/s00334-007-0101-2.

Dupont, L.M. and Kuhlmann, H., 2017, Glacial-interglacial vegetation change in the Zambezi catchment. *Quaternary Science Reviews*, **155**, pp. 127–135, 10.1016/j.quascirev.2016.11.019.

Elenga, H., Peyron, O., Bonnefille, R., Jolly, D., Cheddadi, R., Guiot, J., Andrieu, V., Bottema, S., Buchet, G., De Beaulieu, J.L., Maley, J., Marchant, R., Perezbiol, R., Reille, M., Riollet, Scott, L., Strake, H., Taylor, D., Van Campo,E., Vincens, A., Laarif, F., Jonson, H. and Hamilton, A.C., 2000, Pollen-based biome reconstruction for southern Europe and Africa 18,000 yr BP. *Journal of Biogeography*, **27**(3), pp. 621–634, 10.1046/j.1365-2699.2000.00430.x.

Gaillard, M.-J., Morrison, K.D., Madella, M. and Whitehouse, N., 2018, Past land-use and land-cover change: the challenge of quantification at subcontinental to global scale. *PAGES Magazine*, **26** (1), p. 3, 10.22498/pages.26.1.3.

Gajewski, K., Lézine, A.-M., Vincens, A., Delestan, A. and Sawada, M., 2002, Modern climate–vegetation–pollen relations in Africa and adjacent areas. *Quaternary Science Reviews*, **21** (14–15), pp. 1611–1631, 10.1016/S0277-3791(01)00152-4.

Hedberg, O., 1952, A pollen-analytical reconnaissance in tropical East Africa. *Oikos* **5**(2), pp. 137–166, 10.2307/3565157.

Hély, C., Lézine, A.-M. and APD contributors, 2014, Holocene changes in African vegetation: tradeoff between climate and water availability. *Climate of the Past* **10**, pp. 681–686, 10.5194/cp-10-681-2014.

Hessler, I., Dupont, L., Bonnefille, R., Behling, H., González, C., Helmens, C.F., Hooghiemstra, H., Lebamba, J., Ledru, M.-P., Lézine, A.-M., Maley, J., Marret, F. and Vincens, A., 2010. Millennial-scale changes in vegetation records from tropical Africa and South America during the last glacial. *Quaternary Science Reviews*, **29** (21–22), pp. 2882–2899, 10.1016/j.quascirev.2009.11.029.

Ivory, S. J., Early, R., Sax, D. F., and Russell, J., 2016, Niche expansion and temperature sensitivity of tropical African montane forests. *Global Ecology and Biogeography*, **25**(6), pp. 693–703, 10.1111/geb.12446.

Ivory, S. J., Russell, J., Early, R. and Sax, D. F., 2019, Broader niches revealed by fossil data do not reduce estimates of range loss and fragmentation of African montane trees. *Global Ecology and Biogeography*, **28**(7), pp. 992–1003, 10.1111/geb.12909.

Ivory, S.J., McGlue, M.M., Ellis, G.S., Boehlke, A., Lézine, A.-M., Vincens, A. and Cohen, A.S., 2017, East African weathering dynamics controlled by vegetation-climate feedbacks. *Geology*, **45**(9), pp. 823–826, 10.1130/G38938.1.

Izumi, K. and Lézine, A.-M., 2016, Pollen-based biome reconstructions over the past 18,000 years and atmospheric CO_2 impacts on vegetation in equatorial mountains of Africa. *Quaternary Science Reviews*, **152**, pp. 93–103, 10.1016/j.quascirev.2016.09.023.

Jolly, D., Prentice, C., Bonnefille, R., Ballouche, A., Bengo, M., Brenac, P., Buchet, G., Burney, D., Cazet, J.-P., Cheddadi, R., Edorh, T., Elenga, H., Elmoutaki, S., Guiot, J., Laarif, F., Lamb, H., Lézine, A.-M., Maley, J., Mbenza, M., Peyron, O., Reille, M., Reynaud-Farrera, I., Riollet, G., Ritchie, J.C., Roche, E., Scott, L., Ssemmanda, I., Straka, H., Umer, M.,

Van Campo, E., Vilimumbalo, S., Vincens, A. and Waller, M., 1998, Biome reconstruction from pollen and plant macrofossil data for Africa and the Arabian Peninsula at 0 and 6000 years. *Journal of Biogeography*, **25**, pp. 1007–1027, 10.1046/j.1365-2699.1998.00238.x.

Joussaume, S. and Taylor, K.E., 1995, *Status of the Paleoclimate Modeling Intercomparison Project (PMIP), WCRP-92, WMO/TD 732*, (Geneva: World Climate Research Programme), pp. 425–430.

Kröpelin, S., Verschuren, D., Lézine, A.-M., Eggermont, H., Cocquyt, C., Francus, P., Cazet, J.P., Fagot, M., Rumes, B., Russell, J.M. and Darius, F., 2008, Climate-driven ecosystem succession in the Sahara: the past 6000 years. *Science*, **320**(5877), pp. 765–768, 10.1126/science. 1154913.

Lebamba, J., Vincens, A. and Maley, J., 2012, Pollen, vegetation change and climate at Lake Barombi Mbo (Cameroon) during the last ca. 33,000 cal yr BP: a numerical approach. *Climate of the Past*, **8**(1), pp. 59–78, 10.5194/cp-8-59-2012.

Lebrun, J.-P. and Stork, A.L., 1991-1997. *Enumération des plantes à fleurs d'Afrique tropicale*. Vol. 1–4, (Geneva : Conservatoire et Jardin botaniques de la Ville de Genève).

Lézine, A.-M., Izumi, K., Kageyama, M. and Achoundong, G., 2019, 90,000-year record of Afromontane forest responses to climate change. *Science*, **363**, pp. 177–181, 10.1126/science.aav6821.

Lézine, A.-M., Watrin, J., Vincens, A., Hély, C. and APD contributors, 2009, Are modern pollen data representative of west African vegetation? *Review of Palaeobotany and Palynology*, **156**(3-4), pp. 265–276, 10.1016/j.revpalbo.2009.02.001.

Marchant, R., Richer, S., Boles, O., Capitani, C., Courtney-Mustaphi, C.J., Lane, P., Prendergast, M.E., Stump, D., De Cort, G., Kaplan, J.O. and Phelps, L., 2018, Drivers and trajectories of land cover change in East Africa: Human and environmental interactions from 6000 years ago to present. *Earth-Science Reviews*, **178**, pp. 322–378, 10.1016/j.earscirev.2017. 12.010.

Miller, C.S. and Gosling, W.D., 2014, Quaternary forest associations in lowland tropical West Africa. *Quaternary Science Reviews*, **84**, pp. 7–25, 10.1016/j.quascirev.2013.10.027.

Nash, D.J., De Cort, G., Chase, B.M., Verschuren, D., Nicholson, S.E., Shanahan, T.M., Asrat, A., Lézine, A.M. and Grab, S.W., 2016, African hydroclimatic variability during the last 2000 years. *Quaternary Science Reviews*, **154**, pp. 1–22, 10.1016/j.quascirev.2016.10.012.

Nolan, C., Overpeck, J.T., Allen, J.R., Anderson, P.M., Betancourt, J.L., Binney, H.A., Brewer, S., Bush, M.B., Chase, B.M., Cheddadi, R., Djamali, M., Dodson,J.E., Edwards, M.E., Gosling,W.D., Haberle,S., Hotchkiss, S.C., Huntley, B., Ivory, S.J., Kershaw, A.P., Soo-Hyun, K., Latorre, C., Leydet, M., Lézine, A.-M., Liu, K-B., Liu, Y., Lozhkin, A.V., McGlone, M.S., Marchant, R., Momohara, A., Moreno, P., Müller, S., Otto-Bliesner, B., Shen, C., Stevenson, J., Takahara, H., Tarasov, P.,Tipton, J., Vincens, A., Weng, C. Xu, Q., Zheng, Z. and Jackson, S.T., 2018, Past and future global transformation of terrestrial ecosystems under climate change. *Science*, **361**(6405), pp. 920–923, 10.1126/science.aan5360.

Novello, A., Lebatard, A.E., Moussa, A., Barboni, D., Sylvestre, F., Bourlès, D.L., Paillès, C., Buchet, G., Decarreau, A., Duringer, P. and Ghienne, J.F., 2015, Diatom, phytolith, and pollen records from a 10Be/9Be dated lacustrine succession in the Chad basin: Insight on the Miocene–Pliocene paleoenvironmental changes in Central Africa. *Palaeogeography, Palaeoclimatology, Palaeoecology*, **430**, pp. 85–103, 10.1016/j.palaeo.2015.04.013.

Peyron, O., Jolly, D., Braconnot, P., Bonnefille, R., Guiot, J., Wirrmann, D. and Chalié, F., 2006, Quantitative reconstructions of annual rainfall in Africa 6000 years ago: Model-data comparison. *Journal of Geophysical Research: Atmospheres*, **111**, 10.1029/2006JD007396, article: D24110.

Prentice, I. C., D. Jolly and BIOME 6000 Participants, 2000, Mid-Holocene and glacial-maximum vegetation geography of the northern continents and Africa. *Journal of Biogeography*, **27**, pp. 507–519, 10.1046/j.1365-2699.2000.00425.x.

Prentice, I.C and Webb T., III, 1998, BIOME 6000: reconstructing global mid-Holocene vegetation patterns from palaeoecological records. *Journal of Biogeography*, **25**(6), pp. 997–1005,
10.1046/j.1365-2699.1998.00235.x.

Ritchie, J.C., 1995, Current trends in studies of long-term plant community dynamics. *New
Phytologist*, **130**(4), pp. 469–494, 10.1111/j.1469-8137.1995.tb04325.x.

Scott, L., 2016, Fluctuations of vegetation and climate over the last 75,000 years in the
Savanna Biome, South Africa: Tswaing Crater and Wonderkrater pollen sequences reviewed.
Quaternary Science Reviews, **145**, pp. 117–133, 10.1016/j.quascirev.2016.05.035.

Scott, L., Neumann, F.H., Brook, G.A., Bousman, C.B., Norström, E. and Metwally, A.A., 2012,
Terrestrial fossil pollen evidence of climate change during the last 26 thousand years in southern
Africa. *Quaternary Science Reviews*, **32**, pp. 100–118, 10.1016/j.quascirev.2011.11.010.

Sugita, S., 2007a, Theory of quantitative reconstruction of vegetation I: Pollen from large
sites REVEALS regional vegetation composition. *The Holocene*, **17**(2), pp. 229–241,
10.1177/0959683607075837.

Sugita, S., 2007b, Theory of quantitative reconstruction of vegetation II: All you need is LOVE.
The Holocene, **17**(2), pp. 243–257, 10.1177/0959683607075838.

van Zinderen Bakker. E.M. and Coetzee, J.A., 1952, Pollen spectrum of the southern middleveld
of the Orange Free State. *South African Journal of Science*, **48**, pp. 275–281.

Vincens, A., Bremond, L., Brewer, S., Buchet, G. and Dussouillez, P., 2006, Modern pollen-based
biome reconstructions in East Africa expanded to southern Tanzania. *Review of Palaeobotany
and Palynology*, **140**(3-4), pp. 187–212, 10.1016/j.revpalbo.2006.04.003.

Vincens, A., Lézine, A.-M., Buchet, G., Lewden, D. and Le Thomas, A., 2007, African pollen
database inventory of tree and shrub pollen types. *Review of Palaeobotany and Palynology*,
145(1-2), pp. 135–141, 10.1016/j.revpalbo.2006.09.004.

Watrin, J., Lézine, A.-M. and Hély, C., 2009, Plant migration and plant communities
at the time of the 'green Sahara'. *Comptes Rendus Geoscience*, **341**(8–9), pp.656–670,
10.1016/j.crte.2009.06.007.

Wu, H., Guiot, J., Brewer, S., Guo, Z., Peng, C., 2007, Dominant factors controlling glacial and
interglacial variations in the treeline elevation in tropical Africa. *Proceedings of the National
Academy of Sciences*, **104**(23), pp. 9720–9724, 10.1073/pnas.0610109104.

CHAPTER 3

Preliminary evidence for green, brown and black worlds in tropical western Africa during the Middle and Late Pleistocene

William D. Gosling, Crystal N.H. McMichael, Zoë Groenewoud & Eleonora Roding

Institute for Biodiversity and Ecosystems Dynamics, University of Amsterdam, Amsterdam, The Netherlands

Charlotte S. Miller

Leeds Trinity University, Leeds, United Kingdom

Adele C.M. Julier

Plant Conservation Unit, University of Cape Town, Cape Town, South Africa

ABSTRACT: Modern ecological studies indicate that the degree of openness in African vegetation cover is determined, in part, by the presence of herbivores and fire as consumers of vegetation. Where herbivores are the dominant consumer of vegetation the resultant open state is described as a 'brown' world. Where fire is the dominant consumer of vegetation the resultant open state is described as a 'black' world. While if neither consumer is dominant then a more closed canopy states arises that is described as a 'green' world. Here we use palaeoecological data obtained from Lake Bosumtwi (Ghana) to characterize green, brown, and black worlds during two short sections of around 1000 years each, deposited around 200,000 and 100,000 years ago (Middle and Late Pleistocene). We characterize the vegetation cover using pollen and phytoliths, herbivory using *Sporormiella* and fire using micro-charcoal. We find that during *c.* 1000 years of the Middle Pleistocene fire was the major consumer of vegetation, while during *c.* 1000 years in the Late Pleistocene herbivores were relatively more important consumers of vegetation. We therefore suggest that the Middle Pleistocene section represents a black world, while in the Late Pleistocene section we capture a combination of green, brown and black worlds. The duration of these states seems to range from centuries to millennia and transitions are observed to occur in both an abrupt and a stepwise fashion. These preliminary data demonstrate how palaeoecological information can be used to gain insights into past landscape scale processes over thousands of years. Further work is required to test the robustness of these findings and to provide a higher temporal resolution to aid the link with modern ecological studies.

DOI 10.1201/9781003162766-3

3.1 INTRODUCTION

The theory of 'alternative stable states' proposes that biotic and abiotic feedbacks can determine the composition and structure of an ecosystem regardless of external factors, such as climate (May 1977; Scheffer *et al.* 2001). The presence of herbivory and fire activity in African landscapes has long been suggested to be a key factor in determining the openness of the vegetation cover (Stebbing 1937). To characterize the role of herbivory and fire as mechanisms determining the degree of vegetation openness (alternative stable state) William Bond developed the concept of 'green', 'brown' and 'black' worlds (Figure 1; Bond 2005; Bond 2019). This 'multi-coloured' world model comprises three vegetation states based on differences in herbivory and fire as consumers of vegetation: (i) a 'green' world where consumers (herbivores and fires) are at sufficiently low occurrence levels as to not influence vegetation growth and consequently vegetation cover is more closed and primarily controlled by plant resource acquisition, (ii) a 'brown' world where an abundance of herbivores consuming seeds, seedlings and saplings means a more open vegetation structure, and (iii) a 'black' world where the frequent occurrence of fires consuming vegetation means a more open vegetation structure.

Modern studies in African savannas have demonstrated that herbivory and fire play an important role in the complex interactions that determine vegetation openness (Sankaran *et al.* 2008) and consequently the regulation of both are routinely used in landscape management (Roques *et al.* 2001). Evidence from modern exclosure experiments has demonstrated that the removal of animals from ecosystems can result in changes to vegetation state on timescales of years to decades (e.g. Riginos *et al.* 2012; Staver *et al.* 2009). While the manipulation of fire regimes, through experimental burns and fire suppression initiatives, have demonstrated vegetation state can also be altered through these processes on decadal timescales (e.g. Enslin *et al.* 2000; King *et al.* 1997). Over the longer-term (>100s of years) palaeoecological data have revealed that past changes in herbivory and fire activity have altered African vegetation composition and structure (e.g. Ivory and Russell 2016; Runcia *et al.* 2009; Shanahan *et al.* 2016). Palaeoecological data therefore provide an opportunity to search for evidence of the green, brown and black worlds proposed by Bond (2005), test the timescales of their stability, and discover how transitions between them occur.

Based on Bond's multi-coloured world model it can be suggested that the addition, or subtraction, of either herbivory or fire as a consumer of vegetation in a landscape would result in a shift in the degree of openness (Figure 1). Here we use palaeoecological data to parametrize vegetation openness (Poaceae pollen/phytoliths), herbivory (*Sporormiella*) and fire (charcoal) to provide 'snap shots' of vegetation states in tropical western Africa during the Middle and Late Pleistocene. We use these data to identify periods of time when consumers (both animals and fires) did, and did not, exert a dominant control over the vegetation openness.

Figure 1. Conceptual panorama of vegetation change across an African landscape dependent on the dominant consumer of vegetation. Green = absence of consumers. Brown = herbivores are the dominant consumers. Black = fire is the dominant consumer.

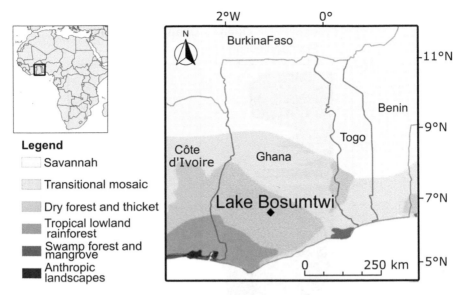

Figure 2. Major vegetation types of western Africa relative to Lake Bosumtwi (Ghana). Vegetation classifications follow White (1983).

3.2 MATERIALS AND METHODS

3.2.1 Study site

Lake Bosumtwi (06° 30' 06.04" N, 01° 24' 52.14" W, 97 metres above sea level; Figure 2) was formed by a meteorite impact *c*. 1 million years ago and is a hydrologically closed system (Koeberl *et al.* 1998). The International Continental Scientific Drilling Program (ICDP) Lake Bosumtwi drilling project extracted a *c*. 295 m sediment core from the centre of the basin in 2004 (named BOS04-5B) (Koeberl *et al.* 2005; Koeberl *et al.* 2007).

Lake Bosumtwi lies within the seasonal migration path of the Intertropical Convergence Zone (ITCZ) and moisture is also delivered to the site by the West African monsoon (Nicholson 2009). This climate configuration results in a strong gradient of decreasing precipitation from south to north across western Africa. Concomitant with the climate gradient is a vegetation gradient which means that today Lake Bosumtwi is surrounded by semi-deciduous forest, with evergreen forest to the south, and savanna to the north (Figure 2; White 1983). Analysis of ancient pollen contained within the BOS04-5B core revealed multiple transitions between forest and savanna vegetation during the last *c*. 520,000 years which have been linked to multi-millennial, orbitally forced, climate changes (Miller and Gosling 2014; Miller *et al.* 2016). In this study, two short (*c*. 5 m) sections of the core, in which vegetation transitions had previously been identified, were sub-sampled for further palaeoecological analysis to explore the impact of shorter-term landscape-scale process on past vegetation dynamics (Figure 3; Gosling *et al.* 2021).

3.2.2 Chronological control

Previously published radiometric dates indicate that the meteorite which formed Lake Bosumtwi impacted the Earth *c*. 1.07 ± 0.05 million years ago (Koeberl *et al.* 1998). Subsequent sediment

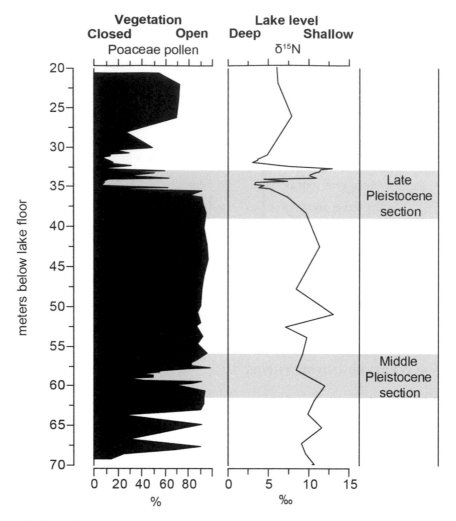

Figure 3. Position of the two sections focused on in this study (grey bars) from Lake Bosumtwi relative to previous palaeoecological work (modified from Miller *et al.* 2016).

accumulation through the core has been constrained by radiocarbon, U-series and optically stim-ulated luminance dating (Shanahan *et al.* 2013a). The two sections of the BOS04-5B core focused on here were recovered from 61.25–56.05 and 38.85–33.45 metres below lake floor (mblf) (Figure 3). Based on the radiometric dating these two sections were likely deposited during the Middle Pleistocene (*c.* 200,000 years ago) and Late Pleistocene (*c.* 100,000 years ago) respectively (Miller *et al.* 2016).

Previous studies into sediment deposition within Lake Bosumtwi indicated that the light-dark couplets of laminations within the sediments likely represent seasonal cycles (Shanahan *et al.* 2013b). Therefore, to gain a minimum estimate of the timescale over which these sections were deposited, the number of laminations within each section was counted on the basis that the sediment core images were included in the initial core descriptions.

3.2.3 Palaeoecological analysis

Twenty sub-samples were prepared for palaeoecological analysis across each of the two sections using standard techniques including density separation, acetolysis and sieving at 180 µm (Moore *et al.* 1991). An exotic marker (*Lycopodium*) was added to each sample to allow the concentration of microscopic remains to be calculated (Stockmarr 1971); University of Lund, batch #483216, containing 18,583 grains ± 4.1%. Each sub-sample was analysed for the concentration of Poaceae pollen, *Sporormiella*, and micro-charcoal (< 180 µm); identifications followed Gosling *et al.* (2013), van Geel and Aptroot (2006), and Whitlock and Larsen (2001) respectively. Phytoliths were also present in the prepared slides. Phytolith concentration from the same sub-samples were analysed across the Middle Pleistocene section; phytolith identifications were based on Piperno (2006).

3.2.4 Statistical analysis and data presentation

To assess the robustness of the palaeoecological datasets 95% confidence intervals were calculated (following Maher 1981). Bayesian Change Point (BCP) analysis (Barry and Hartigan 1993), which determines if there is a significant change in the mean of the variable at a given point in a time series, was used on: (i) *Sporormiella* (herbivore) concentration, and (ii) micro-charcoal (fire) concentration data. The posterior probabilities of the BCP analysis indicated the likelihood of significant change of those variables at each time step sampled (Barry and Hartigan 1993; Blois *et al.* 2011). Where the posterior probability of the BCP analysis exceeded 0.8 for *Sporormiella* or charcoal, a zone boundary was defined, indicating a significant change in herbivory or fire activity. Data were plotted using C2 (Juggins 2005).

3.3 RESULTS

Previously published radiometric dating indicates that the two sections studied here were deposited during the Middle and Late Pleistocene (Figure 3; Shanahan *et al.* 2013a). The results of the younger section are presented in the upper panel (Figure 4A) and the results of the older section are presented in the lower panel (Figure 4B).

3.3.1 Lamination counting

Approximately 1200 light-dark couplets of laminations were identified in the Middle Pleistocene section; however, no laminations could be identified in three parts of the section at: 60.25–59.85, 58.65–58.45, and 57.05–56.85 mblf (Figure 4B). Approximately 750 light-dark couplets of laminations were identified for the Late Pleistocene section; however, no laminations could be identified between 34.45 and 33.45 mblf (Figure 4A).

3.3.2 Palaeoecological analysis

For the Middle Pleistocene section two abrupt changes in vegetation consuming processes were detected by the BCP analysis at 59.25 (micro-charcoal) and 56.65 (*Sporormiella*) mblf defining three zones (Zones i, ii, and iii; Figure 4B). Within Zone i Poaceae pollen occurs in high concentrations (mean of 38,000 grains/cm^3), Poaceae phytoliths, tree phytoliths and micro-charcoal are ever present but are highly variable in concentration (1100–46,000 Poaceae phytoliths/cm^3, 28,000–240,000 tree phytoliths/cm^3, and 250,000–1,600,000 fragments/cm^3), and *Sporormiella* is absent. In Zone ii Poaceae pollen is continually present at medium concentration (mean 26,000 grains/cm^3), Poaceae phytoliths, tree phytoliths and micro-charcoal are usually present and highly variable in concentration (4000–127,000 Poaceae phytoliths/cm^3, 0–360,000 tree phytoliths/cm3, 199,000–6,194,000 fragments/cm^3), and *Sporormiella* is absent. In Zone iii Poaceae pollen and phytoliths occur at high concentrations (mean 35,000 grains/cm^3 and 34,000 phytoliths/cm^3 respectively), but tree phytoliths are absent; micro-charcoal is variable (395,000–978,000 fragments/cm^3), and *Sporormiella* is continually present (395–1133 spores/cm^3).

For the Late Pleistocene section two abrupt changes in vegetation consuming processes were detected by the BCP analysis at 36.45 (*Sporormiella* and micro-charcoal) and 34.25 (*Sporormiella*) mblf defining three zones (Zones I, II, and III; Figure 4A). Within Zone I Poaceae pollen (17,000 grains/cm^3), *Sporormiella* (mean 1200 spores/cm^3), and micro-charcoal (mean 291,000 particles/cm^3) are all relatively highly abundant within the section. In Zone II Poaceae pollen (mean 14,000 grains/cm^3) are abundant, while *Sporormiella* and micro-charcoal vary widely (0–8400 spores/cm^3 and 15,000-496,000 fragments/cm^3 respectively). In Zone III Poaceae pollen (mean 2700 grains/cm^3) and micro-charcoal (mean 4400 fragments/cm^3) are relatively low, while *Sporormiella* varies (0–4000 spores/cm^3) in concentration.

3.4 DISCUSSION

3.4.1 Timescales of deposition

Radiometric dating indicates that the Middle Pleistocene section was deposited *c.* 200,000 years ago during a period of high eccentricity in the Earth's orbital cycle (Miller *et al.* 2016). This orbital configuration of northern hemisphere insolation maximum, during a precession minimum and an eccentricity maximum, extenuates seasonal variations at Lake Bosumtwi (6°N). Therefore, this period of time likely represents ecosystem functioning under high seasonal stress, and a consistently low level of Lake Bosumtwi during this period (Figure 2). The Late Pleistocene section is indicated to have been deposited *c.* 100,000 years ago as part of the longer transition out of the last interglacial period and includes a period of relatively higher lake levels (Figure 2).

Lamination counting within the two sections revealed 1200 light-dark couplets in 5.2 m of sediment in the Middle Pleistocene section, and 750 light-dark couplets in 5.4 m in the Late Pleistocene section. These counts were derived from visual inspection of digital core scan images and consequently likely represent an under-estimate of the true number of bands present because some may not have been captured in the digital imagery of the core surface. To reveal all bands, and develop a robust internal chronology, further analysis of the sediments in thin section is required. The preliminary analysis presented here shows this is a promising line of investigation and, based on the visible bands, it can be roughly estimated that the period of deposition represented by each of these sections is more than 1000 years. The relatively higher number of bands in the older section could be related to the enhanced seasonality and lower sedimentation rates during this period.

3.4.2 Evidence for green, brown and black worlds

Fire (charcoal) was ever present, and highly abundant, in the Middle Pleistocene section (Zone i, ii and iii in Figure 4B), while herbivores (*Sporormiella*) only appear in the uppermost samples (Zone iii in Figure 4B). The first transition (Zone i to Zone ii) is defined by a change in fire activity that is coincident with a slight drop in the average concentration of Poaceae pollen, however, no directional change in the concentration of Poaceae or tree phytoliths is discernible at this time. This suggests that vegetation state did not change significantly. The second transition

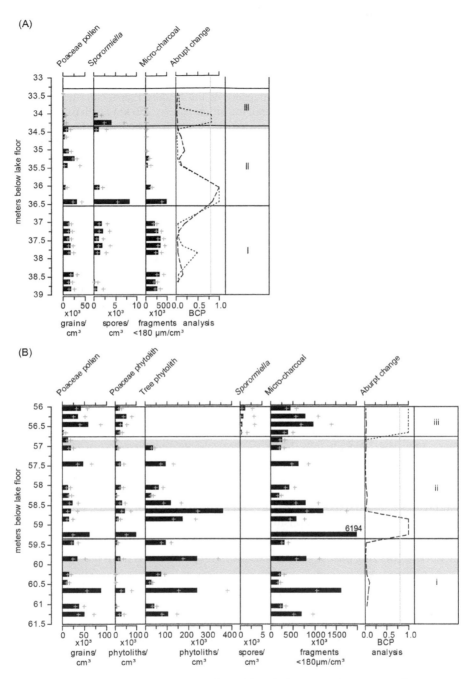

Figure 4. Palaeoecological diagram from two portions of the Lake Bosumtwi (BOS05-3B) sedimentary record: (A) Late Pleistocene (*c.* 39.0-33.0 mblf), and (B) Middle Pleistocene (*c.* 56.0–61.5 mblf). Dark grey plus symbols indicate upper and lower confidence intervals (95%). Horizontal light grey bands indicate portions of the sediment without laminations. For 'Abrupt change' curves: black dotted line indicates BCP analysis of *Sporormiella* concentration data, black dashed line indicated BCP analysis of micro-charcoal concentration data, vertical solid grey line indicates 0.8 significance threshold. In the case of one sample, at 59.25 mblf, the concentration of charcoal fragments exceeds the scales shown and the upper value is indicated numerically. Raw data available from Gosling *et al.* (2021).

(Zone ii to Zone iii) is defined by the appearance of herbivores, and is associated with the loss of tree phytoliths and the rise of Poaceae pollen and phytolith concentrations. This suggests that increased herbivory did diminish, or alter, the woody component of the vegetation. Throughout this section the high concentration of Poaceae (indicated by both pollen and phytoliths) with a variable woody component (indicated by the tree phytoliths) suggests that the landscape around Lake Bosumtwi during this short snapshot of the Middle Pleistocene was similar to the open vegetation found in the forest-savannah transition zone *c*. 250 km north of Lake Bosumtwi today (Julier *et al.* 2018), rather than the closed vegetation found close to the site today (Julier *et al.* 2019; Figure 1). Today in the transition zone fires during the dry season (November-March) limit juvenile establishment and sapling recruitment (Armani *et al.* 2018). The enhanced seasonality during this part of the Middle Pleistocene likely resulted in two prolonged and intense dry seasons during which burning could have occurred. The enhanced seasonality likely meant that fire was an even greater consumer of vegetation then than it is today in the transition zone, suggesting that this entire section likely represents 'black' world open vegetation.

Fire activity and herbivory both change through the Late Pleistocene section (Figure 4A). The first transition (Zone I to II) sees first a spike and then a persistent decline in both herbivory and fire activity. Prior to the spike (Zone I) herbivory levels and fire activity are relatively high and stable, and this is mirrored in the Poaceae concentration. After the spike (Zone II) herbivory and fire activity fluctuate, and are periodically absent, which is coincident with a slight drop in Poaceae concentration. *Sporormiella* and micro-charcoal concentrations relative to the Middle Pleistocene 'black' world section are higher and lower respectively. This suggests that in the Late Pleistocene section herbivores likely played a relatively more important role in determining vegetation structure. Therefore, it can be tentatively suggested that where *Sporormiella* is abundant a 'brown' world persisted. Although it should be noted that both herbivory and fires were clearly present and no doubt both helped to maintain the vegetation openness. Furthermore, it seems that the role of both herbivory and fire as consumers of vegetation (promoting higher Poaceae concentration and inferred openness) is diminishing through Zones I and II. The concomitant decline in consumers and Poaceae perhaps suggests a move from a consumer controlled stable state, towards a more closed canopy 'green(er)' world. The second transition (Zone II to III) is defined by a rise in herbivory, and is coincident with a drop (to near absence) in fire activity, low concentrations of Poaceae pollen, a sedimentological change (loss of laminations), and a drop in lake level (Figure 2). It is consequently unclear if the lower Poaceae pollen concentration (and inferred decreased openness) observed in Zone III are a direct result of diminishing numbers of herbivores and fire activity leading us into a 'green' world, or if other, possibly climatological or taphonomic controls are responsible.

3.4.3 Insights into past vegetation states in tropical western Africa

The preliminary data from the two sections examined here provide evidence for the presence of, and switches between, green, brown and black worlds in tropical western Africa during the Middle and Late Pleistocene. It is also apparent that different combinations of herbivores and fire in the landscape resulted in different degrees of openness (stable states) persisting. It seems likely that the background climatic configuration and seasonality played a significant role in determining the relative importance of these processes; an observation that aligns with modern studies which indicate that fire prevalence is most strongly linked to previous years' rainfall (van Wilgen *et al.* 2004). The presence of both herbivores and fire within the landscape is coincident with elevated concentration of Poaceae pollen and phytoliths in line with expectations from modern studies that suggest these vegetation consumers promote more open ecosystems (Bond 2005; Bond 2019; Warman and Moles 2009). The timescales over which the green, brown and black alternative 'stable states' persisted appears to be at least on the order of centuries to millennia.

It can be tentatively suggested that transitions between green, brown and black worlds occur over decades to centuries; however, it should also be noted that both gradual transitions seem to exist (i.e. in the Late Pleistocene section herbivory, fire activity and Poaceae all decline in a step wise fashion) as well as abrupt shifts (i.e. in the Middle Pleistocene section herbivory appears and tree phytoliths disappear at the same time). The different modes of transition suggest a complex series of interactions are responsible for initiating and driving vegetation dynamics during both time windows. This finding is in line with more recent data from savanna in eastern African that suggest that over the last 1400 years periodic fluctuations in woody cover driven by landscape processes are the rule, not the exception (Gillson 2004). Complex landscape scale processes have also been observed to be important over longer timescales during the transition out of the last interglacial period at Lake Bosumtwi (Shanahan *et al.* 2016) and over shorter timescales in the modern forest-savanna transition zone to the north (Ametsitsi *et al.* 2020). These findings emphasize the need for further empirical data to contextualize the role of processes, such as herbivory and fire activity, in determining vegetation states in Africa across space and through time.

3.4.4 Future perspectives

To determine the relative importance of herbivory and fire as consumers of vegetation, and how this changes through time, at Lake Bosumtwi further research is required. Firstly, a robust internal chronology should be established through detailed examination of the nature and frequency of the laminations, and secondly further details should be extracted from the palaeoecological and palaeoclimatic record. Specifically, detailed pollen and phytolith analyses are required at a higher temporal resolution to characterize trajectories of vegetation change. These vegetation data need to be supported by: (i) further charcoal analyses (micro- and macro-charcoal and charcoal chemistry) to reveal the biomass consumption and temperature of burning (Gosling *et al.* 2019; Whitlock and Larsen 2001), (ii) examination of a full suite of coprophilous fungal spores (not just *Sporormiella*) to provide insights into changing animal assemblages (Loughlin *et al.* 2018; van Geel and Aptroot 2006), and (iii) further independent evidence for climatic change should be sought, including further δ^{15}N analysis to determine changes in moisture balance (Talbot 2001; Talbot and Johannessen 1992), and pollen chemical analysis to identify fluctuations in solar irradiance (Jardine *et al.* 2016; Jardine *et al.* 2021).

3.5 CONCLUSIONS

The examination of Middle and Late Pleistocene sections of the Lake Bosumtwi sediment core provide evidence for, and switches between, 'green', 'brown' and 'black' worlds. The different coloured worlds (alternative stable states) each seem to persist for hundreds to thousands of years. The addition, and subtraction, of herbivory and fire activity from the landscape around Lake Bosumtwi seems to have resulted in vegetation change occurring in both an abrupt and complex gradual fashion over decades to centuries. Although the mechanisms behind the changes in herbivory and fire activity remain largely ambiguous, it seems likely that seasonality modulated by orbital forcing plays a strong role in governing fire activity. To further our understanding of the persistence of, and transition between, alternative stable states around Lake Bosumtwi we now need to improve the chronological control, enhance our sampling resolution, and expand the range of proxies analysed to provide a comprehensive insight into past ecosystem dynamics. Armed with these data we will be able to better anticipate the impact of projected climate changes for tropical western Africa.

ACKNOWLEDGEMENTS

Palaeoecological data were generated by ZG and ER as part of their BSc Biology dissertation projects supervised by WDG and CNHM. CSM and AJ assisted with the pollen identifications. WG wrote the manuscript with contributions from all authors.

DATA AVALIABILITY

All data presented in this manuscript can be downloaded from DOI: 10.6084/m9.figshare.127 38131

REFERENCES

Ametsitsi, G.K., van Langevelde, F., Logah, V., Janssen, T., Medina-Vega, J.A., Issifu, H., Ollivier, L., den Hartogh, K., Adjei-Gyapong, T. and Adu-Bredu, S., 2020, Fixed or mixed? Variation in tree functional types and vegetation structure in a forest-savanna ecotone in West Africa. *Journal of Tropical Ecology* **36**, pp. 133–149, 10.1017/S0266467420000085.

Armani, M., van Langevelde, F., Tomlinson, K.W., Adu-Bredu, S., Djagbletey, G.D. and Veenendaal, E.M., 2018, Compositional patterns of overstorey and understorey woody communities in a forest–savanna boundary in Ghana. *Plant Ecology and Diversity* **11**, pp. 451–463, 10.1080/17550874.2018.1539133.

Barry, D. and Hartigan, J.A., 1993, A Bayesian analysis for change point problems. *Journal of the American Statistical Association* **88**, pp. 309–319, 10.2307/2290726.

Blois, J.L., Williams, J.W.J., Grimm, E.C., Jackson, S.T. and Graham, R.W., 2011, A methodological framework for assessing and reducing temporal uncertainty in paleovegetation mapping from late-Quaternary pollen records. *Quaternary Science Reviews* **30**, pp. 1926–1939, 10.1016/j.quascirev.2011.04.017.

Bond, W.J., 2019, *Open Ecosystems: Ecology and Evolution Beyond the Forest Edge* (Oxford: Oxford University Press).

Bond, W.J., 2005, Large parts of the world are brown or black: A different view on the 'Green World' hypothesis. *Journal of Vegetation Science* **16**, pp. 261–266, 10.1658/1100-9233(2005) 016[0261:LPOTWA]2.0.CO;2.

Enslin, B., Potgieter, A., Biggs, H. and Biggs, R., 2000, Long term effects of fire frequency and season on the woody vegetation dynamics of the Sclerocarya birrea/Acacia nigrescens savanna of the Kruger National Park. *Koedoe* **43**, pp. 27–37, 10.4102/koedoe.v43i1.206.

Gillson, L., 2004, Testing non-equilibrium theories in savannas: 1400 years of vegetation change in Tsavo National Park, Kenya. *Ecological Complexity* **1**, pp. 281–298, 10.1016/j.ecocom. 2004.06.001.

Gosling, W.D., McMichael, C.N.H., Groenewoud, Z., Roding, E., Miller, C.S. and Julier, A.C.M., 2021, Data from: Preliminary evidence for green, brown and black worlds in tropical western Africa during the Middle and Late Pleistocene. *FigShare Data Repository*, 10.6084/ m9.figshare.12738131.

Gosling, W.D., Cornelissen, H. and McMichael, C.N.H., 2019, Reconstructing past fire temperatures from ancient charcoal material. *Palaeogeography, Palaeoclimatology, Palaeoecology* **520**, pp. 128–137, 10.1016/j.palaeo.2019.01.029.

Gosling, W.D., Miller, C.S. and Livingstone, D.A., 2013, Atlas of the tropical West African pollen flora. *Review of Palaeobotany and Palynology* **199**, pp. 1–135, 10.1016/j.revpalbo. 2013.01.003.

Ivory, S.J. and Russell, J., 2016, Climate, herbivory, and fire controls on tropical African forest for the last 60ka. *Quaternary Science Reviews* **148**, pp. 101–114, 10.1016/j.quascirev.2016. 07.015.

Jardine, P.E., Fraser, W.T., Lomax, B.H., Sephton, M.A., Shanahan, T.M., Miller, C.S. and Gosling, W.D., 2016, Pollen and spores as biological recorders of past ultraviolet irradiance. *Scientific Reports* **6**, article 39269, 10.1038/srep39269.

Jardine, P.E., Hoorn, C., Beer, M.A.M., Barbolini, N., Woutersen, A., Bogota-Angel, G., Gosling, W.D., Fraser, W.T., Lomax, B.H., Huang, H., Sciumbata, M., He, H. and Dupont-Nivet, G., 2021, Sporopollenin chemistry and its durability in the geological record: An integration of extant and fossil chemical data across the seed plants. *Palaeontology*. 10.1111/pala.12523

Juggins, S., 2005, *C2 Program version 1.5* (Newcastle upon Tyne: University of Newcastle).

Julier, A.C.M., Jardine, P.E., Adu-Bredu, S., Coe, A.L., Duah-Gyamfi, A., Fraser, W.T., Lomax, B.H., Malhi, Y., Moore, S., Owusu-Afriyie, K. and Gosling, W.D., 2018, The modern pollen-vegetation relationships of a tropical forest-savannah mosaic landscape, Ghana, West Africa. *Palynology* **42**, pp. 324–338, 10.1080/01916122.2017.1356392.

Julier, A.C.M., Jardine, P.E., Adu-Bredu, S., Coe, A.L., Fraser, W.T., Lomax, B.H., Malhi, Y., Moore, S. and Gosling, W.D., 2019, Variability in modern pollen rain from moist and wet tropical forest plots in Ghana, West Africa. *Grana* **58**, pp. 45–62, 10.1080/00173134.2018.15 10027.

King, J., Moutsinga, J. and Doufoulon, G., 1997, Conversion of anthropogenic savanna to production forest through fire-protection of the forest-savanna edge in Gabon, Central Africa. *Forest Ecology and Management* **94**, pp. 233–247, 10.1016/S0378-1127(96)03925-4.

Koeberl, C., Peck, J., King, J.W., Milkereit, B., Overpeck, J.T. and Scholz, C.A., 2005, The ICDP Lake Bosumtwi drilling project: A first report. *Scientific Drilling* **1**, pp. 23–27, 10.5194/sd-1-23-2005.

Koeberl, C., Milkereit, B., Overpeck, J.T., Scholz, C.A., Amoako, P.Y.O., Boamah, D., Danuor, S.K., Karp, T., Kueck, J., Hecky, R.E., King, J.W. and Peck, J.A., 2007, An international and multidisciplinary drilling project into a young complex impact structure: The 2004 ICDP Bosumtwi Crater Drilling Project - An overview. *Meteoritics and Planetary Science* **42**, pp. 483–511, 10.1111/j.1945-5100.2007.tb01057.x.

Koeberl, C., Reimold, W., Blum, J. and Chamberlain, C.P., 1998, Petrology and geochemistry of target rocks from the Bosumtwi impact structure, Ghana, and comparison with Ivory Coast tektites. *Geochimica et Cosmochimica Acta* **62**, pp. 2179–2196, 10.1016/S0016-7037(98)00137-9.

Loughlin, N.J.D., Gosling, W.D. and Montoya, E., 2018, Identifying environmental drivers of fungal non-pollen palynomorphs in the montane forest of the eastern Andean flank, Ecuador. *Quaternary Research* **89**, pp. 119–133, 10.1017/qua.2017.73.

Maher, J.L.J., 1981, Statistics for microfossil concentration measurements employing samples spiked with marker grains. *Review of Palaeobotany and Palynology* **32**, pp. 153–191, 10.1016/0034-6667(81)90002-6.

May, R.M., 1977, Thresholds and breakpoints in ecosystems with a multiplicity of stable states. *Nature* **269**, pp. 471–477, 10.1038/269471a0.

Miller, C.S. and Gosling, W.D., 2014, Quaternary forest associations in lowland tropical West Africa. *Quaternary Science Reviews* **84**, pp. 7–25, 10.1016/j.quascirev.2013.10.027.

Miller, C.S., Gosling, W.D., Kemp, D.B., Coe, A.L. and Gilmour, I., 2016, Drivers of ecosystem and climate change in tropical West Africa over the past ∼540 000 years. *Journal of Quaternary Science* **31**, pp. 671–677, 10.1002/jqs.2893.

Moore, P.D., Webb, J.A. and Collinson, M.E., 1991, *Pollen Analysis*, 2nd edition (Oxford: Blackwell Scientific).

Nicholson, S.E., 2009, A revised picture of the structure of the "monsoon" and land ITCZ over West Africa. *Climate Dynamics* **32**, pp. 1155–1171, 10.1007/s00382-008-0514-3.

Piperno, D.R., 2006, *Phytoliths: A Comprehensive Guide for Archaeologists and Paleoecologists* (Lanham: Alta Mira Press).

Riginos, C., Porensky, L.M., Veblen, K.E., Odadi, W.O., Sensenig, R.L., Kimuyu, D., Keesing, F., Wilkerson, M.L. and Young, T.P., 2012, Lessons on the relationship between live-stock husbandry and biodiversity from the Kenya Long-term Exclosure Experiment (KLEE). *Pastoralism: Research, Policy and Practice* **2**, 10.1186/2041-7136-2-10, article 10.

Roques, K.G., O'Connor, T.G. and Watkinson, A.R., 2001, Dynamics of shrub encroachment in an African savanna: Relative influences of fire, herbivory, rainfall and density dependence. *Journal of Applied Ecology* **38**, pp. 268–280, 10.1046/j.1365-2664.2001.00567.x.

Rucina, S.M., Muiruri, V.M., Kinyanjui, R.N., McGuiness, K. and Marchant, R., 2009, Late Quaternary vegetation and fire dynamics on Mount Kenya. *Palaeogeography, Palaeoclimatology, Palaeoecology* **283**, pp. 1–14, 10.1016/j.palaeo.2009.08.008.

Sankaran, M., Ratnam, J. and Hanan, N., 2008, Woody cover in African savannas: The role of resources, fire and herbivory. *Global Ecology and Biogeography* **17**, 236–245, 10.1111/j.1466-8238.2007.00360.x.

Scheffer, M., Carpenter, S., Foley, J.A., Folke, C. and Walker, B., 2001, Catastrophic shifts in ecosystems. *Nature* **413**, pp. 591–596, 10.1038/35098000.

Shanahan, T.M., Peck, J.A., McKay, N., Heil, C.W., King, J., Forman, S.L., Hoffmann, D.L., Richards, D.A., Overpeck, J.T. and Scholz, C., 2013a, Age models for long lacustrine sediment records using multiple dating approaches - An example from Lake Bosumtwi, Ghana. *Quaternary Geochronology* **15**, pp. 47–60, 10.1016/j.quageo.2012.12.001.

Shanahan, T.M., McKay, N., Overpeck, J.T., Peck, J.A., Scholz, C., Heil, C.W. and King, J., 2013b, Spatial and temporal variability in sedimentological and geochemical properties of sediments from an anoxic crater lake in West Africa: Implications for paleoenvironmental reconstructions. *Palaeogeography, Palaeoclimatology, Palaeoecology* **374**, pp. 96–109, 10.1016/j.palaeo.2013.01.008.

Shanahan, T.M., Hughen, K.A., McKay, N.P., Overpeck, J.T., Scholz, C.A., Gosling, W.D., Miller, C.S., Peck, J.A., King, J.W. and Heil, C.W., 2016, CO_2 and fire influence tropical ecosystem stability in response to climate change. *Scientific Reports* **6**, 29587, 10.1038/srep29587.

Staver, A.C., Bond, W.J., Stock, W.D., van Rensburg, S.J. and Waldram, M.S., 2009, Browsing and fire interact to suppress tree density in an African savanna. *Ecological Applications* **19**, pp. 1909–1919, 10.1890/08-1907.1.

Stebbing, E.P., 1937, *The Forests of West Africa and the Sahara* (London and Edinburgh: W. and R. Chambers Ltd.).

Stockmarr, J., 1971, Tablets with spores used in absolute pollen analysis. *Pollen et Spore* **XIII**, pp. 615–621.

Talbot, M.R., 2001, Nitrogen isotopes in palaeolimnology. In *Tracking Environmental Change Using Lake Sediments. Volume 2. Physical and Geochemical Methods* edited by Last, W.M. and Smol, J.P., (Dordrecht, Boston and London: Kluwer Academic Press), pp. 401–439.

Talbot, M.R. and Johannessen, T., 1992, A high resolution palaeoclimatic record for the last 27 500 years in tropical West Africa from the carbon and nitrogen isotopic composition of lacustrine organic matter. *Earth and Planetary Science Letters* **110**, pp. 23–37, 10.1016/0012-821X(92)90036-U.

van Geel, B. and Aptroot, A., 2006, Fossil ascomycetes in Quaternary deposits. *Nova Hedwigia* **82**, pp. 313–329, 10.1127/0029-5035/2006/0082-0313.

van Wilgen, B., Govender, N., Biggs, H., Ntsala, D. and Funda, X., 2004, Response of savanna fire regimes to changing fire-management policies in a large African national park. *Conservation Biology* **18**, pp. 1533–1540, 10.1111/j.1523-1739.2004.00362.x.

Warman, L. and Moles, A.T, 2009, Alternative stable states in Australia's wet tropics: A theoretical framework for the field data and a field-case for the theory. *Landscape Ecology* **24**, pp. 1–13, 10.1007/s10980-008-9285-9.

White, F., 1983, *The Vegetation of Africa: A Descriptive Memoir to Accompany the Unesco/AETFAT/UNSO vegetation map of Africa* (UNESCO).

Whitlock, C. and Larsen, C., 2001, Charcoal as a fire proxy. In *Tracking Environmental Change Using Lake Sediments Volume 3: Terrestrial, Algal and Siliceous Indicators*, edited by Smol, J.P., Birks, H.J.B., and Last, W.M., (Dordrecht, Boston and London: Kluwer Academic Press), pp. 75–98.

CHAPTER 4

Holocene high-altitude vegetation dynamics on Emi Koussi, Tibesti Mountains (Chad, Central Sahara)

Michèle Dinies[1], Lena Schimmel & Philipp Hoelzmann

Freie Universität Berlin, Institute of Geographical Sciences, Berlin, Germany

Stefan Kröpelin

University of Cologne, Institute of Prehistoric Archaeology, Africa Research Unit, Cologne, Germany

Frank Darius

Technische Universität Berlin, Institute for Ecology, Berlin, Germany

Reinder Neef

German Archaeological Institute (DAI), Scientific Department of the Head Office, Berlin, Germany

ABSTRACT: High mountains are sensitive to climate change. Holocene lake remnants witness to the last Saharan Humid Period (SHP) in the Tibesti, the highest mountains of the Sahara (>3000 m asl). The impact of Holocene climatic change on the terrestrial montane ecosystems, however, is largely unknown. Palynological investigations on a palaeolake sequence from the summit caldera of the Emi Koussi, the highest volcanic complex of the Central Sahara, were performed to address this issue. Based on charred plant particle concentrations, the age depth model indicates a Holocene age of the palaeolake (*c.* 9500-5500 cal yr BP). Shifts in the vegetation composition indicate altered climatic conditions: high proportions of Poaceae and fern as well as highly fluctuating *Artemisia* frequencies characterize the SHP, strongly dominating *Ephedra distachya*-type proportions reflect the present vegetation, evidencing distinctly decreased humidity. The missing mass expansion of tropical tree taxa into the summit zone meanwhile points to persisting low temperatures during the Holocene. The absence of Mediterranean tree taxa during the SHP indicates that migration from the Mediterranean to the southern Sahara was not possible. Hence, a survival of Mediterranean tree taxa within micro-refugia may also be excluded. An exception is *Erica (arborea)* which expanded only during the Middle Holocene, relying on relict populations. Increased precipitation at persisting lower temperatures thus shaped the

[1] Other affiliation: German Archaeological Institute (DAI), Scientific Department of the Head Office, Berlin, Germany

high altitudinal ecosystems during the SHP, triggering modifications in the vegetation composition. The ecosystem type itself, a montane steppe, remained. This is in clear contrast to the contemporaneous marked biome changes in the surrounding Saharan plains.

4.1 INTRODUCTION

Due to their positions amidst today's largest hot desert, the mountain complexes Hoggar and Tibesti in the central Sahara are exceptional regions, resembling islands of increased biodiversity. Even though the altitudinal change in precipitation and temperature leads to more mesic conditions at higher elevations, there is no montane tree belt in these Central Saharan massifs, distinguishing them from the tropical mountains in East Africa as well as from the Mediterranean mountain ranges in the north. A combination of aridity and low temperatures is claimed to prevent major tree growth in the Hoggar and Tibesti, fostering at their summit a vegetation formation dominated by species of the genera *Ephedra, Artemisia* and *Pentzia*, called 'pseudo-steppe' or 'steppe culminale' (e.g. Maire and Monod 1950; Messerli and Winiger 1992; Ozenda 1991; Quézel 1965). The steppe culminale fits the criteria of palearctic steppes characterized by (a) the dominance of herbs, mainly grasses with high chamaephyt proportions in some regions (b) more than 10% vegetation cover (c) occurring in climates too dry for taller woody vegetation and at least occasionally affected by frost (Wesche *et al.* 2016). This definition thus includes the naturally occurring grasslands of the Mediterranean and of the subtropical-tropical arid zones as steppe, similar to the definition of Le Houérou (2001, 2009).

Because of the strong dominance of chamaephyts, the steppe culminale of the Tibesti may be assigned to the desert steppes, characterized by vegetation cover of only 10–25% with often increased dwarf shrub proportions, a vegetation formation mediating to subtropical deserts (cf. Wesche *et al.* 2016). The present heterogenous biogeographical composition of the Central Saharan montane ecosystems points to multiple floristic relations: inner Saharan, Sahelian, Afro-alpine East African, and links to the Mediterranean phytoregion. The latter are documented by Mediterranean woody plants at altitudes higher than 1500 m asl (*Cupressus dupreziana, Erica arborea, Olea europaea* subsp. *laperrinei* and *Rhus tripartite*) and numerous annuals and perennials. However, most of these woody taxa only occur in the Hoggar mountains, compliant with higher proportions of Mediterranean dwarf shrubs and annuals. The more northerly position of the Hoggar massif in comparison to the Tibesti and thus (closer) proximity to the Mediterranean phytoregion accounts for this difference (Maire 1935; Maire and Monod 1950; Médail and Quézel 2018; Ozenda 1991; Quézel 1965).

As to the African Humid Period, several dated palynological and anthracological investigations in the surrounding plains of the Eastern and Central Sahara provide evidence of profound ecosystem changes, with tropical plant taxa migrating some 500 km northwards compared to their present day distribution, thus as far north as the Tibesti or even further north (Gabriel 1977; Haynes *et al.* 1989; Hély and Lézine 2014; Jahns 1995; Lézine 2017; Maley 1981; Ritchie *et al.* 1985; Ritchie 1987; Ritchie 1994; Ritchie and Haynes 1987; Schulz 1987, 1991; Schulz 1991; Mercuri *et al.* 1998; Mercuri 1999; Mercuri and Grandi 2001; Neumann 1989; Trevisan *et al.* 1998; Watrin *et al.* 2009). There is, however, no clear evidence of a similar distinct southward migration of Mediterranean plant taxa into the Central Saharan plains during the Holocene.

An intensified and northwards expanded summer monsoon is supposed to be the main humidity source (Kutzbach 1981). Other atmospheric patterns are assumed to have contributed as well, such as e.g. seasonal tropical plumes and extratropical troughs (Dallmeyer *et al.* 2020; Knippertz 2005; Maley 2000; Skinner and Poulsen 2016), and interactions of extratropic cyclones with the monsoonal system, either with or without overlaps (e.g. Baumhauer 2004; Rogerson *et al.* 2019) or with additional interferences of moisture from the Indian Ocean (Mercuri *et al.* 2018) are discussed. Feedback mechanisms additionally seem to play a role. In models, interactive coupling between atmospheric and oceanic dynamics as well as land surface amplify the African monsoon (e.g. Braconnot *et al.* 1999, 2007; Claussen and Gayler 1997; Claussen *et al.* 1999; Kutzbach and Liu 1997; Pausata *et al.* 2016). The influence of land cover feedbacks is controversial (e.g. Braconnot *et al.* 1999 vs. Braconnot *et al.* 2007), however a positive feedback seems probable (e.g. Krinner *et al.* 2012).

Palaeolake remnants in the montane belt of the Tibesti and Hoggar massifs document the impact of increased precipitation during the African humid period (e.g. Jäkel and Geyh 1982; Maley 1981; Médail and Quézel 2018). The vegetation response however remains vague, because to date only single or few palynological samples scattered throughout the Central Saharan Mountains have been investigated, allowing only very punctual insights. Samples of the Hoggar tentatively dated to the beginning of the Pleistocene show high proportions of temperate and plant taxa, interpreted as relict woodlands of the Tertiary. During the Middle and Late Pleistocene, the proportions of tropical plant taxa as well as types representing the Saharan desert vegetation increased, while Mediterranean and temperate taxa persisted, however with decreasing importance. This led to the hypothesis that the desert vegetation established successively in the plains and thus acted as a barrier, isolating the mountain ranges, which however were connected by hydrogeographic networks activated during periods of increased humidity (Rossignol and Maley 1969). As to the African Humid Periods, the involvement of Mediterranean taxa in the Hoggar/Tassili and Tibesti mountain complexes remains ambiguous. Some authors suggest an expansion of Mediterranean or even temperate tree taxa in the Central Saharan mountains, while a re-investigation in the Hoggar and analyses on lake remnants in the Tibesti revealed only the few Mediterranean tree taxa still present today (e.g. Alimen *et al.* 1968; Gabriel 1977; Quézel 1958; Quézel 1965; Maley 1981; Maley 2000; Prentice *et al.* 2000; Medail and Quézel, 2018 versus Thinon *et al.* 1996 and own investigations).

Here, we present the first pollen diagram encompassing the period *c.* 9500-5500 cal yrs BP, based on palynological investigations of a ^{14}C-dated sediment sequence from the deepest subcaldera of the highest volcanic shield complex, the Emi Koussi at about 3000 m asl. This analysis complements existing single sample investigations from various different geomorphological contexts. Relying on the new sequence, we will: a) present a chronology of the investigated lake remnant based on ^{14}C-dates of organic material, b) indicate Holocene vegetation fluctuations at these highest altitudes and c) deduce climatic configurations from these vegetation reconstructions.

4.2 REGIONAL SETTING

4.2.1 Geomorphology

The Tibesti massif is one of the world's major and most significant examples of intracontinental volcanic provinces and covers more than 100,000 km^2 (Ball *et al.* 2019). Intense volcanic activity began as early as the Oligocene and intensified during the Quaternary (Deniel *et al.* 2015; Gourgaud and Vincent 2004). The Emi Koussi is the largest volcano of the entire Tibesti Volcanic Complex, with a basal diameter from 60 to 80 km and a summit caldera of 9–12 km in diameter. Altitude differences vary between *c.* 3000 m compared to the Bodélé depression in the south and *c.* 2300–2900 m to the Serir Tibesti in the north (Pachur and Altmann 2006). A succession of three volcanic sequences between 2.4 and 1.3 Ma formed the Emi Koussi volcano and each is related to the collapse of a caldera. Era Kohor is the youngest caldera that was formed during the last sequence (Gourgaud and Vincent 2004). Basalts overlaying Terminal Pleistocene to Holocene limnic sediments in the Emi Koussi caldera indicate volcanic eruptions until the recent past (Messerli 1972). Seismic activities are recorded even nowadays. Groundwater-coupled processes like solfatars/fumaroles and hot springs are still active on the eastern slopes of Emi Koussi (Pachur and Altmann 2006).

4.2.2 Climate

In the plains surrounding the Tibesti, hot desert climates (BWh; Köppen 1918) prevail, whereas in the mountains the climatic conditions change due to the adiabatic lapse rate to cold desert climates (BWk) with mean annual temperatures below 18°C. Climate data are not available for the Emi Koussi, but some meteorological data from lower locations indicate the 'classical' precipitation increase and temperature decrease with increasing altitudes. Rainfall at the Trou au Natron at 2450 m asl, thus 1000 m below the sampled Emi Koussi lake remnant, amounts to an average annual precipitation of 93.3 mm and a mean annual temperature of 13.5° C (Gavrilovic 1969). The few data allowing vague insights into the distribution throughout the year indicate a precipitation maximum during May and another during August, with highest monthly means during July (Maire and Monod 1950). For the highest altitudes of the Emi Koussi, precipitation estimations based on single measurements and deduced from altitude, geomorphology and vegetation range from 50 to 150 mm per year (Wininger 1972, Figure 1, Vegetation map). In contrast to the lower mountains and plains that are mainly affected by monsoonal summer precipitation, the high altitudes seem to profit from orographic, frequent but minor precipitations. Coupled with clouds, this effective moisture source propagates plant growth (Quézel 1958).

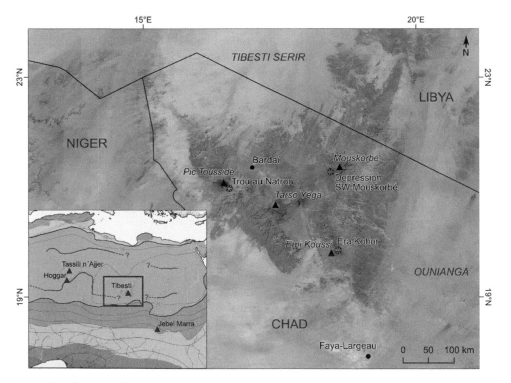

Figure 1. Satellite image (Esri World Imagery) of the Tibesti massif and its surrounding plains, Central Sahara. Inlet map: Schematic map of the vegetation zones (centres of endemism: III – Sudanian, VII – Mediterranean, and the regional transition zones: XI – Guinea-Congolia/Sudania, XVI – Sahel, XVII – Sahara, XVIII – Mediterranean/Sahara), following White (1983). The reconstructed northern limits of Sudanian and Sahelian plant taxa during the SHP are indicated as green lines (Neumann 1989; Schulz 1991). Isohyetes (in grey) are from Railsback *et al.* (2016).

4.2.3 Today's vegetation of the Tibesti

After Maire and Monod (1950), Quézel (1958), Ozenda (1991) and Médail and Quézel (2018), three main altitudinal belts can be distinguished in the Tibesti and Hoggar Mountains: a Saharan-Tropical belt similar to the contracted vegetation surrounding the mountains, followed by a montane Saharan-Mediterranean belt with increasingly diffuse plant cover, and a summit zone. The generalized vegetation map after Quézel (1958, 1964, 1965) shows this altitudinal controlled vegetation distribution on the Emi Koussi volcano itself (Figure 2).

(1) Saharan-Tropical desert vegetation of the plains: contracted desert vegetation with *Panicum turgidum* and *Acacia ehrenbergiana* dominates.

(2) Saharan-Tropical belt up to *c*. 1700 m asl: the vegetation resembles the desert vegetation of the surrounding plains, with Sahelian plant taxa such as acacias (*A. tortilis* var. *raddiana, A. laeta, A. stenocarpa*), *Maerua crassifolia* and *Capparis decidua* in the wadis, and *Hyphaene thebaica* and *Salvadora persica* bordering them. Together with a grassy vegetation formation in depressions or runnels (*Aristidia funiculata – Indigofera sessiliflora* formation), they are supposed to represent remnants of plant formations that expanded northwards during the African Humid Period (Médail and Quézel 2018).

(3a) Lower montane zone of the Saharan-Mediterranean belt, *c*. 1700–2500 m asl: in this zone the Sahelian vegetation fades out, and the plant cover becomes increasingly diffuse. *Acacia stenocarpa* and *A. laeta*, different *Ficus* species, *Grewia tenax, Rhus, Capparis spinosa* and *Boscia salicifolia* grow in the wadis. Montane *Aristida* grasslands (3b1) cover the surfaces between the wadis, with *Linaria* and *Helianthemum* as characteristic taxa. The two *Acacia* species form the upper tree line at altitudes of 1800–2300 m asl, depending on the exposition, while dwarf grown acacias may even reach 200 m higher (Messerli 1972; Quézel 1965). As to the Sahelian affinity of the acacias, temperature is claimed as the limiting factor for tree growth. Yet, aridity seems to be a relevant factor as well: in the Hoggar mountains relict stands of *Olea europaea* subsp. *laperrinei* trees grow in caves at 2700 m asl, and at the Emi Koussi *Erica arborea* grows in ravines between 2500 and 3000 m asl, indicating that theoretically these Mediterranean trees should constitute the upper timber line at higher altitudes or even cover the highest summit. Quézel (1965) thus assumed that these and other Mediterranean (and Temperate) elements disappeared 4000–5000 years ago as a result of aridification.

(3b2) Upper montane zone of the Saharan-Mediterranean belt (steppe culminale), above *c*. 2500 m asl: dwarf shrubs prevail at these highest altitudes. While in the Hoggar mountains *Artemisia herba-alba* (*Seriphidium herba-alba* (Asso) Sojak), *Artemisia campestris* (subsp. *glutinosa* (Besser) Blatt.), *Ephedra major* Host and *Pentzia monodiana* Maire form these summit steppes, endemic counterparts, *Artemisia tilhoana* Quézel and *Ephedra tilhoana* Maire, replace them, with *Pentzia monodiana, Ballota hirsuta* and *Lotus tibesticus* accompanying them as characteristic species. *Euphorbia granulata, Erodium oreophilum, Campanula bordesiana, Helianthemum lippi* and *Helichrysum monodianum* are also common in these steppes. This is supposed to represent a vegetation formation with a long-lasting Quaternary history (Quézel 1958, 1965).

4.3 MATERIALS AND METHODS

4.3.1 Sampling

Within the Era Kohor caldera, palaeolake remnants are preserved at the NW slopes, covered by volcanic talus material (Figures 2 and 3). These diatomaceous sediments exhibit a thickness of 145 cm and were sampled at an altitude of 2796 m asl (top of the sampled sequence;

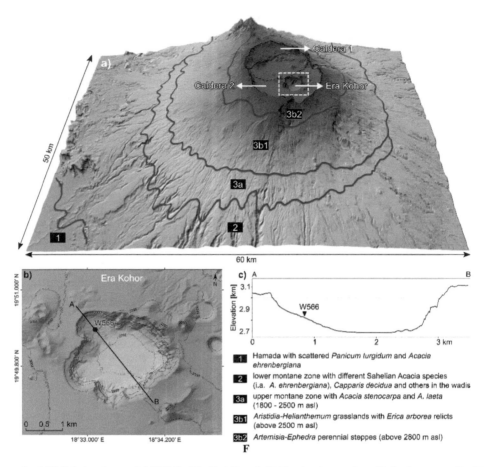

Hamada with scattered *Panicum turgidum* and *Acacia ehrenbergiana*

lower montane zone with different Sahelian Acacia species (i.a. *A. ehrenbergiana*), *Capparis decidua* and others in the wadis

upper montane zone with *Acacia stenocarpa* and *A. laeta* (1800 - 2500 m asl)

Aristidia-Helianthemum grasslands with *Erica arborea* relicts (above 2500 m asl)

Artemisia-Ephedra perennial steppes (above 2800 m asl)

Figure 2. a) Digital elevation model (DEM) of the Emi Koussi shield volcanic complex with the three summit calderas after Gourgand and Vincent (2004) and the generalized vegetation zones after Quezel (1958; 1964; 1965) and Maire and Monod (1935, 1950). b) DEM of the Era Kohor caldera with the position of section W566 c) NW-SE profile across the Era Kohor caldera.

19° 50.376' N, 18° 33.177' E), some 130 m above the present crater floor. The desiccated sediments (diatomite) were sampled in subsequent monoliths (W566-1 to W566-7; Figure 2). Sub-sampling of the monoliths was undertaken at 1 cm sample intervals where possible, after export of the monoliths to the laboratory. However, some monoliths did not reach the laboratory intact and for these segments a single bulk sample was taken.

4.3.2 Preparation of pollen samples and charred plant particles concentrates for ^{14}C-dating

In a first step, samples at intervals of about 10 cm were prepared. Pollen preservation proved highly variable throughout the profile. Subsequently the intervals with pollen preservation were chosen for further palynological investigations. Samples were treated following common acid/base, heavy liquid and acetolysis protocols, thoroughly washing the samples with distilled water after each treatment (Eisele *et al.* 1994; Faegri and Iversen 1989; Moore *et al.* 1991;

Nakagawa *et al.* 1998). A volume of $10cm^3$ sediment (diatomite) was prepared, adding marker spores (Stockmarr 1971).

Pollen concentrations were prepared applying a combination of different protocols (Chester and Prior 2004; Brown *et al.* 1992; Regnéll and Everitt 1996; Nakagawa *et al.* 1998; Vandergoes and Prior 2003; Piotrowska *et al.* 2004) with the following preparation steps: (1) Treatment with $\sim$10% HCl (heated), (2) Treatment with $\sim$10% KOH (heated), (3) Dense-media separation with sodium-polytungstate (SPT; $\sim$2.1 g/cm^3), (4) Treatment with 12 M H_2SO_4 for 2 h, with 1.2 M H_2SO_4 for 3 h (heated), (5) Microscopic analysis of the residues to decide on the next steps for the processing procedure, (6) Density separation (1.6-1.2 g/cm^3) with CsCl or repeated with 2.1 g/cm^3 SPT, with microscopic control of the potential different phases, (7) Sieving (<6 μm).

Pollen slides were analysed using a transmitted light microscope (Leica DMRB) with 400–1000× magnification. Pollens were identified by using the reference collection of the DAI (Deutsches Archäologisches Institut, Berlin, Germany), electronic photographs of the African Pollen Database (downloaded in 2012, www://medias.obs-mip.fr/apd) and published pollen atlases (Bonnefille and Riollet 1980; El-Ghazali 1993; Reille 1992).

4.4 RESULTS

4.4.1 Lithostratigraphy and radiocarbon dating of W566

The diatomite is faintly laminated in its entity, structured by oxidized bands, mainly ochre, some of them sustaining (Figure 3).

Six samples of the W566 sequence were radiocarbon dated (Table 1). The total carbon content of the diatomaceous sediments is generally extremely low and in consequence the datable amount of organic carbon in the prepared concentrates is also low. Macroscopic charred plant remains were not found in the sampled sequence. Thus, concentrates of microscopic charred plant particles were prepared for ^{14}C-dating to enrich organic carbon for the necessary minimum mass of *c*. 0.1 mg to produce reliable ages. However, below 0.1mg C uncertainty increases correspondingly (Kutschera 2016), in addition the risk of contamination during sample preparation rises with decreasing sample mass. The sample with only 0.04 mg C (Poz-12905) has been excluded from calculations of the age depth model because of the methodological uncertainties, as in other investigations (e.g. Parker *et al.* 2004). To establish an age depth model, rbacon, a Bayesian approach with an implemented outlier analysis was chosen (version 2.4.3; Blaauw and Christen 2011). The basal 10 cm were devoid of datable organic material, the ages for these 10 cm were thus extrapolated. With these first absolute datings of the diatomite of the Era Kohor caldera, we prove a Holocene age of the lake remnant.

4.4.2 Pollen sum and diagram

The pollen sum varies between 63 and 910, with 304 on average. It includes all terrestrial types as well as monolete fern spores. Ferns are supposed to have been common on terrestrial sites during the African Humid Period. Similarly, ferns were included in the main sum in earlier investigations in the Tibesti (Maley 1981) and in Yemen (Lézine *et al.* 2007). Aquatics and other spores are excluded. The pollen sum strongly depends on the pollen concentration and preservation of the respective sample: very low pollen concentrations and/or higher proportions of corroded pollen resulted in part in low pollen sums. The diagram was drawn with Tilia (Grimm 2020). The pollen types are grouped according to their ecology in representatives of desert vegetation (of the plains), altitudinal steppes, thorn (and deciduous) savanna, and riparian and water vegetation.

Table 1. Radiocarbon dates from section W566 from the Era Kohor caldera. The ^{14}C dates were calibrated with OxCal. The coverage of the different dated organic components is indicated as percentage values for each sample respectively.

Lab-Nr.	Depth (cm) from top of W566	^{14}C yrs BP	cal yrs BP (2sigma)	Material dated	mg C
Beta441716	0–15	4720±30	5320–5580	90% charred plant particles, 10% tissues and pollen	no data
Poz122314	33–36	7480±50	8180–8380	90% charred plant particles, 10% probably limnic tissues	0.3
Poz122366	95–96	7700±40	8400–8590	85% charred plant particles; 10% minerogenic particles; 5% algae remains	0.8
Beta441715	123–125.5	8360±40	9150–9490	90% charred plant particles, 10% tissues and pollen	no data
Poz125905	132–134	7330±170	7790–8460	80% charred plant particles; 10% minerogenic particles; 10% tissues and pollen	0.04 not considered
Poz122367	136–137	7510±50	8190–8400	80% charred particles, 20% minerogenic particles	0.35

4.5 DISCUSSION

4.5.1 Age-depth model

The sequence of the five ^{14}C-dates clearly indicates a Holocene age of the lake sediments in the Era Kohor subcaldera of the Emi Koussi. Considering the distribution of the ^{14}C-dates in respect to age versus depth, two possibilities can be proposed: (a) The ^{14}C-date at 33–36 cm (Poz-122314) exhibits nearly the same age as the lowest sample (Poz-122367). This can be interpreted as indicating very high sedimentation rates, about 1 m during a few centuries around 8200 cal yr BP. The uppermost sample at 0–15 cm dating to about 5400 cal yr BP, would thus point to a marked hiatus between the basal metre and the upper 30 cm. (b) Assuming approximately constant sedimentation rates, the dates at 33–36 cm (Poz-122314) has to be considered as too old. As the lithology seems uniform and there are no distinct features indicating a hiatus in the uppermost 30 cm, we decided to consider possibility (b).

Age-depth models can be approached by linear inter- and extrapolation or by using e.g. Bayesian statistics. The latter method is based on a constrained ordering of the ^{14}C-dates and simulating numerous different accumulations in between dated depths. Especially for sequences with low dating densities as well as dating scatter and outliers, a Bayesian approach is recommended (Blaauw *et al.* 2018). We therefore chose rbacon to calculate the age depth model. Because no ^{14}C-dates are available for the basal 10 cm, ages have been extrapolated. The resulting age-depth model is shown in Figure 3: two radiocarbon dates (Poz-122314 and Poz-122367) were omitted during model simulation.

Despite the scattering of the ^{14}C-dates, our new radiocarbon datings of organic material that is not susceptible to hard water/reservoir effect indicate an Early and Middle Holocene age of the lake remnant. The Bayesian approach, adapted to scattered data, resulted in an age-depth model indicating a deposition of lake sediments between *c*. 9650 and 5500 cal yr BP.

Based on the published data in Lézine *et al.* (2011b) and complemented by ^{14}C-dates published in Jäkel and Geyh (1982), an extended dataset of published ^{14}C-dates for the Tibesti mountains was compiled. Following a modified classification scheme by Lézine *et al.* (2011b), the dates were assigned to the categories lacustrine, fluviatile terraces, playas, palustrine, arid and cultural and summed per millennium (Figure 6). The frequency distribution of dates of lacustrine contexts peaks during the Early Holocene, as do dates from fluviatile terraces. Considering only ^{14}C-datings other than calcareous crusts, which are possibly susceptible to the hard water effect and volcanic CO_2-gazing, sharpens the peak to the Early Holocene. The newly dated lacustrine sequence from the Emi Koussi caldera, which covers approximately the period 9500–5500 cal yr BP, thus fits into the palaeohydrological scheme of the Tibesti mountain range.

4.5.2 Surface pollen sample and present vegetation

The surface sample, a salt crust from the present-day bottom of the sub-caldera, shows a strong dominance of *Ephedra distachya*-type pollen, followed by high *Artemisia* frequencies and some *Anthemis*-type pollen, probably representing *Pentzia* (see Figure 4; *Pentzia* pollen belong to the *Anthemis*-type, cf. Vezey *et al.* 1994). The surface pollen spectrum thus represents the common chamaephyte steppe with the endemic species *Ephedra tilhoana* and *Artemisia tilhoana*, occurring at altitudes between 2400 m asl and the summit (Quézel 1959). The relationship of the endemic *Ephedra tilhoana* is not investigated genetically. Based on morphological traits alone, two *Ephedra* species are assumed to be closely related (or identical) to the Tibestian endemite: *Ephedra major* Host var. *suggarica* Maire (if not flowering/fruiting) and *E. distachya* L. (H. Freitag, pers. comm. 2020), both in accordance with the dominating *Ephedra distachya* pollen type we identified in our samples. The distribution of both taxa indicates the above-stated relation to the (altitudinal) Mediterranean and Asian steppe (see Figure 4). The affinities of the endemic *Artemisia tilhoana* are less clear. The global distribution and core area of *Artemisia* coincide, however, with the distribution of *Ephedra*, pointing similarly to close relations to the oro-Mediterranean and Asian steppes (e.g. Meusel and Jäger 1992).

4.5.3 Early to Middle Holocene vegetation at the summit of Emi Koussi

Migration of plant taxa in altitude, changes in the plant composition and substitution of the respective ecosystem are possible vegetation responses to climate fluctuations in mountains (e.g Messerli and Winiger 1992).

4.5.3.1 Altitudinal plant migration
In order to detect altitudinal migration, we define plant taxa characteristic for the lower montane belt between about 1700 and 2500 m asl which can be traced palynologically. Common genera with distinct pollen types are: *Acacia, Rhus, Fluggea, Capparis, Boscia, Cordia, Commicarpus/Boerhavia, Grewia* (*tenax*) and *Fagonia* (Maire and Monod 1950; Quézel 1959). Some of these types are present in the Early and Middle Holocene spectra: *Acacia, Capparis/Cadaba* and *Grewia*. However, in spite of the known under-representation of these pollen types (e.g. Ritchie 1994), the low frequencies point to the presence of these thorn savanna taxa close to highest altitudes, but do not indicate a distinct invasion of tropical plant taxa. The aforementioned and postulated persistence of lower temperatures during the Early and Middle Holocene are a plausible explanation of why there is no marked upward migration of tropical taxa at highest altitudes during the Holocene.

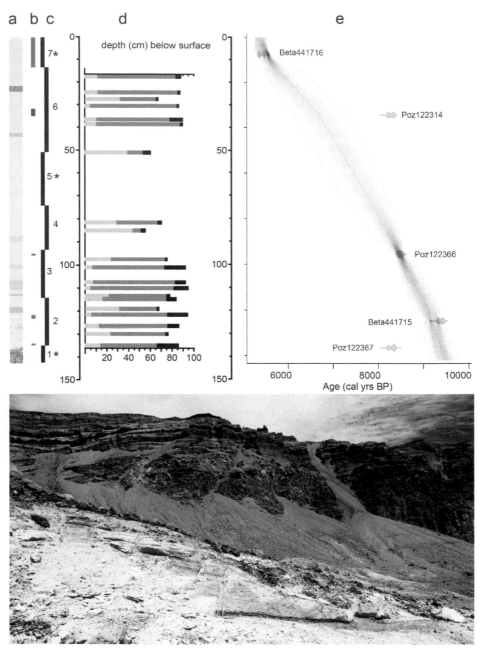

Figure 3. Stratigraphy, summary pollen diagram and age-depth plot of section W566. (a) Schematic lithography of the sampled section. (b) The grey lines indicate the position and thickness of the ^{14}C-sample. (c) The black lines indicate the position and thickness of the sampled monoliths. (*) sections constitute a single sample (damaged during transport). (d) Percentage values of a few selected vegetation formations/pollen types are summed up (light grey – altitudinal steppe, dark grey – grasslands, black – ferns) and plotted against the depth, thus indicating the position of the analysed pollen samples. The same graph is plotted against age in Figure 5. The depth scale is the same in all graphs. (e) Age-depth model (simulated with rBacon).

Figure 4. Above: Surface sample. The percentage values of the recorded pollen types are shown. Middle: Photo of the montane steppe at the Emi Koussi. Below: Geographic distribution of the assumed two closest relatives of the endemic *Ephedra tilhoana*, *E. major* Host and *E distachya* L. Only occurrences documented by plant samples are plotted (GBIF.org).

4.5.3.2 Mediterranean plant taxa migration?

Lower temperatures in combination with increased moisture should have allowed the spread of Mediterranean tree taxa. Today, as mentioned in the introduction, *Erica arborea* is the only woody taxa with Mediterranean distribution in the Tibesti, occurring in a few small populations in ravines at the Emi Koussi at altitudes between 2500 and 3000 m asl, interpreted as relict. Genetic studies on *Erica arborea* indicate a range of expansion during the Pleistocene from the Eastern Africa/Arabia centre westward, pointing to its oro-African relations while colonizing the Mediterranean region and North Africa. Range fragmentation in Northern Africa resulted in the relicts in the Tibesti mountains (Bruneau de Miré and Quézel 1959; Médail and Quézel 2018) and is connected to the expansion of the Saharan desert. Genetic studies of *Erica arborea* imply a spread of this plant taxa during the African Humid Period (Désamoré *et al.* 2011). Few sporadic pollen records of *Erica* before 6000 cal yr BP and higher frequencies about 6000 cal yr BP may indicate the anticipated expansion near today's relict on the Emi Koussi. With proportions of 6.5% *Erica* pollen in the youngest samples of a Trou au Natron diatomite (Maley 1981), larger *Erica* stands are probable in the Tibesti mountains, corroborated by continuous *Erica* pollen input during the 6th–4th millennia BP in Lake Yoa in the plains E of the Tibesti (Kröpelin *et al.* 2008; Lézine *et al.* 2011a). Except for these *Erica* pollen, no Mediterranean nor Temperate plant taxa (excl. the steppic chamaephytes) are recorded in the sequence of twenty samples from the Era Kohor lake deposit. The few and low frequencies of, e.g. *Pinus* and *Tilia*, are interpreted as long-distance transport. In the palynological investigations of the scattered palaeolake sediments of the Tibesti, neither Mediterranean nor Temperate pollen types are recorded (Maley 1981), in line with palynological re-investigations of Mid-Holocene hyrax midden in the Tassili mountains, evidencing only Mediterranean trees as still present today (Thinon *et al.* 1996). This argument is strengthened by the lack of these elements in the charcoal spectra and only lowest pollen proportions of temperate trees, interpreted as long distance transported, in a joint pollen and charcoal investigation in the Tassili mountains (Amrani 2018). The samples from terraces as well as probably dammed up sediments near Bardai in the Tibesti revealed spectra dominated by temperate and Mediterranean pollen types (compiled in Gabriel 1977). Considering the depositional settings, a re-working of older, Pleistocene sediments bearing Mediterranean and Temperate pollen (see Introduction, Rossignol and Maley 1969) can explain these spectra. Relying on the first sequence from a palaeolake, and underpinned by the above-mentioned investigations (Maley 1981; Thinon *et al.* 1996) and our own ongoing analyses of the Early to Middle Holocene sediments of Lake Yoa (Dinies *et al.* 2019), an expansion of Mediterranean woodlands in the Tibesti can be ruled out, though. Two circumstances or a combination of both can explain the lack of Mediterranean woody taxa. First, the Early and Middle Holocene climatic conditions at high altitudes in the Tibesti mountains were unsuitable (too dry). Second, a migration of Mediterranean (woody) taxa from the source area was not possible during the Holocene, because aridity persisted in the plains north of the Tibesti, and relicts that could expand were not present in the mountains, with the exception of the present *Erica*-relict. The few records of Mediterranean taxa with low frequencies in the Fezzan in the area northwest of the Tibesti mountains and the Central Saharan plains corroborate the latter explanation (e.g. Mercuri 2008; Mercuri *et al.* 1998; Schulz 1991; Trevisan *et al.* 1998).

4.5.3.3 Shift in the plant composition

The actual pollen spectrum is dominated by the chamaephytes *Ephedra* and *Artemisia*. In contrast Poaceae prevail in the Early and Middle Holocene spectra, together with at times high *Artemisia* proportions and high fern proportions in the lower part of the sequence. This distinct difference between the basal pollen spectra of the palaeolake and the surface sample, though biased by the different type of samples, evidences that at about 9500 cal yrs BP more humid conditions than today were already established at high altitudes in the Tibesti.

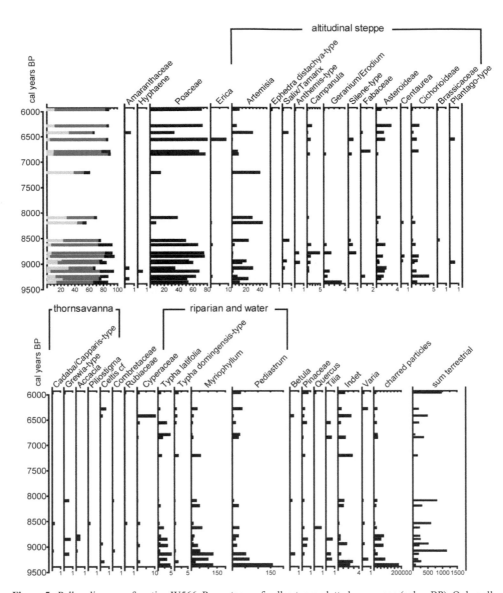

Figure 5. Pollen diagram of section W566. Percentages of pollen types plotted versus age (cal yr BP). Only pollen types present in more than one sample are shown. The first summary graph shows the fluctuations of: light grey – altitudinal steppe, grey – grasslands, black – ferns. The pollen types are grouped according to their ecology.

Ferns still occur today in the Tibesti mountains, however only very rarely and in specific habitats such as shaded fissures and fumerolles at high altitudes, thus in humidity-advantaged localities (Maire and Monod 1950; Quézel 1958). The high fern proportions during the Early and Middle Holocene therefore point to increased humidity, most probably especially to an extended wetter period that fostered their growth, as indicated in the Lake Yoa record in the southeastern foreland (Lézine *et al.* 2011a). Some scattered palynological investigations of palaeolake sediments in the Tibesti revealed high fern spore proportions which are similarly interpreted as mirroring wetter conditions (Maley 1981). Ferns were also abundantly recorded in an Early

Holocene lacustrine deposit in the present-day desert of Yemen (Lézine *et al.* 2007). In an analysis of surface samples throughout the African continent, higher fern proportions are assigned to mean annual precipitation amounts >100 mm, peaking about 1200 mm, corroborating the connection to increased humidity (Gajewski *et al.* 2002). The simultaneously high Poaceae frequencies in our Era Kohor sequence fit into this assumption. Following the interpretation of shifts between *Artemisia* (and Amaranthaceae) versus Poaceae by El Moslimany (1987, 1990), higher grass proportions indicate an increase in humidity during the growing season of grasses. The highest altitudinal pollen spectra in the Early and Middle Holocene thus show a distinct switch in plant composition, indicating increased humidity, probably encompassing a large part of the year. However, despite the marked shift in plant taxa composition, the steppe ecosystem persisted. This is corroborated by the records of additional plant taxa typical for the altitudinal steppes such as *Campanula filicaulis* or *C. bordesiana*, included in the *Campanula* type, or e.g. *Erodium malacoides* included in *Geranium/Erodium* pollen type.

4.5.3.4 Shallow water and riparian vegetation

Myriophyllum-type includes the tropical genus *Laurembergia* and *Myriophyllum*. Due to the temperature considerations, an occurrence of *Myriophyllum* is more probable. *Myriophyllum spicatum* occur on the Djebel Marra at elevations between 1160 and 2450 m asl (Wickens 1967), commonly growing in shallow water habitats (Hegi 1975). The continuous record of *Myriophyllum*-type pollen indicates the persistence of a shallow water zone throughout the Early and Middle Holocene. A continuous riparian belt is similarly corroborated by the continuous records of *Typha domingensis* and *Typha latifolia* (Quézel 1958). While *Myriophyllum spicatum* and *Typha* species tolerate brackish water, at least the recent genus *Pediastrum* is mainly restricted to fresh-water habitats, and is very rarely recorded in brackish biotopes (Hegi 1975; Komarek and Jankovska 2001). However, freshwater influxes into a brackish lake may as well explain the occurrence of *Pediastrum*. The riparian and water vegetation thus indicate a shallow and freshwater-to-brackish lake throughout the Holocene, at least at the position of the sampled lake remnants on the slopes of the Era Kohor crater, some 145 m above the caldera floor.

4.5.4 Early and Middle Holocene high altitude climate settings derived from montane vegetation

Ferns, grasses and the records of some tropical taxa at highest altitudes suggest humid conditions. In contrast, higher *Artemisia* proportions can be connected to an increase in aridity. Though the genus *Artemisia* is characterized by a wide range of bio-climatic requirements, the opposed progression of grasses and *Artemisia* points to the validity of this assumption for the Tibesti (e.g. Langgut *et al.* 2011; Lézine *et al.* 2011a; Subally and Quézel 2002). *Ephedra distachya*-type is another indicator for aridity: low proportions of *Ephedra distachya*-type in the Middle Holocene spectra (cf. also Maley 1981) as compared to its highest frequencies in the surface sample, as well as the general bioclimatic interpretation, evidence its positive correlation with aridity (Gajewski *et al.* 2002; Qin *et al.* 2015).

Based on these assumptions, a humidity progression including ferns and Poaceae versus *Artemisia* and *Ephedra distachya*-type may be postulated (Figure 5). Highly fluctuating, but basically more humid conditions characterize the period between 9500 and 8500 cal yrs BP, while three samples about 8000 cal yr BP point to decreased humidity. These more arid conditions may be concomitant with the arid '8.2' identified elsewhere in Africa event (e.g. Gasse 2000). Whether more arid conditions prevailed until about 7000 cal yr BP or whether results are biased by relying on only one sample remains an open question. During the period from about 7000 to 6000 cal yrs BP humidity increased again, however tentatively less distinctly than during the Early Holocene. Interestingly, the humidity index is roughly in line with the frequency distribution of the hydrological indicators: with the peak in lacustrine dated deposits, the humidity index,

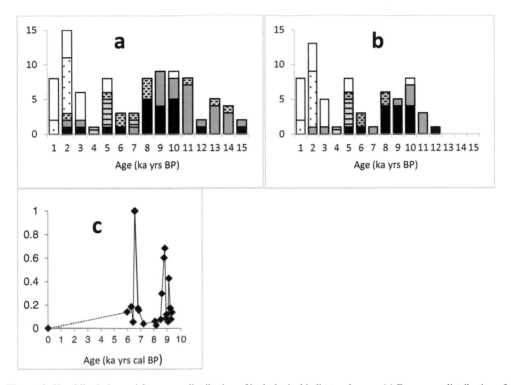

Figure 6. Humidity index and frequency distribution of hydrological indicators by age: (a) Frequency distribution of ^{14}C-dates from different geomorphological and archaeological formations including all ^{14}C-datation, (b) excluding the calcareous crusts. Black = lacustrine sediments, grey = fluviatile terraces, grey/dashes = palustrine sediments, grey/lines = playa sediments, white/points = arid settings, white = cultural layers (c) Humidity index: The proportions of Poaceae and ferns versus *Artemisia* and *Ephedra distchya*-type, scaled to 1, are plotted against age. The surface sample mirroring the present climatic conditions is shown as a separate point.

though distinctly fluctuating, reaches high values during the 9th and 8th millennia BP. Both proxies show a depression during the 7th millennium, pointing to reduced humidity, while in the 6th millennium wetter conditions were established again, with a maximum of the humidity index and (augmented) lacustrine and fluviatile sedimentation.

As to precipitation estimates, the major issues are the distinct shift in vegetation composition and the persistence of the steppe ecosystem at high altitudes. To assess rainfall amounts, the bioclimatic requirements of Mediterranean, North African steppes as well as altitudinal, continental dry steppes dominated by dwarf shrubs may be considered. The precipitation marge for the Mediterranean, North African steppes varies between 100 and 400 mm/a (Le Houérou 1995, 2001). To get a notion of the altitudinal factor, altitudinal, continental dry steppes dominated by *Artemisia* in SW Asia (Iran) may additionally be considered an equivalent. In an altitudinal vegetation-pollen-rain transect in SW Iran, rainfall amounts of about 100–300 mm/a are assigned to *Artemisia*-dominated steppes at altitudes about 800-1500 m asl mirrored by high *Artemisia* pollen proportions. The subsequent, treeless montane and alpine grass-dominated belt at >1500 m asl, characterized in the pollen spectra by high Poaceae frequencies, profits from orographic rainfall increase of about 300–500 mm/a (Dehghani *et al.* 2017). Because of the recurrent high *Artemisia* proportions in the W566 sequence, underpinned by high *Artemisia* frequencies in a sample from another Early Holocene lake remnant of Emi Koussi (Maley 1981), a range near the threshold between *Artemisia* – Poaceae dominated zones may be assumed at about 300–400 mm/a, and up to 500 mm/a for the grass-dominated sections. By applying the actual adiabatic lapse rate (roughly 50 mm per 1000 m), rainfall similarly would amount to 370–540 mm/a. Precipitation estimates for the Middle Holocene in the Ounianga Basin southwest of the Tibesti massif based on transfer functions resulted in 250–420 mm/a (Kröpelin *et al.* 2008; Lézine *et al.* 2011a). At present <5 mm/a are indicated for the Ounianga Basin and 80–150 mm/a are estimated for the high altitudes of the Emi Koussi. Indications concerning temperature fluctuations are less distinct. The lack of a mass expansion of tropical trees points to (persisting) low temperatures. Diatom spectra of the Emi Koussi similarly speak in favour of persisting low or even lower temperatures at the beginning of lake formation (Messerli 1972). Both assumptions are in line with continental scale pollen-based climate reconstructions at 6000 years BP for the region Northern Chad and Southern Libya, indicating an increase of precipitation >200 mm/a with an inconsistent pattern for temperature, however also in parts indicating a temperature decrease, resulting from an increased monsoon (Bartlein *et al.* 2011).

4.6 SUMMARY AND CONCLUSIONS

The age-depth model of a lake remnant based on ^{14}C-dates of charred plant material and the investigation of a sequence of 20 pollen samples from this palaeolake in the Emi Koussi crater, SE Tibesti mountains, allow inferences on some pending issues:

There is no palynological evidence of an Early and Middle Holocene expansion of Mediterranean (and Temperate) plant taxa in the Tibesti mountains (except for the relict *Erica*). For the first time, a palynological sequence has been analysed, improving the previously erratic data basis of single pollen samples. None of the newly investigated pollen samples shows higher frequencies of Mediterranean or Temperate taxa. This is in line with the single-sample palynological investigation of lake remnants in the Trou au Natron and the SE Tibesti (Emi Koussi, Mouskorbé, Tarso Yega, Maley 1981). Neither a re-migration of Mediterranean tree taxa from the north, nor an expansion from possible refuge habitats was thus possible during the Holocene. A rather poorly developed palaeohydrological network reconstructed for the surroundings of the Tibesti (HydroSHEDS; https://www.hydrosheds.org/) seems to corroborate its more isolated situation when compared to the Hoggar mountains in the NW, where there are more Mediterranean tree relicts (*Cupressus dupreziana, Olea europaea* subsp. *laperrinei* e.g. Quézel 1965). An exception is the spread of *Erica arborea* at the end of the African Humid Period in the Tibesti. However, relict stands of *Erica* continue to exist today at the Emi Koussi. Similarly, *Erica arborea* may have outlasted the last Glacial and expanded under favourable conditions about 6000 years ago.

Lake formation at the highest altitudes of the Tibesti and the surrounding plains is synchronous. The radiocarbon-dated charred plant particles from the diatomite W566 of the Era Kohor caldera clearly indicate its Holocene age. The lake formation at highest altitudes thus fits into the main phase of lake and terraces formation in the Tibesti mountains, as shown by the frequency distribution of ^{14}C-dates of these geomorphological formations – and is synchronous to the well-defined palaeolake phase in the Saharan plains (e.g. Cremaschi and Zerboni 2009; Hoelzmann *et al.* 1998; Holmes and Hoelzmann 2017; Kuper and Kröpelin 2006; Lézine *et al.* 2011b; Lécuyer *et al.* 2016; van Neer *et al.* 2020). This does not exclude an increase in humidity pre-dating the formation of lakes: the high Poaceae frequencies in the basal pollen samples of the Emi Koussi sequence indicate wetter conditions already at the beginning of lake formation.

Distinct changes in plant composition with the persistence of the montane steppe ecosystem indicate increased humidity and probably continued low temperatures. Palynological investigations on the W566 lake sediments in the Era Kohor sub-caldera document the persistence of a steppe at high altitudes. Changed climatic conditions, however, are indicated by high Poaceae and fern frequencies and strongly fluctuating *Artemisia* proportions during the African Humid period versus strongly dominating *Ephedra distachya*-type proportions, together with *Artemisia,* reflecting the present vegetation. The resultant calculated humidity index indicates wetter conditions during the Early Holocene, more arid conditions about 8200–8000 cal yr BP presumably lasting until about 7000 cal yr BP, and an increase in humidity during the 6th millennium.

In analogy to Mediterranean and Asian chamaephyte steppes, a three- to fivefold increase in precipitation (*c.* 400–550 mm/a) may be estimated. The lack of a mass expansion of tropical taxa at high altitudes points to persisting low temperatures during the African Humid period. Holocene climate fluctations thus impacted the surrounding plains more strongly, triggering large-scale ecosystem changes. High altitudinal ecosystems in the Tibesti were less affected, probably in consequence of the persisting specific climatic conditions (low temperatures) that still prevail today due to the singular geographic position of the Tibesti between the Mediterranean and the tropical climate regimes.

With its additional habitats such as riparian vegetation and additional grass-dominated plant formations at highest altitudes, the mountain ranges of the Tibesti may have attracted humans and wild animal populations. Due to the more stable conditions than in the plains, the Tibesti has acted (and acts) as a refuge, e.g. for *Ephedra* during the African Humid period or today for *Erica* relict stands – which of course holds true for humans and animals as well.

4.7 ACKNOWLEDGEMENTS

This study is supported by a grant from the DFG priority program SPP 1234 'Entangled Africa'. DEM data were provided by the DLR (Deutsches Zentrum für Luft- und Raumfahrt/German Aerospace Center). Fieldwork was supported by DFG project 57444011-SFB806. We are grateful to Dr. Baba Mallaye, Director of the Centre National de la Recherche pour le Développement (CNRD) in N'Djaména, for his comprehensive support of the missions. We also thank our team members Jan Kuper and Mahadi Chaha for their help with the sampling. We thank an anonymous reviewer and Jean Maley for their most fruitful reviews and suggestions which most significantly improved the manuscript.

REFERENCES

Alimen, H., Beucher, F. and Lhote, H., 1968, Les gisements néolithiques de Tan-Tartait et d'I-n-Itinen Tassili-n-Ajjer (Sahara Central). *Bulletin de la Société préhistorique francaise*, **LXV**, pp. 421–458, 10.3406/bspf.1968.4160.

Amrani, S., 2018, The Holocene flora and vegetation of Ti-n Hanakaten (Tassili n'Ajjer, Algerian Sahara). *Plants and People in the African Past* edited by Mercuri, A., d'Andrea, A., Fornaciari, R. and Höhn, A., Springer, pp. 123–145.

Ball, P., White, N., Masoud, A., Nixon, S., Hoggard, M., Maclennan, J., Stuart, F., Oppenheimer, C. and Kröpelin, S., 2019, Quantifying asthenospheric and lithospheric controls on ma?c magmatism across North Africa. *Geochemistry, Geophysics, Geosystems*, **20**, pp. 3520–3555, 10.1029/2019GC008303.

Bartlein, P. J., Harrison, S. P., Brewer, S., Connor, S., Davis, B. S. A., Gajewski, K., Guiot, J., Harrison-Prentice, T. I., Henderson, A., Peyron, O., Prentice, J. C., Scholze, M., Seppä, H., Shuman, B., Sugita, S., Thompson, R. S., Viau, A. E., Williams, J. and Wu, H., 2011, Pollen-based continental climate reconstructions at 6 and 21 ka: A global synthesis. *Climate Dynamics*, **37**, pp. 775–802, 10.1007/s00382-010-0904-1.

Baumhauer, R., 2004, Some new insights into palaeoenvironmental dynamics and Holocene landscape evolution in the Nigrian Central Sahara (Ténéré, Erg of Ténéré, Erg of Fachi-Bilma). *Zentralblatt für Geologie und Paläontologie*, **1**, pp. 387–403, 10.1127/zgpI/2014/0387-0403.

Blaauw, M. and Christen, J.A., 2011, Flexible paleoclimate age-depth models using an autogressive gamma process. *Bayesian Analysis* **6**(3), pp. 457–474, 10.1214/ba/1339616472.

Blaauw, M., Christen, J. A., Bennett, K. D. and Reimer, P. J., 2018, Double the dates and go for Bayes — Impacts of model choice, dating density and quality on chronologies. *Quaternary Science Reviews*, **188**, pp. 58–66, 10.1016/j.quascirev.2018.03.032.

Bonnefille, R. and Riollet, G., 1980, *Pollens des Savanes d'Afrique Orientale*. Paris : CNRS.

Braconnot, P., Joussaume, S., Marti, O., and de Noblet, N., 1999, Synergistic feedbacks from ocean and vegetation on the African monsoon response to mid-Holocene insolation. *Geophysical Research Letters*, **26**, pp. 2481–2484, 10.1029/1999GL006047.

Braconnot, P., Otto-Bliesner, B., Harrison, S., Joussaume, S., Peterchmitt, J.-Y., Abe-Ouchi, A., Crucifix, M., Driesschaert, E., Fichefet, Th., Hewitt, C. D., Kageyama, M., Kitoh, A., Loutre, M.-F., Marti, O., Merkel, U., Ramstein, G., Valdes, P., Weber, L., Yu, Y. and Zhao, Y., 2007, Results of PMIP2 coupled simulations of the Mid-Holocene and Last Glacial Maximum – Part 2: feedbacks with emphasis on the location of the ITCZ and mid- and high latitudes heat budge. Climate of the Past **3**(2), 279–296, 10.5194/cp-3-279-2007.

Brown, T. A., Farwell, G., Grootes, P. and Schmidt, F., 1992, Radiocarbon AMS dating of pollen extracted from peat samples. *Radiocarbon* **43**(3), pp. 550–556, 10.1017/S00338222000 63815.

Bruneau de Miré, P. and Quézel, P., 1959, Sur la présence de la bruyère en arbre (Erica arborea L.) sur les sommets de l'emi Koussi (massif du Tibesti). *Compte rendu sommaire des séances de la Société de biogéographie*, **36**, pp. 66–70.

Chester, P. and Prior, C., 2004, An AMS ^{14}C pollen-dated sediment and pollen sequence from the Late Holocene, southern coastal Hawke's Bay, New Zealand. *Radiocarbon* **46**(2), pp. 721–731, 10.1017/S0033822200035761.

Claussen, M. and Gayler, V., 1997, The greening of the Sahara during the mid-Holocene: results of an interactive atmosphere-biome model. *Global Ecology and Biogeography Letters*, **6**, pp. 369–377, 10.2307/2997337.

Claussen, M., Kubatzki, C., Brovkin, V., Ganopolski, A., Hoelzmann, P. and Pachur, H. J., 1999, Simulation of an abrupt change in Saharan vegetation in the mid-Holocene. *Geophysical Research Letters*, **26**, pp. 2037–2040, 10.1029/1999GL900494.

Cremaschi, M. and Zerboni, A., 2009, Early to Middle Holocene landscape exploitation in a drying environment: Two case studies compared from the central Sahara (SW Fezzan, Libya). *Comptes Rendus – Geoscience*, **341**, pp. 689–702, 10.1016/j.crte.2009.05.001.

Dallmeyer, A., Claussen, M., Lorenz, S. J. and Shanahan, T., 2020, The end of the African humid period as seen by a transient comprehensive Earth system model simulation of the last 8000 years. *Climate of the Past*, **16**, pp. 117–140, 10.5194/cp-16-117-2020.

Dehghani, M., Djamali, M., Gandouin, E. and Akhani, H., 2017, A pollen rain-vegetation study along a 3600 m mountain-desert transect in the Irano-Turanain region; implications for the reliability of some pollen ratios as moisture indicators. *Review of Palaeobotany and Palynology*, **247**, pp. 133–148, 10.1016/j.revpalbo.2017.08.004.

Désamoré, A., Laenen, B., Devos, N., Popp, M., Gonzalez-Mancebo, J., Carine, M. and Vander-poorten, A., 2011, Out of Africa: north-westwards Pleistocene expansions of the heather *Erica arborea*. *Journal of Biogeography* **38**, 164–176, 10.1111/j.1365-2699.2010.02387.x.

Deniel, C., Vincent, P. M., Beauvilain, A. and Gourgaud, A., 2015, The Cenozoic volcanic province of Tibesti (Sahara of Chad): major units, chronology, and structural features. *Bulletin of Volcanology* **77**(74), pp. 1–21, 10.1007/s00445-015-0955-6.

Dinies, M., Hoelzmann, P., Karls, J., Melles, M., Wennrich, V., Claussen, M., Neef, R. and Kröpelin, S., 2019, Continental records for the 'African Humid Period': lake sediment archives from the Ounianga Basin and the Tibesti Mountains, N Chad. *Abstract, INQUA20[th] Congress*, O-3016.

Eisele, G., Haas, K. and Liner, S., 1994, Methode zur Aufbereitung fossilen Pollens aus minerogenen Sedimenten. *Göttinger Geographische Abhandlungen*, **95**, pp. 165–166.

El-Ghazali, G., 1991, A study on the pollen flora of Sudan. *Review of Palaeobotany and Palynology* **76**(2–4), pp. 99–345, 10.1016/0034-6667(93)90077-8.

El-Moslimany, A.P., 1987, The Late Pleistocene climates of the lake Zeribar region (Kurdistan, Western Iran) deduced from the ecology and pollen production of non-arboreal vegetation. Vegetatio, **72**(3), pp. 131–139.

El-Moslimany, A.P., 1990, Ecological significance of common non-arboreal pollen: examples from drylands of the Middle East. *Review of Palaeobotany and Palynology* **64**, pp. 343–350, 10.1016/0034-6667(90)90150-H.

Faegri, K. and Iversen, J., 1989, *Textbook of Pollen Analysis*. Chichester: John Wiley & Sons Ltd.

Faure, H., 1969, Lacs quaternaires du Sahara. *Mitteilungen des Internationalen Vereins für Limnologie*, **17**, pp. 131–146, 10.1080/05384680.1969.11903878.

Gabriel, B., 1977, Zum ökologischen Wandel im Neolithikum der östlichen Zentralsahara. *Berliner Geographische Abhandlungen*, **27**, pp. 1–111.

Gajewski, K., Lézine, A.-M., Vincens, A., Delestan, A., Sawada, M. and the African Pollen Database, 2002, Modern climate-vegetation-pollen relations in Africa and adjacent areas. *Quaternary Science Reviews*, **21**, pp. 1611–1631, 10.1016/S0277-3791(01)00152-4.

Gasse, F., 2000, Hydrological changes in the African tropics since the Last Glacial Maximum. *Quaternary Science Reviews*, **19**, pp. 189–211, 10.1016/S0277-3791(99)00061-X.

Gavrilovic, D., 1969, Klimatabellen für das Tibesti-Gebirge. *Arbeitsberichte aus der Forschungsstation Bardai/Tibesti* **8** edited by J. Höevermann, G. Jensch, H. Valentin, W. Woehlke and H. Hagedorn, H., pp. 47–48.

Gourgaud, A. and Vincent, P., 2004, Petrology of two continental alkaline intraplate series at Emi Koussi volcano, Tibesti, Chad. *Journal of volcanology and geothermal research*, **129**(4), pp. 261–290, 10.1016/S0377-0273(03)00277-4.

Grimm, E., 2020, *Tilia version 2.6.1* [software]. https://www.tiliait.com/

Haynes, C., Eyles, C., Pavlish, L., Ritchie, J. and Rybak, M., 1989, Holocene palaeoecology of the Eastern Sahara; Selima oasis. *Quaternary Science Reviews*, **8**, pp. 109–136, 10.1016/0277-3791(89)90001-2.

Hegi, G., 1975, *Illustrierte Flora von Mitteleuropa*. V/2, Berlin-Hamburg: Paul Parey.

Hély, C. and Lézine, A.-M., 2014, Holocene changes in African vegetation: Tradeoff between climate and water availability. *Climate of the Past*, **10**, pp. 681–686, 10.5194/cp-10-681-2014.

Hoelzmann, P., Jolly, D., Harrison, S.P., Laarif, F., Bonnefille, R. and Pachur, H.J., 1998, Mid-Holocene land-surface conditions in northern Africa and the Arabian peninsula: A data set for the analysis of biogeophysical feedbacks in the climate system. *Global Biogeochemical Cycles*, **12**(1), pp. 35–51, 10.1029/97GB02733.

Holmes, J. and Hoelzmann, P., 2017, The Late Pleistocene-Holocene African Humid Period as Evident in Lakes. *The Oxford Research Encyclopedia of Climate Science; Regional and Local Climates*, 10.1093/acrefore/9780190228620.013.531.

Jäkel, D. and Geyh, M.A., 1982, ^{14}C-Daten aus dem Gebiet der Sahara, hervorgegangen aus Arbeiten der Forschungsstation Bardai und des Niedersächsischen Landesamtes für Bodenforschung in Hannover. *Berliner Geographische Abhandlungen*, **32**, pp. 143–166.

Jahns, S., 1995, A Holocene pollen diagram from El Atrun, northern Sudan. *Vegetation History and Archaeobotany*, **4**, pp. 23–30, 10.1007/BF00198612.

Knippertz, P., 2005, Tropical-extratropical interactions associated with an Atlantic tropical plume and subtropical jet streak. *Monthly Weather Review*, **133**, pp. 2759–2776, 10.1175/MWR2999.1.

Köppen, W., 1918, Klassifikation der Klimate nach Temperatur, Niederschlag and Jahreslauf. *Petermanns Geographische Mitteilungen*, **64**, pp.193–203.

Komarek, J. and Jankovska, V., 2001, *Review of the green algal genus Pediastrum; Implication for pollen-analytical research.*, Berlin, Stuttgart: Cramer.

Krinner, G., Lézine, A.-M., Braconnot, P., Ramstein, G., Grenier, C. and Gouttevin, I., 2012, Strong timescale-dependent feedbacks on the North African Holocene climate by lakes and wetlands. *Geophysical Research Letters*, **39**(7), L07701, 10.1029/2012GL050992.

Kröpelin, S., Verschuren, D., Lézine, A.-M., Eggermont, H., Cocquyt, C., Francus, P., Cazet, J.-P., Fagot, M., Rumes, B., Russell, J., Darius, F., Conley, D., Schuster, M., Suchodoletz, H. and Engstrom, D., 2008, Climate-driven ecosystem succession in the Sahara: The past 6000 years. *Science*, **320**, pp. 765–768, 10.1126/science.1154913.

Kröpelin, S., Dinies, M., Sylvestre, F. and Hoelzmann, F., 2016, Crater palaeolakes in the Tibesti mountains (Central Sahara, North Chad) - New insights into past Saharan climates. *Geophysical Research Abstracts*, **18**, EGU2016-6557.

Kuper, R. and Kröpelin, S., 2006, Climate-Controlled Holocene Occupation in the Sahara: Motor of Africa's Evolution. *Science*, **313**, pp. 803–807, 10.1126/science.1130989.

Kutschera, W., 2016, Accelerator mass spectrometry: state of the art and perspectives.*Advances in Physics* **X**(1,4), pp. 62–65.

Kutzbach, J. E., 1981, Monsoon climate of the early Holocene: Climate experiment with the Earth's orbital parameters for 9000 years ago. *Science*, **214**(4516), pp. 59–61, 10.1126/science.214.4516.59.

Kutzbach, J. E. and Liu, Z., 1997, Response of the African Monsoon to Orbital Forcing and Ocean Feedbacks in the Middle Holocene. *Science*, **278**(5337), pp. 440–443, 10.1126/science.278.5337.440.

Langgut, D., Almogi-Labin, A., Bar-Matthews, M. and Weinstein-Evron, M., 2011, Vegetation and climate changes in the South Eastern Mediterranean during the Last Glacial-Interglacial cycle (86 ka): new marine pollen record. *Quaternary Science Reviews*, **30**, pp. 3960–3972, 10.1016/j.quascirev.2011.10.016.

Lécuyer, C, Lézine, A.-M., Fourel, F., Gasse, F., Sylvestre, F., Pailles, C., Grenier, C., Travi, Y. and Barral, A., 2016, In-Atei palaeolake documents past environmental changes in central Sahara at the time of the "Green Sahara": Charcoal, carbon isotope and diatom records. *Palaeogeography, Palaeoclimatology, Palaeoecology*, **441**, pp. 834–44, 10.1016/j.palaeo.2015.10.032.

Le Houérou, H. N., 1995, Bioclimatologie et biogéographie des steppes arides du nord de l'Afrique. *Options méditerranéennes*, **10**, pp. 1–396.

Le Houérou, H. N., 2001, Biogeography of the arid steppeland north of the Sahara. *Journal of Arid Environments*, **48**, pp. 103–128, 10.1006/jare.2000.0679.

Le Houérou, H. N., 2009, *Bioclimatology and biogeography of Africa*. Berlin, Heidelberg: Springer.

Lézine, A.-M., Tiercelin, J.-J., Robert, C., Saliège, J.-F., Cleuziou, S., Inzian, M.-L. and Braemer, F., 2007, Centennial to millennial-scale variability of the Indian monsoon during the early Holocene from a sediment, pollen and isotope record from the desert of Yemen. *Palaeogeography, Palaeoclimatology, Palaeoecology*, **243**(3–4), pp. 235–249, 10.1016/j.palaeo.2006.05.019.

Lézine, A.-M., Zheng, W., Braconnot, P. and Krinner, G., 2011a, Late Holocene plant and climate evolution at Lake Yoa, northern Chad: pollen data and climate simulations. *Climate of the Past*, **7**, pp. 1351–1362, 10.5194/cp-7-1351-2011.

Lézine A.-M., Hély, C., Grenier, C., Braconnot P., Krinner G., 2011b, Sahara and Sahel vulnerability to climate changes, lessons from Holocene hydrological data. *Quaternary Science Reviews*, **30**, pp. 3001–3012, 10.1016/j.quascirev.2011.07.006.

Lézine, A.-M., 2017, Vegetation at the Time of the African Humid Period. *Oxford University Press*, 10.1093/acrefore/9780190228620.013.530.

Maire, R., 1935, Contribution à l'étude de la flore du Tibesti. *Mémoires de l'Académie des Sciences de l'institut de France*, **62**, pp. 1–39.

Maire, R. and Monod, T., 1950, Étude sur la flore et la végétation du Tibesti. *Mémoire de lIinstitut francais d'Afrique noire*, **8**, pp. 1–140.

Maley, J., 1981, Etudes palynologiques dans le bassin du Tchad et paléoclimatologie de l'Afrique nord-tropicale de 30000 ans à l'époque actuelle. *Travaux et Documents de l'ORSTOM* **129**, pp. 1–586.

Maley, J., 2000, Last Glacial Maximum lacustrine and fluviatile Formations in the Tibesti and other Saharan mountains, and large-scale climatic teleconnections linked to the activity of the Subtropical Jet Stream. *Global and Planetary Change*, **26**, pp. 121–136, 10.1016/S0921-8181(00)00039-4.

Médail, F. and Quézel, P., 2018, *Biogéographie de la flore du Sahara: Une biodiversité en situation extrême*. Marseille : IRD Éditions, Conservatoire et Jardin botaniques de la Ville de Genève.

Mercuri, A., Grandi, G., Mariotti Lippi, M. and Cremaschi, M., 1998, New pollen data from the Uan Muhuggiag rockshelter (Libyan Sahara, VII-IV millennia BP). In: *Wadi Teshuinat – Palaeoenvironment and prehistory in south-western Fezzan (Libyan Sahara). Survey and excavations in the Tadrart Acacus, Erg Uan Kasa, Messak Settafet and Edeyen of Murzuq, 1990–1995*, edited by Cremaschi, M. and di Lernia, S., pp. 107–122, Firenze: All'Insegna del Giglio.

Mercuri, A., 1999, Palynological analysis of the Early Holocene sequence. *The Uan Afuda cave, Hunter-Gatherer societies of Central Sahara, Arid Zone Archaeology*, **1**, pp. 149–253.

Mercuri, A. M., 2008, Human influence, plant landscape evolution and climate inferences from the archaeobotanical records of the Wadi Teshuinat area (Libyan Sahara). *Journal of Arid Environments*, **72**, pp. 1950–1967, 10.1016/j.jaridenv.2008.04.008.

Mercuri, A. M., Fornaciari, R., Gallinaro, M., Vanin, S. and di Lernia, S., 2018, Plant behaviour from human imprints and the cultivation of wild cereals in Holocene Sahara. *Nature plants*, **4**, pp. 71–81, 10.1038/s41477-017-0098-1.

Messerli, B., 1972, Formen und Formungsprozesse in der Hochgebirgsregion des Tibesti, *Hochgebirgsforschung/High mountain research* 2, Tibesti – Zentrale Sahara, Arbeiten aus der Hochgebirgsregion, Wagner, Innsbruck-München, pp. 23–86.

Messerli, B. and Winiger, M., 1992, Climate, Environmental Change, and Resources of the African Mountains from the Mediterranean to the Equator. *Mountain Research and Development*, **12**(4), pp. 1–315, 10.2307/3673683.

Meusel, H. and Jäger, E., 1992, *Vergleichende Chorologie der zentraleuropäischen Flora*. Stuttgart New York: Fischer.

Moore, P., Webb, J. and Collinson, M., 1991, *Pollen analysis – the treatment of samples*, London: Blackwell Science.

Nakagawa, T., Brugiapaglia, E., Digerfeldt, G., Reille, M., de Beaulieu, J.-L. and Yasuda, Y., 1998, Dense-media separation as a more efficient pollen extraction method for use with organic sediment/deposit samples: comparison with the conventional method. *Boreas*, **27**, pp. 15–24, 10.1111/j.1502-3885.1998.tb00864.x.

Neumann, K., 1989, Holocene vegetation of the Eastern Sahara: charcoal from prehistoric sites. *The African Archaeological Review*, **7**, pp. 97–116, 10.1007/BF01116839.

Ozenda, P., 1991, *Flore et vegetation du Sahara*. Paris: CNRS.

Pachur, H.-J. and Altmann, N., 2006, *Die Ostsahara im Spätquartär*. Heidelberg: Springer.

Parker, A.G., Eckersley, L., Smtih, M. M., Goudie, M. M., Stokes, S., Ward, S., White, K. and Hodson, M.J., 2004, Holocene vegetation dynamics in the northeastern Rub' al-Khali desert, Arabian Peninsula: a phytolith, pollen and carbon isotope study. *Journal of Quaternary Science*, **19**(7), pp. 665-676, 10.1002/jqs.880.

Pausata, F.S.R., Messori, G. and Zhang, Q., 2016, Impacts of dust reduction on the northward expansion of the African monsoon during the Green Sahara period. *Earth and Planetary Science Letters*, **434**, pp. 298–307, 10.1016/j.epsl.2015.11.049.

Piotrowska, N., Bluszcz, A., Demske, D., Granoszewski, W. and Heumann, G., 2004, Extraction and AMS radiocarbon dating of pollen from Lake Baikal sediments. *Radiocarbon*, **46**(1), pp. 181–187, 10.1017/S0033822200039503.

Prentice, I., Jolly, D. and BIOME 6000, 2000, Mid-Holocene and glacial-maximum vegetation geography of the northern continents and Africa. *Journal of Biogeography*, **27**(3), pp. 507–519, 10.1046/j.1365-2699.2000.00425.x.

Qin, F., Wang, Y.-F., Ferguson, D., Chen, W.-L., Li, Y.-M., Zhe Cai, Z., Wang, Q., Ma, H.-Z. and Li, C.-S., 2015, Utility of surface pollen assemblages to delimit Eastern Eurasian steppe types. *PLos ONE*, **10**(3), pp. e0119412, 10.1371/journal.pone.0119412.

Quézel, P., 1958, Contribution à l'étude de la flore et de la végétation du Borkou et du Tibesti. *Mission botanique au Tibesti. Université d'Alger, Institut de recherches Sahariennes*, pp. 99–303.

Quézel, P., 1959, La végétation de la zone nord-occidentale du Tibesti. *Travaux de l'Institut de Recherches sahariennes, Université d'Alger*, **18**, pp. 75–107.

Quézel, P., 1965, *La végétation du Sahara, du Tchad à la Mauritanie*. Stuttgart : Fischer.

Quézel, P., 1964, *Largeau*. Carte international du tapis végétal. Institut géographique national, Tchad.

Railsback, L.B., Brook, G.A., Liang, F., Marais, E., Cheng, H. and Edwards, R.L., 2016, A multi-proxy stalagmite record from northwestern Namibia of regional drying with increasing global-scale warmth over the last 47 kyr: The interplay of a globally shifting ITCZ with regional currents, winds, and rainfall. *Palaeogeography, Palaeoclimatology, Palaeoecology*, **461**, pp. 109–121, 10.1016/j.palaeo.2016.08.014.

Regnéll, J. and Everitt, E., 1996, Preparative centrifugation – a new method for preparing pollen concentrates suitable for radiocarbon dating by AMS. *Vegetation History and Archaeobotany*, **5**, pp. 201–205, 10.1007/BF00217497.

Reille, M., 1992, *Pollen et spores d'Europe et d'Afrique du Nord*, Marseille : Laboratoire de Botanique Historique et Palynologie.

Ritchie, J., Eyles, C. and Haynes, C., 1985, Sediment and pollen evidence for an early to mid-Holocene humid period in the eastern Sahara. *Nature*, **314**, pp. 352–355, 10.1038/314352a0.

Ritchie, J., 1987, A Holocene pollen record from Bir Atrun, northwest Sudan. *Pollen et Spores*, **24**(4), pp. 391–410.

Ritchie, J., 1994, Holocene pollen spectra from Oyo, northwestern Sudan: problems of interpretation in a hyperarid environment. *The Holocene*, **4**(1), pp. 9–15, 10.1177/095968369400400102.

Ritchie, J. and Haynes, C., 1987, Holocene vegetation zonation in the eastern Sahara. *Nature*, **330**, pp. 645–647, 10.1038/330645a0.

Rogerson, M., Dublyansky, Y., Hoffmann, D., Luetscher, M., Töchterle, P. and Spötl, C., 2019, Enhanced Mediterranean water cycle explains increased humidity during MIS 3 in North Africa. *Climate of the Past*, **15**, pp. 1757–1769, 10.5194/cp-15-1757-2019.

Rossignol, M. and Maley, J., 1969, L'activité hors de France des palynologues et paléobotanistes français du Quaternaire. *Etudes Francaises sur le Quaternaire*, pp. 265–274.

Schulz, E., 1987, Die holozäne Vegetation der Zentralen Sahara (N-Mali, N-Niger, SW-Libyen). *Palaeoecology of Africa*, **18**, pp. 143–161.

Schulz, E., 1991, Holocene environments in the Central Sahara. *Hydrobiologia*, **214**, pp. 359–365, 10.1007/BF00050971.

Schulz, E., Hachicha, T., Marquer, L., Pomel, S., Salzmann, U. and Abichou, A., 2014, The distant chant. Climate reconstruction and landscape history. The last two millennia in southeast Tunisia. *Zentralblatt für Geologie und Paläontologie*, **1**(1), pp. 355–386, 10.1127/zgpI/2014/0355-0386.

Skinner, C. B and Poulsen, C. J., 2016, The role of fall season tropical plumes in enhancing Saharan rainfall during the African Humid Period. *Geophysical Research Letters*, **43**, pp. 349–358, 10.1002/2015GL066318.

Soulié-Märsche, I., Bieda, S., Lafond, R., Maley, J., Baitoudji, M., Vincent, P. and Faure, H., 2010, Charophytes as bio-indicators for lake level high satdn at 'Trou au Natron', Tibesti, Chad, during the Late Pleistocene. *Global and Planetary Change*, **72**, pp. 334–340, 10.1016/j.gloplacha.2010.05.004.

Stockmarr, J., 1971, Tablets with spores used in absolute pollen analysis. *Pollen et Spores*, **13**, pp. 615–621.

Subally, D. and Quézel, P., 2002, Glacial or interglacial: *Artemisia,* a plant indicator with dual responses. *Review of Palaeobotany and Palynology*, **120**(1–2), pp. 123–130, 10.1016/S0034-6667(01)00143-9.

Thinon, M., Ballouche, A. and Reille, M., 1996, Holocene vegetation of the central Saharan mountains: the end of a myth. *The Holocene*, **6**(4), pp. 457–462, 10.1177/095968369600600408.

Vandergoes, M. J., Prior, C. A., 2003, AMS dating of pollen concentrates – A methodological study of Late Quaternary sediments from South Westland, New Zealand. *Radiocarbon*, **45**(3), pp. 479–491, 10.1017/S0033822200032823.

van Neer, W., Alhaique, F., Wouters, W., Dierickx, K., Gala, M., Goffette, Q., Mariani, G., Zerboni, A. and di Lernia, S., 2020, Aquatic fauna from the Takarkori rock shelter reveals the Holocene central Saharan climate and palaeohydrography. *PLoS ONE*, **15**(2), pp. e0228588, 10.1371/journal.pone.0228588.

Vezey, E. L., Watson, L. E., Skvarla, J. J. and Estes, J.R., 1994, Plesiomorphic and apomorphic pollen structure characteristics of Anthemideae (Asteroideae: Asteraceae). *American Journal of Botany*, **81**, pp. 648–657, 10.1002/j.1537-2197.1994.tb15496.x.

Wesche, K., Ambarli, D., Kamp, J., Török, P., Treiber, J. and Dengler, J., 2016, The Palaearctic steppe biome: a new synthesis. *Biodiversity and Conservation*, **25**, pp. 2197–2231, 10.1007/s10531-016-1214-7.

Watrin, J., Lézine, A.-M., and Hély, C., 2009, Plant migration and ecosystems at the time of the "green Sahara". *Comptes Rendus Geosciences*, **341**, pp. 656–670, 10.1016/j.crte.2009.06.007.

Wickens, G., 1967, Jebel Marra. The flora of Jebel Marra (Sudan Republic) and its geographical affinities. *Royal Botanic Gardens,Kew. Kew Bulletin additional series*, **5**, pp. 1–368.

Wininger, M., 1972, Die Bewölkungsverhältnisse der zentralsaharischen Gebirge aus Wettersatellitenbildern. *Hochgebirgsforschung/High mountain research* **2**, Tibesti – zentrale Sahara, Arbeiten aus der Hochgebirgsregion, Wagner, Innsbruck-München, pp. 87–120.

CHAPTER 5

Timing and nature of the end of the African Humid Period in the Sahel: Insight from pollen data

Kévin Lemonnier & Anne-Marie Lézine

Laboratoire d'Océanographie et du Climat, Expérimentation et Approche Numérique/IPSL, Sorbonne Université, CNRS-IRD-MNHN, Paris, France

ABSTRACT: Pollen, spores, and algae from a 4m-long sediment core at the Mboro-Baobab site (15°8′58.49″N, 16°54′34.37″W), in the Niayes region of Senegal were used to provide a record of the end of the African Humid Period (AHP) in the western Sahel. We show that the drying of the Mboro-Baobab landscape was gradual, starting from 3750 cal yr BP then culminating at 1300 cal yr BP. In contrast, the response of the lake system and the gallery forest developed in two main phases: the sharp decline of tropical humid forest elements at 3200 cal yr BP followed by the almost complete collapse of the gallery forests at 2500 cal yr BP. Our results are consistent with those from the central Sahel, which show a gradual transition from the AHP to the modern landscape.

5.1 INTRODUCTION

At the end of the African Humid Period (AHP), tropical North Africa experienced a major environmental crisis. The southward displacement of the desert margin (Kuper and Kröpelin 2006) with the related loss in biodiversity (Hély *et al.* 2014; Watrin *et al.* 2009) along with the decline of wetlands (Lézine *et al.* 2011) had dramatic consequences for human populations through a decline in population density, cultural adaptations with the development of pastoralism and migrations toward the main rivers (Nile, Senegal, Niger) (Manning and Timpson 2014 and references therein) and the Equatorial forest block (Lézine *et al.* 2013). How long was this transition from a "green Sahara" to the hyper-arid desert of today? This question remains largely unanswered, as the number of palaeorecords with adequate temporal resolution and age control is low and often contradictory. Based on marine records, deMenocal *et al.* (2000) showed that an abrupt increase in dust transport to the ocean between 5579 and 5299 cal yr BP marked the end of the AHP in the Saharan desert. Studies in the central Sahara suggested that the modern desert was definitely established at 2700 cal yr BP after a period of gradual drying likely originating from the mid-Holocene and intensifying at 4300 cal yr BP (Kröpelin *et al.* 2008; Lézine *et al.* 2011; Lécuyer *et al.* 2016).

In the Sahel, interdunal depressions of Senegal and Nigeria provided detailed records of environmental change during the AHP (Lézine 1988; 1989; Salzmann and Waller 1998; Waller *et al.* 2007). The presence of a water table near the surface favored the development of forest galleries along water bodies. However, pollen studies suggest a contrasted situation: in Senegal, an abrupt change *c.* 2500 cal yr BP, interpreted as the crossing of a biological threshold, brought about the destruction of the forests and the decline of tropical humid elements (Lézine 1989).

DOI 10.1201/9781003162766-5

51

In Nigeria, the modern vegetation was established earlier, *c.* 3300 cal yr BP (Salzmann and Waller 1998). However, new data from Jikaryia Lake (Waller *et al.* 2007) suggest that "rather than a single abrupt event, Late Holocene aridification appears to have occurred progressively" from 4700 cal yr BP onward. Here we present a new and high-resolution pollen record from Mboro-Baobab in the "Niayes" coastal region of Senegal (15°8′58.49″N, 16°54′34.37″W). We focus on the period between 4300 cal yr BP (the base of the Mboro-Baobab pollen sequence) and the last millennium, which has been the subject of a separate publication (Lézine *et al.* 2019). Our goal is to determine whether the transition from AHP to the present was abrupt or gradual in this area, to discuss the turnover of species and the response times of forest trees to changes in hydrological conditions.

5.2 GENERAL SETTING

Niayes are interdunal depressions with accumulations of organic sediments, formed between linear fixed and partially fixed coastal dune ridges, located along the Atlantic coast between 15° N and 16° N, at roughly 17° W. The mean annual rainfall (369 mm) is characteristic of the Sahel, but in response to local soil moisture conditions, the depressions support an array of plant communities that include an azonal extension north of Guinean humid forests and Sudanian dry forests and wooded grasslands (Trochain 1940) (Figure 1A).

The Mboro-Baobab depression is located 2.3 km from the seashore west of the Mboro paleoriver (Figure 1B), at the edge of the central zone of the Niayes where the sub-surface aquifer is the most developed (Putallaz, 1964). The Mboro-Baobab depression is 570 m long and 310 m wide. To the south, it is covered with vegetation and partially exploited for agricultural purposes.

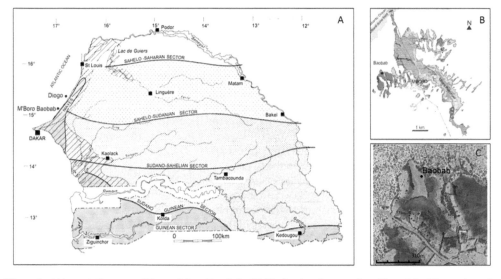

Figure 1. (A) vegetation map of Senegal from Trochain (1940) showing the azonal position of the Niayes along the Atlantic Ocean, north of the Dakar peninsula. Trochain described the vegetation of the Niayes as belonging to a "sub-Guinean domain" owing to its strong relation with the Guinean (rain) forests today found in the Casamance region, south of Senegal. (B) Location map of the Mboro-Baobab interdunal depression west of the Mboro paleoriver channel. In gray, inundated sectors (ORSTOM 1962). (C) Satellite image showing the Mboro- Baobab depression surrounded by dunes fields covered with Sahelian vegetation. The localization of the core in the northern part of the depression is shown.

Its northern part is an open water basin of 270 m by 90 m wide and 0.4 m maximum depth (Figure 1C). Mboro-Baobab is currently filled by brackish waters and occasionally desiccated.

5.3 MATERIAL AND METHODS

A 4 m long core was collected near the northeastern shore of the open water basin using a manual Russian corer (Jowsey 1966) in 2016 (Lézine *et al.* 2019). It was divided into 8 sections of 50 cm each. The sediment consists of homogeneous organic clay. Samples (1 cm^3 each) were taken at 5 cm intervals for pollen and non-pollen palynomorphs (NPP = freshwater algae (*Pediastrum, Botryococcus, Coelastrum, Scenedesmus, Tetraedron* and undifferentiated fungi) analyses. They were processed using the standard HF method described by Faegri and Iversen (1977). Pollen identification was based on the reference slide and photo collections of the LOCEAN laboratory and the Musée National d'Histoire Naturelle in Paris as well as on pollen atlases of tropical African flora from e.g. Maley (1970), APLF (1974), Ybert (1979) and Bonnefille and Riollet (1980). Pollen nomenclature, which follows the standard defined by the African Pollen Database (Vincens *et al.* 2007), was based on Lebrun and Stork (1991-2015). The pollen spectra was composed of 105 taxa among which 57 trees, shrubs, lianas and palms, 21 herbs, 16 undifferentiated (trees, shrubs or herbs), 7 aquatics and 4 ferns (Table 1). Pollen taxa have been grouped according to the phytogeographical affinity of the parent plants following Trochain (1940) and White (1983) (Table 1).

Table 1. Mboro-Baobab pollen taxa. A: trees; AL: tree/lianas; L: lianas; I: undifferentiated; N: herbs; Nq: aquatics. Nomenclature from the African Pollen Database. According to Vincens *et al.* (2007) a "type" is added to the name of the taxon (1) to the genus when several plant genera display similar pollen morphology. The species name can be added in case of monospecificity or when the species is dominant or characteristic in the study area (2) to the species when, the genus being clearly identified, several species can be concerned. SH: Saharan. SL: Sahelian; SU: Sudanian; GU: Guinean.

Family	Taxon	Life form	SH	SL	SU	GU
ACANTHACEAE	*Hygrophila*	N			x	
	Justicia-type	I		x	x	
	Justicia-type *flava*	N		x	x	
AMARANTHACEAE	*Achyranthes*-type *aspera*	N	x	x		
	Aerva-type *lanata*	N	x	x		
	Alternanthera	N	x	x		
	Amaranthaceae undiff.	I	x	x		
	Celosia-type *trigyna*	N	x	x		
	Chenopodium-type	N	x	x	x	
	Gomphrena	N	x	x		
	Suaeda	I	x	x		
ANACARDIACEAE	Anacardiaceae undiff.	A			x	x
	Lannea-type	A		x	x	x
APIACEAE	Apiaceae undiff.	I		x	x	

(*continued*)

Table 1. Continued.

Family	Taxon	Life form	SH	SL	SU	GU
APOCYNACEAE	Apocynaceae undiff.	I	x	x	x	
	Tabernaemontana	AL			x	x
ARECACEAE	*Elaeis guineensis*	PA				x
	Phoenix reclinata-type	PA	x	x	x	x
ASCLEPIADACEAE	*Gymnema*	AL	x			
ASTERACEAE	Asteraceae undiff.	I	x	x	x	
	Centaurea-type *perrottetii*	N	x	x		
	Cichorieae undiff.	I	x	x	x	
BOMBACACEAE	*Adansonia digitata*	A			x	x
BURSERACEAE	*Commiphora africana*-type	A		x	x	
CAESALPINIACEAE	*Detarium senegalense*	A			x	x
CAPPARACEAE	*Capparis fascicularis*-type	AL		x	x	
	Cleome-type *gynandra*	N	x	x		
CARYOPHYLLACEAE	Caryophyllaceae undiff.	I	x	x		
	Cerastium-type	N	x	x		
	Polycarpaea-type	N	x	x	x	
	Polycarpon-type	N	x	x	x	x
CASUARINACEAE	*Casuarina equisetifolia*-type	A				
CELASTRACEAE	Celastraceae undiff.	AL	x	x	x	
	Salacia	AL			x	x
CHRYSOBALANACEAE	*Chrysobalanus/Parinari*	A			x	x
COMBRETACEAE	Combretaceae undiff.	A		x	x	x
COMMELINACEAE	*Commelina*-type *benghalensis*	N		x	x	x
	Commelina-type *forskalaei*	N	x	x		
CONVOLVULACEAE	*Convolvulus*-type	N	x	x	x	
	Ipomoea-type	I	x	x	x	
CYPERACEAE	Cyperaceae undiff.	Nq				
DILLENIACEAE	*Tetracera*	AL			x	x
EBENACEAE	*Diospyros*	A			x	x
	Ebenaceae undiff.	A				
EPHEDRACEAE	*Ephedra*	AL	x			

(*continued*)

Table 1. Continued.

Family	Taxon	Life form	SH	SL	SU	GU
EUPHORBIACEAE	*Acalypha*	I			x	x
	Alchornea	A			x	x
	Anthostema-type	A				x
	Antidesma-type	A			x	x
	Bridelia-type	A			x	x
	Erythrococca-type	A			x	x
	Hymenocardia	A		x	x	x
	Macaranga-type	A			x	x
	Mallotus-type	A			x	x
	Uapaca	A			x	x
FABACEAE	*Aeschynomene*	I			x	x
HALORRHAGACEAE	*Laurembergia tetrandra*	Nq				
LAMIACEAE	*Basilicum*-type *polystachyon*	N			x	
LORANTHACEAE	*Tapinanthus*-type	L		x	x	x
MALVACEAE	*Hibiscus*	I	x	x	x	
MELIACEAE	*Khaya*-type *senegalensis*	A				x
MENISPERMACEAE	*Cissampelos*-type	AL			x	x
	Cocculus	L	x	x	x	
	Tinospora bakis	L	x	x	x	
MIMOSACEAE	*Acacia*	AL	x	x	x	
	Prosopis-type *africana*	A		x	x	x
MORACEAE	*Ficus*	A		x	x	
	Musanga-type	A		x	x	x
	Myrianthus-type	A				x
MYRTACEAE	*Syzygium*-type *guineense*	A			x	x
NYMPHAEACEAE	*Nymphaea*	Nq				
OCHNACEAE	*Lophira*	A			x	x
ONAGRACEAE	*Ludwigia*-type *ascendens*	Nq				
POACEAE	Poaceae undiff.	N	x	x	x	x
POLYGALACEAE	*Polygala*-type	I	x	x	x	
POLYGONACEAE	*Polygonum senegalense*	Nq				
POTAMOGETONACEAE	*Potamogeton*	Nq				

(*continued*)

Table 1. Continued.

Family	Taxon	Life form	SH	SL	SU	GU
RANUNCULACEAE	*Clematis*-type	AL	x	x	x	
RUBIACEAE	*Leptactina*	A				x
	Hymenodictyon-type	A			x	x
	Ixora-type	A			x	x
	Mitracarpus villosus	N		x	x	
	Mitragyna-type *inermis*	A		x	x	x
	Morelia senegalensis	A		x	x	x
	Pavetta	AL			x	x
	Rubiaceae undiff.	I	x	x	x	x
	Spermacoce-type *radiata*	N		x	x	
RUTACEAE	*Zanthoxylum*-type *zanthoxyloides*	A			x	x
SALVADORACEAE	*Salvadora persica*-type	A	x	x		
SAPINDACEAE	*Allophylus*	AL		x	x	x
	Paullinia pinnata	L			x	x
	Sapindaceae undiff.	AL		x	x	x
SAPOTACEAE	Sapotaceae undiff.	AL		x	x	x
	Vitellaria-type *paradoxa*	A				x
SOLANACEAE	*Solanum*-type	I	x	x	x	
STERCULIACEAE	*Dombeya*-type	A			x	x
TAMARICACEAE	*Tamarix*	A	x	x		
TILIACEAE	*Triumfetta*-type	I		x	x	
TYPHACEAE	*Typha*	Nq				
ULMACEAE	*Celtis*	A		x	x	x
ZYGOPHYLLACEAE	*Tribulus*	N	x	x		

Data are presented as percentages and influx values (number of grains/cm^2/year) and diagrams are drawn using Tilia and CONISS (Grimm 1991). Pollen percentages were calculated against a sum excluding aquatics and ferns, the percentages of which were calculated separately against a sum including all the pollen grains and fern spores counted. Algae and fungi percentages were calculated against the sum of NPP. Correspondence statistical analyses (CA) were performed on the raw pollen counts using the Package 'Rcmdr/FactoMinR' (Husson *et al.* 2020).

An age model was derived from three Accelerator Mass Spectrometry (AMS) radiocarbon dates and an additional control point given by the first occurrence of *Casuarina equisetifolia* pollen grains, the parent trees of which were planted in 1948 along the coast (Ndiaye *et al.* 1993), using the Bacon age-depth modeling (Blaauw and Christen 2011) from calibrated ^{14}C ages (Reimer *et al.* 2013) (Table 2).

Table 2. Mboro-Baobab ages. Radiocarbon dating (99.5 cm, 299.5 cm, and 399.5 cm) was conducted at UMS-ARTEMIS AMS Facilities (France). At 30.5 cm downcore, the first appearance of *Casuarina equisetifolia* pollen grains provide a date of 1948 CE. The age of the surface is that of coring (from Lézine *et al.* 2019).

Lab. number	Mean depth (cm)	Nature	Radiocarbon Age BP	Error	$\delta^{13}C$	Age cal yr BP
	0	Date of coring				−66
	30.5	Casuarina pollen				−2
49827	99.5	Bulk sediment	580	30	−22.40‰	604
49828	299.5	Bulk sediment	3265	30	−28.60‰	3496
49829	399.5	Bulk sediment	3865	30	−29.10‰	4298

5.4 POLLEN DIAGRAM

Based on the variations in percentages of the main pollen taxa and NPP types, four zones corresponding to the main phases of landscape evolution over the last 4300 years can be distinguished. Zones 3 and 4 were presented in detail in Lézine *et al.* (2019, Figure 2).

5.4.1 Zone 1A: 4300 – 3700 cal yr BP

Guinean tree taxa such as *Mallotus* (3%), *Zanthoxylum* (2.4%), *Uapaca* (2.5%), *Alchornea* (15%) and *Macaranga* (11%) characterized this zone. Lakeshore populations are reduced with *Typha* percentages less than 1%.

Among the NPPs, *Pediastrum* and *Coelastrum* are largely dominant with percentages that can reach values above 80% for Pediastrum and 30% for *Coelastrum*.

5.4.2 Zone 1B: 3700 – 3050 cal yr BP

There is little change from the previous zone. However, the forest taxa listed above decreased significantly (e.g. *Uapaca* (0.3%)). Other tree taxa such as *Lannea* (1.2%) and *Mitragyna* (1.2%) increased, along with taxa from lakeshore populations, *Syzygium* (7.2%) and *Elaeis* (1.2%). Upland herbaceous taxa, such as Amaranthaceae and *Aerva*, which were occasionally present in zone 1A, also increased. However, their percentages remained low (less than 1.5%).

Among the NPPs, this zone is characterized by the rapid development of *Tetraedron*, which reached up to 80% reflecting the onset of the lake level lowering.

5.4.3 Zone 2A: 3050 – 1600 cal yr BP

This zone is characterized by the sharp increase in *Typha* percentages (15%) and the more gradual increase of Cyperaceae (22.8%) showing the development of reed swamp populations around the lake. Among the trees, *Mallotus* or *Uapaca* gradually decrease to less than 0.5%, whereas *Alchornea* increased (16.5%). Dry herbaceous taxa, such as *Mitracarpus* and *Spermacoce*, already present previously, increased with percentages up to 4%.

NPP are characterized by the progressive increase in the percentages of *Botrycoccus*, (maximum = 17%).

5.4.4 Zone 2B: 1600 – 700 cal yr BP

Cyperaceae (23%) then *Elaeis* (2.6%) increased in this zone. Some tree taxa of Guinean affinity (*Mallotus*, *Anthostema*, *Zanthoxylum*, *Celtis*) or wetlands decreased or even disappeared

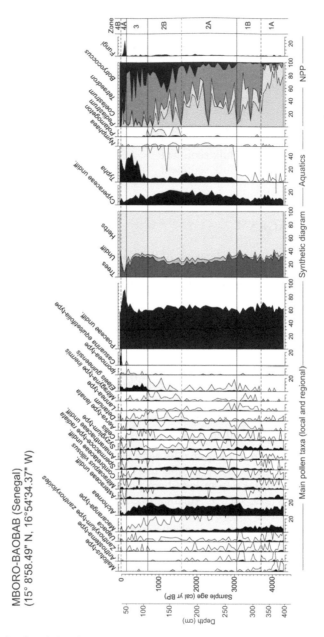

Figure 2. Figure showing from left to right, a simplified diagram of pollen percentages for Mboro-Baobab (Senegal). Black shading shows selected taxa. Light-gray shading shows the same results, multiplied by 10 for highlighting. On the right, the main NPP (Algae and Fungi) are shown.

almost completely, while others were stable (*Alchornea* 19%) or even increased (*Detarium* 2.1%, *Mitragyna* 1.6%). *Typha* percentages decreased significantly (5%) while another aquatic plant, *Potamogeton* (1.9%), appeared.

Among the NPPs, *Tetraedron* and *Botryococcus* continued to increase until reaching 85% and 47%, respectively.

5.4.5 Zone 3: 700 – 129 cal yr BP

Alchornea and *Elaeis* dominated this zone though following an opposite trend. *Elaeis* reached its highest value (12%) while *Alchornea* decreased steadily (minimum at 4.5%). *Syzygium* remained well represented (2%). Among read swamp populations, *Typha* increased remarkably, up to 35% and Cyperaceae decreased (7%). NPPs were dominated by *Tetraedron* (up to 90%) and *Botryococcus* (up to 60%).

5.4.6 Zone 4A: 129 –14 cal yr BP

This zone is characterized by the massive increase in Poaceae (90%) and the correlative decrease in *Typha* (17%). The percentages of *Elaeis* remained at the level reached in the previous area (7%). *Botryococcus* almost exclusively dominated the algae assemblage. This zone is also characterized by the noticeable presence of fungi.

5.4.7 Zone 4B: 14 cal yr BP – Modern

This zone differs from the previous one by the appearance of *Casuarina equisetifolia*, which reaches up to 16%. Poaceae decreased to 55%, in contrast to *Typha*, which increased to values above 30%. *Botryococcus* totally dominated the assemblages of algae.

5.5 ENVIRONMENTAL RECONSTRUCTION AND DISCUSSION

5.5.1 The decline in tropical humid plant taxa

Figure 3 is the graphical representation of a correspondence analysis using the microfossil assemblages for the environmental evolution of the Mboro-Baobab area. Three groups of taxa emerge from this analysis. The first is dominated by tropical humid (Guinean) tree taxa (e.g. *Dictyandra, Macaranga, Mallotus, Syzygium, Uapaca*), the second group is a mixture of taxa from trees and herbs of drier, Sudanian, phytogeographical affinity or having a wide spatial distribution (e.g. *Alchornea*, Asteraceae, *Celtis, Mitracarpus*). The third group is dominated by herbs found today in dry places in the Sahara and Sahel (e.g. *Aerva lanata, Tribulus*, Amaranthaceae). Only few trees are found in this group (e.g. *Acacia, Adansonia digitata*).

The most striking result of the analysis of these groups is the major environmental change revealed at 3200 cal yr BP (Figure 4A). The most humid tree taxa (group 1) which dominated earlier in the record, and particularly around 3875 cal yr BP (Figure 4B), dramatically declined in favor of drier plant taxa (group 2). This decline continued into the late Holocene in two main stages respectively around 2500 and 1300 cal yr BP. Contrasting with this step-like evolution of forest diversity, the regional, dry elements (group 3) exhibit a slow increasing trend starting from 3700 cal yr BP and peaking around 1600-1300 cal yr BP.

5.5.2 Changes in the hydrological regime

The 3200 cal yr BP event closely matches a major change in the hydrological regime (Figure 4C): changes in the algal community show that the lowering of the water level was initiated as soon as 3750 cal yr BP (cf. expansion of *Tetraedron*, Figure 2). At 3200 cal yr BP, the development of the oligotrophic algae *Botryococcus* (Figure 2 and 4C) mark the reduction in nutrient supply due to increasingly dry conditions, which culminated during the last millennium. The strong decline in the freshwater algae *Pediastrum* and *Coelastrum*, followed by the development of *Typha* populations, confirms this major change in the hydrological regime and the lowering of the lake level.

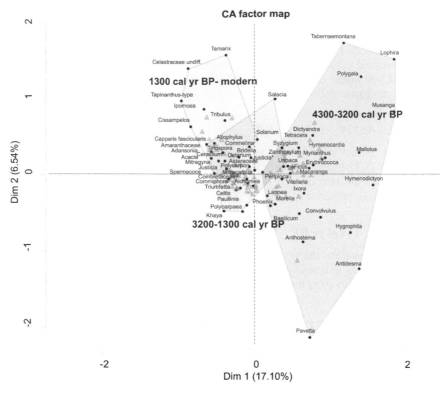

Figure 3. Correspondence analyses on the Mboro-Baobab pollen counts allowing distinction of tree main groups of taxa according to a decreasing humidity gradient. These groups define three periods (in bold).

5.5.3 Timing of the end of the AHP in the Sahel

Earlier work by Lézine (1988, 1989) in the Niayes suggested the sudden disappearance of the gallery forests around 2500 cal yr BP as illustrated at Diogo, only a few kilometers north of Mboro (Figure 1A). Diogo illustrates the magnitude of this environmental crisis by showing the drop in the water level allowing for the development of a Cyperaceae swamp contemporaneous with the collapse of most of the trees, which established during the AHP (Figure 5). Only swamp forest (*Syzygium*) and pioneer (*Alchornea*) trees were able to persist in a context of increased dry conditions (Braconnot *et al.* 2019).

Our Mboro-Baobab record does not show such an abrupt shift from a humid to a dry environment. The tree cover (AP%, cf. Lézine *et al.*, this issue) was never highly developed, probably due to the distal position of the site with respect to the core of the Niaye groundwater system (Diogo) (Putallaz 1962) and seems to have only slightly varied over the last 4300 years (Figure 2). Forest composition however strongly changed. Guinean taxa underwent a clear phase of expansion centered at 3875 cal yr BP. A similar phase of development of tropical trees is seen at Jikariya (Waller *et al.* 2007) in the Manga grasslands of northern Nigeria between 3800 and 3500 cal yr BP that points to the regional character of this short phase of forest recovery. At Mboro-Baobab, this forest phase ended rapidly with steppic conditions increasing as soon as 3750 cal yr BP. Interestingly, this increase in aridity did not have an immediate effect on the composition of the forest gallery but probably only on its density. It is only at 3200 cal yr BP that the replacement of the wettest forest elements by forest taxa more adapted to drier climatic

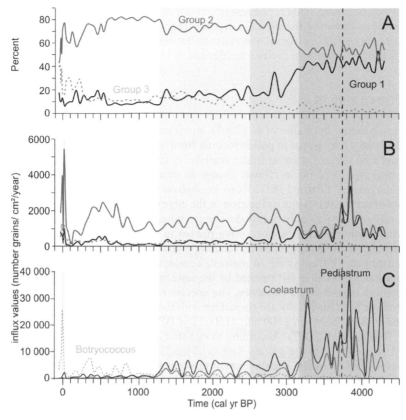

Figure 4. Summary of the evolution of the environment at Mboro-Baobab over the last 4300 years. (A) Percentages of the three groups of pollen taxa revealed by the CA statistical analysis; (B) influx values of these groups; (C) influx values of the algae *Pediatrum, Coelastrum* and *Botryococcus*. The gray bars (from dark to light) show the progressive decline of the most humid elements. The dotted line shows the onset of dryness in a still humid context. Note that curves are superimposed and not cumulative.

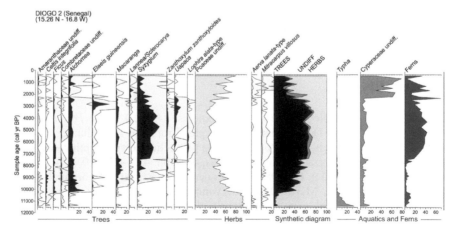

Figure 5. Percentage pollen diagram of main taxa from Diogo in the Niaye area of Senegal (Lézine 1988).

conditions occurred. This replacement was not as abrupt as at Diogo but its amplitude was not negligible as shown by the decrease in percentages of group 1 by about 10% (Figure 4A). Taking into account the chronological uncertainties between sites, our Mboro-Baobab record closely fits with that of Jikaryia, which dates the decline of tropical humid trees at 3150 cal yr BP.

5.6 CONCLUSION

As already noted by Waller *et al.* (2007), significant differences in the timing and amplitude of similar events may occur in pollen records from the same region. The size of the sites and their location with respect to groundwater availability are among the parameters that can influence the local response of plants to climate change in terms of amplitude and timing. Compared to the other Niaye sites (Lézine 1988) Mboro-Baobab contains the most complete and detailed record of the end of the AHP. Despite its location on the edge of the wettest zone of the Niayes region, where the forest gallery was only poorly developed, it has undergone profound environmental changes allowing distinction of the successive phases in the establishment of the modern landscape. The drying of the Mboro-Baobab landscape was gradual, starting from 3750 cal yr BP then culminating at 1300 cal yr BP. In contrast, a major shift in the lake system and the forest diversity occurred at 3200 cal yr BP, marked by the lowering of the water level and the sharp decline of the tropical humid forest elements. The species turnover initiated at that time continued up to 2500 cal yr BP. Our results are consistent with those obtained in the central Sahel, which place the end of AHP between 3300 and 3150 cal yr BP depending on the site (Salzmann and Waller 1998; Waller *et al.* 2007). According to our study, the end of AHP (i.e. the retreat of the most humid elements from the gallery forests of the Niayes region) lasted about 700 years between 3200 and 2500 cal yr BP. The duration of this transition period was likely favored by the presence of a water table close to the surface, which was able to maintain humid conditions for several centuries after the major event of 3200 cal yr BP.

ACKNOWLEDGEMENTS

This work is an "ACCEDE" Belmont Forum contribution (18 BELM 0001 05). Thanks are due to the Institut de Recherche pour le Développement (IRD) and the UCAD Geological Department (Dakar, Senegal) for logistic support in the field and authorizations and to the ECLAIRS International Laboratory (Dakar) and the LOCEAN Laboratory (Paris) for funding. KL and AML are funded by CNRS, France.

REFERENCES

Association des Palynologues de Langue Française (APLF) 1974, *Pollen et spores d'Afrique tropicale*. Travaux et Documents de Géographie Tropicale **16** (Talence : CEGET).

Blaauw, M. and Christen, J. A., 2011, Flexible paleoclimate age-depth models using an autoregressive gamma process. *Bayesian analysis*, **6**(3), pp. 457–474, 10.1214/ba/1339616472.

Bonnefille, R. and Riollet, G., 1980, Pollens des savanes d'Afrique orientale (Paris :CNRS).

Braconnot, P., Zhu, D., Marti, O. and Servonnat, J., 2019, Strengths and challenges for transient Mid- to Late Holocene simulations with dynamical vegetation. *Climate of the Past*, **15**, pp. 997–1024, 10.5194/cp-15-997-2019.

De Menocal, P., Ortiz, J., Guilderson, T., Adkins, J., Sarnthein, M., Baker, L. and Yarusinsky, M., 2000, Abrupt onset and termination of the African Humid Period: Rapid climate responses to gradual insolation forcing. *Quaternary Science Reviews*, **19**(1–5), pp. 347–361, 10.1016/S0277-3791(99)00081-5.

Faegri, K., Kaland, P. E. and Krzywinski, K., 1989, *Textbook of pollen analysis*, 4th edition, (Chichester: John Wiley & Sons Ltd).

Grimm, E. C., 1991, *Tilia and Tiliagraph*, (Springfield: Illinois State Museum).

Hély, C., Lézine, A.-M. and APD contributors, 2014, Holocene changes in African vegetation: tradeoff between climate and water availability. *Climate of the Past*, **10**, pp. 681–686, 10.5194/cp-10-681-2014.

Husson, F., Josse, J. and Le, S., 2020, Package 'RcmdrPlugin. FactoMineR'. http://factominer. free.fr/graphs/RcmdrPlugin.html.

Jowsey, P. C., 1966, An improved peat sampler. *New Phytologist*, **65**(2), pp. 245–248, 10.1111/ j.1469-8137.1966.tb06356.x.

Kröpelin, S., Verschuren, D., Lézine, A.-M., Eggermont, H., Cocquyt, C., Francus, P., Cazet, J. P., Fagot, M., Rumes, B., Russell, J. M., Darius, F., Conley, D. J., Schuster, M., von Suchodoletz, H. and Engstrom, D. R. 2008, Climate-driven ecosystem succession in the Sahara: the past 6000 years. *Science*, **320**(5877), pp. 765–768, 10.1126/science.1154913.

Kuper, R. and Kröpelin, S., 2006, Climate-controlled Holocene occupation in the Sahara: motor of Africa's evolution. *Science*, **313**(5788), pp. 803–807, 10.1126/science.1130989.

Lécuyer, C., Lézine, A.-M., Fourel, F., Gasse, F., Sylvestre, F., Pailles, C., Grenier, C., Travi, Y. and Barral, A., 2016, In-Atei palaeolake documents past environmental changes in central Sahara at the time of the "Green Sahara": Charcoal, carbon isotope and diatom records. *Palaeogeography, Palaeoclimatology, Palaeoecology*, **441**, pp. 834–844, 10.1016/j.palaeo. 2015.10.032.

Lézine, A.-M., 1988, Les variations de la couverture forestière mésophile d'Afrique occidentale au cours de l'Holocène. *Comptes rendus de l'Académie des sciences*, Série 2, **307**(4), pp. 439–445.

Lézine, A.-M., 1989, Late Quaternary vegetation and climate of the Sahel. *Quaternary Research*, **32**(3), pp. 317–334, 10.1016/0033-5894(89)90098-7.

Lézine, A.-M., Hély, C., Grenier, C., Braconnot, P. and Krinner, G., 2011, Sahara and Sahel vulnerability to climate changes, lessons from Holocene hydrological data. *Quaternary Science Reviews*, **30**(21–22), pp. 3001–3012, 10.1016/j.quascirev.2011.07.006.

Lézine, A.-M., Holl, A.F.C., Lebamba, J., Vincens, A., Assi-Khaudjis, C., Février, L. and Sultan, E., 2013, Temporal relationship between Holocene human occupation and vegetation change along the northwestern margin of the Central African rainforest. *Comptes Rendus Geoscience*, **345**(7–8), pp. 327–335, 10.1016/j.crte.2013.03.001.

Lézine, A.-M., Lemonnier, K. and Fofana, C.A.K., 2019, Sahel environmental variability during the last millennium: Insight from a pollen, charcoal and algae record from the Niayes area, Senegal. *Review of Palaeobotany and Palynology*, **271**, 10.1016/j.revpalbo.2019.104103, p. 104103.

Maley, J., 1970, Contributions à l'étude du Bassin tchadien. Atlas de pollens du Tchad. *Bulletin du Jardin botanique National de Belgique*, **40**(1), pp. 29–48, 10.2307/3667543.

Ndiaye, P., Mailly, D., Pineau, M. and Margolis, H.A., 1993, Growth and yield of Casuarina equisetifolia plantations on the coastal sand dunes of Senegal as a function of microtopography. *Forest Ecology and Management*, **56**(1–4), pp. 13–28, 10.1016/0378-1127(93)90100-2.

Manning, K. and Timpson, A., 2014, The demographic response to Holocene climate change in the Sahara. *Quaternary Science Reviews*, **101**, pp. 28–35, 10.1016/j.quascirev.2014.07.003.

ORSTOM, 1962, M Boro. *Carte pédologique des Niayes*, 1:10000. (Dakar: ORSTOM).

Putallaz, J., 1962, *Hydrogéologie de la région des Niayes*. Technical Report BRGM DAK 62 A 12.

Reimer, P. J., Bard, E., Bayliss, A., Beck, J. W., Blackwell, P. G., Ramsey, C. B., Buck, C.E., Cheng, H., Edwards, R.L., Friedrich, M., Grootes, P. M., Guilderson, T.P., Haflidason, H.,

Hajdas, I., Hatté, C., Heaton, T.J., Hoffman, D.L., Hogg, A.G., Hughen, K.A., Kaiser, K.F., Kromer, B., Manning, S.W., Niu, M., Reimer, R.W., Richards, D.A., Scott, E.M., Southon, J.R., Staff, R.A., Turney, C.S.M. and van der Plitch, J., 2013, IntCal13 and Marine13 radiocarbon age calibration curves 0–50,000 years cal yr BP. *Radiocarbon*, **55**(4), pp. 1869–1887, 10.2458/azu_js_rc.55.16947.

Salzmann, U. and Waller, M., 1998, The Holocene vegetational history of the Nigerian Sahel based on multiple pollen profiles. *Review of Palaeobotany and Palynology*, **100**(1–2), pp. 39–72, 10.1016/S0034-6667(97)00053-5.

Trochain, J., 1940, Contribution à l'étude de la végétation du Sénégal (Paris: Larose).

Vincens, A., Lézine, A.-M., Buchet, G., Lewden, D. and Le Thomas, A., 2007, African pollen database inventory of tree and shrub pollen types. *Review of Palaeobotany and Palynology*, **145**(1–2), pp. 135–141, 10.1016/j.revpalbo.2006.09.004.

Waller, M. P., Street-Perrott, F. A. and Wang, H., 2007, Holocene vegetation history of the Sahel: pollen, sedimentological and geochemical data from Jikariya Lake, north-eastern Nigeria. *Journal of Biogeography*, **34**(9), pp. 1575–1590, 10.1111/j.1365-2699.2007.01721.x.

Watrin, J., Lézine, A.-M. and Hély, C., 2009, Plant migration and plant communities at the time of the "green Sahara". *Comptes Rendus Geoscience*, **341**(8–9), pp. 656–670, 10.1016/j.crte.2009.06.007.

White, F., 1983, The vegetation of Africa (Paris: UNESCO).

Ybert, J.-P., 1979, *Atlas de pollens de Côte d'Ivoire*. Initiations-Documentations Techniques 40 (Paris: ORSTOM).

CHAPTER 6

Changes in the West African landscape at the end of the African Humid Period

Anne-Marie Lézine & Kévin Lemonnier

Laboratoire d'Océanographie et du Climat, Expérimentation et Approche numérique/IPSL, Sorbonne Université, CNRS-IRD-MNHN, Paris, France

Martyn P. Waller

Department of Geography and Geology, Kingston University London, Kingston upon Thames, Surrey, United Kingdom

Ilham Bouimetarhan[1] & Lydie Dupont

Center for Marine Environmental Sciences, MARUM, University of Bremen, Bremen, Germany

African Pollen Database contributors[2]

ABSTRACT: Existing pollen datasets from northern Africa stored in the African Pollen Database were used to assess changes in landscape physiognomy at the end of the African Humid Period (AHP) from 5000 cal yr BP to the present using arboreal pollen percentages. The thirty-six sites available were used to map changes in arboreal cover at a sub-continental scale. Based on their location in present-day forested and non-forested areas and their relatively higher temporal resolution eight of them were selected to examine the timing and amplitude of the vegetation response in more detail, and particularly in the Sahel. In spite of low pollen production and dispersal of many tropical plants, which lead to the under representation of most of the trees relative to their abundance in the landscape, we were able to distinguish the geographical pattern and timing of vegetation changes. The landscape response to the end of the AHP was far from homogeneous particularly in the Sahel where a clear east-west gradient of changing tree cover is indicated with the central Sahel being notably species poor. In areas where forests were well developed during the AHP, i.e. in the south and west, the establishment of the modern landscape was abrupt with a threshold crossed between 3300 and 2500 cal yr BP according to local conditions. Elsewhere in northern Africa the switch from tree (C_3) to grass (C_4) dominated landscapes occurred more gradually during the same period. This review shows that the timing of the ecosystem response at the end of the AHP was remarkably synchronous throughout northern Africa.

[1] Other affiliation: *Faculté des Sciences appliquées, CUAM, Université Ibn Zohr, B.P. 8106 Agadir, Morocco*
[2] Akaegbobi, I.M., Assi-Kaudjhis, C., Ballouche, A., Buchet, G., Kadomura, C., Lebamba, J., Maley, J., Marchant, R., Mariotti Lippi, M., Médus, J., Mercuri, A.M., Njokuocha, R.C., Roche, E., Salzmann, U., Schulz, E., Sowunmi, A., Tossou, M., Vincens, A.

DOI 10.1201/9781003162766-6

6.1 INTRODUCTION

Reconstructing vegetation changes in tropical North Africa during the Holocene and understanding how the landscape has evolved from a 'humid' state characterized by the development of numerous lakes and wetlands into the today's hyper-arid desert continues to be the subject of extensive research (e.g. Dallmeyer *et al.* 2020; Krinner *et al.* 2012; Lézine *et al.* 2011; Shanahan *et al.* 2015). Pollen data have been successfully used to assess biodiversity changes that occurred during the African Humid Period (AHP) (deMenocal *et al.* 2000) such as the penetration of tropical trees into the Sahara and the Sahel at its onset or their retreat at its termination (e.g. Hély *et al.* 2014 and references therein). Pollen data are however extremely complex and changes in plant cover and related land-surface albedo remain difficult to be quantified. A large part of this complexity is due to the uneven geographical distribution of the data, difficulties in establishing robust chronologies, and the incomplete and/or discontinuous nature of the sedimentary series (Watrin *et al.* 2009). The desiccation of the ancient Saharan lakes and the deflation of exposed lacustrine deposits, still continuing today, have eroded lake sediments and considerably hampered the preservation of pollen grains. Therefore, preserved sediments provide limited time windows mainly concentrated in wettest periods. Only a single, continuous, sequence covering a time interval from the end of the AHP to the present is available for the entire Sahara (Lézine *et al.* 2011). Continuous pollen series with high temporal resolution are however more numerous to the south, from the Sahel to the Congo basin (e.g. Lézine *et al.* 2013a; Maley and Brenac 1998; Ngomanda *et al.* 2009; Salzmann *et al.* 2002; Vincens *et al.* 2010; Waller *et al.* 2007).

Differences in pollen production and dispersal between tropical plants means that interpretation of vegetation cover within the landscape from ancient pollen records is further complicate, especially in the driest regions. Most of the trees have a low pollen productivity, disperse pollen entomophilously, and are thus under-represented in the pollen spectra (Ritchie 1995). While the presence of tree pollen, even a single grain, may be indicative of the presence of the corresponding plants (Watrin *et al.*, 2006), it is extremely difficult given the current state of knowledge, to quantify the importance of these trees in the landscape. Therefore, the timing and magnitude of the environmental change related to the end of the AHP remains poorly documented.

Here we use a simple index: the percentage of tree pollen (AP %) from selected high resolution pollen records from both continental and marine environments in northern Africa, from 5° to 19°N, in order to characterize the environmental change at the end of the AHP. This index is commonly used to qualitatively evaluate woody cover in temperate regions where most of the tree species are wind-pollinated. Our goal is to discuss how well this index can be applied to northern Africa, given the particularities of the pollen production and dissemination syndromes of tropical tree species.

6.2 ENVIRONMENTAL SETTING

The climatic features of northern Africa are related to the Atlantic monsoon circulation and the seasonal migration of the Intertropical Convergence Zone (ITCZ). In January, the monsoons' influence is confined to the periphery of the Gulf of Guinea and the northernmost areas are subject to the hot, dry continental trade winds. In July, the ITCZ reaches its northernmost position at 22°N allowing moist monsoon air to extend over north tropical Africa up to the southern fringe of the Sahara.

The progressive decrease of the monsoon influence and related rainfall on the continent is correlated with the latitudinal distribution of the vegetation. Five major floristic regions can be delineated from south to north: Guineo-Congolian forests of different types (evergreen and semi-evergreen forests) and secondary wooded grasslands (mean annual rainfall more than 1200 mm), Sudanian dry forests and wooded grasslands (1500–500 mm yr^{-1}), Sahelian wooded grasslands

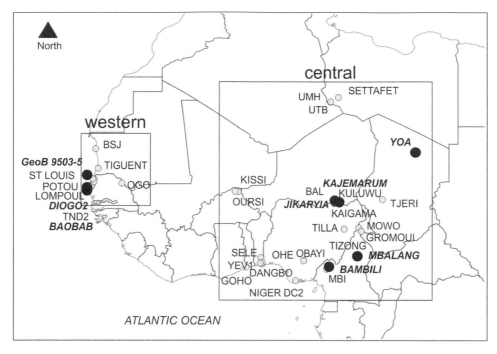

Figure 1. Location map of the Holocene pollen sites used in this study (Table 1). In bold and italic, pollen sites detailed in the text (Figure 3). Boxes define the two sectors shown in Figure 5.

and grasslands (500–100 mm yr^{-1}) and Saharan steppes (less than 100 mm yr^{-1}) (White 1983) (Figure 1). In addition, Afromontane forests occur in the Cameroon Volcanic Line between roughly 1800–2300 m asl (annual rainfall: 2000 mm yr^{-1}; mean temperature: 18 °C) bordered at upper elevations by Afroalpine grasslands and at lower elevations by sub-montane forests and wooded grasslands.

Along the coast, mangrove forests expand at the mouth of rivers where they benefit from both brackish waters and large muddy areas. Their optimal growth is attained under a rain forest climate but two main components expand northward along the coast of West Africa: *Rhizophora racemosa* reaches its northern limit at the mouth of the Senegal River near 16°N and *Avicennia germinans* extends up to Tidra at 19°50 N.

6.3 DATA USED

In this study, we use all the available Holocene pollen data from northern Africa stored in the African Pollen Database (APD) (Figure 1; Table 1). We focus on changes in landscape physiognomy, i.e. the shift from a C_3 plant dominated landscape to a C_4 plant dominated one at the end of the AHP. Therefore, pollen records from equatorial lowland forests (e.g. Elenga *et al.* 1994, 1996; Reynaud-Farrera *et al.* 1996) were excluded, since the impact of the environmental crisis, which occurred at the end of the AHP in the Congo Basin, was more a change in forest composition and density than the replacement of lowland forests by grasslands or wooded grasslands (Vincens *et al.* 1999). The dataset consists of 36 sites among which seven continental (Bambili, Mbalang, Diogo, Mboro-Baobab, Kajemarum, Jikaryia, Yoa) and one marine (GeoB 9503-5) pollen records have been selected in order to characterize the evolution of the landscape in the

Table 1. List of the pollen sites used in this study according to latitude from north to south in the two sectors defined in this study. In bold and italic = selected sites, see black dots on Figure 1.

Site name	Latitude	Longitude	Altitude (m asl)	Country	Time frame (cal yr BP)	Reference
Central sector						
SETTAFET	25.35	11.43	1100	Libya	5000	Schulz (1980)
UAN TABU [UTB]	24.86	10.52	915	Libya	4000–4250	Mercuri, and Grandi (2001)
UAN MUHUGGIAG [UMH]	24.84	10.51	915	Libya	4000–4250	Mercuri *et al.* (1998)
YOA	19.03	20.31	380	Chad	0–5000	Lézine *et al.* (2011)
OURSI	14.65	−0.49	290	Burkina Faso	0–3500	Ballouche and Neumann (1995)
KISSI	14.62	−0.14	280	Burkina Faso	500–750	Ballouche (1997)
TJERI	13.73	16.50	275	Chad	500–4750	Maley (2004)
JIKARYIA	13.31	11.08	343	Nigeria	0–5000	Waller *et al.* (2007)
BAL	13.30	10.94	300	Nigeria	250–5000	Salzmann and Waller (1998)
KAJEMARUM	13.30	11.02	300	Nigeria	2750–5000	Salzmann and Waller (1998)
KAIGAMA	13.25	11.57	330	Nigeria	3500–5000	Salzmann and Waller (1998)
KULUWU	13.22	11.55	330	Nigeria	4750–5000	Salzmann and Waller (1998)
MOWO	10.60	13.99	574	Cameroon	500–750	Delneuf and Médus (1997)
TILLA	10.39	12.12	690	Nigeria	0–5000	Salzmann *et al.* (2002)
GROMOUI	10.32	14.32	430	Cameroon	1000–1250	Delneuf and Médus (1997)
MBALANG	7.32	13.73	1110	Cameroon	0–5000	Vincens *et al.* (2010)
TIZONG	7.25	13.58	1160	Cameroon	0–4000	Lebamba *et al.* (2016)
LAC SELE	7.15	2.43	11	Benin	0–5000	Salzmann and Hoelzmann (2005)
OBAYI	6.82	7.37	550	Nigeria	1750–2000	Njokuocha (2012)
OHE POND	6.80	7.37	1502	Nigeria	1250–1500	Njokuocha and Akaegbobi (2014)

(*Continued*)

Table 1. *Continued.*

Site name	Latitude	Longitude	Altitude (m asl)	Country	Time frame (cal yr BP)	Reference
DANGBO	6.60	2.59	40	Benin	0–5000	Tossou (2002)
YÉVIÉDIÉ	6.53	2.37	53	Benin	1250–5000	Tossou (2002)
GOHO	6.44	2.57	35	Benin	750–4750	Tossou (2002)
MBI	6.08	10.35	2018	Cameroon	0–5000	Lézine *et al.* (2021)
BAMBILI 1	5.94	10.24	2273	Cameroon	0–5000	Lézine *et al.* (2013a)
NIGER DELTA [DC2]	4.55	6.43	0	Nigeria	500–3750	Sowunmi (1981)
Western sector						
BAIE SAINT JEAN [BSJ]	19.47	−16.30	1	Mauritania	1750–3750	Unpublished (Lézine A.M.)
TIGUENT	17.25	−16.02	8	Mauritania	3000–3250	Médus and Barbey (1979)
GeoB 9503-5	16.07	−16.65	−50	Atlantic Ocean	1500–4250	Bouimetharhan *et al.* (2009)
ST LOUIS	16.03	−16.48	2	Senegal	0–1000	Fofana *et al.* (2020)
POTOU	15.75	−16.50	11	Senegal	0–5000	Lézine (1988)
OGO [OS2]	15.57	−13.28	15	Senegal	750–1000	Feller *et al.* (1981)
LOMPOUL	15.42	−16.72	3	Senegal	2000–5000	Lézine (1988)
DIOGO 2	15.27	−16.80	8	Senegal	250–5000	Lézine (1988)
TOUBA N'DIAYE 2 [TND2]	15.17	−16.87	6	Senegal	1000–5000	Lézine (1988)
MBORO-BAOBAB	15.15	−16.91	4	Senegal	0–4500	Lemonnier and Lézine (this volume)

two main vegetation types which expanded in northern Africa during the AHP: (i) continuous forests in the Cameroon Highlands, and (ii) gallery forests around water bodies in the Sahara and Sahel within a dry, open regional environment. The whole dataset was then used to map landscape changes at a sub-continental scale.

The chronology is given in calendar years BP (cal yr BP) after conversion of ^{14}C dates to calendar age according to Stuiver *et al.* (2020). The age models were retrieved from the published articles or from the APD (i.e. based on linear interpolation between the dated samples) taking into account possible discontinuities in sediment deposition. Arboreal pollen types (AP) include trees, shrubs, palms and lianas. Undifferentiated pollen types (UP) include taxa corresponding to plants with various life forms (trees, shrubs, lianas or herbs) and taxa determined at a low taxonomic level, typically the family level (Vincens *et al.* 2007). Percentages are calculated against a sum of total pollen excluding those of aquatics (including Cyperaceae), cultivated and Mediterranean plant types.

6.4 POLLINATION SYNDROMES OF THE TROPICAL PLANTS

The pollination syndromes of the families represented at the seven selected sites (Table 2) are derived from Watson and Dallwitz (1992) and the gymnosperm database (www.conifers.org). This reveals the major challenge of representing landscapes from pollen data in northern Africa. Most of the families present in our pollen records are entomophilous or zoophilous (mainly cheiropterophilous and ornithophilous), whereas 16% are anemophilous or entomophilous and only 5% are strictly anemophilous. Most of the entomophilous/zoophilous plants produce small amounts of pollen, which may be exceptionally large. As a result, they are under-represented (e.g. *Adansonia*) or rarely (e.g. *Isoberlinia, Daniellia*) represented in the pollen assemblages (Watrin *et al.*, 2006). In contrast, members of the Poaceae family, which is exclusively anemophilous, produce large quantities of pollen, and therefore largely dominate the pollen assemblages, particularly those derived from wooded grasslands and grasslands. Taking into account these specifications and the difficulties of measuring the quantity of pollen grains produced by the plants, the most widely used method for describing the pollen signature of the modern vegetation in terms of diversity, biomes or vegetation types is that of surface samples, either from soil, mosses or pollen traps. This method has been successfully developed in tropical Africa since Hedberg (1954). Gajewski *et al.* (2002) have validated the APD dataset, which covers all of sub-Saharan Africa.

6.5 POLLEN REPRESENTATION OF THE MODERN LANDSCAPE

In northern Africa the pollen-plant and pollen-biome relationships have been extensively studied (e.g. Lézine *et al.* 2009; Watrin *et al.* 2007 and references therein). These studies have identified the pollen signature of common plants and biomes along latitudinal (Lézine *et al.* 2009) or altitudinal gradients (Verlhac *et al.* 2018) in order to provide the basis for reliable vegetation reconstructions in the past and for quantitative estimates of climate parameters from pollen data. However, the physiognomy of the vegetation is only rarely addressed (Lézine and Hooghiemstra 1990; Vincens *et al.* 2000). Based on arboreal pollen percentages (AP%) and undifferentiated pollen percentages (UP%), Figure 2A shows that the two extreme vegetation types along a south-north transect in northern Africa, the Guineo-Congolian forest in the south and the desert steppe in the north, are easily identified (Lézine and Hooghiemstra 1990). The Guineo-Congolian forest is represented by AP higher than 80% and UP lower than 8% while the desert steppe is represented by AP lower than 10%. In desert steppe, UP (mainly herbs, sub-shrubs or shrubs from the Amaranthaceae, Brassicaceae, Caryophyllaceae, Asteraceae families) values are exceptionally high, representing more than 50%. While AP% consistently range between 20% and 50% in the Sudanian dry forests of southern Senegal, they strongly vary between 5 and 35% in the Sahelian woodlands and wooded grasslands, making the pollen signature of the tree cover in these grass-dominated landscapes difficult to detect. The difficulty in detecting tree cover is mainly due to the uneven distribution of the trees within the landscape. This is particularly the case in the transition zone between forest and desert where trees are mostly concentrated along rivers and wetlands (Trochain 1940; White 1983). In addition, these vegetation formations share numerous tree species, the distribution of which is denser in the woodlands and sparser in the wooded grasslands (e.g. Combretaceae, *Acacia*). The characteristics of their pollen production and dispersal hampers any reliable reconstruction of vegetation types.

In the Equatorial regions, the forest-wooded grassland transect of Kandara (Vincens *et al.* 2000) shows AP higher than 60% in the forest. AP dramatically decrease to less than 10% in the adjacent wooded grassland and the transition between the two is typified by intermediate values (45%) (Figure 2B). However, as stated by the authors, the different kind of forests (*Albizia* forest and *Rinorea* forest) cannot be distinguished, and no decrease in AP% is observed in the forest

Table 2. List of the families encountered in the Holocene pollen sites used in this study (Table 1) and included in the pollen sum with their pollination syndromes (after Watson and Dallwitz (1992) and the gymnosperm database (www.conifers.org)).

Family	Pollination system	Family	Pollination system
Acanthaceae	entomophilous	Liliaceae	entomophilous
Aizoaceae	entomophilous	Lobeliaceae	entomophilous
Alangiaceae	entomophilous	Loganiaceae	entomophilous
Amaranthaceae	entomophilous	Loranthaceae	entomophilous, or ornithophilous
Anacardiaceae	entomophilous	Lythraceae	entomophilous
Annonaceae	entomophilous	Malpighiaceae	entomophilous
Apiaceae	entomophilous	Malvaceae	entomophilous
Apocynaceae	entomophilous	Melastomataceae	entomophilous, or ornithophilous, or cheiropterophilous
Aquifoliaceae	entomophilous	Meliaceae	entomophilous (usually)
Araliaceae	entomophilous	Melianthaceae	entomophilous, or ornithophilous
Asclepiadaceae	entomophilous	Menispermaceae	unknown
Asparagaceae	entomophilous	Mimosaceae	entomophilous, or ornithophilous, or cheiropterophilous, or anemophilous
Asteraceae	entomophilous (mostly), or anemophilous	Moraceae	anemophilous, or entomophilous
Balanitaceae	entomophilous (mostly), or anemophilous	Myricaceae	anemophilous
Balsaminaceae	entomophilous	Myristicaceae	entomophilous
Begoniaceae	entomophilous	Myrsinaceae	entomophilous
Bignoniaceae	entomophilous, or ornithophilous, or cheiropterophilous	Myrtaceae	entomophilous, or ornithophilous
Bombacaceae	cheiropterophilous	Nyctagynaceae	entomophilous
Boraginaceae	entomophilous	Ochnaceae	entomophilous
Brassicaceae	anemophilous, or entomophilous	Olacaceae	entomophilous
Buddleyaceae	entomophilous, or ornithophilous	Oleaceae	anemophilous (mostly), or entomophilous
Burseraceae	entomophilous	Orobanchaceae	entomophilous
Caesalpiniaceae	entomophilous;	Palmae	entomophilous (mostly), or anemophilous
Campanulaceae	entomophilous	Pandanaceae	anemophilous (mostly), or entomophilous, or ornithophilous, or cheiropterophilous
Capparidaceae	entomophilous	Passifloraceae	entomophilous

(*Continued*)

Table 2. *Continued.*

Family	Pollination system	Family	Pollination system
Caryophyllaceae	entomophilous	Phyllanthaceae	entomophilous
Casuarinaceae	anemophilous	Plantaginaceae	anemophilous, or anemophilous and entomophilous
Celastraceae	entomophilous	Poaceae	anemophilous (exclusively)
Chrysobalanaceae	entomophilous, or cheiropterophilous	Podocarpaceae	anemophilous
Clusiaceae	entomophilous, or ornithophilous	Polygalaceae	entomophilous
Cochlospermaceae	entomophilous	Polygonaceae	anemophilous, or entomophilous
Combretaceae	entomophilous, or ornithophilous	Primulaceae	entomophilous
Commelinaceae	entomophilous	Proteaceae	entomophilous, or ornithophilous, or cheiropterophilous (?), or by unusual means (notably by small marsupials and rodents)
Connaraceae	autogamous or entomophilous	Ranunculaceae	anemophilous, or entomophilous
Convolvulaceae	entomophilous	Resedaceae	entomophilous
Crassulaceae	entomophilous	Rhamnaceae	entomophilous
Cucurbitaceae	entomophilous	Rhizophoraceae	anemophilous, or entomophilous
Cupressaceae	anemophilous	Rosaceae	anemophilous (occasionally), or entomophilous (usually)
Dichapetalaceae	entomophilous	Rubiaceae	entomophilous
Dilleniaceae	entomophilous	Rutaceae	entomophilous
Dioscoreaceae	entomophilous	Salicaceae	anemophilous, or entomophilous
Dipsacaceae	entomophilous	Salvadoraceae	entomophilous
Ebenaceae	entomophilous	Sapindaceae	entomophilous
Ephedraceae	anemophilous, or entomophilous	Sapotaceae	entomophilous
Ericaceae	entomophilous (usually?)	Scrophulariaceae	entomophilous, or ornithophilous
Euphorbiaceae	entomophilous	Simaroubaceae	entomophilous, or ornithophilous
Fabaceae	entomophilous, or ornithophilous, or cheiropterophilous	Solanaceae	entomophilous
Flacourtiaceae	entomophilous	Sphenocleaceae	entomophilous
Gentianaceae	entomophilous	Sterculiaceae	entomophilous

(Continued)

Table 2. *Continued.*

Family	Pollination system	Family	Pollination system
Geraniaceae	entomophilous	Tamaricaceae	entomophilous, or anemophilous
Gesneriaceae	entomophilous, or ornithophilous, or cheiropterophilous	Thymeleaceae	entomophilous
Hippocrateaceae	entomophilous	Tiliaceae	enthomophilous
Hymenocardiaceae	entomophilous (mostly), or anemophilous	Tribulaceae	entomophilous
Hypericaceae	entomophilous	Ulmaceae	entomophilous
Icacinaceae	entomophilous	Urticaceae	anemophilous
Irvingiaceae	entomophilous	Verbenaceae	entomophilous
Lamiaceae	entomophilous or ornithophilous	Vitaceae	entomophilous (mostly), or anemophilous
Lecythidaceae	entomophilous	Zygophyllaceae	entomophilous
Portulaceae	entomophilous		

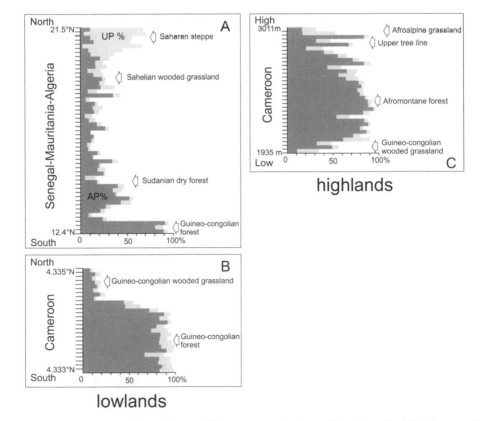

Figure 2. Modern pollen deposition in three different sectors of northern Africa: (A): a South-North transect in Senegal-Mauritania-Algeria (Lézine and Hooghiemstra 1990), (B): the Kandara forest-savanna transect in Southern Cameroon (Vincens *et al.* 2000), and (C): an altitudinal transect in the Cameroon highlands (Verlhac *et al.* 2018). Percentages calculated against a sum excluding fern spores, aquatics and Cyperaceae pollen. Dark grey: AP%. Light grey: UP%.

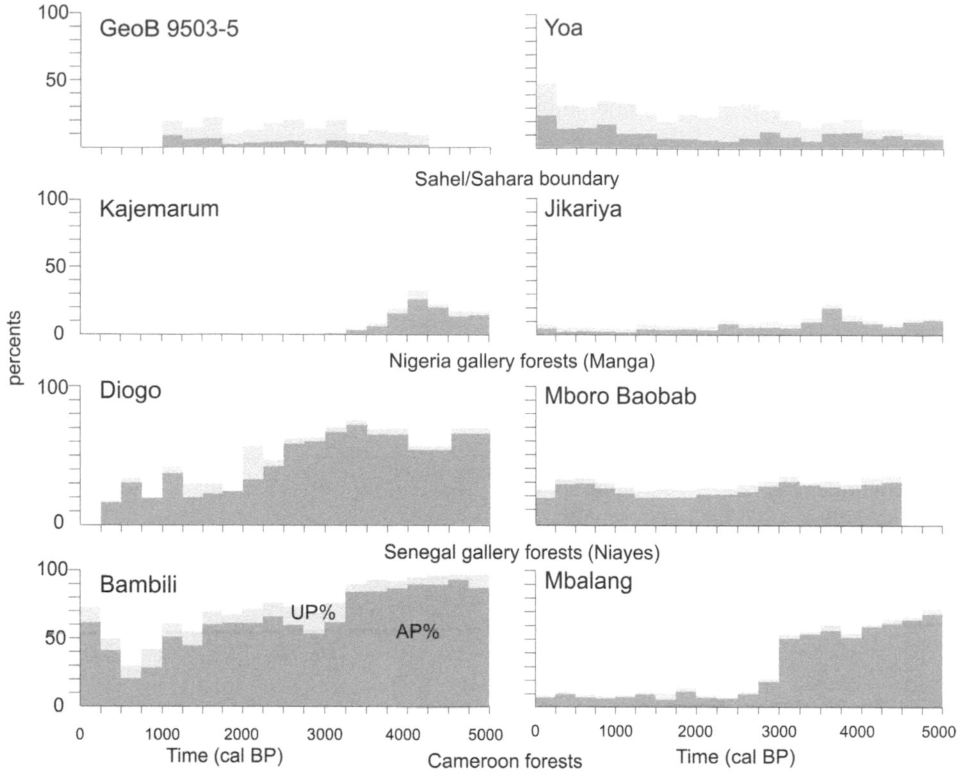

Figure 3. Percentages of Arboreal (AP) and Undifferentiated pollen types (UP) in selected sites from northern Africa from 5000 cal yr BP to the present. In order to facilitate the comparison between sites, data are averaged by 250-years' time bins.

gaps linked to recent settlements. A slightly different pattern is observed in the montane forests Cameroon (Figure 2C; Verlhac *et al.* 2018). As expected, AP reach high values, up to 90%, within the forest. However, they strongly vary, reflecting the highly heterogeneous forest cover (Momo Solefack 2009) (Figure 2C). This is particularly the case at higher elevations where the trees are irregularly distributed and the upper tree line is often patchy due to marked topography. Afroalpine grasslands at the top of the mountains and wooded grasslands below the forest block show AP averaging 15% and 10%, respectively.

6.6 THE HOLOCENE POLLEN SEQUENCES

Vincens *et al.* (1999) first reviewed the equatorial forest response at the end of the AHP. They showed that the lowland forests changed in nature from rain or swamp forests to woodlands or secondary forests characterized by an increase in light demanding trees. While the timing of this change is remarkably coherent in the lowlands of Cameroon at around 2400 cal yr BP (Lézine *et al.* 2013b and reference therein), it varies in Congo within a time interval from 3800 and 1800 cal yr BP according to local hydrological conditions (Vincens *et al.* 1999). In mountain areas, a clear signal of forest disruption occurred at 3300–3000 cal yr BP (Figure 3): AP values abruptly fell from 50 to 9% in the Adamawa plateaus (Mbalang) and then remained below 11% up to the present day. At higher elevations (Bambili), AP decreases from 84% to 54% at 3300 cal yr

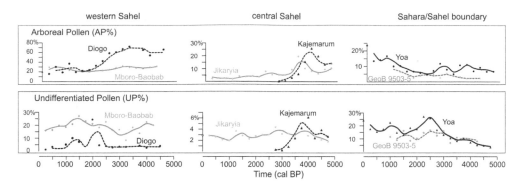

Figure 4. Arboreal pollen percentage (AP%) and undifferentiated pollen percentage (UP%) from selected sites from the Sahara and the Sahel. Dashed lines show the second order running average.

BP and then remained at this level up to 2500 cal yr BP. AP subsequently increased again to a maximum of 65% indicating a later forested phase, albeit less developed compared to that of the AHP. This second phase of forest development ended during the last millennium.

In the Sahel, the pollen diagrams testify to a contrasting situation: in the west, near the Atlantic coast, AP% were remarkably high during the Holocene forest phase indicating dense gallery forests in the coastal region (Diogo). AP decreased from 72% to 19% between 3250 and 1250 cal yr BP through a well-marked threshold at 2500 BP characterized by a drop of about 20%. In the same region however, Mboro-Baobab did not record such an evolution and AP values displayed rather constant percentages during the last 5000 years, only slightly fluctuating between 30 and 20%. In the central Sahel, AP never reached percentages as high as those recorded in the west, even during the AHP reflecting poorly developed gallery forests around water bodies. The maximum of AP at Kajemarum and Jikaryia in the Manga region of northern Nigeria did not exceed 26% and 19%, respectively. At Kajemarum, AP definitively disappeared at 3000 cal yr BP. At Jikaryia, the transition between the Holocene and the present landscape was less contrasted though a slight decrease in AP% was observed at the same time. At the Sahara/Sahel transition (Yoa, GeoB 95013-5) AP were constantly less than 23% during the AHP. At Yoa, the retreat of tropical plant species (e.g. *Celtis, Piliostigma, Grewia*) and the establishment of desert conditions at the end of the AHP occurred in two phases at 4700 and 2700 cal yr BP, respectively (Lézine *et al.* 2011). This transition occurred in a context of poorly developed tree cover and is therefore not reflected by our AP index. Interestingly, AP% slightly increased during the last few centuries which may reflect increased anthropogenic influence, for instance the establishment of palm tree plantations at Yoa (Lézine *et al.* 2011), or the occurrence of episodic humid events in the Senegal River Basin (Bouimetarhan *et al.* 2009). In order to refine the timing and structure of the end of the AHP in the Sahel, we plotted the undifferentiated pollen types (UP) (Figure 4) that are especially abundant in the most arid regions as shown in the modern samples (Figure 2). Three sectors of the Sahara and the Sahel emerge from the examination of UP %: at the Sahara/Sahel boundary, the UP % gradually increased from 5000 cal yr BP (or even before) to the present with a peak at 2700 cal yr BP. A roughly similar trend is observed in the Central Sahel. Here they were present from 4000 cal yr BP and then remained constant throughout the last millennia. In contrast, UP % increased much later, since 2700 cal yr BP in the western Sahel.

6.7 DISCUSSION

Our study illustrates one of the major problems of tropical palynology: can the abundance of trees and grasses in the landscape be described using a simple index: the percentage of arboreal pollen? Open landscapes dominate in Africa today between 30°S and 30°N with the exception of the equatorial mangrove, and afromontane forests. These open landscapes are of various types, ranging from dry forests, to woodlands, wooded grasslands and grasslands (White 1983). The particularities of pollen production and dispersal of tropical plants mean that trees are often under-represented in pollen spectra, making distinction between these types of vegetation difficult. Based on the modern samples (Figure 2), we have considered three AP% categories to characterize the late Holocene landscape of northern Africa (Figures 5 and 6): (i) AP higher than 45% corresponding to a dense tree cover, (ii) AP lower than 15% indicating grass-dominated landscapes, and (iii) intermediate values (15 < AP < 45%), secondary/dry forests. Figure 5 shows that dense forest stands remained developed in mangroves along the coast of the Gulf of Guinea, in the western Sahel and in the Cameroon highlands well after the establishment of xeric conditions at the end of the AHP. Surprisingly, it suggests that there was no transition from forested to open landscapes from south to north throughout northern Africa with open landscapes dominating largely from 5000 to *c*. 3000 cal yr BP north of 10°N. The absence of intermediate tree cover between forests and open landscapes is likely an artefact due to the difficulties evoked in distinguishing woodlands from wooded grasslands and grasslands. While different kinds of open environments are difficult to infer from AP%, it is even more difficult to derive related climatic factors. For instance, modern AP% are higher at Yoa in the Saharan steppe, where the annual rainfall does not exceed 7 mm (23%) than at Mbalang in the Sudanian wooded grassland, where it averages 1500–1600 mm (6.5%). Environmental reconstructions in such grass-dominated landscapes are thus particularly challenging and new methodological approaches such as those developed by Sugita (2007) are crucial for a better estimation of regional vegetation proportions (See Gaillard *et al.*, this volume). Despite this major limitation and the purely qualitative aspect of our approach, two key results emerge from our review (6.7.1 and following).

6.7.1 Abrupt forest collapses or gradual establishment of xeric conditions?

In densely forested areas from western Sahel (Diogo) and the Cameroon Highlands (Bambili, Mbalang), AP% permits identification of the tipping points between forested and non-forested (or less-forested) states that occurred at times, varying according to altitude and hydrological conditions, between 3300 and 2500 cal yr BP (Figure 3). However, the determination is only qualitative since the evaluation of its magnitude depends mainly on the pollen productivity and dispersion of tropical trees (see Gaillard *et al.*, this volume). Abrupt forest collapse most certainly originated from the middle Holocene, when an increased seasonality in rainfall was established (Vincens *et al.* 2010). At Bambili, the progressive destabilization of the forest is thought to have originated from the '8.2 event', which could thus be considered as the 'early warning signal' (Lenton 2011) of the environmental crisis that took place several millennia later, at 3300 cal yr BP (Lézine *et al.* 2013a). Unlike these sites, where a threshold was abruptly crossed leading to the collapse of the forests, the establishment of an open landscape at the end of the AHP was gradual (Figure 4); tree cover decreased gradually in already arid areas, particularly in the central Sahel. This supports the interpretations by Kröpelin *et al.* (2008) that the transition to xeric conditions at the end of the AHP was a gradual one. Consequently, pollen data from the Sahel do not support the earlier hypotheses of an abrupt environmental change at the end of the AHP as proposed by de Menocal *et al.* (2000).

6.7.2 A complex pattern of environmental change

Changes in distribution of trees across the landscape was not merely latitudinal following a south-north gradient of decreasing precipitation (Figure 5). The most striking feature is the difference between the central and western Sahel. We therefore consider two separate transects, one along the west coast and the other in central northern Africa. Forests widely developed in the west.

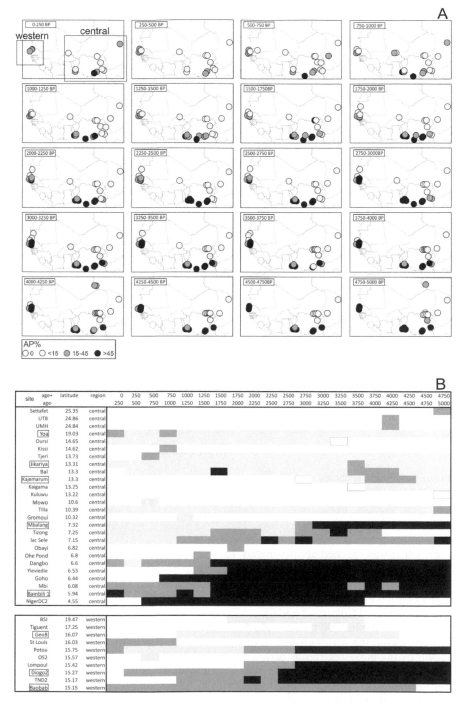

Figure 5. (A) Spatial distribution of tree pollen percentages from 5000 cal yr BP to the present in northern Africa. Three categories are considered with AP > 45% showing dense tree cover, 15 < AP < 45% intermediate tree cover and AP < 15% sparse tree cover. (B) Patterns of tree cover change according to latitude in the two sectors defined in Figure 1. The selected sites are indicated on the left column by a box.

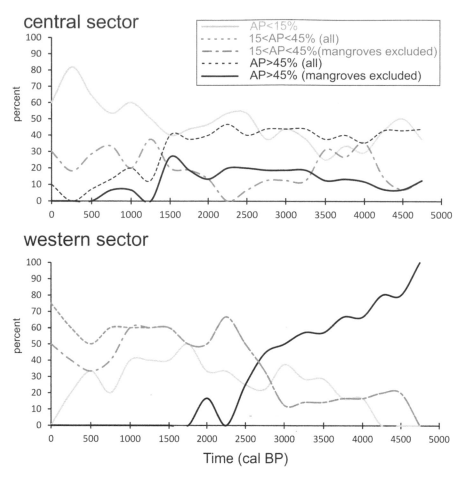

Figure 6. Timing of landscape change in the western (on the left) and central (on the right) sectors of northern Africa. The curves show the number of sites in the categories defined by AP% in northern Africa (AP = O, AP < 15%, 15 < AP < 45%, >45%). Data are expressed in per cent of the total of sites present in each time bin (cf. Figure 5B).

Their expansion reached approximately 16°N, i.e. the mouth of the Senegal River. In contrast, forests only occurred along the coast and in the highlands in central Africa. Trees were rare north of 10°N in the Sahel, except during a time interval spanning from 4500 to 3500 cal yr BP in the Manga region (13°N) and in the central Saharan massifs in southern Libya (25°N) where they slightly increased (Figures 5 and 6).

Despite the limited number of sites in both sectors clear patterns can be discerned. In the western sector, the major change in forest cover occurred from 3000 to 2250 cal yr BP. This is shown by the opposing trends between the number of sites with AP values above 45%, which were gradually decreasing in number and those with AP values ranging from 15 to 45%, which were gradually increasing. Sparse tree cover became gradually commoner over the western sector from 4250 cal yr BP with the modern landscape composed of woodlands and wooded grasslands being definitively established after 1750 cal yr BP. Unlike the western sector, sites with sparse tree cover dominated the central sector throughout the last 5000 years. The number of such sites progressively increased in two phases respectively dated 3500 and 1500 cal yr BP. Sites with dense tree cover were restricted to the Cameroon Highlands and the mangrove areas along the Gulf of Guinea. In both sectors, they declined from 1500 cal yr BP onwards. The most striking result of our study is the behaviour of sites with intermediate tree cover. As already noted, trees increased in the Manga sector of the Sahel around 3800 cal yr BP due to the recovery of tropical taxa during a short wet phase (Waller *et al.* 2007). After 3500 cal yr BP, trees disappeared from the Manga sector. Elsewhere in the central sector of northern Africa, the number of sites with intermediate and sparse tree cover increased from 2250 cal yr BP onwards.

6.8 CONCLUSIONS

Reconstructing vegetation cover from AP% is highly challenging, particularly given the predominantly open vegetation types that are widely distributed in northern Africa from the Sudanian dry forests to the Sahelian wooded grasslands and Saharan grasslands. Despite this caveat, our review shows that the landscape response to the end of the AHP was far from homogeneous particularly in the Sahel where a clear east-west gradient of tree cover change is indicated, the central Sahel being notably poor in tree species as shown for instance in the Lake Chad (Tjeri) and the Niger bend (Oursi, Kissi) areas. In areas where forests were well developed during the AHP, i.e. in the south and west, the establishment of the modern landscape was abrupt with a threshold crossed between 3300 and 2500 cal yr BP according to local conditions. Elsewhere in northern Africa the switch from tree (C_3) to grass (C_4) dominated landscapes occurred at approximately the same time, but more gradually. This review allows the timing of the ecosystem response at the end of the AHP be identified, which was remarkably synchronous throughout northern Africa, in contrast to hydrological data, which suggest a time transgressive end to the AHP (Shanahan *et al.* 2015).

ACKNOWLEDGEMENTS

This work contributes to the ACCEDE ANR Belmont Forum project ((18 BELM 0001 05). Thanks are due to M.J. Gaillard (Kalmar University, Sweden) for constructive comments on the manuscript and the African Pollen Database for data access. AML and KL are funded by CNRS, LD by the University of Bremen and IB by the BMF grant PMARS2015-100.

REFERENCES

Anthony, E., 1989, Chenier plain development in Northern Sierra Leone, West Africa. *Marine Geology* **90**, pp. 297–309, 10.1016/0025-3227(89)90132-1.
Ballouche, A., 1997, Dynamique des paysages végétaux sahélo-soudaniens et pratiques agro-pastorales à l'Holocène: exemples du Burkina Faso. Monographies-Bulletin de l'Association de Géographes Français, Paris.
Ballouche, A. and Neumann, K., 1995, A new contribution to the Holocene vegetation history of the West African Sahel: pollen from Oursi, Burkina Faso and charcoal from three sites in northeast Nigeria. *Vegetation History and Archaeobotany*, **4**, pp. 31–39, 10.1007/BF00198613.
Bouimetarhan, I., Dupont, L., Schefuß, E., Mollenhauer, G., Mulitza, S. and Zonneveld, K., 2009, Palynological evidence for climatic and oceanic variability off NW Africa during the late Holocene. *Quaternary Research*, **72**, pp. 188–197, 10.1016/j.yqres.2009.05.003.
Dallmeyer, A., Claussen, M., Lorenz, S.J. and Shanahan, T., 2020, The end of the African Humid Period as seen by a transient comprehensive Earth system model simulation of the last 8000 years. *Climate of the Past*, **16**, pp. 117–140, 10.5194/cp-16-117-2020.

Delneuf, M. and Médus, J., 1997, Comparaison de deux environnements anthropisés de la période protohistorique du Nord-Cameroun. In *L'Homme et le milieu végétal dans le bassin du Lac Tchad: Séminaire du Réseau Méga-Tchad*, edited by Barreteau, D., Dognin, R. and von Graffenried, C. (Paris: ORSTOM), pp. 145–170.

De Menocal, P., Ortiz, J., Guilderson, T., Adkins, J., Sarnthein, M., Baker, L. and Yarusinsky, M., 2000, Abrupt onset and termination of the African Humid Period: rapid climate responses to gradual insolation forcing. *Quaternary Science Reviews*, **19**(1–5), pp. 347–361, 10.1016/S0277-3791(99)00081-5.

Elenga, H., Schwartz, D. and Vincens, A., 1994, Pollen evidence of late Quaternary vegetation and inferred climate changes in Congo. *Palaeogeography, Palaeoclimatology, Palaeoecology* **109**, pp. 345–356, 10.1016/0031-0182(94)90184-8.

Elenga, H., Schwartz, D., Vincens, A., Bertaux, J., de Namur, C., Martin L., Wirrmann, D. and Servant M., 1996, Diagramme pollinique Holocène du lac Kitina (Congo): mise en évidence de changements paléobotaniques et paléoclimatiques dans le massif forestier du Mayombe. *Comptes-Rendus de l'Académie des Sciences, Paris* **323**, IIA, pp. 403–410.

Feller, C., Médus, J., Paycheng, C. and Chavane, B., 1981, Etude pédologique et palynologique d'un site protohistorique de la moyenne vallée du fleuve Sénégal. *Palaeoecology of Africa*, **13**, pp. 235–247.

Fofana, C.A.K., Sow, E. and Lézine, A.-M., 2020, The Senegal River during the last millennium. *Review of Palaeobotany and Palynology*, **275**, p. 104175, 10.1016/j.revpalbo.2020.104175.

Gaillard, M.-J., Githumbi, E., Achoundong, G., Lézine, A.-M., Hély, C., Lebamba, J., Marquer, L., Mazier, F., Li, F., Sugita, S., this volume, The challenge of pollen-based quantitative reconstruction of Holocene plant cover in subtropical and tropical regions – a review and a pilot study in West Africa. *Palaeoecology of Africa*, **35**, chapter 12, 10.1201/9781003162766-12.

Gajewski, K., Lézine, A.-M., Vincens, A., Delestan, A. and Sawada, M., 2002, Modern climate–vegetation–pollen relations in Africa and adjacent areas. *Quaternary Science Reviews*, **21**(14–15), pp. 1611–1631, 10.1016/S0277-3791(01)00152-4.

Hedberg, O., 1954, A pollen analytical reconnaissance in tropical East Africa. *Oikos*, **5**, pp. 137–165, 10.2307/3565157.

Hély, C., Lézine, A.-M. and APD contributors, 2014, Holocene changes in African vegetation: tradeoff between climate and water availability. *Climate of the Past*, **10**, pp. 681–686, 10.5194/cp-10-681-2014.

Krinner, G., Lézine, A.-M., Braconnot, P., Sepulchre, P., Ramstein, G., Grenier, C. and Gouttevin, I., 2012, A reassessment of lake and wetland feedbacks on the North African Hlocene climate. *Geophysical Research Letters*, **39**(7), 10.1029/2012GL050992), article: L07701.

Kröpelin, S., Verschuren, D., Lézine, A.-M., Eggermont, H., Cocquyt, C., Francus, P., Cazet, J.P., Fagot, M., Rumes, B., Russell, J.M., Conley, D.J., Schuster, M., Von Suchodoletz, H., Engstrom, D.R., 2008, Climate-driven ecosystem succession in the Sahara: the past 6000 years. *Science*, **320**(5877), pp. 765–768, 10.1126/science.1154913.

Lebamba, J., Vincens, A., Lézine, A.-M., Marchant, R. and Buchet, G., 2016, Forest-savannah dynamics on the Adamawa plateau (Central Cameroon) during the "African humid period" termination: A new high-resolution pollen record from Lake Tizong. *Review of Palaeobotany and Palynology*, **235**, 129–139, 10.1016/j.revpalbo.2016.10.001.

Lemonnier, K., and Lézine, A.-M., this volume, Timing and nature of the end of the Holocene Humid Period in the Sahel: Insight from pollen data. *Palaeoecology of Africa*, **35**, chapter: 5, 10.1201/9781003162766-5.

Lenton, T.M., 2011, Early warning of climate tipping points. *Nature Climate Change*,**1**(4), pp. 201–209, 10.1038/nclimate1143.

Lézine, A.-M., 1988, Les variations de la couverture forestière mésophile d'Afrique occidentale au cours de l'Holocène. *Comptes Rendus de l'Académie des Sciences, Paris*, **307**(2), pp. 439–445.

Lézine, A.-M., 1997, Evolution of the West African mangrove during the late Quaternary: a review. *Géographie Physique et Quaternaire*, **51**, pp. 405–414, 10.7202/033139ar.

Lézine, A.-M. and Hooghiemstra, H., 1990. Land-sea comparisons during the last glacial-interglacial transition: pollen records from West Tropical Africa. *Palaeogeography, Palaeoclimatology, Palaeoecology* **79**(3–4), pp. 313-331, 10.1016/0031-0182(90)90025-3.

Lézine, A.-M., Watrin, J., Vincens, A. and Hély, C., 2009, Are modern pollen data representative of West African vegetation? *Review of Palaeobotany and Palynology*, **156**(3–4), pp. 265–276, 10.1016/j.revpalbo.2009.02.001.

Lézine, A.-M., Zheng, W., Braconnot, P. and Krinner, G., 2011, Late Holocene plant and climate evolution at Lake Yoa, northern Chad: Pollen data and climate simulations. *Climate of the Past*, **7**, pp. 1351–1362, 10.5194/cp-7-1351-2011.

Lézine, A.-M., Assi-Kaudjhis, C. Roche, E., Vincens, A. and Achoundong, G., 2013a, Towards an understanding of West African montane forest response to climate change. *Journal of Biogeography*, **40**(1), pp. 183–196, 10.1111/j.1365-2699.2012.02770.x.

Lézine, A.-M., Holl, A.F.C., Lebamba, J., Vincens, A., Assi-Khaudjis, C., Février, L. and Sultan, E, 2013b, Temporal relationship between Holocene human occupation and vegetation change along the northwestern margin of the Central African rainforest. *Comptes Rendus Geoscience*, **345**(7–8), pp. 327–335, 10.1016/j.crte.2013.03.001.

Lézine, A.-M., Izumi, K. and Achoundong, G., 2021, Mbi Crater (Cameroon) illustrates the relations between mountain and lowland forests over the past 15,000 years in Western Equatorial Africa. *Quaternary International*, 10.1016/j.quaint.2020.12.014.

Maley, J., 2004, Le bassin du Tchad au Quaternaire récent: formations sédimentaires, paléoenvironnements et préhistoire. La question des paléotchads. In *L'évolution de la végétation depuis deux millions d'années*, edited by Sémah, A.-M. and Renault-Miskovsky, J. (Paris: Errance), pp. 179–217.

Maley, J. and Brenac, P., 1998, Vegetation dynamics, palaeoenvironments and climatic change in the forests of western Cameroon during the last 28,000 years BP. *Review of Palaeobotany and Palynology* **99**, pp. 157–187, 10.1016/S0034-6667(97)00047-X.

Médus, J. and Barbey, C., 1979, Deux analyses polliniques de sédiments minéraux de Mauritanie méridionales. *Association sénégalaise pour l'Etude du Quaternaire, Bulletin de Liaison*, **54–55**, pp. 75–79.

Mercuri A.-M., Grandi G.T., Mariotti Lippi M. and Cremaschi M. (1998). New pollen data from the Uan Muhuggiag rockshelter (Libyan Sahara, VII–IV millennia BP). In *Wadi Teshuinat – Palaeoenvironment and prehistory in south-western Fezzan (Libyan Sahara). Survey and excavations in the Tadrart Acacus, Erg Uan Kasa, Messak Settafet and Edeyen of Murzuq, 1990–1995*, edited by Cremaschi, M. and di Lernia, S. (Firenze, Edizioni All'Insegna del Giglio), pp. 107–122.

Mercuri, A.-M. and Grandi, G.T., 2001, Palynological analyses of the Late Pleistocene, Early Holocene and Middle Holocene layers. In *Uan Tabu in the settlement history of the Libyan Sahara*, Arid Zone Archaeology, Monographs 2, Ch. 10, edited by Garcea, E.A.A. (Firenze: Edizioni All'Insegna del Giglio), pp. 161–188.

Momo Solefack, M.C., 2009, *Influence des activités anthropiques sur la végétation du Mont Oku (Cameroun)*. PhD Thesis, Université de Picardie, Amiens, France.

Ngomanda, A, Neumann, K, Schweizer, A and Maley, J., 2009, Seasonality change and the third millennium BP rainforest crisis in southern Cameroon (Central Africa). *Quaternary Research*, **71**(3), pp.307–18, 10.1016/j.yqres.2008.12.002.

Njokuocha, R.C., 2012, Holocene pollen deposits and recent vegetation distribution of Obayi Lake, Nsukka. *Quaternary international*, **262**, pp. 20–24, 10.1016/j.quaint.2011.10.033.

Njokuocha, R.C. and Akaegbobi, I.M., 2014, A contribution to the Holocene vegetation history of Nigeria: Pollen from Ohe Pond Nsukka, southeastern Nigeria. *Quaternary International*, **338**, pp. 28–34, 10.1016/j.quaint.2013.07.036.

Reynaud-Farrera, I., Maley, J. and Wirrmann, D., 1996. Végétation et climat dans les forêts du Sud-Ouest Cameroun depuis 4 770 ans BP: analyse pollinique des sédiments du Lac Ossa. *Comptes Rendus de l'Académie des Sciences, Paris*, **322**(II A), pp. 749–755.

Ritchie, J.C., 1995, Current trends in studies of long-term plant community dynamics. *New Phytologist*, **130**(4), pp. 469–494, 10.1111/j.1469–8137.1995.tb04325.x.

Salzmann, U. and Hoelzmann, P., 2005, The Dahomey Gap: an abrupt climatically induced rain forest fragmentation in West Africa during the late Holocene. *The Holocene*, **15**(2), pp. 190–199, 10.1191/0959683605hl799rp.

Salzmann, U. and Waller, M., 1998, The Holocene vegetational history of the Nigerian Sahel based on multiple pollen profiles. *Review of Palaeobotany and Palynology*, **100**, pp. 39–72, 10.1016/S0034-6667(97)00053-5.

Salzmann, U., Hoelzmann, P. and Morczinek, I., 2002, Late Quaternary climate and vegetation of the Sudanian zone of northeast Nigeria. *Quaternary Research*, **58**(1), pp. 73–83, 10.1006/qres.2002.2356.

Schulz, E., 1980, Zur Vegetation der östlichen zentralen Sahara und zu ihrer Entwicklung im Holozän. Würzburger Geographische Arbeiten 51.

Shanahan, T.M., McKay, N.P., Hughen, K.A., Overpeck, J.T., Otto-Bliesner, B., Heil, C.W., King, J., Scholz, C.A. and Peck, J., 2015, The time-transgressive termination of the African Humid Period. *Nature Geoscience*, **8**(2), pp. 140–144, 10.1038/ngeo2329.

Sowunmi, M.A., 1981, Nigerian vegetational history from the Late Quaternary to the Present day. *Palaeoecology of Africa*, **13**, pp. 217–234.

Stuiver, M., Reimer, P.J., and Reimer, R.W., 2020, CALIB 7.1 [WWW program] at http://calib.org, accessed 2020-6-9.

Sugita, S., 2007, Theory of quantitative reconstruction of vegetation I: pollen from large sites REVEALS regional vegetation composition. *The Holocene*, **17**(2), pp. 229–241, 10.1177/0959683607075837.

Tossou, M.G., 2002, Recherche palynologique sur la végétation Holocène du Sud-Bénin (Afrique de l'Ouest). PhD Thesis, Université de Lomé, Togo.

Trochain, J., 1940, Contribution à l'étude de la végétation du Sénégal. (Paris: LAROSE)

Verlhac, L., Izumi, K., Lézine, A.-M., Lemonnier, K., Buchet, G., Achoundong, G. and Tchiengué, B., 2018, Altitudinal distribution of pollen, plants and biomes in the Cameroon highlands. *Review of Palaeobotany and Palynology*, **259**, pp. 21–28, 10.1016/j.revpalbo.2018.09.011.

Vincens, A., Buchet, G., Servant, M. and ECOFIT Mbalang collaborators, 2010, Vegetation response to the "African Humid Period" termination in Central Cameroon (7°N) – new pollen insight from Lake Mbalang. *Climate of the Past*, **6**, pp. 281–294, 10.5194/cp-6-281-2010.

Vincens, A., Dubois, M.A., Guillet, B., Achoundong, G., Buchet, G., Beyala, V.K.K., De Namur, C. and Riera, B., 2000, Pollen-rain–vegetation relationships along a forest–savanna transect in southeastern Cameroon. *Review of Palaeobotany and Palynology*, **110**(3–4), pp. 191–208, 10.1016/S0034-6667(00)00009-9.

Vincens, A., Lézine, A.-M., Buchet, G., Lewden, D. and Le Thomas, A., 2007, African pollen database inventory of tree and shrub pollen types. *Review of Palaeobotany and Palynology*, **145**(1–2), pp. 135–141, 10.1016/j.revpalbo.2006.09.004.

Vincens, A., Schwartz, D., Elenga, H., Reynaud-Farrera, I., Alexandre, A., Bertaux, J., Mariotti, A., Martin, L., Meunier, J.D., Nguetsop, F. and Servant, M., 1999, Forest response to climate changes in Atlantic Equatorial Africa during the last 4000 years BP and inheritance on the modern landscapes. *Journal of Biogeography*, **26**(4), pp. 879–885, 10.1046/j.1365-2699.1999.00333.x.

Waller, M.P., Street-Perrott, F.A. and Wang, H., 2007, Holocene vegetation history of the Sahel:pollen, sedimentological andgeochemical data from Jikariya Lake,north-eastern Nigeria. *Journal of Biogeography*, **34**, pp. 1575–1590, 10.1111/j.1365-2699.2007.01721.x.

Watrin, J., Lézine, A.-M., Gajewski, K. and Vincens, A., 2007, Pollen–plant–climate relationships in sub-Saharan Africa. *Journal of Biogeography*, **34**(3), pp. 489–499, 10.1111/j.1365-2699.2006.01626.x.

Watrin, J., Lézine, A.-M., and Hély, C., 2009. Plant migration and ecosystems at the time of the "green Sahara". *Comptes Rendus Geosciences* **341**, 656–670, 10.1016/j.crte.2009.06.007.

Watson, L. and Dallwitz, M.J., 1992 onwards, The families of flowering plants: descriptions, illustrations, identification, and information retrieval. Version: 2nd May 2020. delta-intkey.com.

White, F., 1983, *The Vegetation of Africa* (Paris: UNESCO).

CHAPTER 7

Reconstructing vegetation history of the Olorgesailie Basin during the Middle to Late Pleistocene using phytolith data

Rahab N. Kinyanjui[1]

Department of Earth Sciences, National Museums of Kenya, Nairobi, Kenya

Michael Meadows[2]

Department of Environmental and Geographical Science, University of Cape Town, Rondebosch, South Africa

Lindsey Gillson

Plant Conservation Unit, Botany Department, University of Cape Town, Rondebosch, South Africa

Marion. K. Bamford

The Evolutionary Studies Institute, University of the Witwatersrand, Johannesburg, South Africa

Anna K. Behrensmeyer

Department of Paleobiology, Smithsonian Institution, Washington, DC USA

Richard Potts

Human Origins Program, Smithsonian Institution, Washington, DC USA

ABSTRACT: The Olorgesailie basin, located in the East African Rift System (EARS), southern Kenya (1.5–1.6°S, 36.4–36.5°E, 940–1040 m asl), is an important site for palaeoanthropological, palaeontological and geological research, with sediments dating back more than 1 Ma. Little is known about the palaeovegetation and palaeoenvironmental context of this important site and how this varied through space and time. Here we use phytolith data to reconstruct the vegetation history through the Middle and Late Pleistocene (*c.* 670 ka to *c.* 64 ka). The analysis of 24 samples from

[1] Other affiliation: *Department of Environmental & Geographical Science, University of Cape Town, South Africa; School of Geographical Sciences, East China Normal University, Shanghai, China; and College of Geography and Environmental Sciences, Zhejiang Normal University, Jinhua, China*
[2] Other affiliation: *School of Geographic Sciences, East China Normal University, Shanghai, China; and College of Geography and Environmental Sciences, Zhejiang Normal University, China*

DOI 10.1201/9781003162766-7

palaeosols within the Olorgesailie and Oltulelei Formations uses diagnostic phytoliths to plot relative phytolith abundance diagrams documenting temporal and spatial vegetation variations. We use three phytolith indices (climate index, aridity index and tree density index) to estimate temperature, precipitation, and vegetation structure from sediments that are well-constrained chronologically by $^{40}Ar/^{39}Ar$ dating. Spatial vegetation variation is captured through examining phytoliths from three distinct localities (Loc. OLT, Loc. B and Loc. G) over a distance of about 5 km. Results suggest that local vegetation changes approximately correspond with stratigraphic units. Phytolith indices reflect warm and moist conditions *c.* 670 ka, with cool and dry conditions from *c.* 650 ka during which time riparian and a variety of other habitats were present. An unconformity, which extends over *c.* 180,000 years (500 to 320 ka), is interpreted as largely caused by tectonic processes coupled with precipitation variability. A shift from woody vegetation to C$_4$ grasslands is evident following the unconformity, and riparian habitats were again present. Considerable climate variability is apparent thereafter, with inconsistent precipitation conditions until *c.* 220 ka when more stable and moist conditions set in until *c.* 64 ka. The study documents vegetation dynamics directly associated with the Middle to Late Pleistocene palaeontological and archaeological record of Olorgesailie.

7.1 INTRODUCTION

East Africa is well known for its archive of human evolutionary history and the long-term history of ecological interactions between faunal and floral species and their environments (Bobe and Behrensmeyer 2002; Potts 1998). Tectonic processes created the episodic fossilization and preservation of records of ancient life in the East Africa Rift system (EARS) (Brown and Feibel 1991; Feibel 1999; Campsiano and Feibel 2008). Extensively collected and documented EARS fossils have addressed a variety of research questions including human and faunal evolutionary history, extinction and speciation events, species-environment interactions and the impact of climate change on past ecosystems (Bobe and Behrensmeyer 2004; deMenocal 2004; Feakins *et al.* 2007; Lepre *et al.* 2007; Potts and Faith 2015; Trauth *et al.* 2007).

Previous studies suggest that African climate and vegetation cover have been highly variable since the early Pliocene, oscillating from warm and humid to cool and dry episodes that can be correlated with orbital oscillations (Cerling *et al.* 2011; deMenocal 1995; Feakins *et al.* 2007; Lepre *et al.* 2007; Potts 2013; Trauth *et al.* 2007). These oscillations resulted in habitats that varied between woodland, wooded grassland and open grassland. Such vegetation dynamics influenced hominin diets and ecomorphology in the region, such as in the Awash and Olduvai basins (Albert *et al.* 2009; Ashley *et al.* 2010a; Bamford *et al.* 2006; Basell 2007; Cerling *et al.* 2010; White *et al.* 2010; WoldeGabriel *et al.* 2009).

In the Olorgesailie basin, vegetation cover has previously been reconstructed in only a single palaeosol dated *c.* 1 Ma, based on carbon isotope data and faunal remains (Sikes *et al.* 1999). The rich palaeontological, archaeological and geological data published from the basin have otherwise lacked a parallel body of evidence on vegetation cover and how its structure changed over time and across the landscape. Initial steps geared to reconstructing the vegetation cover and structure were hindered by poor preservation of fossil pollen and other organic plant remains due to high salinity in the basin (Livingstone and Mworia 1999). A subsequent feasibility study showed the potential for using fossil phytoliths to reconstruct vegetation cover and structure during the Middle to Late Pleistocene of the basin (Kinyanjui 2012).

Phytoliths are plant silica bodies that form when silica is deposited within and around plants cells during evapotranspiration, resulting in cell casts/replicas that are morphologically distinct. The silica cells are preserved in the soils after the decomposition of the parent plants (Mulholland and Rapp 1992; Piperno 1988; 2006). They are inorganic in nature, preserve well in a variety of depositional regimes, and are particularly useful because of good preservation in regions where organic plant remains are susceptible to oxidation and thus less likely to accumulate (Piperno 2006). Phytolith research in East Africa has gained popularity as a reliable proxy for reconstructing past vegetation cover (Alexandre *et al.* 1997; Bremond *et al.* 2005a, b; 2008; Fredlund and Tieszen 1994) geared toward understanding past environments (Albert *et al.* 2012; 2015; Bamford *et al.* 2006; Barboni *et al.* 2009; 2010; Estaban *et al.* 2020; Kinyanjui 2013). In addition, phytolith indices can be used to determine changes in precipitation, vegetation structure dynamics, and provide coarse estimates of variations in temperature through time (Aleman *et al.*, 2008; Alexandre *et al.* 1997; Bremond *et al.* 2008; Fredlund and Tieszen 1997).

This paper presents phytolith data from the upper Olorgesailie Formation dated between *c.* 670 ka and 500 ka (Deino *et al.* 2018; Isaac 1978) and the Oltulelei Formation dated between *c.* 320 ka to *c.* 36 ka (Behrensmeyer *et al.* 2018; Deino *et al.* 2018). The stratigraphic sequence covers a significant period when hominins acquired new adaptive strategies associated with changing environments, which also coincided with replacement of mega-herbivore species by smaller herbivore taxa in the basin (Potts 1998; Potts and Faith 2015; Potts *et al.* 2018) associated with climate and local tectonic processes (Behrensmeyer *et al.* 2018; Potts *et al.* 2020).

7.2 STUDY SITE: OLORGESAILIE BASIN

The Olorgesailie basin is located in the southern part of the Kenyan Rift Valley (1°3′S and 36°2′E; 940–1040 m asl) (Figure 1). It is renowned for its high concentration of Acheulean hand axes dated *c.* 1.2 Ma to 500 ka (Deino and Potts 1990; Isaac 1977; 1978; Potts 1989; Potts *et al.* 1999). The basin lies near the southern end of the Eastern Arc of the East African Rift System (EARS). To the south it is bounded by lavas associated with Mt. Olorgesailie, an extinct Pliocene-Early Pleistocene age volcano, while to the north, east and west the basin is delimited by faulted ridges of trachytes, phonolites, and basalts extruded during the Plio-Pleistocene (Baker and Mitchell 1976; Isaac 1978; Shackleton 1978). The basin's present drainage system flows from north to south following the slope of the rift floor (Baker 1958), and from east to west along the northern foothills of Mt. Olorgesailie, although tectonic movements and volcanism altered these drainage patterns in the past (Behrensmeyer *et al.* 2002; 2018). The main outcrops of the basin north of Mt. Olorgesailie cover an area of *c.* 65 km^2.

7.2.1 Present-day climate and vegetation

Climate in the Olorgesailie region is largely controlled by the annual north-south oscillation of the Intertropical Convergence Zone (ITCZ) (Hills 1978; Kenworthy 1966; Nicholson 1996) resulting in two distinct rainy seasons: (i) long rains from March to June, and (ii) short rains from October to December (Asani and Kinuthia 1979; Griffiths 1958; 1972; Nicholson 1996; Sansom 1954). There are, however, anomalies in the rainfall pattern influenced by local topography. In this regard, the southern Kenya Rift Valley receives variable mean annual rainfall, for example 1000–1500 mm in the rift-margin highlands and 250–500 mm in the rift-axis lowlands (Griffiths 1972; Mutai and Ward 2000). The mean annual rainfall around the Olorgesailie basin is less than 500 mm which, combined with an evapo-transpiration rate exceeding 2400 mm/year, results in the negative hydrological balance responsible for the semi-arid condition of the basin (Damnati and Taieb 1995).

Vegetation in the Olorgesailie basin today is classified as a semi-arid variant of *Commiphora-Acacia* bushland (White 1983) or as a Northern *Acacia-Commiphora* bushland (WWF Eco-regions) (Figure 1). The dominant woody species include *Acacia tortilis* (Forssk.) Hayne, *A. senegal* (L.) Willd., *A. mellifera (Vahl)Benth,* (Figure 2B), *Commiphora africana* (A. Rich) Engl., *C. campestris* Engl., *C. samharensis* Schweinf., *C. schimperi* (O. Berg) Engl. *Terminalia* sp., *Balanites* spp., *Grewia bicolor* Juss., *G. villosa* Willd., *Boscia coriacea* Pax, (Figure 2B)

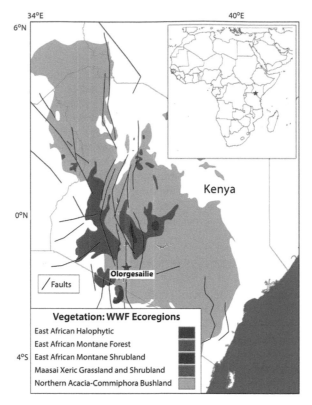

Figure 1. Map showing the location of the Olorgesailie basin and WWF vegetation designated ecoregions (White 1983).

and *Salvadora persica* L. The shrub component comprises *Sericocomopsis hildebrandtii* Schinz, *Barleria* sp., *Aerva* sp. and *Indigofera* sp. Most of the trees are deciduous and are in leaf during and after rains (Livingstone and Mworia-Maitima 1999; Mworia *et al.* 1988). Grasslands are C_4-dominated, mostly Chloridoideae family such as *Chloris roxburghiana* Schult., *Dactyloctenium bogdanii* S.M Phillips, *Eragrostis cilianesis* (All.) F.T. Hubb., *Tetrapogon cenchriformis* (A. Rich.) W.D Clayton, *Sporobolus jacquemontii* Kunth. and *Brachyachne* spp. (Livingstone and Mworia-Maitima 1999; Mworia *et al.* 1988). Elements of riverine gallery forest are also present along the seasonal river channels (lagas); they include *Syzygium cordatum* Krauss, *S. guineense* (Willd.) DC., *Delonix elata* (L.) Gamble and *Terminalia* spp., while *Kyllinga alba* Nees and *K. welwitschii* are C_4 sedges common in the floodplains (Livingstone and Mworia-Maitima 1999; Mworia *et al.* 1988).

7.2.2 Geoarchaeology

Current lithostratigraphic description distinguishes two units: the Olorgesailie and Oltulelei Formations (Behrensmeyer *et al.* 2002; 2018; Isaac 1978; Shackleton 1978), dated primarily by the single-crystal $^{40}Ar/^{39}Ar$ method (Behrensmeyer *et al.* 2018; Deino and Potts 1990; 1992; Deino *et al.* 2018; Potts *et al.* 2018).

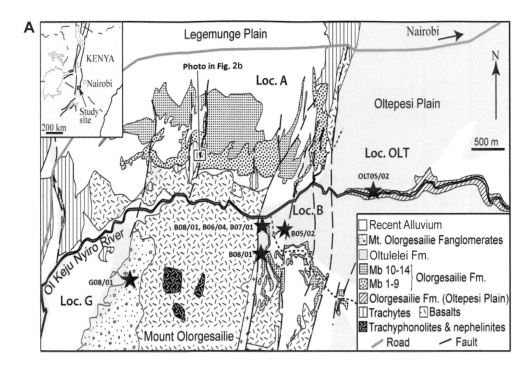

Figure 2. (A) Map showing Olorgesailie basin, localities, geo-trenches (marked with star) and the basin's geology. (B) *Acacia melliferae* and *Boscia coriaceae* visible in the photograph are among the dominant woody species in the basin.

7.2.2.1 Olorgesailie formation

These are deposits comprising lacustrine, wetland/swampy, fluvial and colluvial sediments form-ing a composite thickness of *c*. 80 m (Behrensmeyer *et al.* 2002). There are fourteen members (Isaac 1978; Shackleton 1978) with the base of the oldest member (Member 1) dated *c*. 1.2 Ma and the youngest Member (Member 14) *c*. 499 ka (Behrensmeyer *et al.* 2002; Deino and Potts 1990, 1992; Deino *et al.* 2018). Erosion and excavations have exposed a rich archaeological record including Acheulean stone artifacts associated with extinct mammals, butchery sites and

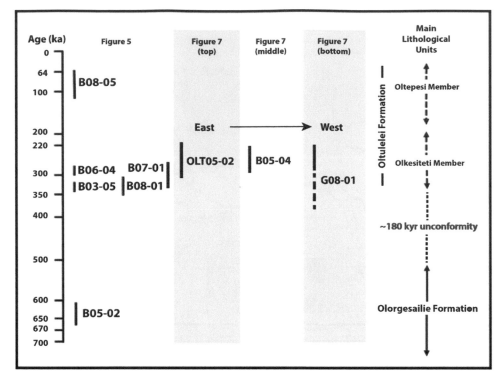

Figure 3. Schematic diagram showing lithostratigraphic/age relationship between sections.

hominin remains of *Homo erectus* (Potts 1994; Potts *et al.* 1999; 2004). Stable isotope data suggest vegetation structure around 1 Ma was predominantly C_4 grasslands (Sikes *et al.* 1999).

7.2.2.2 Oltulelei formation
These are younger deposits that overlie the Olorgesailie Formation, dated between *c.* 320 ka and *c.* 36 ka (Figure 3). The sediments rest on a major erosional unconformity, representing a hiatus of *c.* 180,000 years (Behrensmeyer *et al.* 2002; 2018; Deino *et al.* 2018) and consist of colluvial, alluvial, freshwater and saline lake deposits preserved in a series of erosional and channel filling regimes (Behrensmeyer *et al.* 2002; 2007; 2018). There are three successive members; Olkesiteti (*c.* 320–190 ka), Oltepesi (*c.* 190–50 ka) and Tinga (*c.* 50–36 ka) (Behrensmeyer *et al.* 2018; Deino *et al.* 2018). Exposures of these sediments have yielded stone tools of Middle Stone Age (MSA) technology and vertebrate fossils (Brooks *et al.* 2018; Potts *et al.* 2018).

7.3 METHODS

7.3.1 Laboratory methods, microscopy and identification

Phytolith extraction followed Albert's (1999) protocol with additional Acid Insoluble Fraction (AIF) steps (Mercader *et al.* 2010). All samples reached the minimum of 250 phytoliths counted for inclusion in the analysis, of these all but three samples yielded >500 phytoliths. Phytoliths

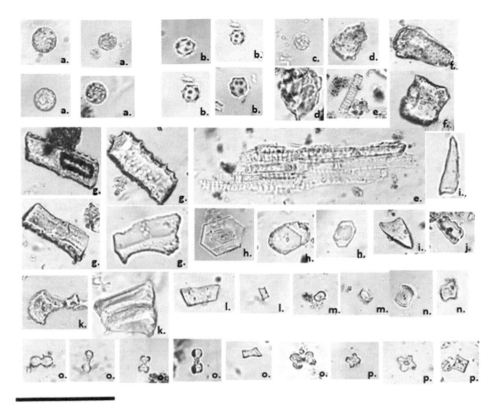

Scale bar-2mm

Figure 4. Micro-photographs of key fossil phytolith morphotypes identified. (a) Globular granulate, (b) globular echinate (Palm phytoliths), (c) Globular echinate (d) Globular verrucate, (e) Tracheid types, (f) Blocky types, (g) Sclereid bodies, (h) Achene (sedge phytoliths), (i) Scutiform/trichome types, (j) Crenate type, (k) Bulliform types, (l) Trapeziforms, (m) rondels, (n) saddles, (o) bilobates, (p) crosses.

were identified at 400× magnification using an Olympus BX52 light microscope. Micropho-tographs were taken and viewed using image processing software; Image-Pros plus 5.1 and Infinity Capture 2. Images were stored as TIFF/JPEG files.

Morphological description and classification were based on phytolith shape and size based on the International Codes for Phytolith Nomenclature (ICPN) (Madella *et al.* 2005), while additional literature was consulted for specific taxa. Grass short cell phytoliths (GSCPs) assignments were based largely on the following studies (Barboni and Bremond 2009; Bremond *et al.* 2008; Esteban *et al.* 2016; Mercader *et al.* 2010; Neumann *et al.* 2009; 2017; Novello *et al.* 2012; Piperno 2006; Rossouw 2009; Strömberg 2004; Twiss 1992; Twiss *et al.* 1969), Woody (tree and shrub) (Albert and Weiner 2001; Mercader *et al.* 2009; Neumann *et al.* 2009; Piperno 2006; Runge 1999; Strömberg 2004) and Palm phytoliths (Figures 4b) (Albert *et al.* 2009; Ashley *et al.* 2010b; Neumann *et al.* 2009), Cyperaceae morphotypes (Honaine *et al.* 2006; Ollendorf 1992; Piperno 2006; Strömberg 2004), while the non-diagnostic types (trichomes, bulliforms, elon-gates) followed (Albert and Weiner 2001; Mercader *et al.* 2009; Neumann *et al.* 2009; Piperno 2006; Runge 1999; Strömberg 2004) (see Table 1).

Table 1. Summary of the diagnostic morphotypes used to identify various taxa presented herein.

Major Taxa	Diagnostic morphotype (s)	Morphotype description	Comparable references
Aristidoideae	• Bilobate convex outer margin long shaft.	• Bilobate with rounded lobes and long shaft ($>10\,\mu m$).	Piperno and Pearsall 1998; Piperno 2006; Neumann et al., 2009.
Panicoideae	• Bilobates panicoid type. • Bilobates concave outer margin long/short shaft. • Bilobates convex outer margin outer margin short shaft. • Quadra-lobates/crosses.	• Bilobate that is symmetrical, with rounded lobes. Shaft $<10\,\mu m$). • Bilobates with lobes curving inwards with either long shaft ($>10\,\mu m$) or short shaft ($<10\,\mu m$). • Bilobate with rounded lobes and short shaft ($<10\,\mu m$). • Equidimensional bodies with four lobes	Fredlund and Tieszen 1994; Pieperno 2006; Fahmy 2008; Rossouw 2009; Neumann et al. 2009; Mercader et al. 2010; Novello et al., 2012.
Chloridoideae	• Saddle (Figure 4n). • Saddle squat.	• Saddle bodies with two opposite convex edges and two opposite concave edges in top view. • Saddle with side notches and a much longer axis than its tangential line.	Piperno 2006; Rossouw 2009; Neumann et al. 2009; Novello et al. 2012; Mercader et al., 2010.
Undifferentiated grasses	• Bulliforms (Figure 4k).	• Parallelepipedal bulliform cells.	Madella et al. 2005; Neumann et al. 2009; Novello et al., 2012.
Palm trees	• Globular echinate (palm type) (Figure 4c).	• Globular echinate, with defined spines.	Albert et al. 2006; 2009; Piperno 2006; Barboni et al., 2007; 2009.
Woody taxa (Dicot)	• Globular granulate (Figure 4a). • Globular verrucate (Figure 4d). • Tracheids (Figure 4g). • Sclereids (Figure 4e).	• Spheroid/ spherical with fine knobs or knots, grainy. • Spheroid/spherical with irregularly shaped wart-like processes. • Tracheids. • Silicified elongate to slightly curved treachery tissue.	Strömberg 2004; Madella et al. 2005; Albert et al. 2005; Piperno 2006; Neumann et al. 2009.
Herbaceous taxa	• Globular psilate, elongate spiny.	• Spheroid/spherical with smooth to sub-smooth surface.	Madella et al. 2005; Piperno 2006; Neumann et al., 2009.
Cyperaceae	• Achene types, hat shapes (Figure 4h).	• Pentagonal or hexagonal shape, with psilate surface and a central rounded cone.	Ollendorf 1992; Piperno 2006; Neumann et al., 2009.

7.3.2 Data analyses

Morphotypes with >2% occurrence were plotted against sampling depths and age estimates using TILIA software (Grimm 2007).The zonation is determined statistically by constrained incremental sum-of-squares analysis (CONISS), a component of the TILIA and TILIA GRAPH software (Grimm 2007), based on sequential samples that coincide with the geological age bracket from which they were sampled.

We calculated three phytolith indices: (i) tree cover density (D/P) to distinguish grassland versus wooded vegetation, whereby a high D/P ratio indicates woody vegetation, and a low D/P ratio indicates open grassland, (ii) climate index ((I_C) to estimate the temperature gradient, whereby high I_C indicates cool climates, and low I_C indicates warmer climates, and (iii) aridity index (Iph) to estimate aridity-humidity gradient across time (Alexandre *et al.* 1997; Diester-Haas *et al.* 1973), whereby a high Iph indicates arid while low Iph indicates moist-conditions (Alexandre and Bremond 2009; Bremond *et al.* 2005a, Bremond *et al.* 2005b; 2008; Neumann *et al.*, 2009).

The D/P index is calculated as the ratio of diagnostic woody dicotyledons morphotypes (globular granulate/ornate) versus diagnostic grass short cell phytoliths (GSCPs) (saddles+bilobates+crosses+polylobates+trapeziform rondels) (Alexandre *et al.* 1997; Bremond *et al.* 2005b; Bremond *et al.* 2008; Neumann *et al.* 2009). The Iph index is calculated as the ratio of Chloridoideae (saddle) versus the sum of Chloridoideae (saddle) and Panicoideae (cross + bilobates + polylobates) (Bremond *et al.* 2005b; Neumann *et al.* 2009). The I_C index is the ratio of Pooideae C_3 grasses (trapeziform and keeled rondels) versus C_4 Panicoideae and Chloridoideae grasses (saddles + bilobates + crosses + polylobates+trapeziform) (Bremond *et al.* 2008).

7.4 RESULTS

7.4.1 Temporal changes

Temporal variation in phytolith assemblages are presented in summary diagrams which shows changes in the proportion of the various morphotypes (Figure 5). Five main taxonomic categories are identified and classified into: (i) C_3 grasses (Pooideae), (ii) C_4 grasses (Panicoideae, Chloridoideae and Aristidoideae), (iii) other grasses (undifferentiated grasses), (iv) tree-shrub-herbs (woody and other taxa), and (v) aquatic taxa (palms and sedges).

A total of 18,679 phytoliths were assigned to 41 different morphotypes . Concentrations of phytoliths averaged 747 per sample, of which around 61% were grasses, 12% trees and shrubs, while palms and sedges morphotypes accounted for 2% each. Non-diagnostic morphotypes accounted for 0.01% only. Of the 11,400 grass phytoliths 91% are diagnostic. Grass phytoliths were the most prominent component in all samples except for three from the Olorgesailie Fm. (R9, R16 and R17) which were dominated by trees/shrubs morphotypes. Variations in the phytolith assemblage where clustered into zones using CONNISS; these vegetation based divisions were found to correspond closely with the chronostratigraphic units (Figure 5).

7.4.1.1 Upper Olorgesailie formation zone (c. 670 ka to 500 ka)
The phytolith assemblages in the Upper Olorgesailie Fm. zone show greater abundance of trees/shrubs morphotypes relative to the grass morphotypes (Figure 5). C_4 Aristidoideae morphotypes are consistently low in this unit. Palm and sedge phytoliths are present in moderate abundance in nearly all of the six samples from this zone. Phytolith indices, I_C, Iph and D/P in the Olorgesailie Fm zone are as follows:- mean and max. respectively I_C (0.3, 0.6), Iph (0.5, 0.7) and D/P (0.4, 1.0). The I_C maximum is recorded in the sample prior to 650 ka while Iph and D/P maxima are recorded around 650 ka (Figure 6).

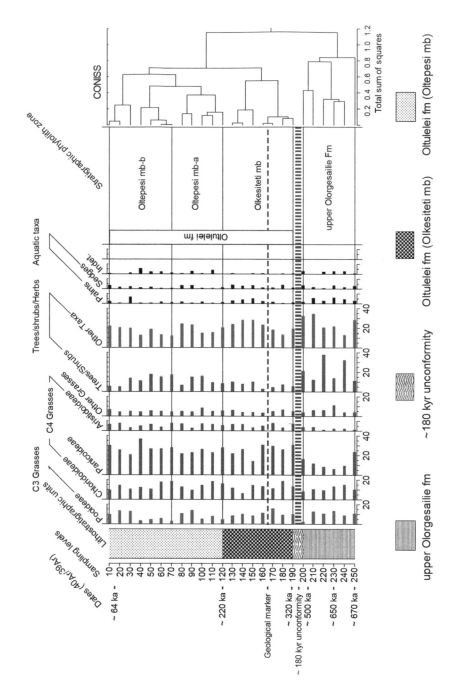

Figure 5. Phytolith relative abundance plotted against sample levels 10 – 250, representing the combined stratigraphic sequence through the Oltulelei and upper Olorgesailie formations. Note that actual sample intervals are variable in thickness but are equalized for this diagram.. See text for explanation of phytolith zones. CONISS = constrained incremental sum-of-squares analysis, a component of the TILIA and TILIA GRAPH software (Grimm 2007).

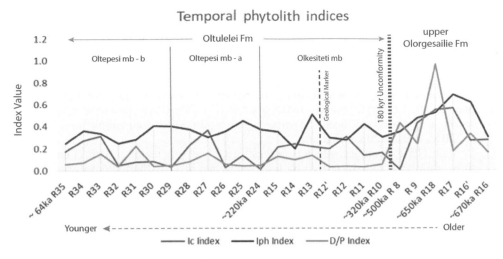

Figure 6. Phytolith indices plotted against samples. Climate index (I_C) increases with cool conditions, Aridity index (Iph) increases with low precipitation and Tree cover density (D/P) increases with more wooded vegetation structure. Graph plotted with older to younger going from right to left. Geological markers indicate with dotted red lines.

7.4.1.2 Olkesiteti mb zone (c. 320 ka to 220 ka)

This zone is separated from the underlying Olorgesailie Fm. zone by the *c.* 180,000 year unconformity (Behrensmeyer *et al.* 2002; 2018). Phytolith assemblages are mostly dominated by grasses, especially C_4 Panicoideae. The Olkesiteti mb zone represents phytolith data for samples R10 to R24 and marks the beginning of reduced frequencies of trees/shrubs phytoliths with increased GSCPs dominated by C_4Panicoideae morphotypes (Figure 5). Through the zone there is a significant increase in palm phytoliths, which had virtually disappeared following the unconformity. Interestingly, this variation, along with increased GSCPs and reduction of trees/shrubs morphotypes, appears to correspond with the geological marker in the unit. Phytolith preservation is good, as indicated by the low count in the indeterminate category. Phytolith indices, I_C, Iph and D/P in the zone are as follows:- mean and max. respectively I_C (0.2, 0.3), Iph (0.3, 0.5) and D/P (0.1, 0.1) (Figure 6).

7.4.1.3 Oltepesi mb zone (c. 220 ka to 64 ka)

The Oltepesi mb zone is marked by increased GSCPs, especially C_4 Panicoideae morphotypes, similar to the Olkesiteti mb zone. However, in this zone a slight increase in trees/shrubs morphotypes is evident, exceeding 15% in most of the samples. The zone is further sub-divided into two sub-zones, *viz.* Oltepesi mb zone-a and -b following the CONNISS clustering.

    a) Sub-zone Oltepesi mb-a exhibits a continuation of C_4 Panicoideae and C_4 Chloridoideae grass dominance. Palm phytoliths are rare but an increase in tree/shrub phytoliths > 15% occurs, especially in samples R26 (100) and R27 (90) (Figure 5).

    b) Sub-zone Oltepesi mb-b is characterized by trees/shrubs and C_4 Panicoideae grasses with the exception of sample R33 (30), which has a high occurrence of C_3 Pooideae grasses, while both C_4 Panicoideae and Chloridoideae grasses are relatively low in this sub-zone. Phytolith indices, I_C, Iph and D/P in the entire zone are as follows:- mean and max. respectively I_C (0.1, 0.4), Iph (0.3, 0.4) and D/P (0.1, 0.2) (Figure 6).

7.4.2 Spatial variation

We documented variations in phytolith assemblages across the three localities sampled from three geological sections, OLT05-02, B05-04 and G08-01 (Figure 2). The distance between Localities OLT and B is *c*. 2.2 km and between B and G *c*. 2.5 km, in the east-west direction (Behrensmeyer *et al*. 2018). Summary phytolith diagrams (Figure 7) show percentage occurrence of selected morphotypes identified and classified by taxonomic affiliation, which are then grouped into five major categories: (i) C_3 grasses, (ii) C_4 grasses, (iii) woody and herbaceous taxa, (iv) Other taxa, and (v) riparian/aquatic taxa. The percentages are plotted for each sample for the period between *c*. 320 ka and *c*. 220 ka (Figure 7).

OLT05-02 samples, which is farthest east, show two major clusters (Figure 7). The lower three samples are characterized by low occurrence of C_4 Chloridoideae grasses (<10%) and relatively higher occurrence of trees/shrubs and dicots phytoliths at >20%. The upper cluster is dominated by grass morphotypes, derived especially from C_4 Panicoideae and Chloridoideae. Chloridoideae morphotypes increase to more than 10% in this cluster. C_3 Pooideae morphotypes are more or less consistent across all six samples. Palm and sedge phytoliths are rare in all samples. Phytolith indices, mean and max. respectively for the OLT05-02 samples are as follows: I_C (0.7, max. 1.6), Iph (0.3, 0.4) and D/P (0.1, 2.1) (Figure 8).

B05-04 samples are similar except for sample number R19 (*c*. 320 ka), which is dominated by tree/shrub morphotypes with a high occurrence of 60% dropping sharply to 20% or less in the subsequent samples towards *c*. 220 ka. Also present in low frequencies in this sample (R-19) are palm phytoliths, which are absent in the other samples in this section. Sedges are present in low percentages only across all five samples. Phytolith indices, mean and max. respectively for the B05-04 samples are as follows: I_C (0.5, max. 0.6), Iph (0.5, 0.6) and D/P (0.5, 0.6) (Figure 8).

G08-01 samples from the site situated farthest to the west, are divided into two major clusters that correspond with the red soil zone , which is geologically significant as the oldest unit of the Oltulelei Fm. in this locality (Behrensmeyer, *et al*. 2018) (Figure 7); the top of the red soil forms the 'geological marker' indicated in Figure 5. The lower cluster consists of samples R35, R36 and R37. Sample R37 differs from the rest in that it has the highest occurrence grass morphotypes coupled with lowest occurrence of tree/shrub morphotypes (10%). The upper cluster (R39, R41, and R42) is characterized by the presence of phytolith morphotypes from all categories. Palm phytoliths are present in significant frequencies except for R38 (*c*. 320 ka), which is notable for its high proportion of dicots, trees and shrubs.

We calculated phytolith indices for only five samples because it was not possible to do so for sample R38; C_4 grass morphotypes in this sample were absent, and extremely low C_3 Pooideae and tree/shrub morphotypes did not permit calculation of indices. Phytolith indices, mean and max. respectively for the B05-04 samples are as follows: I_C (0.6, max. 0.8), Iph (0.3, 0.5) and D/P (2.3, 7.5) (Figure 8).

7.5 DISCUSSION

7.5.1 Temporal vegetation change in the Olorgesailie basin (*c*. 670 ka to 64 ka)

Temporal changes in the vegetation cover and structure correspond to a relatively stable tectonic intervals in the basin around 670 ka to 500 ka, when vegetation structure was characterized by wooded grasslands as indicated by a higher D/P index (>0.5). The landscapes during this time window were characterized by riparian habitats and fresh water resources indicated by palms and sedges (Albert *et al*. 2009; Ashley *et al*. 2010 a, b; Bamford *et al*. 2006). Iph index suggests moderately moist conditions in the upper Olorgesailie Formation. After the *c*. 180,000 years unconformity, vegetation cover transitioned to C_4 Panicoideae dominated grasslands, that

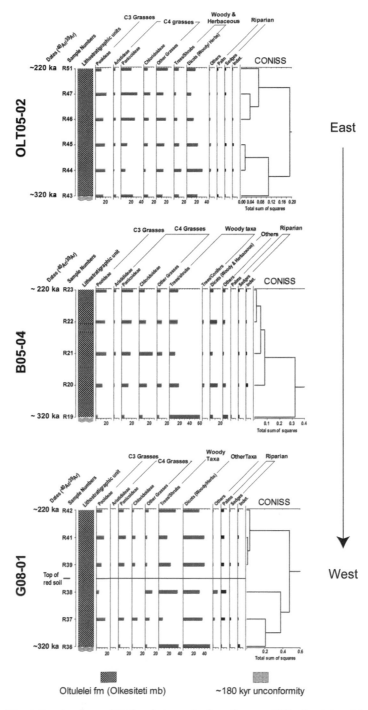

Figure 7. Phytolith relative abundance plotted against samples from the same 100 ka time interval for Localities OLT, B and G, which shows kilometre-scale landscape variability in local vegetation across the Olorgesailie basin. In G08-01, 'Top of red soil' = 'Geological Marker' in Figures 5 and 6. See text and caption to Figure 5 for further information.

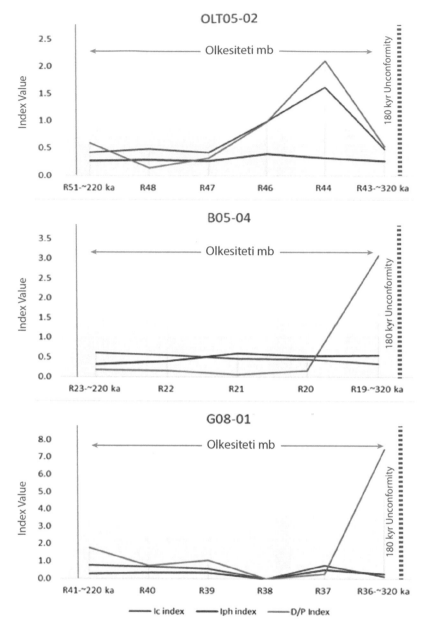

Figure 8. Olorgesailie phytolith indices plotted against samples for localities OLT, B and G. Graphs plotted with older to younger going from right to left. Climate index (I$_C$) increases with cool conditions, Aridity index (Iph) increases with low precipitation and Tree cover density (D/P) increases with more wooded vegetation structure. Geographically, OLT05-02 is farthest east, B05-04 in the middle, and G08-01 is farthest west (Figure 2).

fluctuates subtly through to *c.* 64 ka. I$_C$ and Iph indices suggest warm and moist conditions. Vegetation cover between *c.* 220 ka and 64 ka, was stable and dominated by C4 Panicoideae grasslands with low proportions of woody elements. Wetlands were present through to *c.* 220 ka. As mentioned earlier, variation corresponds to lithostratigraphic units.

7.5.2 Vegetation heterogeneity across the Olorgesailie basin (*c*. 320 ka to 220 ka)

Vegetation heterogeneity existed across the landscapes of the Olorgesailie basin between *c*. 320 ka and *c*. 220 ka. Within locality OLT, toward the eastern part of the rift basin (Figure 2), vegetation cover was initially dominated by woody elements that shifted towards C_4 Panicoideae grasslands. Palms and sedges were rarely present. Locality B, near the northern shoulder of Mt. Olorgesailie, was also dominated by C_4 Panicoideae grasslands except around *c*. 320 ka, which was marked by high dominance in woody vegetation. Sedges were rarely present and palms completely absent. Locality G on the western side of Mt. Olorgesailie was consistently dominated by woody and dicot vegetation, with abundant palms. C_3 Pooideae grasses across the three localities follows an east-west declining trend and are more prominent across the eastern and central localities (B and OLT) than at Locality G. This suggests that topographic complexity in the basin also influenced vegetation variation across the palaeolandscapes at least 5 km distance. Mt. Olorgesailie, which is today characterized by C_3 vegetation, likely also was a source of the C_3 Pooideae phytoliths input into the basin during the Oltulelei Fm. The evidence overall indicates considerable spatial variation in vegetation structure across the basin during the period of sedimentation.

7.5.3 Palaeoanthropological significance of vegetation data

The phytolith data presented here represent temporal resolution of 10^{4-5} years over a total interval of *c*. 600,000 years and are currently the only comprehensive outcrop-based data in East Africa for inferring vegetation during the Middle to Late Pleistocene transition in early hominin technology from Acheulean to Middle Stone Age (MSA) tools, by *c*. 350 ka to 320 ka (Brooks *et al.* 2018; Potts *et al.* 2018). Before the *c*. 180,000-year erosional phase, the vegetation structure associated with the late Acheulean technology at Olorgesailie was more wooded, and included wetlands/riparian resources, contrasting with the vegetation structure associated with MSA technology, which was dominated by warm and moist-adapted C_4 Panicoideae grasslands. This corresponds with a marked change in the geology and lithology of the basin between *c*. 500 and 320 ka, which exhibited a shift to a landscape with fluvial channels, flood plains and spring deposits (Behrensmeyer *et al.* 2018; Potts *et al.* 2018). The vegetation change was accompanied by large mammal turnover, evidenced by the paleontological evidence in the southern Kenyan rift (Potts *et al.* 2018). Herbivores may have played a significant role in engineering the vegetation structure as it shifted from more wooded to open grasslands in the basin (Potts *et al.* 2020). These dynamics could have influenced human adaptive strategies that led to new technological innovations (Brooks *et al.* 2018; Potts *et al.* 2020).

Vegetation heterogeneity, specifically the occurrence of a mosaic of riparian, grasslands and woodland patches in the time interval between 320 ka and 220 ka, suggests a variety of habitats that availed key resources for hominins and offered attractive spots for food resources. Such mixed habitats have been inferred at other prehistoric sites such as Olduvai and Awash and proved to be of palaeoanthropological significance (Albert *et al.* 2009; Ashley *et al.* 2010; Bamford *et al.* 2006; Barboni *et al.* 1999).

7.6 CONCLUSIONS

Phytolith data show that vegetation structure and cover in the Olorgesailie basin was more wooded prior (*c*. 670 ka to 500 ka) to the *c*. 180,000 year unconformity. Riparian habitats were present with subtle variation between palaeolandscapes. Trees and shrubs fluctuated between landscapes while grasslands remained stable. From 320 ka to 64 ka vegetation cover shifted to grasslands dominated by C_4 Panicoideae with some riparian habitats. Between *c*. 320 ka and 220 ka, our evidence shows that relatively adjacent habitats in the basin included a diversity of open grasslands, woodland, and riparian elements.

Phytolith data from this study assist in developing an understanding of local palaeoenvironments with which early hominins interacted during the Pleistocene (*c.* 670 ka to 64 ka). The data suggest that late Acheulean technology in the southern Kenya rift was associated with more woody vegetation structure than early MSA technology, which appears to have coincided with grassy landscapes in the Olorgesailie basin. This technological transition at Olorgesailie represents the oldest example currently known of the replacement of the Acheulean by the MSA (Brooks *et al.* 2018; Deino *et al.* 2018). We propose that this important transition may have occurred in response to changing vegetation and food resources available to the fauna, and may help to explain the replacement of large-bodied by smaller-bodied herbivores that were increasingly reliant on mixed vegetation diets (Potts *et al.* 2020). The varied habitats present in the basin offered critical resources such as fresh water and forage that would have sustained hominins and other animals.

In conclusion, vegetation dynamics in Olorgesailie during the Pleistocene reflected responses to both precipitation variability and tectonic processes. The vegetation data presented here are significant in that they are directly extracted from and related to fossil- and artifact-bearing outcrops and shed new light on the evidence for interactions between hominins and other faunal and floral communities in the basin. We postulate that such interactions likely played a critical role in shaping hominin behavior and technological innovations.

ACKNOWLEDGEMENTS

Our sincere gratitude goes to the following institutions, National Museums of Kenya (NMK), University of Cape Town, Smithsonian Institution, and University of Calgary. Support from the Peter Buck Fund for Human Origins Research (Smithsonian) is much appreciated acknowledged. We also thank Alan Deino for his long-term collaboration on the geological dates discussed in this paper; Jennifer Clark for her assistance in all the logistics that involved the Smithsonian Institution and for assisting with the figures; Drs. Veronica Muiruri and Stephen Rucina for the technical provisions during analyses at the Palynology and Paleobotany lab, NMK.

REFERENCES

Albert, R.M., and Weiner, S., 2001, Study of phytoliths in prehistoric ash layers from Kebara and Tabun caves using a quantitative approach. In: *Phytolith: Applications in Earth sciences and Human History*, edited by Meunier, J.D., Colin, F., (Lisse: A.A Balkema), pp. 251–266.

Albert, R.M., Bamford, M.K. and Cabanes, D., 2006, Taphonomy of Phytolith and macroplants in different soils from Olduvai Gorge (Tanzania) and the application to Plio-Pleistocene Palaeoanthropological samples. *Quaternary International* **148**, pp.78–94, 10.1016/j.quaint. 2005.11.026.

Albert, R.M., Bamford, M.K., and Cabanes, D., 2009, Palaeoecological significance of palms at Olduvai Gorge, Tanzania, based on Phytolith remains. *Quaternary International* **193**, pp. 41–48, 10.1016/j.quaint.2007.06.008.

Alexandre, A. and Brémond, L., 2009, Comment on the Paper in Quaternary International: "Methodological concerns for Analysis of phytolith assemblages: Does count size matter?" (C.A.E. Strömberg). *Quaternary International*, **193**, pp. 141–142, 10.1016/j.quaint. 2008.03.015.

Alexandre, A., Meunier, J.-D., Lézine, A.-M., Vincens, A. and Schwartz, D., 1997, Phytoliths: indicators of grasslands dynamics during the late Holocene in intertropical Africa. *Palaeogeography, Palaeoclimatology, Palaeoecology* **136**, pp. 213–229, 10.1016/S0031-0182(97)00089-8.

Asani, G.C., and Kinuthia, J.H., 1979, Diurnal variation of Precipitations in East Africa- Nairobi, Kenya. *Metrological Department Research Report*, **8**, pp.1–58.

Ashley, G.M., Barboni, D., Dominguez-Rodrigo, M., Bunn, H.T., Mabulla, A.Z.P., Diez-Martin, F., Barba, R. and Baquedano, E., 2010a, A Spring and wooded habitats at FLK Zinj and their relevance to origins of Human behaviour. *Quaternary Research*, **74**(3), pp. 304–314, 10.1016/j.yqres.2010.07.015.

Ashley, G.M., Barboni, D., Dominguez-Rodrigo, M., Bunn, H.T., Mabulla, A.Z.P., Diez-Martin, F., Barba, R., 2010b, Paleoenviromental and Paleoecological reconstruction of a fresh water oasis in Savannah grasslands at FLK North, Olduvai Gorge Tanzania. *Quaternary Research*, **74**(3), pp. 333–343, 10.1016/j.yqres.2010.08.006.

Baker, B.H., 1958, *Geology of the Magadi Area: Geological Survey of Kenya*, Report Number 42, (Nairobi: Government Printer).

Baker, B.H. and Mitchell, J.G., 1976, Volcanic stratigraphy and geochronology of the Kedong-Olorgesailie area and the evolution of the southern Kenya rift Valley. *Journal of the Geological Society of London*,**132**, pp. 467–484, 10.1144/gsjgs.132.5.0467.

Bamford, M.K., Albert, R.M., and Cabanes, D., 2006, Plio-Pleistocene macroplant fossil remains and phytoliths from Lowermost Bed II in the eastern palaeolake margin of Olduvai Gorge, Tanzania. *Quaternary International*, **148**, pp. 95–112, 10.1016/j.quaint.2005.11.027.

Barboni, D. and Bremond, L., 2009 Phytoliths of East African grasses: An assessment of their environmental and taxonomic significance based on floristic data. *Review of Palaeobotany and Palynology*, **158**, pp. 29–41, 10.1016/j.revpalbo.2009.07.002.

Barboni, D., Bonnefille, R., Alexandre, A. and Meunier, J. D., 1999, Phytoliths as paleoenvironmental indicators, West Side Awash Valley, Ethiopia. *Palaeogeography, Palaeoclimatology, Palaeoclimatology*, **152**, pp. 87–100, 10.1016/S0031-0182(99)00045-0.

Barboni, D., Bremond, L. and Bonnefille, R., 2007, Comparative study of modern phytolith assemblages from inter-tropical Africa. *Palaeogeography, Palaeoclimatology, Palaeoclimatology*, **246**, pp. 454–470, 10.1016/j.palaeo.2006.10.012.

Barboni, D., Ashley, G.M., Dominguez-Rodrigo, M. and Bunn, H.T., Mabulla, A. Z.P and Baquedano, E., 2010, Phytoliths infer locally dense and heterogeneous paleovegetation at FLK North and surrounding localities during upper Bed I time, Olduvai Gorge, Tanzania. *Quaternary Research*, **74**, pp. 344–354, 10.1016/j.yqres.2010.09.005.

Behrensmeyer, A.K., Potts, R., Deino, A. and Ditchfield, P., 2002, Olorgesailie, Kenya: a million years in the life of a rift basin. In *Sedimentation in Continental Rifts*, SEPM Special. Publication 73, edited by Renault, R.W. and Ashley, G., pp. 99–106.

Behrensmeyer, A.K., Potts, R., and Deino, A., 2018, The Oltulelei Formation of the southern Kenyan Rift Valley: A chronicle of rapid landscape transformation over the last 500 ky. *Geological Society of America Bulletin*, **130**, pp.1474–1492, 10.1130/B31853.1.

Bremond, L., Alexandre, A., Hély, C., and Guiot, J., 2005a, A Phytolith index as a proxy of tree cover density in tropical areas: calibration with Leaf Area Index along a forest-savanna transect in southeastern Cameroon. *Global and Planetary Change*, **45**(4), pp. 277–293, 10.1016/j.gloplacha.2004.09.002.

Bremond, L., Alexandre, A., Hély, C., and Guiot, J., 2005b, Grass water stress estimated from phytoliths in West Africa. *Journal of Biogeography*, **32**(2), pp. 311–327, 10.1111/j.1365-2699.2004.01162.x.

Bremond, L., Alexandre, A., Wooller, M. J., Hély, C., Williamson, D., Schäfer. P. A., Majule, A. and Guiot, J., 2008, Phytolith indices as proxies of grass subfamilies on the East African tropical mountains. *Global and Planetary Change*, **61**, pp. 209–224, 10.1016/j.gloplacha.2007.08.016.

Brooks, A.S., Yellen, J.E., Potts, R., Behrensmeyer, A.K., Deino, A.L., Leslie, D.E., Ambrose, S.H., Ferguson, J.R., d'Errico, F., and Zipkin, A.M., 2018, Long-distance stone transport and pigment use in the earliest Middle Stone Age. *Science*, **360**, p. 90–94, 10.1126/science.aao2646.

Brown, F., and Feibel, C.S., 1991, Stratigraphy, depositional environments, and palaeogeography of Koobi Fora Formation. In *Koobi Fora Research Project*, Volume 3, edited by Harris, J.M., (Oxford: Clarendon Press), pp. 1–30.

Cerling, T.E., Levin, N.E., Quade, J., Wynn, J.G., Fix, D.L., Kingstone, J.D., Klein, R.G. and Brown, F.H., 2010, Comment on the Paleoenvironments of *Ardipithecus ramidus*. *Science*, **328**, pp. 1105, 10.1126/science.1185274.

Cerling, T.E., Wynn, J.G., Andanje, S.A., Bird, M.I., Korir, D.K., Levin, N.E., Mace, W., Macharia, A.N., Quade, J., and Remien, C.H., 2011, Woody cover and hominin environments in the past 6 million years. *Nature*, **476**, pp. 51–56, 10.1038/nature10306.

Damnati, B. and Taieb, M., 1995, Solar and ENSO signatures in laminated deposits from lake Magadi (Kenya) during the Pleistocene/ Holocene transition. *Journal of African Earth Sciences*, **21**(3), pp. 373–382, 10.1016/0899-5362(95)00094-A.

Deino, A., and Potts, R., 1990, Single-crystal ^{40}Ar/^{39}Ar dating of Olorgesailie Formation, southern Kenya Rift. *Journal of Geophysical Research*, **95**, pp. 8453–8470, 10.1029/JB095iB06p08453.

Deino, A. and Potts, R., 1992, Age-probability spectra for examination of single crystal ^{40}Ar/^{39}Ar dating results: examples from Olorgesailie, southern Kenya rift. *Quaternary International*, **13–14**, pp. 47–53, 10.1016/1040-6182(92)90009-Q.

Deino, A., Dommain, R., Keller, C.B., Potts, R., Behrensmeyer, A.K., Beverly, E.J., King, J., Heil, C.W., Stockhecke, M., and Brown, E.T., 2019, Chronostratigraphic model of a high-resolution drill core record of the past million years from the Koora Basin, south Kenya Rift: Overcoming the difficulties of variable sedimentation rate and hiatuses. *Quaternary Science Reviews*, **215**, pp. 213–231, 10.1016/j.quascirev.2019.05.009.

deMenocal, P., 1995, Plio-Pleistocene African Climate. *Science*, **270**, pp. 53–59, 10.1126/science.270.5233.53.

deMenocal, P., 2004, African Climate Change and Faunal evolution during the Pliocene-Pleistocene. *Earth and Planetary Science Letters*, **220**, pp. 3–24, 10.1016/S0012-821X(04)00003-2.

Esteban, I., Vlok, J., Kotina, E.L., Bamford, M.K., Cowling, R.M., Cabanes, D., Albert, R.M., 2017, Phytoliths in plants from the south coast of the Greater Cape Floristic Region (South Africa). *Review of Palaeobotany and Palynology*, **245**, pp. 69–84, 10.1016/j.revpalbo.2017.05.001.

Fahmy, A. G., 2008, Diversity of lobate phytoliths in grass leaves from Sahel region, West Tropical Africa: Tribe Paniceae. *Plant Systematics and Evolution*, **270**, pp. 1–23, 10.1007/s00606-007-0597-z.

Feibel, C.S. 1999. Basin Evolution, sedimentary dynamics, and hominid habitats in East Africa. In *African Biogeography, Climate Change and Human Evolution*, edited by Bromage, T.G. and Schrenk, F. (Oxford: Oxford University Press), pp. 276–281.

Fredlund, G.G. and Tieszen, L.T., 1994, Modern Phytolith Assemblages from the North American Great Plains. *Journal of Biogeography*, **21**, pp. 321–335, 10.2307/2845533.

Fredlund, G.G. and Tieszen, L.T., 1997, Calibrating grass phytolith assemblages in climatic terms: Application to Late Pleistocene assemblage from Kansas and Nebraska. *Palaeogeography, Palaeoclimatology, Palaeoecology*, **136**, pp. 199–211, 10.1016/S0031-0182(97)00040-0.

Grimm, E.C., 2007, TILIA 1.01 and TILIA Graph. Illinois State University, Illinois.

Griffiths, J.F., 1958, Climatic Zones of East Africa. *East African Agricultural Journal*, **23**, pp. 179–185, 10.1080/03670074.1958.11665143.

Griffiths, J.F., 1972, Climates of Africa. *World Survey of Climatology*, **10**, pp. 313–347.

Hills, R.C., 1978, The structure of the Inter-Tropical Convergence Zone in Equatorial Africa and its relationship to East Africa rainfall. *Transaction of the Institute of British Geographers,* New series **4**(3), pp. 329–352, 10.2307/622055.

Honaine, M.F., Zucol, A.F. and Osterrieth, M.L., 2006, Phytolith Assemblages and Systematic Associations in Grassland species of South-Eastern Pampean Plains, Argentina. *Annals of Botany*, **98**, pp. 1155–1165, 10.1093/aob/mcl207.

Isaac, G.L., 1977, Olorgesailie, Archaeological Studies of a Middle Pleistocene Lake Basin in Kenya, (Chicago: University of Chicago Press).

Isaac, G.L., 1978, The Olorgesailie Formation: stratigraphy, tectonics and the paleogeographic context of the Middle Pleistocene archaeological sites. *Geological Society, London, Special Publications*, **6**, pp. 174–206, 10.1144/GSL.SP.1978.006.01.15.

Kenworthy, J.M., 1966, Temperature conditions in the tropical highland climates in East Africa. *The East African Geographical Review*, **1966**(4), pp. 1–11.

Kinyanjui, R.N., 2012, Phytolith Analysis as a Paleoecological Tool for Reconstructing Mid- Late Pleistocene Environments in the Olorgesailie Basin, Kenya, MSc Dissertation (Environmental and Geographical Sciences Department: University of Cape Town).

Livingstone, D.A. and Mworia-Maitima, J., 1999, *Preliminary Report on Preservation Status of Organic Plant Remains in Olorgesailie Basin*, Unpublished Report, (Niarobi: National Museums of Kenya).

Madella, M., Alexandre, A., and Ball, T., 2005, International Code for Phytolith Nomenclature 1.0. *Annals of Botany*, **96**, pp. 253–260, 10.1093/aob/mci172.

Mercader, J., Bennett, T., Esselmont, C., Simpson, S. and Walde, D., 2009, Phytoliths in woody plants from the Miombo woodlands of Mozambique. *Annals of Botany*, **104**(1), pp. 91–113, 10.1093/aob/mcp097.

Mercader, J., Astudillo, F., Barkworth, M., Bennett, T., Esselmont, C., Kinyanjui, R., Grossman-Laskin, D., Simpson, S. and Walde, D., 2010, Grass and Sedge phytoliths from Niassa, Mozambique. *Journal of Archaeological Science*, **37**(8): pp. 1953–1967, 10.1016/j.jas.2010.03.001.

Mulholland, S.C. and Rapp, G. Jr., 1992, A morphological classification of grass silica- bodies. In *Phytolith Systematic, Emerging Issues: Advances in Archaeological and Museum Science*, Volume 1, edited by Rapp, G. Jr. and Mulholland, S.C. (New York: Plenum Press), pp. 65–89.

Mutai, C.C and Ward, M.N., 2000, East African rainfall and the tropical circulation/convection on interseasonal to interannual timescales. *Journal of Climate*, **13**, pp. 3915–3939, 10.1175/1520-0442(2000)013<3915:EARATT>2.0.CO;2.

Mworia, J., Dallmeijer, A. and Jacobs, B., 1988, Vegetation and modern pollen rain at Olorgesailie, Kenya. *Utafiti*, **1**(1), pp. 1–22.

Neumann, K., Fahmy, A., Lespez. L., Balloche, A., Huysecom, E., 2009, The early Holocene palaeenvironment of Ounjougou (Mali): Phytoliths in multiproxy context. *Palaeogeography, Palaeoclimatology, Palaeoecology*, **276**, pp. 87–106, 10.1016/j.palaeo.2009.03.001.

Neumann, K., Fahmy, A.G., Muller-Scheessel, N., Schmidt, M., 2017, Taxonomic, ecological and palaeoecological significance of leaf phytoliths in West African grasses. *Quaternary International*, **434**, pp. 15–32, 10.1016/j.quaint.2015.11.039.

Nicholson, S.E., 1996, A Review of Climate Dynamics and Climate Variability in eastern Africa. In *The Limnology, Climatology and Paleoaclimatology of the East Africa Lakes*, edited by Johnson, T.C. and Odada, E.O., (Amsterdam: Gordon and Breach Publishers), pp. 25–56.

Novello, A., Barboni, D., Berti-Equille, L.,Mazur, J-C., Poilecot, P., Vignaud, P., 2012, Phytolith signal of Aquatic plants and soils in Chad, Central Africa. *Review of Paleobotany and Palynology*, **178**, pp. 43–58, 10.1016/j.revpalbo.2012.03.010.

Ollendorf, A.L., 1992, Toward classification scheme of sedge (Cyperaceae) phytoliths. In *Phytolith Systematics, Emerging Issues: Advances in Archaeological and Museum Science*, Volume 1, edited by Rapp, G. Jr. and Mulholland, S.C. (New York: Plenum Press), pp. 91–111.

Owen, R.B., Potts, R., Behrensmeyer, A.K. and Ditchfield, P., 2008, Diatomaceous sediments and environmental change in the Pleistocene Olorgesailie Formation, southern Kenya Rift Valley. *Palaeogeography, Palaeoclimatology, Palaeoclimatology*, **269**, pp. 17–37, 10.1016/j.palaeo.2008.06.021.

Piperno, D.R., 1988, *Phytolith Analysis: An Archaeological and Geological Perspective.* (San Diego: Academic Press).

Piperno, D.R., 1989, The occurrence of phytoliths in reproductive structures of selected tropical angiosperms and their significance in tropical paleoecology, paleoenthnobotany, and systematic. *Review of Palaeobotany and Palynology*, **61**, pp. 147–173, 10.1016/0034-6667(89)90067-5.

Piperno, D.R., 2006, Phytoliths: A Comprehensive guide for Archaeologists and Paleoecologists, (Oxford: Altamira Press).

Piperno, D.R. and Pearsall, D.M., 1998, The silica bodies of tropical American grasses: Morphology, taxonomy, and implications for grass systematics and fossil phytolith identification. *Smithsonian Contributions to Botany*, **5**, pp. 1–45, 10.5479/si.0081024X.85.

Potts, R., 1989, Olorgesailie: New excavations and findings in Early and Middle Pleistocene contexts, southern Kenya Rift Valley. *Journal of Human Evolution*, **18**, pp. 477–484, 10.1016/0047-2484(89)90076-6.

Potts, R., 1994, Variables versus models in early Pleistocene hominid land use. *Journal of Human Evolution*, **27**, pp. 7–24, 10.1006/jhev.1994.1033.

Potts, R., 1998, Environmental hypotheses of hominin evolution. *Yearbook of Physical Anthropology*, **41**, pp. 93–136, 10.1002/(SICI)1096-8644(1998)107:27+<93::AID-AJPA5>3.0.CO;2-X.

Potts, R., 2001, Mid-Pleistocene environmental change and human evolution. In *Human Roots: Africa and Asia in the Middle Pleistocene*, edited by Barham, L., Robson-Brown, K., (Bristol: Western Academic Press), pp. 5–21.

Potts, R., 2013, Hominin evolution in settings of strong environmental variability. *Quaternary Science Reviews*, **73**, pp. 1–13, 10.1016/j.quascirev.2013.04.003.

Potts, R., and Deino, A., 1995, Mid-Pleistocene change in large faunas of East Africa. *Quaternary Research*, **43**, pp. 106–113, 10.1006/qres.1995.1010.

Potts, R., and Faith, J.T., 2015, Alternating high and low climate variability: The context of natural selection and speciation in Plio-Pleistocene hominin evolution. *Journal of Human Evolution*, **87**, pp. 5–20, 10.1016/j.jhevol.2015.06.014.

Potts, R., Behrensmeyer, A.K., and Ditchfield, P., 1999, Paleolandscape variation and Early Pleistocene hominid activities: Members 1 and 7, Olorgesailie Formation, Kenya. *Journal of Human Evolution*, **37**, pp. 747–788, 10.1006/jhev.1999.0344.

Potts, R., Behrensmeyer, A.K., Deino, A., Ditchfield, P., and Clark, J., 2004, Small Mid-Pleistocene hominin associated with East African Acheulean technology. *Science*, **305**, pp. 75–78, 10.1126/science.1097661.

Potts, R., Dommain, R., Moerman, J.W., Behrensmeyer, A.K., Deino, A.L., Beverly, E.J., Brown, E.T., Deocampo, D., Kinyanjui, R., Lupien, R.L., Owen, R.B., Rabideaux, N., Riedl, S., Russell, J.M., Stockhecke, M., deMenocal, P., Faith, J.T., Garcin, Y., Noren, A., Scott, J.J., Western, D., Bright, J., Clark, J.B., Cohern, A.S., Heil Jr., C.W., Keller, C.B., King, J., Levin, N., Brady, K., Muiruri, V., Renaut, R., Rucina, S.M., Uno, K., 2020, Increased ecological resource variability during a critical transition in hominin evolution. *Science Advances*, **6**, 10.1126/sciadv. abc8975, article: eabc8975.

Rapp, G. Jr., and Mulholland, S.C., 1992, Phytolith Systematics: Emerging Issues. Advances in Archaeological and Museums Science. (New York: Plenum Press).

Rossouw, L., 2009, The application of fossil grass-phytolith analysis in the reconstruction of Cainozoic environments in the South African interior, PhD Dissertation, (Faculty of Natural and Agricultural Sciences: University of the Free State, Bloemfontein).

Runge, F. 1999, The opal phytolith inventory of soils in central Africa – quantities, shapes, classifications and spectra. *Review of Palaeobotany and Palynology*, **107**, 1–2, pp. 23–53.

Sansom, H.W., 1954, The climate of East Africa based on Thornthwaite's classification. *East Africa Meteorological Department Memoir*, **3**, pp. 1–49.

Shackleton, R.M., 1978, A geological map of the Olorgesailie area. In *Geological Background to Fossil Man*, map insert, edited by Bishop, W.W., (Edinburgh: Scottish Academic Press).

Sikes, N.E., 1994, Early hominid habitat preference in East Africa: Paleosol carbon isotopic evidence. *Journal of Human Evolution*, **27**, pp. 25–45, 10.1006/jhev.1994.1034.

Sikes, N.E., and Ashley, G.M., 2007, Stable isotopes of pedogenic carbonates as indicators of paleoecology in the Plio-Pleistocene (Upper Bed I), western margin of Olduvai Basin, Tanzania. *Journal of Human Evolution*, **53**, pp. 574–594, 10.1016/j.jhevol.2006.12.008.

Sikes, N.E., Potts, R. and Behrensmeyer, A.K., 1999, Early Pleistocene habitat in Member 1 Olorgesailie based on paleosol stable isotopes. *Journal of Human Evolution*, **37**, pp. 721–670, 10.1006/jhev.1999.0343.

Strömberg, C.A.E., 2004, Using Phytolith assemblages to reconstruct the origin and spread of grass-dominated habitats in the Great Plains during the late Eocene to early Miocene. *Palaeogeography, Palaeoclimatology, Palaeoecology*, **207**(3–4), 59–75, 10.1016/j.palaeo.2003.09.028.

Trauth, M.H., Maslin, M., Deino, A., Strecker, M.R., Bergner, A.G.N., and Duhnforth, M., 2007, High-and low-latitude forcing of Plio-Pleistocene East African climate and human evolution. *Journal of Human Evolution*, **53**, pp. 475–486, 10.1016/j.jhevol.2006.12.009.

Twiss, P.C., 1992, Predicted world distribution of C3 and C4 grass phytoliths. In *Phytolith Systematics. Emerging Issues: Advances in Archaeological and Museum Science*, Volume 1, edited by Rapp. G.J. and Mulholland, S.C., (New York: Plenum Press), pp. 113–128.

Twiss, P.C., Suess, E. and Smith, R.M., 1969, Morphological Classification of Grass Phytoliths. *Soil Science Society of America Proceedings*, **33**, pp. 109–115, 10.2136/sssaj1969.03615995003300010030x.

White, F., 1983, The Vegetation of Africa. *Natural Resources Resarch, UNESCO*, **20**, pp. 1–356.

Wolde Gabriel, G., Ambrose, S.H., Barboni, D., Bonnefille, R., Bremond, L., Currie, B., DeGusta, D., Hart, W.K., Murray, A.M., Renne, P.R. Jolly-Saad, M.C., Stewart, K.M. and White, T.D., 2009, The geological, isotopic, botanical, invertebrate, and lower vertebrate surroundings of Ardipithecus ramidus. *Science*, **326**, article: 65-65e5, 10.1126/science.1175817.

CHAPTER 8

Sedimentological, palynological and charcoal analyses of the hydric palustrine sediments from the Lielerai-Kimana wetlands, Kajiado, southern Kenya

Esther N. Githumbi[1]

Department of Physical Geography and Ecosystem Science, Lund University, Lund, Sweden

Colin J. Courtney Mustaphi[2]

Department of Environmental Sciences, Geoecology, University of Basel, Basel, Switzerland

Robert Marchant

Department of Environment and Geography, Institute for Tropical Ecosystems, University of York, York, United Kingdom

ABSTRACT: Intermittent and sometimes ephemeral wetlands found in savannah ecosystems are subject to complex hydrological and geomorphological processes that influence vegetation and fire patterns, and their use by humans, cattle and wildlife. Today these wetlands are impacted by changes in land use, climate, and wildlife use. Small wetlands in semi-arid climates are dynamic ecosystems that respond rapidly to biological, physical and chemical processes and accumulate sedimentary deposits making them excellent archives of past environmental changes. The Lielerai-Kimana wetlands in the Greater Amboseli Ecosystem of southern Kenya are located within the current protected area network. A 384 cm long palustrine sediment core was recovered from the Lielerai-Kimana wetlands, radiometrically dated, and used for sedimentological, palynological and charcoal analyses. The ^{210}Pb (n = 16 ages) and ^{14}C (n = 10 ages) results suggest a complex depositional pattern covering *c.* 1500 years. The mismatches between ^{210}Pb and ^{14}C date estimates suggest some degree of sediment mixing which may prevent centennial-scale interpretation of the palaeoenvironmental record. The sediments are characterised by a high silt content from

[1] Other affiliation: *Department of Biology and Environmental Science, Linnaeus University, Kalmar, Sweden*
[2] Other affiliation: *Water Infrastructure and Sustainable Energy (WISE) Futures, Nelson Mandela African Institution of Science & Technology, Tengeru, Arusha, Tanzania; and York Institute for Tropical Ecosystems, Department of Environment and Geography, University of York, York, United Kingdom*

DOI 10.1201/9781003162766-8

1500–600 cal yr BP with the organic content increasing from 600 cal yr BP to the present. The pollen data show little changes in vegetation composition, suggesting a mosaic of savannah, woody savannah, riparian, and montane forest persisted throughout the period of deposition. Charcoal and organic content suggest the continuous presence of herbaceous fuels at the coring site from 600 cal yr BP to the present. The results highlight the spatiotemporal heterogeneity and complexity of wetland records across the semi-arid landscapes of Amboseli during the Late Holocene. Further studies on hydroclimate and vegetation change, disturbance ecology (fire, erosion, bioturbation), and human-environment interactions would further develop our understanding of the environmental history and drivers of environmental change across these ecosystems that are increasingly under human land use pressures.

8.1 INTRODUCTION

Amboseli is an important ecosystem in southern Kenya due to its high biodiversity, long history of pastoralism, rapidly developing agricultural potential, and subsequent land-use changes over the past decades. High levels of tourism since the 1960s (Western 1973, 2007; Western *et al.* 2020) make Amboseli an important contributor to Kenya's gross domestic product (GDP) and the livelihoods of local communities (Hodgson 2000; Makindi 2016). Amboseli has experienced, and continues to experience, wide-scale yet spatially complex habitat changes such as: *Vachellia xanthophloea* woodland reduction, riparian and wetland fluctuation, introduced agriculture and a proliferation of ruderal plant species (Western and Van Praet 1973). Within semi-arid regions, such as the Amboseli, water is a crucial natural resource and ecosystem service. Changes in livelihood strategies from pastoral to sedentary has led to increased pressure on water resources for farming and irrigation (Okello and Kioko 2011; Wishitemi *et al.* 2015). Amboseli wetlands are intermittently distributed across the semi-arid savannah landscape and have complex responses to hydroclimate, land-use and land cover changes, defaunation and wildlife management (Meijerink and Wijngaarden 1997). Increased densities of wildlife populations outside protected areas due to fencing along migration corridors (Kenya Wildlife Service 2009; Osipova *et al.* 2018) and fencing in some of Amboseli's wetlands (Okello and Kioko 2011) has been reported. The Amboseli Ecosystem Management Plan 2008–2018 (Kenya Wildlife Service 2009) and Kenya Wetlands Atlas (MEMR 2012) identified wetlands as an exceptional resource value for the Amboseli ecosystem that are susceptible to rapid and irreversible change with complex social co-benefits and potential negative impacts on biodiversity and conservation. Palaeoenvironmental records developed from wetland ecosystem sedimentary archives can provide a long-term context for modern ecosystem changes and develop data suitable for characterising the wetlands, and their histories, as part of developing evidenced-based support for management (Gillson and Marchant 2014; Githumbi *et al.* 2020).

The abundance of wetlands in Amboseli enabled five palaeoecological studies, focused on a variety of proxies, to provide information relevant to understanding different aspects of environmental, ecological and human-environment interaction change during the Late Holocene (last *c.* 4200 years). Understanding how the various wetlands responded to large-scale controls (climate variability) and internal controls (hydrological connectivity, topography) is necessary when utilising wetland palaeoenvironmental records to observe temporal change and compare changes across a landscape. Palynological analyses published from Amboseli wetlands, include Namelok (Rucina *et al.* 2010) and Esambu, Kimana and Ormakau (Githumbi 2017) and were presented in summary formats for discussion alongside archaeological insights (Githumbi *et al.* 2018a, Shoemaker 2018) and carbon accumulation in wetlands (Gallego-Sala *et al.* 2018). The oldest radiocarbon dated published record from the Amboseli wetlands is currently from Indoinyo-Esambu, which dates back to *c.* 5000 cal yr BP, while the Namelok record covers the past *c.* 3000 cal yr BP. The radiocarbon dated sediment core presented here from the Lielerai-Kimana dates back to *c*. 1500 cal yr BP. Sedimentological, geochronological, palynological and charcoal data from the Lielerai-Kimana sediment core are used to characterise the sediments, explore the limitations and opportunities of the geochronological results, and interpret the pollen and charcoal results. The findings are discussed in relation to the history of wetland dynamics in the Amboseli area of southern Kenya.

8.1.1 Study site

Amboseli is located in southern Kenya between the northern base of Kilimanjaro, near the Kenya-Tanzania border, Tsavo West National Park and the Chyulu Hills (Figure 1). A semi-arid savannah and scrubland ecosystems, the Amboseli basin occupies *c*. 3000 km^2 and experiences bimodal rainfall distribution. Rainfall patterns are determined by the African monsoon, which is, in turn, driven by the latitudinal migration of the Intertropical Convergence Zone (ITCZ) and its interaction with more local climate systems and topographic feedbacks (Hulme 1996; Marchant *et al.* 2007). Precipitation mainly falls in two rainy seasons: March–May and October–December. Average annual precipitation is 586 mm year^{-1} although with considerable interannual variation (range 226 to 990 mm year^{-1}) from 1979 to 2009 (Githumbi 2017). An overall increase in temperature (daily minimum and maximum) was observed between 1976 and 2001, however, there was little change in precipitation (Altmann *et al.* 2002). Today, Amboseli supports a wide range of wildlife including ungulates, carnivores, and birds, that co-exist with pastoralist and agricultural land uses, making it an important wildlife conservation area and designated important bird area (IBA). Several of the wetlands across Amboseli are crucial for wildlife migrations among the network of protected (Kenya Wildlife Service 2009; Osipova *et al.* 2018), however, with

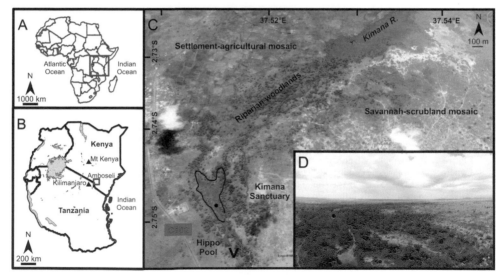

Figure 1. Map of the Lielerai-Kimana wetlands and study site (red boxes). (A) Inset map of Africa and (B) East Africa. (C) The Kimana Sanctuary Lielerai-Kimana wetlands and riparian forests along the Kimana River, which flows north and then east. The Cyperaceae and Poaceae covered riparian wetland is outlined with black and the coring location is shown with a black circle (coordinates: 2.748833 S, 37.515367 E, 1222 m asl). The 'V' symbol indicates the location and perspective of the inset photograph D, and Kenyan road C102. (D) Oblique air photograph of the wetland showing the coring location (black circle) and landscape features around the Kimana Sanctuary protected area (Courtney Mustaphi *et al.* 2014; Githumbi *et al.* 2014). Google Earth image dated 27 April 2014 (copyright CNES/Airbus and Google Inc.) Basemap imagery: Google Earth, 2020, http://earth.google.com/web/.

Figure 2. Photographs of landscapes and features around the Kimana wetland. (A) The coring location between the tall Cyperaceae and Poaceae dominated wetland. (B) Short Poaceae and Cyperaceae that cover parts of the southern end of the Kimana wetland. (C) The Hippo pool (viewed from the west facing southeast). (D) A biogeomorphological riverbank feature heavily influenced by *Hippopotamus amphibius* and *Crocodylus niloticus* entering and exiting the Kimana River (flowing to the left of photograph, northward). (E) The eastern margin of the Kimana Sanctuary viewing northeast showing the mosaic of barren ground, savannah, woody savannahs, and wetlands. (F) An example of crops grown in the Amboseli-Oloitoktok area. Photographs: A, Esther Githumbi, April 2014; B, D, E, Colin Mustaphi, March 2015; and C, F, Rob Marchant, March 2015.

increasing anthropogenic land-use pressures and human-wildlife conflicts are increasing (Okello 2005).

The Amboseli landscape comprises of savannah and scrubland plains, lacustrine plains, wetlands, and volcanic footslopes (Meijerink and Wijngaarden 1997). The woody-grassy savannah landscapes are dominated by *Pennisetum* spp., *Cenchrus* spp., and African Acacia species (*Vachellia* and *Senegalia*, and referred in this paper as Acacias) (Figure 2). The Amboseli wetlands are located on the Kilimanjaro aquifer and they receive groundwater and surface runoff primarily from the slopes of Mount Kilimanjaro (Grossmann 2008; Meijerink and Wijngaarden 1997). Amboseli wetlands are either spring fed (Enkongu, Esambu, Ormakau, Namelok) or riverine (Lielerai-Kimana) and highly modified through climatological, biological, geomorphological and hydrological processes. The wetlands overlie deep, well-drained and neutral pH soils developed from lava flows and partly reworked pyroclastic deposits of Pleistocene age (Williams 1972). These wetland areas are dominated by *Cyperus* spp. and *Sporobolus* spp., while *Vachellia* (Acacias), *Balanites, Commiphora, Euphorbia* and *Syzygium* dominate the adjacent woody savannahs and scrublands.

The Lielerai-Kimana toponym comes from the Maa language, where 'Lielerai' refers to the yellow fever trees (*Vachellia xanthophloea*) and 'Kimana' means 'a continuous circle'. The Lielerai-Kimana wetland is fed by the Kimana River, which flows through the wetlands towards the Chyulu Hills. Parts of the wetland complex was under various forms of protected area status that have included ecotourism and community-based conservation (Githumbi *et al.* 2018a; Meguro and Inoue 2011) and is an ecological hotspot for wildlife migration and is a hippopotami breeding site.

8.2 MATERIALS AND METHODS

8.2.1 Field methods

The Lielerai-Kimana core site was located, within the Kimana Sanctuary (also known as *Sidai Oleng*), at 2.748833°S, 37.5151367°E (1222 m asl). Today the area around Lielerai-Kimana is a Cyperaceae-covered wetland with hydric soils and palustrine sediments (Figures 2A, 2B). In July 2014, the profile of the sediments beneath the wetlands were determined through probing using fiberglass rods. A coring location was then chosen where the organic sediment accumulation seemed to be high and sandy layers appeared to be absent. Then the sediments were cored using a hand operated, 50 cm long, 5 cm diameter hemicylindrical Russian peat corer (Jowsey 1966). Sediment cores from two overlapping parallel boreholes were obtained. Together the parallel cores contained the complete sedimentary sequence and reached a maximum depth of 384 cm (Courtney Mustaphi *et al.* 2014; Githumbi 2017). Upon recovery, the core sections were wrapped in plastic and aluminium foil, and transferred into split plastic PVC pipes for transport. The sediment cores were then transported to the Palynology and Palaeobotany Section of the National Museums of Kenya (NMK; Nairobi, Kenya) where they were placed in cold storage at 4°C. Subsamples were later extracted at NMK, and analyses were undertaken at NMK and the University of York (UK).

8.2.2 Laboratory methods

8.2.2.1 Geochronology
Ten accelerator mass spectrometer (AMS) radiocarbon dates were obtained from the Lielerai-Kimana sediment core (Table 1). Two bulk sediment subsamples were sent to Direct AMS (Bothell, WA, USA) that were acid–base–acid washed, combusted, graphitised and measured using a National Electrostatics Corporation (NEC) 1.5 SDH Compact Pelletron 500 kV AMS. Two bulk sediment subsamples were sent to the [14]CHRONO laboratory (Queen's University Belfast, Northern Ireland) and acid wash pre-treated prior to combustion to CO_2, graphitization and AMS using a NEC 0.5 MV compact accelerator. Six sediment subsamples were wet sieved using a 63 μm mesh and the fine organic detrital fraction (<63 μm) was graphitised and dated at the East Kilbride facility of the SUERC Radiocarbon Laboratory (University of Glasgow) using either the National Electrostatic Corporation 5 MV tandem AMS or 250 kV single-stage AMS. Radiocarbon ages (^{14}C years BP) were calibrated using the IntCal13 curve (Reimer *et al.* 2013). Calibrated ages are presented in years BP (Before Present, CE 1950) by convention.

Table 1. Alpha counter radioactivities (counts per second or Bq) and CRS model calendar age results of ^{210}Pb dating the uppermost sediments of the Lielerai-Kimana sediment core collected in CE2014 (-64 yr BP). Note that samples below ^{210}Pb background radioactivities (34–40 cm) do not produce CRS calendar age estimates. Acronyms: BP, before present AD1950; CRS, constant rate of supply model; NA, not applicable, ages not modelled.

Depth (cm)	Dry weight (g)	Bulk density (g cm^{-3})	Corrected d ^{209}Po (counts)	Corrected d ^{210}Po (counts)	^{210}Po ^{209}Po	^{210}Pb activity (Bq kg^{-1})	^{210}Pb- activity error $\pm 1\sigma$ (Bq kg^{-1})	CRS year BP	CRS 2σ error (years)
4–5	0.2523	0.2678	2073.30	134.25	0.06	23.55	2.16	−63	2.01
5–6	0.3035	0.3109	1654.24	95.75	0.06	17.50	1.89	−62	2.29
6–8	0.502	0.2614	973.14	69.73	0.07	13.10	1.66	−61	2.30
8–10	0.3801	0.2077	1760.26	88.46	0.05	12.13	1.36	−60	2.31
10–12	0.4117	0.2247	1490.22	85.34	0.06	12.77	1.46	−59	2.32
12–14	0.5203	0.2864	1375.20	135.30	0.10	17.35	1.61	−57	2.32
14–16	0.4018	0.2202	1434.21	88.46	0.06	14.09	1.59	−55	2.34
16–18	0.3981	0.2184	1486.22	139.46	0.09	21.63	1.97	−50	2.37
18–20	0.3977	0.2347	1421.21	113.44	0.08	18.42	1.85	−45	2.40
20–22	0.4735	0.2839	1999.29	218.55	0.11	21.19	1.57	−38	2.45
22–24	0.3152	0.2079	1768.26	150.91	0.09	24.85	2.17	−24	2.58
24–26	0.3312	0.1887	1197.18	53.08	0.04	12.28	1.77	−15	2.65
26–28	0.2737	0.1577	1147.17	53.08	0.05	15.51	2.23	4	3.01
28–30	0.1683	0.1607	1163.17	18.73	0.02	8.78	2.09	20	3.36
30–32	0.2761	0.1824	1835.27	38.51	0.02	6.97	1.16	37	3.84
32–34	0.2236	0.1541	1430.21	23.94	0.02	6.87	1.45	79	7.24
34–36	0.2231	0.1345	1341.20	15.61	0.01	4.79	1.24	NA	NA
36–38	0.2042	0.1272	1170.17	9.37	0.01	3.60	1.2	NA	NA
38–40	0.2425	0.1589	1567.23	18.73	0.01	4.52	1.07	NA	NA

Lead (^{210}Pb) dating of the uppermost sediments was done by drying sediment subsamples from intervals between 4–40 cm depth and measured through alpha spectroscopy at the University of Exeter, UK. A constant rate of supply (CRS) model of calendar age estimates with age estimate errors that have a Gaussian distribution (Appleby 2005) was developed with the core top (0 cm) set as −64 years BP, the year of core collection. ^{210}Pb background activities were reached at 34–36 cm depth.

An age-depth model combining both calibrated ^{14}C and ^{210}Pb ages was developed using Bacon (Blaauw and Christen 2011) in RStudio version 1.3.959.

8.2.2.2 Sedimentological characterisation

The Lielerai-Kimana sediment cores were longitudinally split and surfaces cleaned flat with a knife and the composite stratigraphy of overlapping core drives was established in the laboratory. The cores were scanned at BOSCORF National Oceanography Centre (NOC), Southampton, UK, using a molybdenum tube ITRAX™ Core Scanner for μXRF semi-quantification of 36 elements at 500 μm resolution (Croudace and Rothwell 2015). The X-ray beam irradiated the centerline of the core using a 3 kW Mo X-ray tube typically run at 55 kV and 50 ma for x-radiography and 30 kV and 50 ma for the XRF scan. A Geotek multi-sensor core logger (MSCL-XYZ) with a Bartington MS2E magnetic susceptibility point sensor and a Konica Minolta spectrophotometer (quantified greyscale reflectance) scanned the cores at 0.5 cm intervals from 0–384 cm stratigraphic depth.

Loss-on-ignition (LOI) analysis (Heiri *et al.* 2001) was used to estimate organic matter content, carbonate content and residuals mean clastic particle size every 5 cm down core. Wet sediment subsamples were weighed dried at 105°C for 24 hours, then burned at 550°C for 5 hours and weighed, and ashed at 950°C for 3 hours and weighed, to calculate the dry weight (and dry bulk density), organic matter and carbonate contents respectively. Particle size distribution analysis used the Malvern Mastersizer 2000 laser granulometer (MEH/MJG180914). The binned particle size distributions were aggregated into percentage clay (<1–4 μm), silt (4–63 μm), and sand (63–2000 μm) fractions (Blott and Pye 2012). Bulk sediment C:N ratios for the top 111–0 cm at contiguous 1-cm intervals were measured using a CHN analyser at the University of Exeter, UK.

8.2.2.3 Sieved (>125 μm) charcoal analysis

For sieved charcoal analysis, subsamples of 1 cm^3 of wet sediment were removed at a continuous 1 cm resolution. Samples were soaked in a sodium metaphosphate solution (Tsakiridou *et al.* 2020) for >24 hours and then wet sieved through a 125-μm mesh and the larger fraction was transferred to a gridded Petri dish. Charcoal pieces were identified and total charcoal was tallied through visual inspection and manipulated with a metal probing needle under a Zeiss Stemi 2000-C optical stereo microscope at 10–40 $\times$ magnifications and converted to sieved charcoal concentrations (pieces cm^{-3}).

8.2.2.4 Pollen preparation and analysis

Wet sediment subsamples of 1 cm^3 were extracted from the core at 20–cm intervals and processed for pollen analysis using sequential sediment digestions (Fægri and Iversen 1989). One tablet of exotic *Lycopodium* spores was added prior to pollen processing to act as a counter for calculating pollen concentration or estimation of absolute influx (Lund University batch number 3862, exotic count per tablet: 9666 $\pm$ 212.3). Subsamples were chemically digested, centrifuged, decanted, and rinsed with deionised water (or glacial acetic acid) after each digestion step. The digestion step included 10% HCl (room temperature), 10% KOH (room temperature), 50% HF (70°C), dewatering with glacial acetic acid, and acetolysis (Erdtman 1960; Fægri and Iversen 1989). Digested sediment residues were transferred into vials using 96% ethanol, centrifuged and decanted. Glycerine was added to match the volume of residue and then left for the ethanol to evaporate in an oven at 60°C then a droplet was mounted on standard microscope slides with a coverslip (Bennett and Willis 2001).

Pollen counts were conducted across horizontal line scans of the coverslips using a Leica DM4000B microscope at a magnification of 400×. Identifications were confirmed using a reference collection derived from pollen and herbarium specimens from the National Museums of Kenya and published pollen atlases (Bonnefille and Riollet 1980; Hamilton 1976, 1982; Hamilton and Perrott 1980; Schüler *et al.* 2014). Pollen diagrams were plotted using C2 v.1.7.6 (Juggins 2003). The stratigraphically constrained cluster analysis using incremental sum of squares (CONISS, Grimm 1987) was used to delimit statistically significant assemblage zones of the relative abundances of all pollen types using a broken stick test (Bennett 1996). Pollen were grouped using published pollen atlases (Schüler and Hemp 2016), previously published pollen grouping (Githumbi *et al.* 2018a; Rucina *et al.* 2010), field observation of taxa, and ecology and habitat descriptions (Dharani 2011).

8.3 RESULTS

8.3.1 Chronology and stratigraphy

In total sixteen ^{210}Pb (Table 1) and 10 AMS radiocarbon dates (Table 2) were used in the development of the Bacon age-depth model (Figure 3; Blaauw and Christen 2011). The ^{210}Pb samples

Table 2. Reported AMS radiocarbon dates with IntCal13 calibrated ages with maximum 2σ error range (year BP, before present; Reimer *et al.* 2013) from the Lielerai-Kimana sediment core.

Depth (cm)	Lab code	Age (^{14}C years)	Calibrated age (year BP)	Material
20–21	SUERC-64044	578 ± 37	600 ± 73	<63 µm sieved organic detritus
40–41	SUERC-64045	526 ± 37	539 ± 88	<63 µm sieved organic detritus
50–51	SUERC-57339	424 ± 37	486 ± 158	<63 µm sieved organic detritus
50–51	SUERC-58881	500 ± 37	527 ± 96	<63 µm sieved organic detritus
75–76	SUERC-64046	647 ± 37	605 ± 64	<63 µm sieved organic detritus
100–101	UBA-26122	540 ± 24	543 ± 81	bulk sediment
100–101	SUERC-64047	630 ± 35	602 ± 58	<63 µm sieved organic detritus
260–261	D-AMS 009667	946 ± 25	850 ± 67	bulk sediment
310–311	D-AMS 009668	455 ± 24	509 ± 27	bulk sediment
370–371	UBA-26122	1458 ± 31	1340 ± 44	bulk sediment

were from the top 40 cm of the core and the AMS radiocarbon dates ranged from the whole core. There is a high degree of uncertainty in the age-depth model from 650–100 cal yr BP due to the range of dates in this uppermost section, suggesting some degree of movement of old carbon.

The Lielerai-Kimana core was predominantly organic rich (*c.* 20%), poorly sorted, sandy and coarse silt sediments (*c.* 61% silt, *c.* 23% sand and *c.* 11% clay) and three broad lithological units were observed (Figure 4). The base of the core (384 cm) to 380 cm contained very poorly sorted dark grey sandy and coarse silts (mean particle size 70–170 µm) with light coloured sandy laminations. The second unit from 380–103 cm was a massive, visually homogeneous, poorly sorted, very dark brown organic sediment (10–25%), with increased clay content and less sand content (mean particle size 23–250 µm). The uppermost unit, from 103 cm to the top, contained very poorly sorted, medium dark brown sandy silt (mean particle size 80–250 µm) with a very high organic content (40–90%). C/N values were relatively high and variable within the measured section (Figure 3), 111–0 cm depth, carbon ranged in value between 16% and 48% while nitrogen ranged between 0.6% and 1.2%, the three C/N peaks are observed at 32 cm, 44 cm and 86 cm. Magnetic susceptibility ranges from 0 to 617.41 κ with a mean of 191.45 κ.

8.3.2 Charcoal

Sieved charcoal particle (>125 µm) concentrations ranged 2–5700 pieces cm^3 with a mean of 329 (median of 104 pieces cm^3) and 1σ standard deviation of 656. Throughout the core, but particularly in the uppermost section the charcoal concentrations are highly variable. Charcoal concentrations were generally <200 pieces cm^3 from 384–100 cm and averaged 1000 pieces cm^3 from 100–20 cm. Charcoal concentrations were <1000 pieces cm^3 from 20–0 cm (Figure 4). When the charcoal concentrations were converted to charcoal accumulation rates (influx, pieces cm^2 year^1), the pattern is very similar to the concentration values, with the exception of a slight increase in the top 30 cm that is not evident in the concentration values. Several surface sediment samples of 0–1 cm depth from different parts of the swamp (n = 13) were analysed using the same charcoal sieving method and averaged 17 pieces cm^3 with a range of 0–162 pieces cm^{-3}. The conspicuous increase in charcoal around 100 cm is concomitant with increased organic content and sands, and changes to the elemental composition measured by ITRAX (Figure 4).

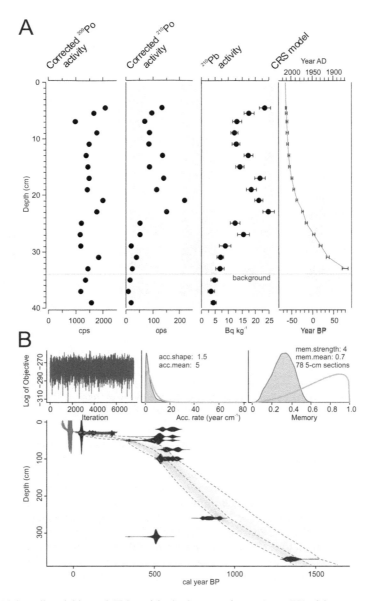

Figure 3. (A) Alpha radioactivities and CRS model calendar age estimates (years BP) of the uppermost sediments of the Kimana sediment core. (B) Age-depth model iteration using IntCal13 calibration curve and default BACON settings.

8.3.3 Pollen

Forty-six pollen taxa were identified from the Lielerai-Kimana sediment core and pollen-slide charcoal was also counted. Although the apparent variability of pollen relative abundances is relatively low, the CONISS zonation identified two significant pollen zones and the dendrogram showed insignificant zone breaks (subzones; Figure 5). The pollen taxa have been ordered by dominant growth form in Figure 5. The pollen assemblages are similar throughout the past 1500

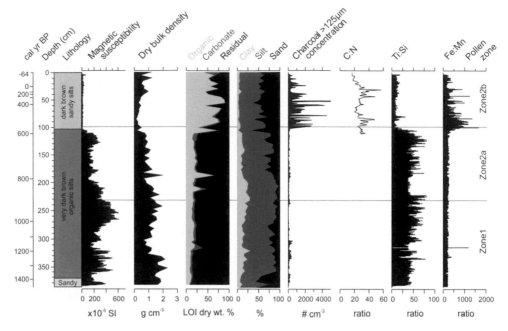

Figure 4. Sedimentological results of the Kimana wetland sediment core and age-depth model estimated downcore ages (at left). The visual lithological description and sediment characterisation measurements of volume magnetic susceptibility, dry sediment bulk density, and loss-on-ignition results, particle size distribution summary and charcoal concentrations. C:N measurements were collected for the top 111-0 cm. Elemental ratios were calculated from ITRAX XRD results (cps units). Pollen zones delineated by horizontal lines from CONISS analysis of the pollen relative abundances (see Figure 5).

years, consistent with the current savannah, wooded savannah, and riparian vegetation surrounding the site. Throughout the record the trees and shrubs have a combined mean value of *c.* 50%, *c.* 24% and *c.* 26%. The herbaceous taxa are *c.* 31%, the aquatics *c.* 12% and the grass *c.* 6%. The dominating taxa occurring above >2% throughout the record are *Vachelia* (Acacia) *c.* 3%, *Carissa c.* 2%, *Polyscias c.* 2%, *Balanites c.* 2%, *Podocarpus c.* 3%, *Euphorbia c.* 3%, *Commelina c.* 3.7%, *Commiphora c.* 2.6%, Asteraceae *c.* 2.8%, Amaranthaceae *c.* 2.6%, Cyperaceae *c.* 7%, *Typha c.* 3.5% and Poaceae *c.* 5%. The lowermost significant pollen assemblage zone 1, from the core bottom to *c.* 900 cal yr BP (385 to 230 cm) has tree pollen abundances of *c.* 24%, shrub pollen *c.* 27%, herbs *c.* 31%, aquatics *c.* 12% and grasses *c.* 6%. The uppermost significant pollen assemblage zone 2, from *c.* 900 cal yr BP to present (230 cm to the top) is divided into two subzones: zone 2a (from *c.* 800–500 cal yr BP, 230–90 cm) with tree pollen abundances of *c.* 24%, *c.* 27% shrub pollen, *c.* 31% herbs, *c.* 12% aquatics and grasses *c.* 6%. The topmost subzone 2b (from *c.* 500 cal yr BP to present, 90–0 cm) shows tree pollen abundances of *c.* 26%, *c.* 25% shrub pollen, *c.* 31% herbs, *c.* 12% aquatics, and *c.* 6% grasses.

8.4 DISCUSSION

8.4.1 Sedimentology and chronology

The poorly sorted sediments and high sand content suggest complex transport and depositional energy regimes influence the sediment deposition into the Lielerai-Kimana wetland. The wetland is under high bioturbation pressure from microorganisms, herbivores and has a river running

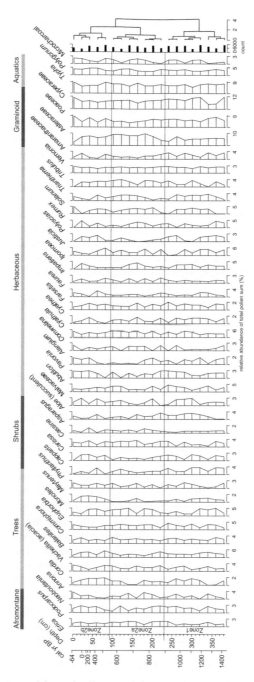

Figure 5. Relative abundance of the total pollen sum. Select taxa shown. Pollen groups based on qualitative predominance of taxa or structural growth forms in the Amboseli area (Githumbi *et al.* 2018a and 2018b; Rucina *et al.* 2010) with succulents shown in brackets. CONISS zonation with 1 statistically significant zone break and 1 insignificant zone bound between zone 2a and 2b (CONISS dendrogram units not shown; total sum of squares ×1000).

through the wetland that ephemerally floods. From the base of the core (384 cm) up to 100 cm, the sediments are characterised by moderate organic, high silt, and higher clay content. This could reflect the ephemeral nature of the catchment that is fed by a river with the headwaters in the nearby Kilimanjaro mountain. The low (<20%) organic content measured in the lower half of the core (384–200 cm; Figure 4) is similar to organic content measured from modern hippopotamus pool margins, trails and mudflats inside the Ngorongoro Crater (Deocampo 2002); although, Ngorongoro crater wetlands are generally much larger and far less shaded by trees than is currently found at Lielerai-Kimana. In the uppermost unit (100 cm to the top; Figure 4), the magnetic susceptibility drops significantly, we observe a high Fe:Mn ratio, low Ti:Si ratio, high organic, charcoal and sand content. The radiocarbon ages suggest a high degree of remobilisation of organic content from within the catchment aged between 650–400 cal yr BP. It is frequently observed that organic matter (including charcoal after fires) and sands at the surface are transported by surface sheet wash during intense rain events on top of the very hard compacted soils (see also Shoemaker 2018, page 61). These materials could be transported into the riverine and palustrine features from the wider Amboseli landscape and trapped by vegetation (Courtney Mustaphi *et al.* 2015; Koff and Vandel 2008). Within the top 100 cm (Figure 3) the C/N values are very high and variable, indicative of terrestrial sources and the high variability may be due to certain layers, such as 30–50 cm and around 100 cm, having high rates of microbial activity.

Bulk sediment radiocarbon dates can produce older ^{14}C dates because of aggregation of remobilised fine organic matter from the catchment and have a lower precision (Barnekow *et al.* 1998; Grimm *et al.* 2009). Bulk sediment offsets have not been explicitly investigated in wetland ecosystems in eastern Africa. The Lielerai-Kimana record at 101–100 cm provides one example of a comparison for a bulk sediment age estimate of 540 ± 24 (UBA-26122; Figure 3) and <63 um sieved organic detritus age of 630 ± 35 (SUERC-64047). In this example, the age of the finer organic detritus was older than the total bulk sediment (with visible rootlets removed). In palustrine sediments, rootlet remains need to be avoided because of vertical movement of carbon in the stratigraphy. The offset could be due to a remainder of younger fine rootlets persisting in the bulk sediment sample, remobilised clay- and silt-sized organics in the catchment or remobilised organics with a variety of radiocarbon ages deposited together – with the limited comparable results, here it is difficult to discern the cause.

The distribution of the radiocarbon samples is less monotonic and not in stratigraphic superposition; but overall, there is a tendency towards older dates (1500–940 cal yr BP) towards the bottom of the core. The ^{210}Pb profile has some evidence of disturbed sediments over the past century, but is not completely mixed or noisy, and there are large offsets of around 400 years, between the ^{210}Pb and ^{14}C radiometric dates. Offsets between radiocarbon dating of organic content and ^{210}Pb dating of the uppermost clastic sediments has been observed in sediments from temperature lacustrine deposits with the potential for remobilisation of older organic detritus in the catchment (Rinta *et al.* 2016). In nearby volcanic Crater Lake Duluti, Mount Meru, century-scale disparities between the ^{210}Pb age estimates and radiocarbon ages occurred in the top sediments (Öberg *et al.* 2012). This could be due to changing sedimentation from within the closed catchment of the crater due to land use and change cover change, and as varying water levels influencing the submerged macrophyte coverage and resuspension and deposition of organic sediments (Öberg *et al.* 2013).

The top of the record presented a reasonable curve of higher ^{210}Pb alpha radioactivities that decreased and reached a background activity by 34 cm depth; yet, the radioactivities suggest a slight some degree of sediment mixing from 23 to 7 cm (Figure 3A). The uppermost radiocarbon ages are several centuries older than the ^{210}Pb age estimates suggesting reworking of carbon from within the catchment for bulk and fine (<63 μm) organic matter. Potential causes include riverine overbank flow (wetland is located close to Kimana river), deposition of sediments and organic matter from the catchment, vegetation bioturbation through root establishment, and mammalian bioturbation (the wetland is a key watering point for Amboseli wildlife and a Hippo pool). This may prevent centennial-scale interpretation of the palaeoenvironmental record. A combination of ^{210}Pb and radiocarbon dates at the top of our record improves the utility of the age-depth model by highlighting the influx of old and material. The proximity of Lielarai-Kimana wetland to Kimana River could be the main driver of the influx observed where flooding of the river and use of the wetland by the resident hippos is a constant source of sediment flow.

8.4.2 Regional palaeoenvironmental records

Several wetlands in the region have basal ages from 3000–1500 cal yr BP, suggesting a phase of wetland establishment and maintenance by hydroclimatic, topographical/geological (Liutkus and Ashley 2003; Owen *et al.* 2004), and biogeomorphological processes (Deocampo 2002; McCarthy *et al.* 1998). Palustrine sediments from the Amboseli and Tsavo areas have basal radiocarbon dates >5000 cal yr BP (Gillson 2004; Githumbi *et al.* 2018a; Rucina *et al.* 2010). Esambu had an older radiocarbon date of *c.* 5000 cal yr BP, and the sediments at this age were highly represented dry savannah pollen types, such as *Vachellia* (acacia) and Amaranthaceae/Chenopodiaceae as opposed to wetland taxa (Githumbi *et al.* 2018b). That suggests the landscape was an open patchy savannah where the lack of herbaceous vegetation or a continuous understory indicates a dry period and that the dimensions and spatial distribution of the hydric palustrine ecosystems within the semi-arid ecosystem varies through time in response to hydroclimatic, topographical, and biogeomorphological processes (Ashley *et al.* 2002, Casanova and Powling 2015; Jolly *et al.* 2008). Sediment cores collected from relatively small, montane wetlands in adjacent highland areas have similar basal ages (5000–1500 cal yr BP), including Kilimanjaro (Coetzee 1967; Courtney Mustaphi *et al.* 2020; Hamilton 1982), North Pare Mountains (Heckmann 2014), and Mount Shengena (Finch *et al.* 2017). The Late Holocene in this region was characterised by intermittent drought phases following the end of the African Humid Period (Verschuren *et al.* 2009) that influenced the vegetation and fire regimes of the savannahs (Gillson 2004; Nelson *et al.* 2012; Urban *et al.* 2015). The stability of wetland vegetation has been described in coastal wetlands of eastern South Africa during the Late Holocene (Turner and Plater 2004).

8.4.3 Land-cover and land-use change

Semi-arid savannah and shrubland ecosystems of eastern African are characterised by high spatiotemporal variability in precipitation regimes that results in a diverse ecosystem composition, structure and distribution. The Amboseli landscape has undergone a series of land-use regimes through time and currently supports pastoralists, agriculture and wildlife. The dynamic nature of the Amboseli local ecosystem is shaped by these interactions between climate, ecosystem, herbivores and humans. Pollen, non-pollen palynomorph (NPP) and charcoal data from Esambu swamp indicated that sparse woodland and shrub taxa characterised the semi-arid Amboseli landscape from *c.* 5000–1700 cal yr BP (Githumbi *et al.* 2018b) with a change from a drier environment with a small aquatic areas that expanded as regional climates become moist. Several of the published archaeological records from Amboseli (Foley 1981; Shoemaker 2018) and Kilimanjaro area (Fosbrooke and Sassoon, 1965) also published radiocarbon ages during the late Holocene or ceramic wares dated to the Holocene and suggest extensive human land uses across Amboseli (Iles *et al.* 2018; Odner 1971). The ecosystem was occupied by hunter-gatherer groups and subsequent pastoral Neolithic communities with regionally archaeological evidence indicating a shift in livelihood strategies with specialised pastoralism emerging *c.* 3000 cal yr BP (Shoemaker 2018). The Namelok record, which covers the last *c.* 3000 cal yr BP, suggests the vegetation was a savannah mosaic dominated by acacias and Amarathaceae/Chenopodiaceae, Asteraceae, *Cissampelos*, Poaceae and *Salvadora* (Rucina *et al.* 2010). Pollen taxa recorded from the Lielerai-Kimana record (Figure 5) concur with the other wetland records where the dominant taxa include *Vachellia* (Acacia), Asteraceae, Amaranthaceae, *Commiphora, Balanites* and *Euphorbia.* The Lielerai-Kimana pollen taxa composition suggests that the landscape supported an open savannah shrubland over the last *c.* 1200 cal yr BP with most of the change experienced over the last *c.* 500 cal yr BP. This is evident from the charcoal record (Figure 4), which shows the continued presence of fires, and low biomass consuming fires until the last 500 years where the charcoal concentration and accumulation rate triple the mean values for the whole record. The Lielerai-Kimana record shows a mix of taxa that occur at low and high elevations including Afromontane taxa such as *Cordia, Podocarpus* and *Polyscias.* This suggests that Amboseli wetlands record some of the wind-blown pollen from vegetation taxa on Kilimanjaro and Chyulu Hills.

During the Late Holocene, the Amboseli landscape has supported pastoralism and hunter-gatherer livelihood strategies (Lane 2004). Caravan trade routes around Kilimanjaro penetrated into Amboseli and the region was exploited for the ivory trade leading to high levels of defaunation (Håkansson 2012); notably during the 1800s and 1900s driven by high ivory prices (Håkansson 2004). The 19th century droughts and disease epidemics decimated both livestock and wild grazer populations in the region. To date, there has been no direct evidence in the archaeological records of agriculture or combination livelihoods of agro-pastoralism in Amboseli during the Late Holocene, although the absence of evidence is inconclusive and does not rule out the possibility (Githumbi *et al.* 2018a; Shoemaker 2018). Land uses outside of parks has placed greater pressure on protected areas as large herbivore populations increase in density. This concentration of elephants in the parks has led to a decline in woody taxa in particular in smaller parks such as Amboseli (Western *et al.* 2020). From the 1980s, conversion of the Amboseli wetlands for subsistence and industrial food crop agriculture was the predominant land use change (Githumbi *et al.* 2018b; MEMR 2012; Okello and Kioko 2011) outside of the national park.

Over the last *c.* 500 cal yr BP, significant changes are observed in the Amboseli ecosystem pollen, fire and geochemical records; the physiochemical signals indicate aridity interspersed with intervals of increased moisture, which could be increased local rainfall or increased water table levels caused by increased precipitation on Kilimanjaro and Chyulu. Namelok (Rucina *et al.* 2010) records a dry period from *c.* 500 cal yr BP indicated by the continuous presence and increase in taxa such as *Vachellia* (Acacia), Amaranthaceae/Chenopodiaceae, Cyperaceae and Poaceae. This is accompanied by a significant increase in charcoal levels attributed to presence of anthropogenic activity such as pastoralism and eventually agriculture at Lielerai-Kimana, Namelok and Esambu. Taxa associated with human activity are not identified from the Lielerai-Kimana record; however, *Ricinus communis* and *Cannabis sativa* are recorded with increasing numbers in Namelok from *c.* 1600 cal yr BP (Rucina *et al.* 2010). The Esambu wetland records a decrease in woody taxa with increasing Cyperaceae and Poaceae, a combination of a dry interval and human presence in the savannah and surrounding highlands. The decreased charcoal content in the top 20 cm (^{210}Pb dates of −38 cal yr BP) is consistent with the low occurrence of fires currently on the landscape observed by satellite observations (Githumbi *et al.* 2018a) and the high level of grazing pressure both by cattle and wild herbivores on the grazing-dominated savannahs (Bond and Keeley 2005; van Langevelde *et al.* 2003).

8.5 CONCLUSIONS

Small tropical wetlands provide information about the local changes in a landscape. However, they have to be utilised with caution on their own due to the inability to discern the driver or scale of ecosystem change. The Lielerai-Kimana wetland allows us to visualise the general Amboseli environment, i.e. a savannah shrubland and the changes within the wetland but not the drivers of the change. Data from Lielerai-Kimana, Esambu, Enkongu, Namelok wetlands highlight the continued maintenance of a semi-arid savannah with varying levels of woodland. Significant social-ecological changes have occurred within the same interval that we observe the significant increase of charcoal *c.* 500 cal yr BP; arrival of the Maa-speaking pastoralists into Amboseli, widespread Swahili settlement increasing, arrival of the colonialist, increased elephant poaching and more recently land management policies. These past and current land cover and land use impacts and the timing of the vegetation, sedimentary and fire regime changes across the wetlands indicate localised controls; this emphasises the differential ecosystem responses of each wetland and highlights that small wetlands can be used to corroborate regional events.

ACKNOWLEDGEMENTS

We thank Stephen Rucina, Veronica Muiruri, Rahab Kinyanjui and Rebecca Muriuki from the National Museums of Kenya, Palaeobotany and Palynology Section for field and laboratory support; Joseph Mutua and Nicholas Gakuu for fieldwork logistics; and the British Institute in Eastern Africa (Nairobi) for their continued support. We thank William Gosling, Anne-Marie Lezine and Louis Scott for their organisation of this special issue. This study contributes to the African Pollen Database (APD), Neotoma Paleoecology Database, and the PAGES supported Global Paleofire Working Group (GPWG2) and PAGES-LandCover6k. Past Global Changes (PAGES) is funded by the Swiss Academy of Sciences and Chinese Academy of Sciences and supported in-kind by the University of Bern, Switzerland. ENG and CCM were supported by a European Commission Marie Skłodowska-Curie Initial Training Network grant to RM and coordinated by Paul Lane through Uppsala University (FP7-PEOPLE-2013-ITN project number 606879). [210]Pb and C/N data presented in this paper was supported by NERC Radiocarbon Allocation 1681.1012 or funded by the Natural Environment Research Council NERC standard grant number NE/I012915/1 to Dan Charman at University of Exeter, UK, and we thank Angela Gallego-Sala and the Millipeat Project. This research formed part of the PhD dissertation of EG, University of York. In memory of our friends and colleagues, Richard Payne and Eric Grimm.

REFERENCES

Altmann, J., Alberts, S.C., Altmann, S.A., and Roy, S.B., 2002, Dramatic change in local climate patterns in the Amboseli basin, Kenya. *African Journal of Ecology*, **40**(3), pp. 248–251, 10.1046/j.1365-2028.2002.00366.x.

Appleby, P.G., 2005, Chronostratigraphic techniques in recent sediments. In *Tracking Environmental Change Using Lake Sediments: Developments in Paleoenvironmental Research*, Volume 1, edited by Last W.M., Smol J.P., (Dordrecht: Kluwer Academic Publishers), pp. 171–203.

Ashley, G.M., Goman, M., Hover, V.C., Owen, R.B., Renaut, R.W. and Muasya, A.M., 2002, Artesian blister wetlands, a perennial water resource in the semi-arid Rift Valley of East Africa. *Wetlands*, **22**(4), pp. 686–695, 10.1672/0277-5212(2002)022[0686:ABWAPW]2.0.CO;2.

Barnekow, L., Possnert, G., and Sandgren, P., 1998, AMS ^{14}C chronologies of Holocene lake sediments in the Abisko area, northern Sweden – a comparison between dated bulk sediment and macrofossil samples. *GFF*, **120**(1), pp. 59–67.

Bennett, K.D., 1996, Determination of the number of zones in a biostratigraphical sequence. *New Phytologist*, **132**(1), pp. 155–170, 10.1111/j.1469-8137.1996.tb04521.x.

Blaauw, M., and Christen, J.A., 2011, Flexible paleoclimate age-depth models using an autoregressive gamma process. *Bayesian Analysis*, **6**(3), pp. 457–474, 10.1214/ba/1339616472.

Blott, S.J., and Pye, K., 2012, Particle size scales and classification of sediment types based on particle size distributions: Review and recommended procedures. *Sedimentology*, **59**(7), pp. 2071–2096, 10.1111/j.1365-3091.2012.01335.x.

Bond, W., and Keeley, J., 2005, Fire as a global 'herbivore': the ecology and evolution of flammable ecosystems. *Trends in Ecology & Evolution*, **20**(7), pp. 387–394, 10.1016/j.tree.2005.04.025.

Bonnefille, R., and Riollet, G., 1980, Pollens des savanes d'Afrique orientale. (Paris: CNRS).

Casanova, M.T., and Powling, I.J., 2015, What makes a swamp swampy? Water regime and the botany of endangered wetlands in western Victoria. *Australian Journal of Botany*, **62**(6), pp. 469–480, 10.1071/BT14119.

Coetzee, J.A., 1967, Pollen analytical studies in eastern & southern Africa. *Palaeoecology of Africa*, **3**, pp. 1–146.

Courtney Mustaphi, C.J., Githumbi, E., Shoemaker, A., Degefa, A.Z., Petek, N., van der Plas, G., Muriuki, R.M., Rucina, S.M., and Marchant, R., 2014, *Ongoing sedimentological and palaeoecological investigations at Lielerai Kimana and Ormakau Swamps, Kajiado District, Kenya*, Report. (Kenya: Resilience in East African Landscapes Project), pp. 1–32.

Courtney Mustaphi, C.J., Davis, E.L., Perreault, J.T., and Pisaric, M.F., 2015, Spatial variability of recent macroscopic charcoal deposition in a small montane lake and implications for reconstruction of watershed-scale fire regimes. *Journal of Paleolimnology*, **54**(1), pp. 71–86, 10.1007/s10933-015-9838-2.

Courtney Mustaphi CJ., Kinyanjui R., Shoemaker A., Mumbi C., Muiruri V., Marchant L., Rucina S., and Marchant R., 2020, A 3000-year record of vegetation changes and fire at a high-elevation wetland on Kilimanjaro, Tanzania. *Quaternary Research*, **99**, pp. 34–62, 10.1017/qua.2020.76.

Croudace, I.W., and Rothwell, R.G., 2015, Micro-XRF Studies of Sediment Cores. In *Tracking Environmental Change Using Lake Sediments*, Volume 2: Physical and Geochemical Methods. (Dordrecht: Springer Netherlands).

Deocampo, D.M., 2002, Sedimentary structures generated by *Hippopotamus amphibius* in a lake-margin wetland, Ngorongoro Crater, Tanzania. *Palaios*, **17**(2), pp. 212–217, 10.1669/0883-1351(2002)017<0212:SSGBHA>2.0.CO;2.

Dharani, N., 2011, *Field Guide to Common Trees and Shrubs of East Africa*. 1st ed. (Cape Town: Struik Publishers).

Erdtman, G., 1960, The acetolysis method-a revised description. *Svensk Botanisk Tidskrift*, **54**, pp. 516–564.

Fægri, K., and Iversen, J., 1989, *Textbook of Pollen Analysis*, IV Edition, (New York: John Wiley and Sons).

Finch, J., Marchant, R., and Courtney Mustaphi, C.J., 2017, Ecosystem change in the South Pare Mountain bloc, Eastern Arc Mountains of Tanzania. *The Holocene*, **27**(6), pp. 796–810, 10.1177/0959683616675937.

Foley, R., 1981, Off-Site Archaeology and Human Adaptation in Eastern Africa:Analysis of Regional Artifact Density in the Amboseli, Southern Kenya. *Cambridge Monographs in African Archaeology 3*. British Archaeological Research International Series 97, 410p.

Fosbrooke, H.A., and Sassoon, H., 1965, Archaeological remains on Kilimanjaro. *Tanganyika Notes and Records* **64**, pp. 62–64.

Gallego-Sala, A.V., Charman, D.J., Brewer, S., Page, S.E., Colin Prentice, I., Friedlingstein, P., Moreton, S., Amesbury, M.J., Beilman, D.W., Bjamp, S., Blyakharchuk, T., Bochicchio, C., Booth, R.K., Bunbury, J., Camill, P., Carless, D., Chimner, R.A., Clifford, M., Cressey, E., Courtney-Mustaphi, C., Vleeschouwer, O., Jong, R., Fialkiewicz-Koziel, B., Finkelstein, S.A., Garneau, M., Githumbi, E., Hribjlan, J., Holmquist, J., M Hughes, P.D., Jones, C., Jones, M.C., Karofeld, E., Klein, E.S., Kokfelt, U., Korhola, A., Lacourse, T., Roux, G., Lamentowicz, M., Large, D., Lavoie, M., Loisel, J., Mackay, H., MacDonald, G.M., Makila, M., Magnan, G.,

Marchant, R., Marcisz, K., Martamp, A., Cortizas, N., Massa, C., Mathijssen, P., Mauquoy, D., Mighall, T., G Mitchell, F.J., Moss, P., Nichols, J., Oksanen, P.O., Orme, L., Packalen, M.S., Robinson, S., Roland, T.P., Sanderson, N.K., Britta Sannel, A.K., Steinberg, N., Swindles, G.T., Edward Turner, T., Uglow, J., Vamp, M., Bellen, S., Linden, M., Geel, B., Wang, G., Yu, Z., Zaragoza-Castells, J., and Zhao, Y., 2018, Latitudinal limits to the predicted increase of the peatland carbon sink with warming. *Nature Climate Change*, **8**, pp. 907–914.

Gillson, L., 2004, Testing non-equilibrium theories in savannas: 1400 years of vegetation change in Tsavo National Park, Kenya. *Ecological Complexity*, **1**(4), pp. 281–298, 10.1016/j.ecocom.2004.06.001.

Gillson, L. and Marchant, R., 2014, From myopia to clarity: sharpening the focus of ecosystem management through the lens of paleoecology. *Trends in Ecology & Evolution*, **29**(6), pp. 317–325, 10.1016/j.tree.2014.03.010.

Githumbi, E.N., 2017, *Holocene Environmental and Human Interactions in East Africa*. PhD Thesis. University of York, UK.

Githumbi, E.N., Courtney Mustaphi, C.J., and Marchant, R., 2014, Natural and anthropogenic causes of environmental change in the Amboseli and Mau Forest regions. *British Institute in Eastern Africa Annual Report 2013–2014*, (Nairobi and London: British Institute in Eastern Africa), pp. 11–12.

Githumbi, E.N., Kariuki, R., Shoemaker, A., Courtney-Mustaphi, C.J., Chuhilla, M., Richer, S., Lane, P., and Marchant, R., 2018a, Pollen, people and place: Multidisciplinary perspectives on ecosystem change at Amboseli, Kenya. *Frontiers in Earth Science*, **5**, article: 113, 10.3389/feart.2017.00113.

Githumbi, E.N., Courtney Mustaphi, C.J., Yun, K.J., Muiruri, V., Rucina, S.M., and Marchant, R., 2018b, Late Holocene wetland transgression and 500 years of vegetation and fire variability in the semi-arid Amboseli landscape, southern Kenya. *Ambio*, **47**(6), pp. 682–696, 10.1007/s13280-018-1014-2.

Githumbi, E.N., Marchant, R., and Olago, D., 2020, Using the Past to Inform a Sustainable Future: Palaeoecological Insights from East Africa. In *African and the Sustainable Development Goals*. (Springer International Publishing), pp. 187–195.

Grimm, E.C., 1987, CONISS: a FORTRAN 77 program for stratigraphically constrained cluster analysis by the method of incremental sum of squares. *Computers & Geosciences*, **13**(1), pp. 13–35, 10.1016/0098-3004(87)90022-7.

Grimm, E.C., Maher, L.J., and Nelson, D.M., 2009, The magnitude of error in conventional bulk-sediment radiocarbon dates from central North America. *Quaternary Research*, **72**(2), pp. 301–308, 10.1016/j.yqres.2009.05.006.

Grossmann, M., 2008. The Kilimanjaro aquifer. In: *Conceptualizing Cooperation on Africa's Transboundary Groundwater Resources*, Volume 32, edited by Herrfahrdt-Pähle, E. and Scheumann, W. (German Development Institute), pp. 91–124.

Håkansson, N.T., 2004, The human ecology of world systems in East Africa: The impact of the ivory trade. *Human Ecology*, **32**(5), pp. 561–591, 10.1007/s10745-004-6097-7.

Håkansson, N.T., 2012, Ivory – Socio-ecological consequences of the East Africa ivory trade. In *Ecology and Power: Struggles over Land and Material Resources in the Past, Present, and Future*, edited by Hornborg, A., Clark, B. and Hermele, K., (London and New York: Routledge), pp. 124–142.

Hamilton, A.C., 1976, Identification of East African Urticales Pollen. *Pollen et Spores*, **18**(1), pp. 27–66.

Hamilton, A.C., 1982, *Environmental History of East Africa: A study of the Quaternary*. (London: Academic Press).

Hamilton, A.C., and Perrott, R.A., 1980, Modern pollen deposition on a tropical African mountain. *Pollen et Spores*, **22**(3–4), pp. 437–468.

Heckmann, M., 2014, Farmers, smelters and caravans: Two thousand years of land use and soil erosion in North Pare, NE Tanzania. *Catena*, **113**, pp. 187–201, 10.1016/j.catena.2013.07.010.

Heiri, O., Lotter, A.F., and Lemcke, G., 2001, Loss on ignition as a method for estimating organic and carbonate content in sediments: Reproducibility and comparability of results. *Journal of Palaeolimnology*, **25**, pp. 101–110, 10.1023/A:1008119611481.

Hodgson, D.L., 2000, Taking Stock: State Control, Ethnic Identity and Pastoralist Development in Tanganyika. *The Journal of African History*, **41**(1), pp. 55–78, 10.1017/S0021853799007574.

Hogg, A.G., Hua, Q., Blackwell, P.G., Niu, M., Buck, C.E., Guilderson, T.P., Heaton, T.J., Palmer, J.G., Reimer, P.J., Reimer, R.W., Turney, C.S.M., and Zimmerman, S.R.H., 2013, SHCal13 Southern Hemisphere Calibration, 0–50,000 Years cal BP. *Radiocarbon*, **55**(4), pp. 1889–1903, 10.2458/azu_js_rc.55.16783.

Hulme, M., 1996, Climate change within the period of meteorological records. *In*: *The physical geography of Africa*, edited by Adams, W.M., Goudie, A.S. and Orme, A.R., (New York: Oxford University Press), pp. 88–102.

Iles, L., Stump, D., Heckmann, M., Lang, C., and Lane, P.J., 2018, Iron Production in North Pare, Tanzania: Archaeometallurgical and Geoarchaeological Perspectives on Landscape Change. *African Archaeological Review*, **35**(4), pp. 507–530, 10.1007/s10437-018-9312-4.

Jolly, I.D., McEwan, K.L., and Holland, K.L., 2008, A review of groundwater–surface water interactions in arid/semi-arid wetlands and the consequences of salinity for wetland ecology. *Ecohydrology*, **1**(1), 43–58, 10.1002/eco.6.

Jowsey, P.C., 1966, An improved peat sampler. *New Phytologist*, **65**(2), pp. 245–248, 10.1111/j.1469-8137.1966.tb06356.x.

Juggins, S., 2003, *C2 user guide: Software for ecological and palaeoecological data analysis and visualization.* (Newcastle upon Tyne: University of Newcastle), pp. 1–73.

Kenya Wildlife Service, 2009, Amboseli Ecosystem Management Plan, 2008–2018. (Kenya Wildlife Service Biodiversity Planning, Assessment & Compliance Department).

Koff, T., and Vandel, E., 2008, Spatial distribution of macrofossil assemblages in surface sediments of two small lakes in Estonia. *Estonian Journal of Ecology*, **57**, pp. 5–20, 10.3176/eco.2008.1.01.

Lane, P.J., 2004, The 'moving frontier' and the transition to food production in Kenya. *Azania*, **39**, pp. 243–264, 10.1080/00672700409480402.

Liutkus, C.M., and Ashley, G.M., 2003, Facies model of a semiarid freshwater wetland, Olduvai Gorge, Tanzania. *Journal of Sedimentary Research*, **73**(5), pp. 691–705, 10.1306/021303730691.

Makindi, S.M., 2016, Local communities, biodiversity conservation and ecotourism: A case study of the Kimana Community Wildlife Sanctuary, Kenya. *African Journal of Hospitality, Tourism and Leisure*, **5**(3), pp. 1–15.

Marchant, R., Mumbi, C., Behera, S., and Yamagata, T., 2007, The Indian Ocean dipole? The unsung driver of climatic variability in East Africa. *African Journal of Ecology*, **45**(1), pp. 4–16, 10.1111/j.1365-2028.2006.00707.x.

McCarthy, T.S., Ellery, W.N., and Bloem, A., 1998, Some observations on the geomorphological impact of hippopotamus (Hippopotamus amphibius L.) in the Okavango Delta, Botswana. *African Journal of Ecology*, **36**(1), pp. 44–56, 10.1046/j.1365-2028.1998.89-89089.x.

Meguro, T., and Inoue, M., 2011, Conservation Goals Betrayed by the Uses of Wildlife Benefits in Community-based Conservation: The Case of Kimana Sanctuary in Southern Kenya. *Human Dimensions of Wildlife*, **16**(1), pp. 30–44, 10.1080/10871209.2011.531516.

Meijerink, A.M.J., and Wijngaarden, W., 1997, Contribution to the groundwater hydrology of the Amboseli ecosystem, Kenya. In *Groundwater/Surface Water Ecotones: Biological and Hydrological Interactions and Management Options*, edited by Gibert, J., Mathieu, J. and Fournier, F., (Cambridge University Press), pp. 111–118.

Ministry of Environment and Mineral Resources, 2012, *Kenya Wetlands Atlas*. (Nairobi: Government of Kenya).

Nelson, D.M., Verschuren, D., Urban, M.A., and Hu, F.S., 2012, Long-term variability and rainfall control of savanna fire regimes in equatorial East Africa. *Global Change Biology*, **18**(10), pp. 3160–3170, 10.1111/j.1365-2486.2012.02766.x.

Öberg, H., Andersen, T.J., Westerberg, L.O., Risberg, J., and Holmgren, K., 2012, A diatom record of recent environmental change in Lake Duluti, northern Tanzania. *Journal of Paleolimnology*, **48**(2), pp. 401–416, 10.1007/s10933-012-9615-4.

Öberg, H., Norström, E., Malmström Ryner, M., Holmgren, K., Westerberg, L.-O., Risberg, J., Eddudóttir, S.D., Andersen, T.J., and Muzuka, A., 2013, Environmental variability in northern Tanzania from AD 1000 to 1800, as inferred from diatoms and pollen in Lake Duluti. *Palaeogeography, Palaeoclimatology, Palaeoecology*, **374**, pp. 230–241, 10.1016/j.palaeo.2013.01.021.

Odner, K., 1971, A Preliminary Report on an Archaeological Survey on the Slopes of Kilimanjaro. *Azania: Archaeological Research in Africa*, **6**(1), pp. 131–149, 10.1080/00672707109511549.

Okello, M.M., 2005, Land use changes and human–wildlife conflicts in the Amboseli Area, Kenya. *Human Dimensions of Wildlife*, **10**(1), pp. 19–28, 10.1080/10871200590904851.

Okello, M.M. and Kioko, J.M., 2011, A Field Study in the Status and Threats of Cultivation in Kimana and Ilchalai Swamps in Amboseli Dispersal Area, Kenya. *Natural Resources*, **2**(4), pp. 197–211, 10.4236/nr.2011.24026.

Osipova, L., Okello, M.M., Njumbi, S.J., Ngene, S., Western, D., Hayward, M.W., and Balkenhol, N., 2018, Fencing solves human-wildlife conflict locally but shifts problems elsewhere: A case study using functional connectivity modelling of the African elephant. *Journal of Applied Ecology*, **55**(6), pp. 2673–2684, 10.1111/1365-2664.13246.

Owen, R.B., Renaut, R.W., Hover, V.C., Ashley, G.M., and Muasya, A.M., 2004, Swamps, springs and diatoms: Wetlands of the semi-arid Bogoria-Baringo Rift, Kenya. *Hydrobiologia*, **518**(1–3), pp. 59–78, 10.1023/B:HYDR.0000025057.62967.2c.

Reimer, P.J., Bard, E., Bayliss, A., Beck, J.W., Blackwell, P.G., Ramsey, C.B., Buck, C.E., Cheng, H., Edwards, R.L., Friedrich, M., 2013, IntCal13 and Marine13 radiocarbon age calibration curves 0–50,000 years cal BP. *Radiocarbon*, **55**(4), pp. 1869–1887, 10.2458/azu_js_rc.55.16947.

Rinta, P., van Hardenbroek, M., Jones, R.I., Kankaala, P., Rey, F., Szidat, S., Wooller, M.J., and Heiri, O., 2016, Land use affects carbon sources to the pelagic food web in a small boreal lake. *PLOS ONE*, **11**(8), article: e0159900, 10.1371/journal.pone.0159900.

Rucina, S.M., Muiruri, V.M., Downton, L., and Marchant, R., 2010, Late-Holocene savanna dynamics in the Amboseli Basin, Kenya. *The Holocene*, **20**(5), pp. 667–677, 10.1177/0959683609358910.

Schüler, L., and Hemp, A., 2016, Atlas of pollen, spores, and their parent taxa of Mt Kilimanjaro and tropical East Africa. *Quaternary International*, **425**, pp. 301–386, 10.1016/j.quaint.2016.07.038.

Schüler, L., Hemp, A., and Behling, H., 2014, Relationship between vegetation and modern pollen-rain along an elevational gradient on Kilimanjaro, Tanzania. *The Holocene*, **24**(6), 702–713, 10.1177/0959683614526939.

Shoemaker, A., 2018, Pastoral pasts in the Amboseli landscape: An archaeological exploration of the Amboseli ecosystem from the later Holocene to the colonial period. Department of Archaeology and Ancient History, PhD thesis. (Sweden: Uppsala University).

Tsakiridou, M., Cunningham, L., and Hardiman, M., 2021, Toward a standardized procedure for charcoal analysis. *Quaternary Research*, **99**(1998), pp. 329-340, 10.1017/qua.2020.56.

Turner, S. and Plater, A., 2004, Palynological evidence for the origin and development of late Holocene wetland sediments: Mdlanzi Swamp, KwaZulu-Natal, South Africa. *South African Journal of Science*, **100**(3–4), pp. 220–229.

Urban, M.A., Nelson, D.M., Street-Perrott, F.A., Verschuren, D., and Hu, F.S., 2015, A late-Quaternary perspective on atmospheric pCO_2, climate, and fire as drivers of C_4-grass abundance. *Ecology*, **96**(3), pp. 642–653, 10.1890/14-0209.1.

van Langevelde, F., Claudius, A.., Kumar, L., Van De Koppel, J., De Ridder, N., Van Andel, J., Skidmore, A.K., Hearne, J.W., Stroosnijder, L., Bond, W.J., Prins, H.H.T., and Rietkerk, M., 2003, Effects of fire and herbivory on the stability of savanna ecosystems. *Ecology*, **84**(2), pp. 337–350, 10.1890/0012-9658(2003)084[0337:EOFAHO]2.0.CO;2.

Verschuren, D., Sinninghe Damsté, J.S., Moernaut, J., Kristen, I., Blaauw, M., Fagot, M., and Haug, G.H., 2009, Half-precessional dynamics of monsoon rainfall near the East African Equator. *Nature*, **462**(7273), pp. 637–641, 10.1038/nature08520.

Western, D., 1973, The Structure, Dynamics and Changes of the Amboseli Ecosystem. PhD thesis. (Kenya: University of Nairobi).

Western, D., and Van Praet, C., 1973, Cyclical changes in the habitat and climate of an East African ecosystem. *Nature*, **241**(5385), pp. 104–106, 10.1038/241104a0.

Western, D., Tyrrell, P., Brehony, P., Russell, S., Western, G., and Kamanga, J., 2020, Conservation from the inside out: Winning space and a place for wildlife in working landscapes. *People and Nature*, **2**(2), pp. 279–291, 10.1002/pan3.10077.

Williams, L.A.J., 1972, *Geology of the Amboseli Area*. (Nairobi: Ministry of Natural Resources).

Wishitemi, B.E.L., Momanyi, S.O., Ombati, B.G., and Okello, M.M., 2015, The link between poverty, environment and ecotourism development in areas adjacent to Maasai Mara and Amboseli protected areas, Kenya. *Tourism Management Perspectives*, **16**, pp. 306–317, 10.1016/j.tmp.2015.07.003.

CHAPTER 9

The new Garba Guracha palynological sequence: Revision and data expansion

Graciela Gil-Romera[1], Mekbib Fekadu[2] & Lars Opgenoorth

Department of Ecology, Philipps-Marburg University, Marburg, Germany

David Grady & Henry F. Lamb[3]

Department of Geography and Earth Sciences, Aberystwyth University, Aberystwyth, United Kingdom

Lucas Bittner & Michael Zech

Heisenberg Chair of Physical Geography, Technische Universität Dresden, Dresden, Germany

Georg Miehe

Department of Geography, Philipps-Marburg University, Marburg, Germany

9.1 SITE DETAILS

In the context of the DFG-funded project 'Mountain Exile Hypothesis' (FOR2358) we retrieved a 15.5 m long sediment core from Garba Guracha (GG) (Bale Mountains National Park, Ethiopia – BMNP) during February 2017. A 16,000 year record of 62 pollen samples has been previously produced from GG by Umer *et al.* (2007). In this data paper, we present a new palynological and macrocharcoal record, with 259 pollen and 1301 charcoal samples covering the last 14 ka BP. Previous versions of the charcoal and *Erica* pollen records can be found in Gil-Romera *et al.* (2019). The depth-age model for the presented record in this new manuscript can be found in Bittner *et al.* 2020.

GG is a small (500 × 300 m), 6 m deep lake (6.9°N, 39.9°E), lying in a cirque at 3950 m asl in the BMNP of the Bale-Arsi Massif, Ethiopia (Figure 1). GG sits in a north-facing valley of the Sanetti plateau that was locally glaciated during the Last Glacial Maximum (Ossendorf *et al.* 2019; Tiercelin *et al.* 2008). High seasonality characterizes the rainfall pattern with a dry season from November to February and a bimodal wet season from March to October. Local rainfall patterns in BMNP present an altitudinal (lowlands to highlands) and latitudinal (N to S) increasing rainfall gradient, defining the vegetation zonation (Miehe and Miehe 1994). GG lies between the Ericaceous and afroalpine vegetation belts, however only dispersed *Erica* spp.

---

[1] Other affiliation: *Department of Geoenvironmental Processes and Global Change, Pyrenean Institute of Ecology-CSIC, Zaragoza, Spain*
[2] Other affiliation: *Department of Biology, Addis Ababa University, Addis Ababa, Ethiopia*
[3] Other affiliation: *Department of Botany, Trinity College Dublin, Dublin, Ireland*

DOI 10.1201/9781003162766-9

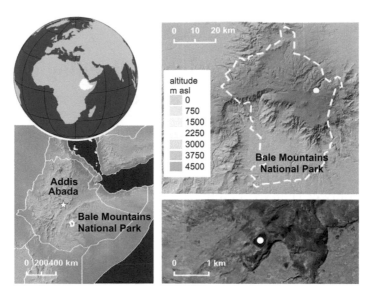

Figure 1. Map of the study area (Bale Mountains National Park) situated east of the Ethiopian Rift Valley in Ethiopia, East Africa (made by Dr. Miguel Sevilla-Callejo). Layout of satellite image is from TerraMetrics© for Google Maps (2019).

stands are present on the south east slopes of the catchment. The dominant plant communities around the lake consist of *Alchemilla haumannii* and *Helichrysum splendidum-H. citrispinum* shrublands. The northern lowland area has been heavily transformed by human activities and the natural vegetation remaining are dry afromontane forests dominated by *Podocarpus falcatus* and *Olea europaea* to 2700 m asl, followed by *Myrica salicifolia* and *Dodonaea viscosa* communities that give way at the upper forest limit to *Hagenia abyssinica-Juniperus procera* and *Schefflera* spp. communities up to 3200 m asl. *Erica trimera* and *E. arborea*, together with *Hypericum revolutum* form the Ericaceous belt found from 3200 m to 4000 m asl. The southern slopes receive rainfall throughout the year presenting a mixed broadleaved Afromontane rainforest dominated by *Pouteria aidssima*, *Ficus mucuso* but without *Juniperus* (Friis *et al.* 2010).

9.2 SEDIMENT DESCRIPTION AND METHODS

We retrieved two overlapping cores at 4.8 m water depth with a Livingstone corer from a raft. We obtained sediments to a maximum sediment depth of 1550 cm. Details of core retrieval, sedimentology and lithofacies can be found in Bittner *et al.* (2020). Charcoal particles and pollen data presented here cover from 13,700 to 94 cal. yr BP, as the bottom most sections were formed by highly inorganic sediments yielding counts too low to be statistically significant. Charcoal was sampled at contiguous intervals (1300 samples, Gil-Romera *et al.* 2019) and we analysed 259 pollen samples using a modified version of the laboratory protocol of Moore *et al.* (1994), adding *Lycopodium* spores (Lund University, batch # 1031, 15,636 spores per tablet, SD 592) to estimate concentration and pollen accumulation rates (PAR). A zonation was applied in order to facilitate the pollen data set interpretation. These pollen assemblage zones were done using a CONISS (Grimm 1987) cluster analysis. We calculated the Principal Curve (PrC) for our pollen assemblage to summarize the palynological compositional turnover along one ordination

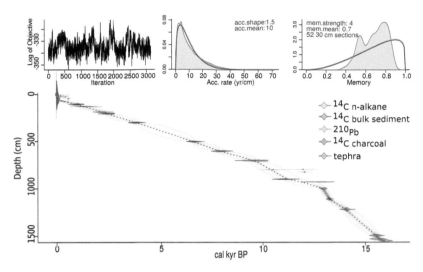

Figure 2. Age-depth model for Garba Guracha (GG) presenting the prior parameters into the Bayesian framework on the upper panels Modified from Bittner *et al.* (2020).

axis (De'ath 1999). Charcoal and pollen plots and numerical analyses were done using several R packages (R Core Team 2012). Full code, data and further details can be downloaded from Gil-Romera's GitHub repository shorturl.at/rxMW1 and from Neotoma database.

9.3 DATING

The 15.9 ka BP GGU17-1B age-depth model was built combining 48 dated samples using the Bayesian framework of the Bacon package in R (Blaauw and Christen 2011; R Core Team, 2017) (Figure 2). Dates used in this model can be found in supporting online material, and chronological methods are described in Bittner *et al.* (2020).

9.4 INTERPRETATION

We identified 98 taxa and counted an average of 456 pollen grains (SD = 120). We define six pollen zones (GGU1-GGU6) representing the last 13.7 ka BP of vegetation dynamics in GG area (Figure 3).

9.4.1 GGU1 Afroalpine vegetation (1170–1031 cm, 13.7–13 ka BP, 30 pollen samples)

The landscape was dominated by afroalpine vegetation reflecting cool, dry conditions (Figure 3). PAR values indicate a clear expansion of afroalpine and dry forest communities, with a minor development of the Ericaceous belt. At *c.* 13.1 ka BP we identify a large pulse of biomass burning that may be related to local fire activity. However, we cannot discard the possibility that charcoal particles accumulated on a frozen lake surface (Tiercelin *et al.* 2008) prior to the beginning of our record, which were deposited on the lake bottom after the ice melted.

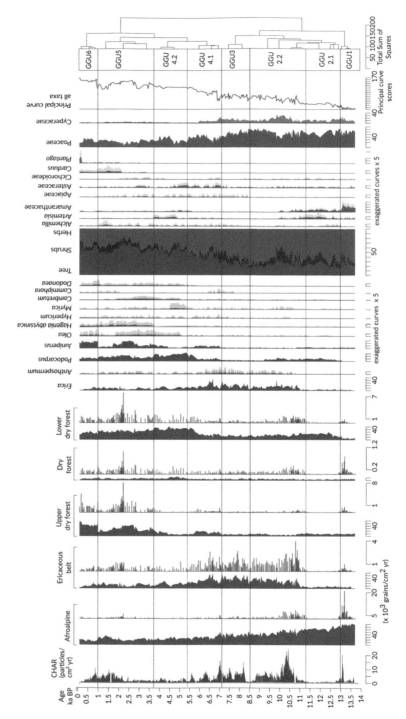

Figure 3. Pollen diagram for lake Garba Guracha at 3950 m asl. Analyst: Graciela Gil-Romera. Vegetation belts are presented in percentages (filled silhouette) and PAR (in bars). Taxa included in each group can be found in the dictionary stored at shorturl.at/rxMW1.

9.4.2 GGU2 Afroalpine-Ericaceous belt turnover (1030–637 cm,13–8.5 ka BP, 80 pollen samples)

9.4.2.1 GGU 2.1 Amaranthacea-Poaceae dominating (1030–900 cm, 13–11.2 ka BP, 26 pollen samples)
Afroalpine herbs (e.g. Amaranthaceae) abruptly decrease at 12.5 ka BP coinciding with Younger Dryas-like conditions (YD). YD conditions have been described in Eastern Africa before (Talbot *et al.* 2007) broadly concurring in time (12.9–11.7 ka BP) with the YD in the Greenland ice cores (GS1, Rasmussen *et al.* 2006). While the YD may not have had a great impact on most lowland African records, afroalpine communities clearly declined, especially the PAR of these communities, except *Artemisia* which does not decrease in abundance. The lower forest seems to expand despite PAR suggesting a poor development. These vegetation dynamics may be pointing to a short and abrupt cold, dry pulse at these altitudes.

9.4.2.2 GGU 2.2 Expansion of Erica (899–637 cm, 11.2–8.5 ka BP, 54 pollen samples)
There is a notable expansion of most vegetation communities, as indicated by all PAR values and increased PrC scores. This expansion is especially important in the Ericaceous belt, potentially responding to increasing temperatures and moisture availability linked to the African Humid Period onset (AHP) (deMenocal *et al.* 2000). Under those conditions *Erica* thickets may have fostered biomass burning during the dry season, activating a fire feedback well known in the local modern-day fire ecology (Johansson and Granström 2014). Such a mechanism would explain the Ericaceous PAR variability – that does not always correspond to that of abundance – where expansion events are followed by burning that leads to new resprouting thickets and therefore new biomass to be burnt. In addition, dry forests keep a relatively constant and low presence despite an initial PAR increase, that together with the *Erica* dominance, may suggests the establishment of a wetter and warmer phase during this period. The relative expansion of hydrophytes such as Cyperaceae would also indicate higher water availability. From 10 to 8 ka BP Poaceae expansion coincides with reduced CHAR and *Erica* PAR indicating a progressive opening of the landscape.

9.4.3 GGU3 Full extension of the Ericacoeus belt (636–549 cm, 8.5–7.2 ka BP, 30 pollen samples)

Erica spreads as indicated by its PAR, with a consequential biomass burning and decline in the herbal community while the dry forest boundaries keep similar PAR and abundances. We infer relatively similar climate conditions at this stage, with no clear community transformations as supported by unchanging PrC scores.

9.4.4 GGU4 Ericaceous belt-dry montane forest turnover (548–291 cm 7.2–3.6 ka BP, 52 pollen samples)

9.4.4.1 GGU4.1 Progressive Erica decline (547–418 cm, 7.2–5.4 ka BP, 33 pollen samples)
The Ericaceous belt slightly contracts at 7.2 ka BP as the Upper dry forest boundary expands, concurring with increasing biomass burning. The Lower dry forest expands at the end of this zone, while the Ericaceous belt progressively contracts as illustrated by increased PrC scores. The GG landscape may have opened up further as *Erica* cover declines from 6 ka BP onwards, enabling a higher deposition of lowland dry forest pollen taxa, indicating progressively drying conditions.

9.4.4.2 GGU4.2: Lower dry Afromontane forest expansion (417–291 cm, 5.4–3.6 ka BP, 19 pollen samples)
The Ericaceous belt and the local biomass burning continue declining while the Lower dry forest boundary, dominated by *Podocarpus* and *Olea*, expands with varying pollen concentrations,

concurring with the spread of *Artemisia* with equivalent values to those of zone GGU2a. This landscape succession suggests drying conditions concurring with the regional AHP termination (de Menocal *et al.* 2000).

9.4.5 GGU5 Upper dry forest expansion, (290–109 cm, 3.6–1 ka BP, 50 pollen samples)

This zone features the constant presence of the dry forest communities, with *Juniperus* dominating until *c.* 2 ka BP. After this *Juniperus* declines in conjunction with new biomass burning and the scarce, consistent presence of ruderals (e.g. *Plantago*) or other herbs in the Afroalpine domain. We hypothesize that this landscape change is likely due to a dry forest upwards shift under similar climate conditions to GGU4b. Increasing human pressure from *c.* 2 ka BP may have triggered more frequent or intense fires to obtain new pastures at the Upper dry forest-Ericaceous Belt limit. The increase in PAR *c.* 2 ka BP in all forest components might be linked to brief climate changes, as increasing temperature or short drought spells (Marshall *et al.* 2009).

9.4.6 GGU6 Upper dry forests and ruderal plants presence (108–60 cm, 1000–94 yr BP, 17 pollen samples)

The last millennium at GG presents a locally open landscape with abundant herbs, especially those indicating human agency (e.g. *Plantago)*, with scarce *Erica* presence and an upwards shift of the Upper dry forest. While we cannot discount the possibility of persistent drying conditions and relative temperature changes, we understand landscape changes in this zone to be mainly human-driven.

ACKNOWLEDGEMENTS

This study was funded by the DFG research unit FOR 2358 'Mountain Exile Hypothesis'.

REFERENCES

Bittner, L., Bliedtner, M., Grady, D., Gil-Romera, G., Martin-Jones, C., Lemma, B., Mekonnen, B., Lamb, H.F., Yang, H., Glaser, B., Szidat, S., Salazar, G., Rose, N.L., Opgenoorth, L., Miehe, G., Zech, W., Zech, M., 2020, Revisiting afro-alpine Lake Garba Guracha in the Bale Mountains of Ethiopia: rationale, chronology, geochemistry, and paleoenvironmental implications. *Journal of Paleolimnology*, **64**, pp. 293–314, 10.1007/s10933-020-00138-w.

De'ath, G., 1999, Principal Curves: A New Technique for Indirect and Direct Gradient Analysis. *Ecology* **80**, pp. 2237–2253, 10.1890/0012-9658(1999)080[2237:PCANTF]2.0.CO;2.

de Menocal, P.B., Ortiz, J., Guilderson, T., Adkins, J., Sarnthein, M., Baker, L., Yarusinski, M., 2000. Abrupt onset and termination of the African Humid Period: Rapid climate response to gradual insolation forcing. *Quaternary Science Reviews*, **19**, pp. 347–361, 10.1016/S0277-3791(99)00081-5.

Friis, I., Demissew, S., van Breugel, P., 2010, *Atlas of the Potential Vegetation of Ethiopia*. (Copenhaguen: The Royal Danish Academy of Sciences and Letters).

Gil-Romera, G., Adolf, C., Benito, B.M., Bittner, L., Johansson, M.U., Grady, D.A., Lamb, H.F., Lemma, B., Fekadu, M., Glaser, B., Mekonnen, B., Sevilla-Callejo, M., Zech, M., Zech, W., Miehe, G., 2019, Long-term fire resilience of the Ericaceous Belt, Bale Mountains, Ethiopia. *Biology Letters*, **15**, 20190367, 10.1098/rsbl.2019.0357.

Grimm, E.C., 1987, CONISS: A FORTRAN 77 program for stratigraphically constrained cluster analysis by the method of incremental sum of squares. *Computers & Geosciences*, **13**, pp. 13–35, 10.1016/0098-3004(87)90022-7.

Johansson, M.U., Granström, A., 2014, Fuel, fire and cattle in African highlands: traditional management maintains a mosaic heathland landscape. *Journal of Applied Ecology*, **51**, pp. 1396–1405, 10.1111/1365-2664.12291.

Marshall, M.H., Lamb, H.F., Davies, S.J., Leng, M.J., Kubsa, Z., Umer, M., Bryant, C., 2009. Climatic change in northern Ethiopia during the past 17,000 years: A diatom and stable isotope record from Lake Ashenge. *Palaeogeography, Palaeoclimatolology, Palaeoecology*, **279**, pp. 114–127, 10.1016/j.palaeo.2009.05.003.

Miehe, S., Miehe, G., 1994, Ericaceous forests and heathlands in the Bale Mountains of South Ethiopia. Ecology and man's impact. (Hamburg: Warnke).

Ossendorf, G., Groos, A.R., Bromm, T., Tekelemariam, M.G., Glaser, B., Lesur, J., Schmidt, J., Akçar, N., Bekele, T., Beldados, A., Demissew, S., Kahsay, T.H., Nash, B.P., Nauss, T., Negash, A., Nemomissa, S., Veit, H., Vogelsang, R., Woldu, Z., Zech, W., Opgenoorth, L., Miehe, G., 2019, Middle Stone Age foragers resided in high elevations of the glaciated Bale Mountains, Ethiopia. *Science*, **365**, pp. 583–587, 10.1126/science.aaw8942.

Rasmussen, S.O., Andersen, K.K., Svensson, A.M., Steffensen, J.P., Vinther, B.M., Clausen, H.B., Siggaard-Andersen, M.-L., Johnsen, S.J., Larsen, L.B., Dahl-Jensen, D., Bigler, M., Röthlisberger, R., Fischer, H., Goto-Azuma, K., Hansson, M.E., Ruth, U., 2006, A new Greenland ice core chronology for the last glacial termination. *Journal of Geophysical Research*, **111**, D06102, 10.1029/2005JD006079.

R Core Team, 2012, *R: A Language and Environment for Statistical Computing*, (Vienna: R Foundation for Statistical Computing).

Talbot, M.R., Filippi, M.L., Jensen, N.B., Tiercelin, J.-J., 2007, An abrupt change in the African monsoon at the end of the Younger Dryas. *Geochemistry, Geophysics, Geosystems*, **8**, Q03005, 10.1029/2006GC001465.

Tiercelin, J.J., Gibert, E., Umer, M., Bonnefille, R., Disnar, J.R., Lézine, A.M., Hureau-Mazaudier, D., Travi, Y., Keravis, D., Lamb, H.F., 2008, High-resolution sedimentary record of the last deglaciation from a high-altitude lake in Ethiopia. *Quaternary Science Reviews*. **27**, pp. 449–467, 10.1016/j.quascirev.2007.11.002.

Umer, M., Lamb, H.F., Bonnefille, R., Lézine, A.-M., Tiercelin, J.-J., Gibert, E., Cazet, J.-P., Watrin, J., 2007, Late Pleistocene and Holocene vegetation history of the Bale Mountains, Ethiopia. *Quaternary Science Reviews*, **26**, pp. 2229–2246, 10.1016/j.quascirev.2007.05.004.

CHAPTER 10

Lower to Mid-Pliocene pollen data from East African hominid sites, a review

Raymonde Bonnefille

Honorary Director of Research, CNRS, CEREGE, Aix Marseille University, CNRS, IRD, INRAE, Coll. France, Technopole Arbois-Méditerranée, Aix en Provence, France

Benjamin Bourel

CEREGE, Aix Marseille University, CNRS, IRD, INRAE, Coll. France, Technopole Arbois-Méditerranée, Aix en Provence, France

ABSTRACT: This paper presents fossil pollen data (39 samples) available for the Lower Pliocene outcrops at five hominid sites in East Africa located within the intertropical region between 3° South to 11° North. They are dated from 4.2 to 2.95 millions years (Ma). Most of these data were obtained many years ago, except one additional pollen assemblage recently obtained in the Woranso-Mille area of the Lower Awash valley, Ethiopia. In East Africa, Plio-Pleistocene sedimentary sequences are internationally renowned for the continuous palaeo-anthropological researches pursued for decades, following the discoveries of preserved remains of fossil hominins associated to a rich, diversified and abundant mammalian fauna. As a result of these new discoveries, geological studies have continued, being connected to following field expeditions and exploration of new sites in the Ethiopian Rift. We present here the revised geological context, adding chronological precision to previous age constraints of the original pollen data. For each terrestrial Pliocene site, we also include some comments regarding the significance of pollen counts and briefly discuss their interpretation for reconstructing past vegetation, environment, and inferred climatic parameters at the studied hominin sites.

10.1 INTRODUCTION

Pollen data provide information dealing with plant biodiversity and composition of past vegetation. Together with other palaeoecological proxies, these results concern regional environmental conditions, habitats, and food resources. All items are abundantly discussed to explain hominin species niches and evolution. Pollen studies from outcrops at hominin sites – that will be presented hereafter, and from deep core sediments off eastern Africa (e.g. Feakins *et al.* 2013) stand as critical contribution to the knowledge of hominin palaeoenvironments. Pollen data presented here mostly concern (with one exception) Pliocene outcrops from northern Ethiopia and Tanzania investigated a few decades ago.

Since the first publications of the original pollen data (Bonnefille 2010; Bonnefille and Riollet 1987; Bonnefille *et al.* 1987), new dated tephra (or tuffs) have been identified in the same outcrops (Table 1) and progress in the ^{40}Ar/^{39}Ar dating method have been made, notably for

Table 1. List of dated tuffs from the Pliocene Hadar Formation (Ethiopia) used to calculate interpolated age of Hadar pollen samples (Campisano 2007).

Tuff and basalts	Abbreviation	Age (Ma)	Original references
Bouroukie Tuff 2U	BKT-2U	2.96±0.006	Campisano 2007
Bouroukie Tuff 2L	BKT-2L	2.96±0.02	Semaw *et al.* 1997
Kada Hadar Tuff	KHT	3.20±0.01	Walter, 1994
Triple Tuff 4	TT-4	3.256±	Campisano 2007
Kada Damum Basalt	KMB	3.30±0.04	Renne *et al.* 1993
Kada Me'e Tuff	KMT	3.36±0.002	Campisano 2007
Sidi Hakoma Tuff	SHT	3.42±0.03	Walter and Aronson 1993

sites located in the Lower Awash Valley. Allowing previous stratigraphic placement of the pollen samples initially positioned versus tephra layers used as marker beds, it is now possible to revise ages of the pollen assemblages and place them in a more accurate chronological framework.

In this paper, we present the updated pollen data as supplementary data available for further research and interpretation (Supporting Online Material [SOM] Tables S1 to S4). We discuss them according to geographical location and chronological order. However, while addressing past vegetation changes through time it is necessary to consider the present and past altitudinal location at each site. Indeed, under the same global climatic conditions, past vegetation can differ between sites at different geographic locations. It is well known that relief modifies local climatic conditions. Altitudinal distribution of the vegetation inside zones (or belts) influences diversity in habitat patterns (Friis *et al.* 2010). A key feature of tectonic activity is the disruption of the river drainage. Geomorphological processes maintain habitats with variable topography through time (Reynolds *et al.* 2011). However, establishing past elevation at each fossil site for different time periods during the Lower Pliocene remains difficult. In this paper it is accepted that Ethiopian rifting and architecture of the Omo–Turkana basin was already attained since the eruption of the Gombe Group Basalt dated at *c.* 4.2 Ma (Erbello and Kidane 2018).

10.2 PRELIMINARY REMARKS FOR FURTHER USE OF POLLEN DATA

Pollen analysis of Plio-Pleistocene deposits from East Africa started in the 1970's with the first discoveries and field exploration in the search of hominin fossil remains documenting our ancestral origin. These initial pollen studies were strongly linked to the pioneer original stratigraphic, geological descriptions and mapping of the sedimentary terrestrial outcrops that yielded these remains. The universal and standardized framework, adopted at that time for collecting faunal remains and pollen samples at Laetoli, and at Hadar (Figure 1), was shared by all scientists involved, and still stands valid. Therefore, each pollen sample was given a label starting with the number corresponding to the fossil fauna locality, including hominin localities such as Afar Locality 333 (AL 333). Pollen sampling done along a section follows with underscore letters (a, b, c, etc.) indicating the stratigraphic succession.

Progress in absolute dating techniques provided further precision to date the fossil hominins remains, the abundant associated fauna and, therefore, pollen data. The initial stratigraphic position of pollen samples can now be integrated into the new geological studies recently carried out at each of the different sites (Campisano 2007; Campisano and Feibel 2008a, 2008b; Deino 2011; Harrison 2011). At the time of collecting pollen samples (except for sample MRD-210), GPS was not available, therefore approximate coordinates for the Hadar samples had to be

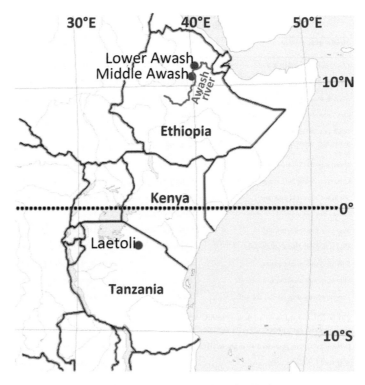

Figure 1. Location map of the Pliocene sites in Tanzania and Ethiopia (Middle and Lower Awash).

calculated by importing the initial map previously established on aerial photographs (Bonnefille *et al.* 1987) into a Geographic Information System (QGIS v3.2.2) (Table 2).

In the early days of pioneer African palynology, fossil pollen studies at Olduvai, Laetoli and Hadar were also associated to preliminary investigation of modern pollen soil samples from various vegetation types and plant collecting in vegetation surrounding the Pliocene sites. Interpretation of pollen data in term of past vegetation changes through time should also consider additional information of the setting in which these data were obtained. Indeed, different environmental contexts such as the occurrences of small or big lakes, existing delta or meandering river and floodplains, locally modify the pattern of the regional vegetation reflected by the composition and percentages of pollen taxa in the assemblages (spectra).

In this paper, sedimentary conditions provided by detailed and new geological studies are briefly summarized while comparing the chronological succession of the pollen data between the different terrestrial sites. Pliocene East African pollen sites are located at various elevations, even in the African Rift. Researchers who will address the vegetation dynamic through time, such as comparing with isotopic studies, have to remember that composition of modern and past vegetation vary according to local climatic patterns, relief and occurrence of water bodies. Moreover, East Africa is an area of tectonic changes and active volcanism. But palaeo-topography and palaeo-geography are far from being established at sites older than 3 Ma.

Pollen data included in this paper inform on terrestrial past vegetation for snapshot windows through the Pliocene. Considering their accurate dating they can be considered as valuable benchmark in the expectation of continuous pollen record from terrestrial cores. Although thoroughly investigated, Pliocene outcrops of tropical East Africa, hardly preserved fossil pollen.

Table 2. List of Pliocene pollen samples from the Lower Awash, Ethiopia. Nb 0 (Bonnefille *et al.* 1987; Saylor *et al.* 2019), N (Bonnefille *et al.* 2004).

Nb 0	N	Stratigraphic position	Locality	Geographical coordinates	Collector	Submember	Age (Ma)
55Nj	(13)	Clay 2 m below tuff BKT-2	Hadar, Kada Hadar	11° 08'31"N; 40°34'49"E	RB, AV	KH-2	2.95
116f	(12)	Clay 10 m above tuff complex TT-4	Hadar, Kada Hadar	11° 08'25"N; 40°35'34"E	RB 1973	DD-1/DD-2	3.22
116e	(11)	Silt 2 m above tuff complex TT-4	Hadar, Kada Hadar	11° 08'25"N; 40°35'34"E	RB 1973	DD-1	3.23
333d	(10)	Clay Ostracod layer in tuff complex TT-4	Hadar, Baruteita	11° 07'39"N; 40°34'38"E	RB, AV	SH-4/DD-1	3.256±0.018
333c	(9)	Black clay, 50 cm below ostracod layer	Hadar, Baruteita	11° 07'39"N; 40°34'38"E	RB, AV	SH-4	3.24
333b	(8)	Sandy clay below tuff complex TT-4 and Pink Marl	Hadar, Baruteita	11° 07'39"N; 40°34'38"E	RB, AV	SH-4	3.25
134b	(7)	Green clay 20 cm below Tuff complex TT-4	Hadar	11° 07'45"N; 40°35'53"E	RB 1974	SH-4	3.25
266b	(6)	Green clay below tuff complex TT-4	Hadar, Kada Hadar	11° 07'40"N; 40°35'44"E	RB 1974	SH-4	3.25
228b	(5)	Silt lower contact of pink marl	Hadar, Sidi Hakoma	11° 06'19"N; 40°34'35"E	RB 1974	SH-3/SH-4	3.28
228a	(4)	Brown clay gastropod, 11 m above SH-3s	Hadar, Sidi Hakoma	11° 06'19"N; 40°34'35"E	RB 1974	SH-3	3.30
280 Nj	(3)	Silty clay 3 m above top SH-3s	Hadar Kada Hadar	11° 07'15"N; 40°37'53"E	RB 1974	SH-3	3.30
KMm	(2)	Clay with plant debris, gasteropods	Hurda	11° 06'36"N; 40°39'03"E	RB, AV	SH-2	3.35
KMl2	(2)	Silty sand with plant remains	Hurda	11° 06'36"N; 40°39'03"E	RB, AV	SH-2	3.35
KMl1	(2)	Silty sand with plant remains	Hurda	11° 06'36"N; 40°39'03"E	RB, AV	SH-2	3.36
KMk	(2)	Laminated clay with intercalated tuff	Hurda	11° 06'36"N; 40°39'03"E	RB, AV	SH-2	3.36
KMj	(2)	Laminated clay	Hurda	11° 06'36"N; 40°39'03"E	RB, AV	SH-2	3.37
KMi	(2)	Grey clay with gastropods	Hurda	11° 06'36"N; 40°39'03"E	RB, AV	SH-2	3.37
KMh	(2)	Ostracods clay with gastropods 10 m above SHT	Hurda	11° 06'36"N; 40°39'03"E	RB, AV	SH-2	3.37
KMg	(2)	Silty sand 2 m above lignite layer	Hurda	11° 06'36"N; 40°39'03"E	RB, AV	SH-2	3.37
KMf	(2)	Clay gastropods and ostracods above lignite layer	Hurda	11° 06'36"N; 40°39'03"E	RB, AV	SH-2	3.37
KMe	(2)	Lignite thick layer within KMT	Hurda	11° 06'36"N; 40°39'03"E	RB, AV	SH-2	3.36±0.02
KMd2b	(2)	Clay with gastropods, fishes below lignite layer	Hurda	11° 06'36"N 40°39'03"E	RB, AV	SH-2	3.37
KMd2a	(2)	Clay with gastropods, fishes below lignite layer	Hadar, Hurda	11° 06'36"N; 40°39'03"E	RB, AV	SH-2	3.37
KMc	(2)	Silty clay 2.5m above sample b	Hurda	11° 06'36"N; 40°39'03"E	RB, AV	SH-2	3.37
KMb	(2)	Brown silt 1 m above sample a	Hurda	11° 06'36"N; 40°39'03"E	RB, AV	SH-2	3.38
KMa	(2)	Black clay 26 m above SHT	Hurda	11° 06'36"N; 40°39'03"E	RB, AV	SH-2	3.38
398b	(1)	Green clay interbedded between two SHT layers	Dikika Oudaleita	11° 06'12"N; 40°36'39"E	RB 1974	BM	3.42
MRD-210		Deltaic sands, below Mille Tuff	Worenso MRD-VP-1	11°32'55"N; 40°28'01"E	Levin 2018	BM	3.80
Meshellu		"Lignite/coal" 20 cm thick above green clay	Dikika Meshellu River	11°05'00"N; 40°40'37"E	RB, JK 1973	BM	4.00

Among the whole set of sediment samples collected from the different geological strata at the different sites, the degree of success in pollen extraction as result of preservation was <10% of the total processed samples, despite preferentially selecting lacustrine deposits (Bonnefille *et al.* 1987). It is hoped that further investigation and pollen studies carried out on sediment obtained from deep terrestrial drilling projects, such as the Hominin Sites and Palaeolakes Drilling Project (Cohen *et al.* 2016) can provide additional results. Following this line, two cores (NAW14-1A, NAO14-1B) (Figure 2) collected from the assumed position of the Hadar palaeo-lake indicate that the core sediments cover the period between 2.91 Ma and 3.33 Ma (Campisano 2012; Campisano *et al.* 2017; Garello 2019). These sediments have provided nine additional pollen assemblages corresponding to Pliocene strata (Bourel 2020).

During the early days of development of tropical palynology, fossil pollen identifications were based upon many botanical papers dealing with plant systematics. Descriptions of pollen

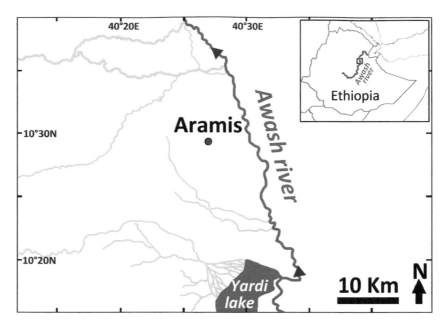

Figure 2. Location map of the lower Pliocene pollen samples from Aramis (dated 4.4 Ma) in the Middle Awash valley.

from modern plants of tropical Africa illustrated in several pollen atlases (Bonnefille 1971a, 1971b; Bonnefille and Riollet 1980) were the basis for fossil pollen identifications. It might also be noted that all pollen identifications and counting of the data presented herein were carried out in the same laboratory though its location and name have changed over the years (pre-1979: Laboratoire de Géologie du Quaternaire (LGQ) CNRS Bellevue-Meudon; 1979–1994 LGQ, Marseille-Luminy University; post-1994: CEREGE, Aix-en-Provence). Researchers and technical staff used the same criteria and nomenclature for their pollen diagnoses and compared fossil pollen to analogues included in the same modern pollen reference collection (n=7000 species) built by the first author, and now stored at CEREGE. This fact points to the consistency and taxonomical homogeneity of the East African Pliocene pollen data provided in this paper. Considering that our pollen studies were initiated before isotopic methods had been developed, the reference collection is focused on tree species more than on herbaceous plants. The homogenous pollen morphology of all genera and species among gramineae does not allow distinction between C_3 and C_4 grasses.

From the phytogeographical point of view, the African Plio-Pleistocene sites discussed here are located within the intertropical region. This zone belongs to the phytogeographical domain 'Somalia-Masai *Acacia–Commiphora* steppe', mostly defined upon species endemism and homogenous ecological climatic conditions. It is named after the abundance of more than hundred trees species for each of these two very common genera (White 1983). This region is now also called and mapped as 'deciduous bushland and thicket' (van Breugel *et al.* 2015). The landscape of the 'Somalia-Masai *Acacia–Commiphora* deciduous bushland and thicket' does not appear homogenous since a great diversity of local ecosystems exists. The density of forest cover varies near rivers, lakes, springs, high water tables, and on mountains. On a broader regional scale, the density of tree and shrub cover within this landscape depends on the monsoon rainfall distribution at the different sites. Moreover, the Pliocene sites investigated in this paper are located between 500 and 1500 m asl. Therefore, temperature, amount of rainfall, and patterns of rainfall distribution vary accordingly to the topographic context of the Rift.

Proximity of mountains and volcanoes allows various types of evergreen forests to be established on their slopes. Within the large Somalia–Masai region, the vegetation belts on different mountains have distinct floristic characteristics (such as different types of forests) and constitute an independent phytogeographical domain called 'Afromontane', with specific plant taxa (White 1983). Pliocene fossil pollen assemblages from fluviatile sediment collected in the lowlands may include pollen signal of different ecosystems mixed together. While reconstructing past vegetation around a studied site, spatial distribution of different ecosystems may be postulated using knowledge of the palaeogeographical and geological contexts, both at the local and regional scales. Such an approach relies on thorough studies of modern pollen deposition (taphonomy), and appropriate assumptions based upon geological studies.

10.3 LOWER PLIOCENE DATA (4.4 TO 3.8 MA)

Palaeobotanical data from this time period come from Ethiopia (four sites) and Tanzania (one site).

10.3.1 Middle Awash Valley, Aramis locality, 565 m asl, 10°30′N, 40°30′E, 4.3 Ma, Ethiopia

Extensive Pliocene floodplain deposits occur in the Middle Awash Valley of the Ethiopian Rift. At Aramis (Figure 2), field prospections provided fossil remains of *Ardipithecus ramidus,* an interesting hominin associated with an abundant fauna dated at 4.4 Ma (White *et al.* 2009). Pollen samples collected directly in the sediment that contained *Ardipithecus ramidus* did not provide consistent pollen results. We found a few grains of sedges (Cyperaceae), grasses and palm tree attributed to *Hyphaene/Borassus*. All these plants were later documented by abundant phytoliths found in the same dated sediment (Barboni *et al.* 2019; WoldeGabriel *et al.* 2009). Fossil wood studies identified trees such as *Ficus* (Jolly-Saad *et al.* 2010), *Syzygium, Cordia* and *Rothmania* (Jolly-Saad and Bonnefille 2012), together with *Celtis*, a tree identified by the abundant calcified stones of its fruits (WoldeGabriel *et al.* 2009). All these trees provide edible fruits. In agreement with the fossil fauna and preserved geological evidences (tufa) for groundwater discharge areas (springs), the palaeobotanical record is interpreted as a spring forest or palm woodland that allowed a forest habitat for this hominin adapted to climbing within a grass-dominated landscape (Barboni *et al.* 2019).

10.3.2 Lower Awash valley, Meshellu/Dikika, 500 m asl, 11°05′N, 40°41′E, 4 – 3.8 Ma, Ethiopia

A pollen sample collected in 1973 by R. Bonnefille and J. Kalb from lacustrine organic sediment exposed along the Meschellu drainage, east of the Dikika area (Figure 3) has provided a pollen spectrum that documents a past vegetation that is fairly different from that of Aramis site (Bonnefille 2010, Table 1) (SOM Table S1). According to its stratigraphic position below the Sidi Hakoma Tuff (SHT), the pollen sample belongs to the Basal Member of the Hadar Formation. It is likely placed between the Ikini Tuff dated to 3.8 Ma (Wynn *et al.* 2006) and the Afar Stratoid Series Basalt dated to 4 Ma (Wynn *et al.* 2006). At Meshellu/Dikika, *c.* 4 Ma ago, a large extensive swamp with abundant aquatic *Typha* and sedges occurred among an herbaceous vegetation including herbs such as *Isoglossa*, Apiaceae, Brassicaceae, Caryophyllaceae, Gentianaceae, Polygonaceae, mostly abundant in current highland swamps or grasslands, such as observed by the first author in the Bale Mountains. All tree pollen taxa (*Alchornea, Macaranga, Hagenia,* Rubiaceae, *Hymenodictyon*, etc.) belong to highland humid forests in which the conifer *Podocarpus* is absent. This pollen assemblage is devoid of pollen taxa from dry evergreen bushland or

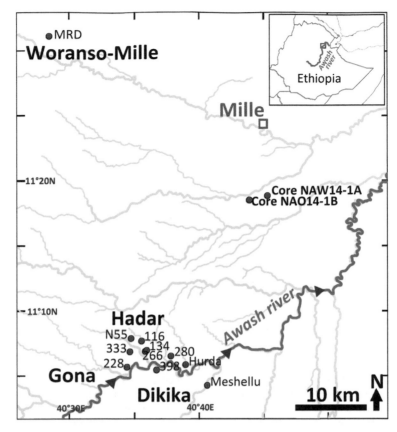

Figure 3. Location map of the Pliocene pollen samples from Woranso/ Mille (3.8 Ma), and Hadar site (3.4 to 2.9 Ma), in the Lower Awash valley.

sub-desert steppe which prevail in present dry climatic conditions of the Ethiopian Rift. Humid and cool local conditions explain the occurrence of an extended swamp among a diversified tropical forest with great affinities with high elevation humid forests and grasslands of the Ethiopian plateau today.

10.3.3 Lower Awash Valley, Woranso-Mille, 660 m asl, 11°33′N, 40°28′E, 3.8 Ma, Ethiopia

The MRD-210 pollen spectrum was recently extracted from a sandy horizon directly providing remains of *Australopithecus anamensis* dated 3.8 Ma (Figure 3) (Saylor *et al.* 2019) (SOM Table S2). Within the geological context of deltaic deposits, abundant pollen of the creeping herb *Tribulus terrestris* and Nyctaginaceae with the forb *Aerva* occupied the alternatively inundated floodplain such as found today in the Okavango Delta. Pollen of *Acacia* (a botanical genus now separated into *Vachellia* and *Senegalia*) is exceptionally abundant. The low pollen representation of this taxon in samples of modern vegetation, and its low dispersal ability (Hamilton 1982) indicate that dense *Acacia* woodland occurred locally. Including *Euclea* and *Combretum*, such woodland seems to have affinities with modern deciduous woodland encountered today on escarpments at *c.* 1300–1500 m asl. *Boscia* and Capparaceae together with *Rhus, Acalypha* and *Trilepisium* could also be found in riparian forest. Other tree pollen, such as *Juniperus,*

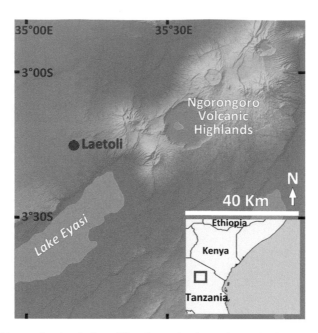

Figure 4. Topographic map showing the Laetoli location on the slopes of several volcanos around the Ngorongoro caldera, East African Rift (Tanzania).

Podocarpus, Olea and *Hagenia* attest to the occurrence of a drier conifer mountain forest, similar in composition to that present on Eastern Ethiopian highlands. Habitat scores based on plant functional types (PFTs) method are highest for steppe (25%) and woodland (34%) whereas riparian dry open *Acacia* woodland is documented, locally, in the palaeovegetation (Saylor *et al.* 2019).

10.3.4 Laetoli site, 1800 m asl, 3°13′S, 35°13′E, 3.83 – 3.8 Ma, Tanzania

In the southern part of the Eastern Rift Valley, the Laetoli site is located at much higher elevation (1800 m asl) than other Middle and Lower Pliocene deposits from the Awash Valley in Ethiopia (Figure 1). Deposits consist of air-fall tuffs provided by the repeated activity of Ngorongoro volcanic highlands with no evidence of permanent water system other than possibly localized springs, and river drainage, south of our sampling area.

Pollen preservation was very poor in most terrestrial samples (n=40) collected in the volcanic fossil-rich deposits of the Upper Laetolil Beds yielding remains of *Australopithecus afarensis,* including tracks of its fossilized footprints (Leakey *et al.* 1987). The Upper Laetolil Beds revealed less productive than the other sites, a conclusion confirmed by further investigation (Rossouw and Scott 2011).

Fossilized termite mounds were so abundant that we decided to investigate their infill in order to obtain some information on past vegetation (Bonnefille and Riollet 1987). It is assumed that termites feeding upon local surrounding vegetation, the ventilation shaft must have been open to the air during the interval in which the termitary was occupied. The hives were sealed shortly after the demise of the colony. Atmospheric pollen trapped in the hives was preserved under anaerobic conditions and the termite structure buried and fossilized by the volcanic ash. Indeed, macroscopic observation during field collecting showed organic matter, vegetal fragment tissues and pollen preserved in the hive chamber infill. Under such taphonomic conditions, possible bias of these pollen assemblages can be expected. In order to test such hypothesis, modern soil samples were collected in different ecosystems of the regional vegetation along a 30km altitudinal transect, together with plant species (appendix A in Leakey *et al.* 1987). Modern pollen assemblages from short and medium grasslands between 1500 to 1800 m asl are characterized by the dominance of pollen from herbaceous plants, pollen of grasses being less abundant than pollen of others herbs or forbs of the Asteraceae, Acanthaceae, Amaranthaceae, Fabaceae and Cyperaceae. Overgrazing and burning by pastoralists may have contributed to increasing forbs and ligneous plants typical of the current short or medium grasslands of Laetoli/Olduvai region. In modern pollen surface samples collected from the wooded grassland or woodland surrounding the Laetoli area today at 1700–1800 m elevation, tree pollen including *Acacia*, *Euclea* and other Sudano-Zambezian taxa account for 20% of the total counts (Bonnefille and Riollet 1987). A homogenous signal of *c.* 2% pollen from forests, now covering the slopes of the 20 to 30 km distant volcanoes correspond to aeolian pollen transport. In the fossil pollen data, the distribution of the various taxa appears consistent with that in modern soil samples, an argument bringing to the fore that past vegetation at the site can be reconstructed from fossil pollen extracted from termite mounts infill.

Altogether, pollen data at Laetoli consist of ten pollen assemblages sampling only the Upper Laetolil Beds between 3.83 to 3.68 Ma. The oldest nine pollen samples collected in a 4 metres stratigraphic interval around Tuff 1 (Hay 1987) span an interval of *c.* 6 ka before 3.83 Ma, on the basis of sedimentation rate (Deino 2011). In between, there is a time span of 150 ka with no pollen information. The youngest pollen assemblage was obtained from a hive chamber collected 3 m below the footprint tuff (Tuff 7) dated 3.66 Ma (Deino 2011). It has an attributed age of 3.68 Ma (youngest sample ML-81-7).

The Laetoli pollen data (SOM Table S3) total 4354 identified and counted pollen grains. We note an important number of pollen taxa from trees and shrubs (26), most of them being regular components of evergreen bushland and mountain forests. Some pollen taxa *Boscia* and *Celtis* being also identified as seeds and fruit stones, macro-botanical remains recovered from the top part of the Laetolil Beds at 3.66 Ma (Bamford 2011). It is likely that deciduous woodland occupied the peneplain between Eyasi and Olduvai divides from 3.8 to 3.6 Ma ago. Percentages of pollen from Afromontane forests are three to five times greater than today indicating that such forests were more extensive or closer to the site during the Mid-Pliocene. The Laetoli herbaceous palynoflora is also remarkable by a great number of pollen taxa (*c.* 54) that belong to tropical families such as Acanthaceae, Amaranthaceae, Euphorbiaceae, Zygophyllaceae, with many Asteraceae, Fabaceae and Solanaceae, etc. The fossil pollen composition points to rich and diversified low and medium grasslands, together with grass pollen being more abundant than in the pollen samples from modern vegetation, a conclusion in agreement with the record of several grass phytoliths (Rossouw and Scott 2011).

At Laetoli, the fossil pollen data show that several different vegetation types, highland forests, woodland and grasslands, were available to the rich fauna and bipedal hominin *Australopithecus afarensis*. The local vegetation at Laetoli was woodland or wooded grassland, whereas highland forests were located on nearby volcanoes. The nine oldest Laetoli pollen spectra appear close in age to that recently obtained at Woranso/Mille, in the Lower Awash valley of the Ethiopian Rift (Saylor *et al.* 2019) at more than a thousand kilometres distance in the northern tropics. The occurrence of *Acacia*, *Boscia, Celtis*, Capparaceae, and several others at both sites despite strongly different sedimentological contexts, indicates a great expansion of this woodland vegetation that was widespread in the southern and northern tropics. Woodland and wooded grasslands close to forests were available both for *A. anamensis* in Ethiopia and *A. afarensis* in Tanzania. At Woranso-Mille, these hominins made use of mixed grassland–wooded vegetation more or less similar in composition to woodlands and grasslands existing in the Laetoli plains, but in Ethiopia such vegetation was associated with a more densely wooded riparian forest along an important river delta.

10.4 MID-PLIOCENE (3.42-2.95 MA), ETHIOPIA

Fossil pollen from this time period are only documented at Hadar site, 500 m asl, along the lower valley of the Awash River (11°06'N, 40°35'E), in the Afar depression of the Ethiopian Rift (Figure 2). The sediments and intercalated volcanic tuffs were deposited in a low topography of a rapidly subsiding sedimentary basin (Roman *et al.* 2008). Pollen data from 27 samples (totalling over 15,000 counted grains) were obtained during the first years of palaeontological exploration of the research area (1973–1977). They were collected from sediments outcropping in the type sections of the Hadar Formation dated from 3.42 to 2.95 Ma (Figure 5). In the original publication, these results were presented according to stratigraphic position and the conventional K-Ar dates available at that time (Bonnefille *et al.* 1987). We revisit here the chronological attribution of former pollen samples using the most recent tephro-chronostratigraphy, including additional ^{40}Ar/^{39}Ar dates and newly identified and dated tephra (tuffs) (Campisano 2007; Campisano and Feibel 2008a, 2008b; Roman *et al.* 2008) (Table 2, Figure 5).

The oldest pollen sample (398b) was collected at Oudaleita, a locality on the right bank of the Awash River (Bonnefille *et al.* 1987), now included in the research area of the Dikika Project, although the outcrops still belongs to the Hadar geological Formation. It was extracted from a lacustrine diatom-rich green clay intercalated between two layers of the Sidi Hakoma Tuff (SHT) that separates the Sidi Hakoma Member from the the Basal Member. SHT is precisely dated 3.42±0.03 Ma at Hadar (Walter and Aronson 1993 revised by Campisano 2007) and correlates with two other tuffs from the Turkana basin (Brown 1982), notably with the Tulu Bor B at East Turkana also dated 3.42±0.03 Ma (McDougall *et al.* 2012). At the lowest level of the Sidi Hakoma Member of the Hadar Formation, the perfectly-dated pollen assemblage 398b shows a remarkable diversity of plants attested by 57 identified pollen taxa. This is partly explained by the high total pollen count (*c.* 1200 grains) providing highly reliable information. But such pollen assemblage fits well with deposition of fluvial deltaic channel, as shown by pollen content of a modern sample mud from the Awash River (Bonnefille 1969). Abundant grass pollen (36%) are associated with Afroalpine herbs such as *Alchemilla, Carduus, Laurembergia, Lythrum, Potamogeton* and *Hypericum,* which today in Ethiopia is largely represented by an abundant small tree (rather than herb species) at the upper limit of the the forest zone. Therefore, such pollen assemblage likely documents mountain grassland above the forest zone, while sedges and *Typha* were scarce. A total of 24% pollen from trees and shrubs indicates an important tree cover, whereas the high pollen taxa diversity attests of extended well-established woodlands. Significant percentages of *Celtis, Acacia, Dombeya, Ekebergia, Euclea, Brucea* and *Juniperus* indicate that components of the present-day evergreen bushlands and deciduous woodlands that occupy the mid-elevation vegetation zone (1500–1800 m asl) on slopes of the northern Ethiopian plateau, were particularly dense at that time. A high percentage of ferns spores (15%) in this oldest Hadar pollen sample is unusual and likely reflect high atmospheric humidity, or more humid conditions such as attested also by the occurrence of *Alangium* and *Garcinia,* these trees being components of more humid forests (Bonnefille *et al.* 1987).

The next samples (n = 14) were collected within a 19 m-thick section exposed at Hurda wadi (previously spelled Ourda) and located *c.* 26 m above the SHT Tuff. This section of dark clay deposits includes a thick coal/lignite layer (sample KMe) intercalated between sandy levels of SH-2 and SH-3 sub-members. Recent geological study found that this lignite likely correlates to the thin darker unit layer within the new Kada Me'e Tuff complex (KMT), newly identified below the SH3 sands and dated 3.36±0.02 Ma (Campisano 2007; Campisano and Feibel 2008b). This date was not available in former publications, although it is in fairly good agreement with previous estimated chronological placement of the samples (Bonnefille *et al.* 2004, Figure 3). The chronology of the numerous pollen data from the lower SH-2 sub-member of the Sidi Hakoma Member is now well constrained. It spans a time interval of *c*. 30 kyr, from 3.38 to 3.35 Ma. Interpolated ages were calculated for sample KMa to m considering that all samples are located below the SH-3 sands. However, this unit defined in the Hadar type section is not clearly expressed in the Hurda section where clay deposits have a greater thickness than in the Hadar type sections (Campisano and Feibel 2008a). Pollen particularly well-preserved in the Hurda section are stratigraphically attributed to the SH-2 sub-member of the Hadar Formation. We attribute their preservation to anoxic conditions attested by the occurrence of *Cyprideis* ostracods indicating paludal environment (Peypouquet *et al.* 1983), and the presence of natrojarosite, a blue mineral attesting of acidic environment without clastic input (Tiercelin 1986).

The pollen assemblages are divided into two pollen zones showing different taxonomic composition and pollen percentages. The lower samples (KMa to i) surrounding the 3.36 Ma lignite and KMT Tuff, are characterized by abundant pollen of grasses associated to sedges and aquatic *Typha* (23% of the total pollen), a plant that requires freshwater. *Typha* pollen are recorded under such high percentages only in modern samples from periodically inundated delta setting, such as the Omo delta (Vincens 1982). Pollen of the aquatic *Polygonum* and abundant Asteraceae (former Compositae) confirm that it was similar to a high elevation swamp. The next five samples (KMj to m) show an important increase in arboreal pollen from humid forest such as *Ilex, Myrica, Prunus* and *Hagenia*, all trees characterizing the upper limit of the forest zone, such as in humid southwestern Ethiopian highlands (Bonnefille *et al.* 1993). These trees have low dispersal ability (Hamilton 1972). A total arboreal pollen percentage of 20%, and the great number of tree taxa, many from mountain forests, a few from evergreen woodland, indicate that a diversified well-established humid forest occurred at a few kilometres distance from the Hurda section located closer to the depocenter of the Hadar lake. At that time, the landscape in this area was densely wooded with grasslands occupying the lake shores, with *Typha* and rare sedges swamp. Hominin remains have not been found at Hurda, but there are many of their fossils extracted from the Sidi Hakoma section in SH-2 and SH-3 sub-members of the Hadar area located a few kilometres west of Hurda. At Hadar, vertisol mudstones are representative of floodplain deposits of a meandering fluvial system in which remains of hominin fossils are found (Campisano and Feibel 2008a).

Following the chronological succession of the Hadar Formation, four pollen samples collected higher up in the stratigraphic section at localities 228, 280, 266 (Figure 2), in sediment yielding gastropod beds and pink marl below the Triple Tuff, are dated between *c*. 3.30 to 3.25 Ma (Table 2). Although sediments deposited during this time period correspond to beach/nearshore and lacustrine environment (Campisano and Feibel 2008a), their pollen spectra strongly differ from those of the KM section (Bonnefille *et al.* 1987). Collected at sites all less than 5km apart (Bonnefille *et al.* 1987), these pollen spectra show consistent results characterized by the predominance of grass pollen (*c*. 80%), whereas *Typha* (5%) and sedges (10%) are recorded under much lower percentages. Pollen of mountain forest (0 to 5%) disappear or occur as one or two taxa only. The herbaceous component of the pollen assemblages varies as expected in open grasslands with rare trees occupying the flat plains surrounding a large fluctuating lake. The occurrence of several herbs such as *Alchemilla, Artemisia, Plantago*, now common in the Afroalpine grassland, together with more abundant Asteraceae, associated with the dry conifer *Juniperus, Phoenix* (palm) tree, and Ericaceae shrubs indicates cooler conditions. At the top of the SH-4 sub-member included in the Mammoth event, drier, colder, and more variable climatic conditions were inferred by a statistical process used to calculate humidity coefficient from pollen data (Bonnefille *et al.* 2004). Sample 228b collected at hominin locality 228 provided the lowest value of humidity coefficient.

Above in the stratigraphy, a group of six pollen samples was extracted from clay deposits in stratigraphic proximity to the Triple Tuff TT-4 initially dated at 3.26 Ma (Walter 1994), and now considered at 3.256±0.018 Ma (Campisano 2007). Tuff TT-4 marks the limit between the Sidi Hakoma and the Denen Dora Members of the Hadar Formation. Abundance of grass pollen together with fluctuating percentages of the aquatic *Typha* and sedges, suggest vegetation typical of periodically flooded lake margins. The occurrence of the shrub *Dodonaea viscosa* observed as a pioneer on basaltic sub-modern lava flows (field observation by RB) and known as a fire-resistant species, appears in sediment contemporaneous with eruption of the basalt. In these six samples, arboreal taxa such as *Commiphora, Acacia, Hyphaene/Borassus*, and *Tamarix* are pollen markers of the arid sub-desert steppe like the one now occupying the Rift Valley. The co-occurrences of these trees indicate that warmer climatic conditions established in the region *c.* 3.25 Ma, during a period of higher global climatic variability attested by the marine isotopic record (Dolan *et al.* 2015). At Hadar, in the lower Awash valley, samples dated 3.25 Ma were collected at three distinct localities 134, 333 and 116, all found in the northern part of the Hadar area, upstream the Kada Hadar wadi (Bonnefille *et al.* 1987). A sample of green clay exposed at locality 134, below the TT-4 (top of SH-4 sub member) produced a total pollen count exceeding 2300 grains distributed among 34 distinct plant taxa. Grasses average 60% whereas tree pollen has increased with the addition of taxa commonly found in mid-elevation woodland such as *Croton, Acacia, Dodonaea, Commiphora, Hyphaene/Borassus* and *Tamarix*. All are indicators of woodland located at lower elevation, under warmer climatic conditions than mountain forest trees. Abundant pollen of herbs such as Amaranthaceae, Acanthaceae and *Tribulus* that grow on delta plains also indicate higher aridity and/or temperature or increased salinity of the lake waters and margins, a characteristic also indicated by ostracods assemblages from the same layer (Peypouquet *et al.* 1983).

The next three samples come from the hominin locality AL333, but are not contemporaneous with hominins remains since they were collected from sediment lower in the stratigraphy. A sample of sandy clay below TT-4 shows high grass percentage (66%), associated with abundant *Typha* (32%) suggesting the return of an extended marsh or a shallow freshwater fluctuating lake. Few tree pollen grains are found in the black clay that lies 50cm below the ostracod layer. In the sample from the ostracod layer corresponding to the boundary between Sidi Hakoma and Denen Dora Members, grasses (98%) dominate the pollen assemblage. We also note pollen from the aquatic herb of Onagraceae.

At locality 116, two kilometres towards the escarpment, grass pollen is still dominant (82%) and sedges increase. A subsequent percentage of olive tree pollen associated with *Combretum* and *Euclea* indicates that dry seasonal woodlands spread and came closer to the lake. The pollen sample 116e, upstream of the Kada Hadar wadi, may have consisted of a wooded grassland environment with olive trees on the northern slope towards the escarpment. In conclusion, the composition of pollen assemblages close in stratigraphy to the 3.25 Ma dated Triple Tuff is in good agreement with lacustrine and episodic sub-aerial exposures of depositional environment at the top of the Sidi Hakoma Member, and the base of the Denen Dora Member (Campisano and Feibel 2008a).

After a gap in the record of *c.* 300 ka with no pollen information for the Kada Hadar Member, the next spectrum come from the brown clay exposed just below the ash fall Bouroukié Tuff (BKT-2), at locality 55N where the first tuff sample was collected and provided the age of 2.88±0.08 Ma (Walter and Aronson 1982). In the upper KH-2 sub-member, the BKT-2 Tuff complex is now considered 2.95 Ma (2.94–2.96 Ma by Campisano 2007; DiMaggio *et al.* 2008). A rich pollen assemblage (n > 1000) extracted from this clay shows more than 30 taxa and 7.5% trees and shrubs. The arboreal taxa are exclusively found in current Afromontane forests (*Myrica, Podocarpus, Juniperus, Hagenia* and Ericaceae) and others are markers of evergreen drier woodland/bushland (*Euclea, Ekebergia, Dodonaea, Acacia* and *Rumex*). Sedges, *Typha* and grasses constitute the herbaceous cover on shores of a shallow lake. A certain degree of salinity can be inferred from the high proportion of Amaranthaceae/Chenopodiaceae herbs or shrubs and the low proportion of *Typha*.

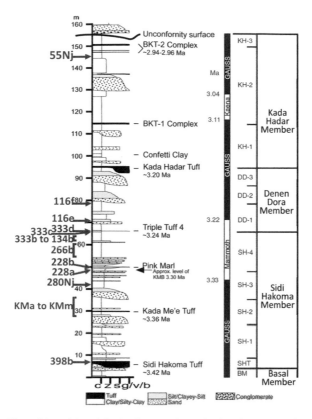

Figure 5. Stratigraphic position of the original pollen samples (red color) from Bonnefille *et al.*, (1987), in the composite section of the Hadar Formation after Campisano and Feibel (2008).

After the deposition of BKT-2 Tuff, the Hadar geological Formation ends with a well-marked unconformity (Figure 5). The Busidima Formation lies stratigraphically above it, after a gap of deposition of *c.* 0.3 Ma. Sediments of the Upper Pliocene (2.9 to 2.6 Ma) lack in the Hadar Formation. A few additional pollen spectra were extracted from archaeological horizons dated 2.35 Ma, i.e. in the Pleistocene Busidima Formation. They are not discussed in this paper restricted to the Pliocene.

10.5 DISCUSSION

10.5.1 Interpretation of pollen data

In the difficult challenge of interpreting pollen assemblages in terms of pattern of past vegetation, landscape or habitat we rely on our field experience and botanical knowledge of tropical vegetation improved by additional exploration of pollen signal from many other modern ecosystems in East Africa (Bonnefille 2011). In pollen studies from temperate vegetation where most pollen is dispersed by wind, taxa with the greatest percentages are assumed to reflect the local vegetation because the greatest amount of pollen is dispersed at proximity of the plant producing them. A similar pattern is suggested for tropical vegetation such as shown in the sub-desert vegetation of the plains surrounding lake Turkana (Bonnefille 1977). For example, pollen assemblages of modern soil samples collected in different vegetation zones along the slopes of the escarpment of the Ethiopian Rift, north of Hadar, provided pollen indicators of the distinct vegetation belts along an altitudinal gradient, the succession comprising samples from sub-desert steppe to evergreen bushland and Afromontane forest (Bonnefille *et al.* 1987). Different vegetation types illustrate a similar gradient on a broader scale in Tanzania (Vincens *et al.* 2006), whereas other samples from southwestern Ethiopia document humid forests from mountains (Bonnefille *et al.* 1993). However, in the tropical region, pollen from grasses, dispersed by wind, is produced in significantly greater amounts than pollen from other tropical plants, notably most forest trees (Hamilton 1982). As a consequence of this discrepancy, pollen percentages will not directly reflect the proportion of plants in one given vegetation type. Moreover, the sedimentological context in which the fossil pollen grains are deposited, modify the distribution pattern of the percentages of different taxa. Fluvial transport over long distances, can mix distinct pollen grains from various vegetation types, that do not necessarily reflect their surrounding ecological conditions (Vincens 1982). Composition of the pollen assemblage varies according to the deposition mode of the sediment. At Hadar, pollen assemblages such as from the mud of shallow fresh water lake or swamps are characterized by the abundance of *Typha*. Pollen assemblages such as sample 134 which is from a river channel deposit (Campisano and Feibel 2008a), includes a great diversity of pollen taxa from various vegetation types covering a larger source area, thus potentially representing the entire river basin (Bonnefille *et al.* 1987).

10.5.2 Environment of hominins through the Mid and Upper Pliocene

Through the million-year time span between 3.8 and 2.9 Ma, the 39 pollen assemblages discussed in this paper clearly indicate that geographically distinct hominin sites are associated with different types of vegetation. Grasslands, wooded grasslands, woodlands and forests (riparian or mountain), all existed, including different species composition and tree cover density. In a same region, pollen studies show that past vegetation types were also changing through time. In the lower Awash valley of Ethiopia, the oldest sample (4–3.8 Ma) from Meshellu now located at *c.* 500 m elevation, indicate proximity of high elevation mountain forest. At Worenso–Mille, the pollen assemblage associated to *Australopithecus anamensis* which fossil remains were found in the same strata dated to 3.8 Ma at the same locality, documents a mid-elevation *Acacia* woodland with abundant *Acacia, Cordia, Croton, Rhus, Tamarix* and various shrub taxa from the Cappari-daceae and also Amaranthaceae that indicate a warm climate. Closer to the northern Ethiopian escarpment, where deciduous and woodland dominated the local landscape some riparian forest and grasslands occupied the delta that also received a pollen signal of distal mountain forest. There are few highland dry conifer forests taxa such as *Olea* and *Juniperus*. A closely related mid-elevation woodland is documented at Laetoli in Tanzania, south of the equator, at about the time when the oldest *A. afarensis* appeared in the East African fossil record. Such drier past vegetation contrasts with that of more humid conditions found in the Lower Awash valley, a few hundred thousand years later when *A. afarensis* prevailed at Hadar.

It is at Hadar that well-preserved, numerous pollen data best document past vegetation. At Hadar from 3.4 to 3.2 Ma, the remarkable feature is the floristic affinities of the fossil taxa to components of the present Afromontane vegetation domain. These include taxonomic attribution of fossil pollen to living plants common in high elevation evergreen bushlands, forests and Afroalpine grasslands. These similarities clearly indicate much cooler and wetter vegetation and a past climate that contrasts sharply with the extremely arid current climate of the Afar region. During the Mid-Pliocene, atmospheric CO_2 concentration was likely similar to today (400 ppm) (Ning Tan *et al.* 2020), which may partly explain forests expansion at that time. From 3.4 Ma to 2.9 Ma, global decrease in CO_2 concentration is postulated to have enhanced polar glaciation. During the Pliocene, oceanic and atmospheric circulations were totally different from the present situation. Some modelling experiments indicate more precipitation over East Africa at 3.3 Ma (Dolan *et al.* 2015).

At 3.4 Ma, in the oldest deposits at Hadar, all the trees belong to woodlands and forests now encountered only in Ethiopian highlands (Friis *et al.* 2010). The fossil pollen data set lacks indicators of the common sub-desert trees (such as Capparidaceae, Salvadoraceae, *Ziziphus, Combretaceae*). Indeed, Mid-Pliocene vegetation at Hadar was markedly different from that of today. *Australopithecus afarensis* and *A. anamensis* were sharing a vegetated landscape that is difficult to reconstruct. Elevation and climatic conditions of the region differed markedly from the current palaeogeographical setting. In a subsiding flat spreading basin, the Hadar Lake could have been closer to the Rift escarpment and its relief. At that time, it could be postulated that the large lake was located at elevation much higher than 500 m. Today forests are established above 1800 m elevation, under a minimum amount of over 1000 mm/yr precipitation with no short dry season, and an average annual temperature of 18°C. These requirements significantly contrast with the low rainfall, high temperature and long dry season of today's climatic conditions in the Rift. Regular characteristics of the sedimentology during the time interval 3.4 to 3.2 Ma, presume of stabilized forest soil and agree with pollen indicating forests occurrences. From 3.32 to 3.22 Ma, the climatically stable and humid lacustrine phase at Hada corresponds to a global period of low variability attested by the astronomically calculated lower insolation and precession amplitude (Campisano and Feibel 2007, Figure 3). In contrast, during the deposition of the KH-2 Member (3.12 to 2.96 Ma), drier conditions were inferred from faunal studies, notably the increase in bovid tribes (wildebeests and gazelles) adapted to grazing arid lands after 3.15 Ma (Campisano and Reed 2007). The drying trend supported by faunal analysis cannot be confirmed by pollen since no data exist for this time range. DD and KH Members were not initially densely sampled for pollen studies. However, at the top of the KH-2 sub-member of the Hadar Formation, the pollen spectrum BK-2 documents open woodland indicating long dry seasons unsuitable for forest growth. Although derived from a single pollen assemblage, this interpretation is in good agreement with sedimentological and faunal interpretation of local environment and global climate patterns around 3 Ma.

10.5.3 Pollen reconstructed climatic conditions

An attempt to reconstruct climatic parameters from Hadar pollen data was performed (Bonnefille *et al.* 2004). Such method uses references of selected taxa frequencies in pollen assemblages obtained from 966 samples collected in different vegetation types at East African sites where corresponding meteorological parameters were interpolated. In 2001, R. Bonnefille collected at Hadar 13 new modern pollen assemblages from soil and river mud. Thirty years had passed since the collection of modern pollen data had been obtained and used as a calibration data set. We processed the new modern samples as if they were fossil ones and apply to them previous methodological approaches developed to calculate climatic parameters from upper Pleistocene and Holocene pollen data (Bonnefille *et al.* 1992; Jolly *et al.* 1998). The results could be compared to measured mean annual rainfall and temperature values provided by a meteorological station located nearby Hadar. Such procedure constituted a control test intended to validate our method. Indeed pollen-reconstructed values obtained for both modern surface soil and river samples collected in 2001 (these likely analogues to fossil pollen deposition) correspond closely to those measured at the nearby meteorological station (Bonnefille *et al.* 2004). Our statistical method was therefore validated. We could apply such likelihood procedure to infer mean annual precipitation and temperature for the Pliocene. Of course, the fact that climatic requirements from plant taxa during the Pliocene remained the same as today has to be an accepted assumption. For Hadar, the results indicate that from 3.42 to 3 Ma, mean annual terrestrial temperature was *c.* 10°C lower than the 25°C current value. Mean annual rainfall values average 1000 mm/yr, which is two to three times the amount registered today (Bonnefille *et al.* 2004). Moreover, a noticeable cold shift towards 15°C associated with a minor rainfall increase is registered at *c*. 3.36–3.35 Ma (samples KMk to KMm section, Figure 3, Table 2) after deposition of the anoxic black organic layer 'lignite' (KMe) dated by the KMT Tuff at 3.36±0.02 Ma (Campisano 2007), and occurring in a stable deposition of lacustrine clay. Below the lower limit of the Mammoth sub-chron (3.33 Ma), the cool and humid event registered at Hadar appears in good contemporaneity with the oceanic first isotopic cooling event (Shackleton *et al.* 1995), therefore connecting a local environmental event to a global climatic event (Campisano and Feibel 2007). Pollen results and inferred climatic parameters clearly show that global climatic variations had a strong impact upon climate and vegetation in East Africa. However, the vegetation changes documented here for a few thousand years concern changing species of trees in the forest, but the forest was maintained in the region throughout this period. At Hadar, fossil remains of *A. afarensis* are found in many localities both before and after the cooling spell (below and above the SH-3 sands). The palaeo-geographical context indicates a permanent but fluctuating large lake, surrounded by periodically flooded flats covered by swampy sedges and cattail tall herbs. Located near a forest with great diversity of tree species, such environmental conditions offered various resources for food subsistence and shelter for this hominin which maintained abundant groups for more than half a million years.

Later, at about 2.95 Ma, when arctic glaciation extended, a single pollen assemblage documents past vegetation at Hadar. At that time a strong contrast was calculated in reconstructed values between mean annual temperature (MAT) inferred by 'best analogue' statistics and mean temperature of the coldest month (MTCO) inferred by the Plant Functional Type (PFT) method (Jolly *et al.* 1998). Mean annual precipitation remained high (>1000 mm), and the humidity coefficient showed its greatest value (Bonnefille *et al.* 2004). These values indicate stronger seasonal contrast in the local terrestrial climate, in good timing correspondence with increased variability of dust content registered in marine cores of the Arabian sea shown by a noticeable peak at 2.9 Ma (deMenocal 1995). In the marine core DSDP 231 from the Gulf of Aden, plant wax biomarkers show increased proportion of C_4 plants, while pollen from the halophytes Chenopodiaceae/Amaranthaceae) are in greater percentages (Feakins *et al.* 2013) synchronously with the lowest percentages of total arboreal pollen (Bonnefille 2010). Several marine indicators point to strong changes over the whole East African region, and more contrasted climatic conditions at Hadar. However, in the Lower Awash Valley, trends towards more aridity did not reach a threshold that enabble forest growth from the Hadar region. Many species of forest and woodland trees (e.g. *Olea, Podocarpus, Hagenia, Myrica, Macaranga, Acacia, Ekebergia, Nuxia, Euclea, Juniperus, Dodonaea*) persisted in the basin, until 2.9 Ma, still offering food and shelter resources. The variety of tree species indicates a diversified forest that does not resemble riparian forests, impoverished in species such as those found nowadays in arid region along the Awash or the Omo Rivers. Riparian forests, although providing a locally wooded environment among arid grassland floodplains, are less diversified in trees and herbs species (Carr 1998) and they are geographically localized. They could not provide food and continuous long-term living conditions for early hominins populations such those found in the Lower Awash Valley for a million years. Composition of the riparian vegetation is known to change very quickly. However, when hominins seem to have disappeared from the region (at 2.9 Ma), greater seasonal temperature (and rainfall) contrasts possibly made living conditions more difficult than those experienced earlier by occupants in the Hadar/Dikika region.

During the middle Pliocene, from 3.8 to 3.2 Ma, multiple, roughly contemporaneous sites, all containing hominins, are present across East Africa, spanning regular occurrence of *Australopithecus afarensis*. At Hadar (Ethiopia), Laetoli (Tanzania), East and West Turkana (Kenya), hominins were rare members of large mammalian fossil communities. Significant discrepancies in hominin abundance versus other mammalian fossils could have been caused by local ecology rather than by taphonomic biases (Villaseñor *et al.* 2020). Indeed, at Hadar, local ecological conditions with diversified woodlands and forests, and local grasslands (strongly differing from modern savannah/steppe) provided favourable conditions for persistence of *A. afarensis* in the region. Stronger climatic changes after 2.9 Ma could have caused their extinction. But the Last Appearance Datum (LAD) of the species is difficult to determine due to a gap in the deposition of the sediments at Hadar after 2.9 Ma.

10.6 CONCLUSIONS

Pollen data from the Pliocene in East tropical Africa document many vegetation types that experienced strong changes in their taxonomic composition. Some of these changes occurred over a few thousand years. However, the resolution of the record is not sufficient to attribute them to fluctuations in precipitation following the monsoon precession cycle, although it is likely. Past vegetation for deposits *c.* 3.8 Ma is reconstructed as dense woodland or wooded grassland both in Tanzania south of the equator and Ethiopia, north of the equator. Such vegetation implies a seasonal distribution of rainfall over a broad region in East Africa. At 3.42 Ma many indicators of cool mountain forests and high elevation grasslands prevailed under humid conditions of higher rainfall, with short dry seasons that maintained forests at Hadar for more than 200 ka. A cold event documented slightly before 3.33 Ma, close to the lower limit of the Mammoth event, occurred in good timing connexion with the known marine isotope stage MIS M2 oceanic cooling and rainfall increase simulated by some model experiments (Dolan *et al.* 2015). At Hadar vegetation changes are indicated by modifications in the tree species composition of the forests. It is remarkable that such noticeable changes (both in vegetation and local climate) had no effect on *A. afarensis* local occurrences. Regarding hominins habitat, existing forests offered abundant and diverse food resources and shelters for *A. afarensis*. At Hadar, hominin fossils are found within sediments deposited before and after the cold spell. From 3.1 to 2.96 Ma, analyses of the fossil fauna, notably the abundance of different bovid tribes, are interpreted as indicating a drying trend in the palaeoenvironmental conditions. Increased body sizes of *A. afarensis* are documented synchronously. A single pollen assemblage confirms the drying trend post 2.9 Ma at a time of greater variability of the global climate following increasing arctic glaciations.

ACKNOWLEDGEMENTS

We are grateful to two anonymous reviewers for their comments on an earlier version of the manuscript. We thank Doris Barboni for stimulating and encouraging this paper, her help to provide the data in digital format and final editing.

REFERENCES

Bamford, M.K., 2011, Fossil Leaves, Fruits and Seeds, In: Harrison, T. (Ed.), Paleontology and Geology of Laetoli: Human Evolution in Context: Volume 1: Geology, Geochronology, Paleoecology and Paleoenvironment, Vertebrate Paleobiology and Paleoanthropology Series. (Springer: Netherlands, Dordrecht), pp. 235–252, 10.1007/978-90-481-9956-3_11.

Barboni, D., Ashley, G.M., Bourel, B., Arráiz, H., Mazur, J.-C., 2019, Springs, palm groves, and the record of early hominins in Africa. *Review Palaeobotany Palynology* **266**, pp. 23–41, 10.1016/j.revpalbo.2019.03.004.

Bonnefille, R., 1969, Analyse pollinique d'un sediment recent: vases actuelles de la riviere Aouache (Ethiopie). *Pollen et Spores*, **11**, pp. 7–16.

Bonnefille, R., 1971a, Atlas des pollens d'Ethiopie: Pollens actuels de la basse vallee de l'omo, recoltes botaniques 1968. *Adansonia*, **2**, pp. 463–518.

Bonnefille, R., 1971b, Atlas des pollens d'Éthiopie: Principales espéces des forêts de montagne. *Pollen et Spores*, **13**, pp. 15–72.

Bonnefille, R., 1977, Représentation pollinique d'environnements arides à l'Est du Lac Turkana (Kenya). Presented at the 9° Congrès INQUA, Bulletin de l'Association française pour l'étude du quaternaire, Birmingham, UK, pp. 235–247.

Bonnefille, R., 2010, Cenozoic vegetation, climate changes and hominid evolution in tropical Africa. *Global Planetary Change*, **72**, pp. 390–411, 10.1016/j.gloplacha.2010.01.015.

Bonnefille, R., 2011, Rainforest responses to past climatic changes in tropical Africa, in: Bush, M., Flenley, J., Gosling, W. (Eds.), *Tropical Rainforest Responses to Climatic Change*, Springer Praxis Books. Springer, Berlin, Heidelberg, pp. 125–184, 10.1007/978-3-642-05383-2_5.

Bonnefille, R., and Riollet, G., 1980, *Pollens des savanes d'Afrique orientale*. Centre national de la recherche scientifique, Paris, France.

Bonnefille, Vincens, A., and Buchet, 1987, Palynology, stratigraphy and palaeoenvironment of a pliocene hominid site (2.9-3.3 M.Y.) at Hadar, Ethiopia. *Palaeogeography Palaeoclimatology Palaeoecology*, **60**, pp. 249–281, 10.1016/0031-0182(87)90035-6.

Bonnefille, R., Riollet, G., 1987, *Palynological spectra from the Upper Laetolil Beds*, in: The Pliocene Site of Laetoli, Northern Tanzania. Leakey, M.D., Harris, J.M. (Eds.), Oxford, pp. 52–61.

Bonnefille, R., Buchet, G., Friis, I.B., Kelbessa, E., Mohammed, M.U., 1993, Modern pollen rain on an altitudinal range of forests and woodlands in South West Ethiopia. *Opera Botanica* **121**, pp. 71–84.

Bonnefille, R., Chalié, F., Guiot, J., Vincens, A., 1992, Quantitative estimates of full glacial temperatures in equatorial Africa from palynological data. *Climate Dynamics* **6**, pp. 251–257, 10.1007/BF00193538.

Bonnefille, R., Potts, R., Chalié, F., Jolly, D., Peyron, O., 2004, High-resolution vegetation and climate change associated with Pliocene Australopithecus afarensis. *Proceedings of the National Academy of Sciences*, **101**, pp. 12125–12129, 10.1073/pnas.0401709101.

Bourel, B., 2020, Pollen use for the spatial high-resolution reconstruction of Plio-Pleistocene Hominini environments in the East African Rift (Ethiopia and Tanzania). Aix-Marseille Université, Aix-en-Provence, France.

Brown, F.H., 1982, Tulu Bor tuff at Koobi Fora correlated with the Sidi Hakoma tuff at Hadar. *Nature*, **300**, pp. 631–633, 10.1038/300631a0.

Campisano, C.J., 2007, Tephrostratigraphy and hominin paleoenvironments of the Hadar Formation, Afar Depression, Ethiopia. Rutgers University – Graduate School-New Brunswick. https://doi.org/10.7282/T3NS0V99

Campisano, C.J., 2012, Geological summary of the Busidima Formation (Plio-Pleistocene) at the Hadar paleoanthropological site, Afar Depression, Ethiopia. *Journal of Human Evolution* **62**, pp. 338–352, 10.1016/j.jhevol.2011.05.002

Campisano, C.J. and Feibel, C.S., 2007, Connecting local environmental sequences to global climate patterns: evidence from the hominin-bearing Hadar Formation, Ethiopia. *Journal of Human Evolution*, African Paleoclimate and Human Evolution **53**, pp. 515–527. 10.1016/j.jhevol.2007.05.015.

Campisano, C.J., and Reed, K.E., 2007, *Spatial and temporal patterns of Austral-opithecus afarensishabitats at Hadar, Ethiopia*, in: Abstracts. Presented at the Paleoanthropology SocietyAnnual Meeting, Philadelphia, Pennsylvania, p. A6.

Campisano, C.J., Feibel, C.S., 2008a, Depositional environments and stratigraphic summary of the Pliocene Hadar Formation at Hadar, Afar Depression, Ethiopia, *Geological Society of America Special Papers*, **446**, pp. 179–2011.

Campisano, C.J., Feibel, C.S., 2008b, *Tephrostratigraphy of the Hadar and Busidima Formations at Hadar, Afar Depression, Ethiopia*, in: The Geology of Early Humans in the Horn of Africa. Geological Society of America, pp. 135–162.

Campisano, C.J., Cohen, A.S., Arrowsmith, J.R., Asrat, A., Behrensmeyer, A.K., Brown, E.T., Deino, A.L., Deocampo, D.M., Feibel, C.S., Kingston, J.D., Lamb, H.F., Lowenstein, T., Noren, A., Olago, D., Owen, R.B., Pelletier, J.D., Potts, R., Reed, K.E., Renaut, R.W., Russell, J.M., Russell, J.L., Schäbitz, F., Trauth, M.H., Wynn, J.G., 2017, The Hominin Sites and Paleolakes Drilling Project: High-resolution paleoclimate records from the East African Rift System and their implications for understanding the environmental context of hominin evolution. *Paleo Anthropology*, **2017**, pp. 1–43, 10.1130/abs/2017am-295426.

Carr, C.J., 1998, Patterns of vegetation along the Omo River in southwest Ethiopia. *Plant Ecology*, **135**, pp. 135–163. 10.1023/A:1009704427916.

Cohen, A., Campisano, C., Arrowsmith, R., Asrat, A., Behrensmeyer, A.K., Deino, A., Feibel, C., Hill, A., Johnson, R., Kingston, J., Lamb, H., Lowenstein, T., Noren, A., Olago, D., Owen, R.B., Potts, R., Reed, K., Renaut, R., Schäbitz, F., Tiercelin, J.-J., Trauth, M.H., Wynn, J., Ivory, S., Brady, K., O'Grady, R., Rodysill, J., Githiri, J., Russell, J., Foerster, V., Dommain, R., Rucina, S., Deocampo, D., Russell, J., Billingsley, A., Beck, C., Dorenbeck, G., Dullo, L., Feary, D., Garello, D., Gromig, R., Johnson, T., Junginger, A., Karanja, M., Kimburi, E., Mbuthia, A., McCartney, T., McNulty, E., Muiruri, V., Nambiro, E., Negash, E.W., Njagi, D., Wilson, J.N., Rabideaux, N., Raub, T., Sier, M.J., Smith, P., Urban, J., Warren, M., Yadeta, M., Yost, C., Zinaye, B., 2016. The Hominin Sites and Paleolakes Drilling Project: inferring the environmental context of human evolution from eastern African rift lake deposits. *Scientific Drilling* **21**, pp. 1–16, 10.5194/sd-21-1-2016.

Deino, A.L., 2011, *40 Ar/39 Ar dating of Laetoli, Tanzania*, in: Paleontology and Geology of Laetoli: Human Evolution in Context. Springer, pp. 77–97.

Deino, A.L., 2011, *40Ar/39Ar Dating of Laetoli, Tanzania*, in: Harrison, T. (Ed.), Paleontology and Geology of Laetoli: Human Evolution in Context. Volume 1: Geology, Geochronology, Paleoecology and Paleoenvironment, Vertebrate Paleobiology and Paleoanthropology. (Springer Science+Business Media B.V.: Netherlands), pp. 77–97.

deMenocal, P., 1995, Plio-Pleistocene African Climate. *Science*, **270**, pp. 53–59, 10.1126/science.270.5233.53.

DiMaggio, E.N., Campisano, C.J., Arrowsmith, J.R., Reed, K.E., Swisher, C.C., Lockwood, C.A., 2008, *Correlation and stratigraphy of the BKT-2 volcanic complex in west-central Afar, Ethiopia*, in: The Geology of Early Humans in the Horn of Africa. Geological Society of America. 10.1130/2008.2446(07).

Dolan, A.M., Haywood, A.M., Hunter, S.J., Tindall, J.C., Dowsett, H.J., Hill, D.J., Pickering, S.J., 2015. Modelling the enigmatic late pliocene glacial event—Marine isotope stage m2. *Global Planetary Change*, **128**, pp. 47–60, 10.1016/j.gloplacha.2015.02.001.

Erbello, A., Kidane, T., 2018, Timing of volcanism and initiation of rifting in the Omo-Turkana depression, southwest Ethiopia: Evidence from paleomagnetism. *Journal of African Earth Sciences* **139**, pp. 319–329, 10.1016/j.jafrearsci.2017.12.031.

Feakins, S.J., Levin, N.E., Liddy, H.M., Sieracki, A., Eglinton, T.I., Bonnefille, R., 2013, Northeast African vegetation change over 12 m.y. *Geology*, **41** G33845.1. 10.1130/G33845.1.

Friis, I., Demissew, S., van Breugel, P., 2010, *Atlas of the potential vegetation of Ethiopia*, The Royal Danish Academy of Sciences and Letters. ed. Det Kongelige Danske Videnskabernes Selskab, Copenhagen.

Garello, D.I., 2019, Tephrostratigraphy of Pliocene Drill Cores from Kenya and Ethiopia, and Pleistocene Exposures in the Ledi-Geraru Research Project Area, Ethiopia: Geological Context for the Evolution of Australopithecus and Homo (Ph.D.). Ann Arbor, United States.

Hamilton, A.C., 1972, The interpretation of pollen diagrams from highland Uganda. *Palaeoecology of Africa*, **7**, pp. 45–149.

Hamilton, A.C., 1982, *Environmental history of East Africa: A study of the Quaternary.* Academic press London.

Harrison, T., 2011, Laetoli Revisited: Renewed Paleontological and Geological Investigations at Localities on the Eyasi Plateau in Northern Tanzania, in: Harrison, T. (Ed.), Paleontology and Geology of Laetoli: Human Evolution in Context: Volume 1: Geology, Geochronology, Paleoecology and Paleoenvironment, Vertebrate Paleobiology and Paleoanthropology Series. (Springer: Netherlands, Dordrecht), pp. 1–15. 10.1007/978-90-481-9956-3_1.

Hay, R., 1987, *Geology of the Laetoli area,* in: Laetoli: A Pliocene Site in Northern Tanzania. Oxford, United Kingdom, pp. 23–47.

Jolly, D., Prentice, I.C., Bonnefille, R., Ballouche, A., Bengo, M., Brenac, P., Buchet, G., Burney, D., Cazet, J.-P., Cheddadi, R., 1998, Biome reconstruction from pollen and plant macrofossil data for Africa and the Arabian peninsula at 0 and 6000 years. *Journal of Biogeography.* **25**, pp. 1007–1027, 10.1046/j.1365-2699.1998.00238.x.

Jolly-Saad, M.-C., Bonnefille, R., 2012, Lower Pliocene Fossil Wood from the Middle Awash Valley, Ethiopia. *Palaeontographica Abt.* B 43–73, 10.1127/palb/289/2012/43.

Jolly-Saad, M.-C., Dupéron-Laudoueneix, M., Dupéron, J., Bonnefille, R., 2010, Ficoxylon sp., a fossil wood of 4.4 Ma (Middle Awash, Ethiopia). *Comptes Rendus Palevol* **9**, pp. 1–4, 10.1016/j.crpv.2009.12.001.

Leakey, M.D. Mary D., Harris, J.M., 1987, *Laetoli, A Pliocene site in northern Tanzania,* (Oxford University Press: Oxford, United Kingdom).

Lisiecki, L.E., Raymo, M.E., 2005, A Pliocene-Pleistocene stack of 57 globally distributed benthic $\delta^{18}O$ records. *Paleoceanography* **20**, 10.1029/2004PA001071.

McDougall, I., Brown, F.H., Vasconcelos, P.M., Cohen, B.E., Thiede, D.S., Buchanan, M.J., 2012, New single crystal 40Ar/39Ar ages improve time scale for deposition of the Omo Group, Omo–Turkana Basin, East Africa. *Journal of the Geological Society,* **169**, pp. 213–226, 10.1144/0016-76492010-188.

Peypouquet, J.P., Carbonel, P., Taieb, M., Tiercelin, J.J., Perinet, G., 1983, Ostracoda and evolution process of paleohydrologic environments in the Hadar Formation (the Afar Depression, Ethiopia). in Maddocks, R.F., ed., Applications of Ostracoda: Houston, Texas, University of Houston Geosciences, pp. 277–285.

Reynolds, S.C., Bailey, G.N., King, G.C.P., 2011, Landscapes and their relation to hominin habitats: Case studies from Australopithecus sites in eastern and southern Africa. *Journal of Human Evaluation* **60**, pp. 281–298, 10.1016/j.jhevol.2010.10.001.

Roman, D.C., Campisano, C.J., Quade, J., DiMaggio, E.N., Arrowsmith, J.R., Feibel, C., 2008, *Composite tephrostratigraphy of the Dikika, Gona, Hadar, and Ledi-Geraru project areas, northern Awash, Ethiopia,* in: The Geology of Early Humans in the Horn of Africa. (Geological Society of America: Boulder, Colorado), pp. 119–134.

Rossouw, L. and Scott, L., 2011, *Phytoliths and pollen, the microscopic plant remains in Pliocene volcanic sediments around Laetoli, Tanzania,* in: Harrison, T. (Ed.), Paleontology and Geology of Laetoli: Human Evolution in Context. Volume 1: Geology, Geochronology, Paleoecology and Paleoenvironment. (Springer Science+Business Media B.V.: Dordrecht).

Saylor, B.Z., Gibert, L., Deino, A., Alene, M., Levin, N.E., Melillo, S.M., Peaple, M.D., Feakins, S.J., Bourel, B., Barboni, D., 2019, Age and context of mid-Pliocene hominin cranium from Woranso-Mille, Ethiopia. *Nature,* **573**, pp. 220–224, 10.1038/s41586-019-1514-7.

Tiercelin, J.J., 1986, The Pliocene Hadar Formation, Afar depression of Ethiopia. *Geological Society of London Special Publication,* **25**, pp. 221–240, 10.1144/GSL.SP.1986.025.01.19.

van Breugel, P., Breugel, P. van, Kindt, R., Lillesø, J.-P.B., Bingham, M., Demissew, S., Dudley, C., Friis, I., Gachathi, F., Kalema, J., Mbago, F.M., Moshi, H.N., Mulumba, J., Namaganda, M., Ndangalasi, H.J., Ruffo, C.K., Védaste, M., Jamnadass, R., Graudal, L., 2015, *Potential Natural Vegetation Map of Eastern Africa (Burundi, Ethiopia, Kenya, Malawi, Rwanda, Tanzania, Uganda and Zambia).* Version 2.0. https://vegetationmap4africa.org.

Villaseñor, A., Bobe, R., Behrensmeyer, A.K., 2020, Middle Pliocene hominin distribution patterns in Eastern Africa. *Journal of Human Evolution*, **147**, 102856. 10.1016/j.jhevol.2020.102856.

Vincens, A., 1982, Palynologie, environnements actuels et plio-pléistocènes l'Est du Lac Turkana (Kenya). PhD thesis, University Aix-Marseille II, Marseille, France.

Vincens, A., Bremond, L., Brewer, S., Buchet, G., Dussouillez, P., 2006, Modern pollen-based biome reconstructions in East Africa expanded to southern Tanzania. *Review of Palaeobotany and Palynology*, **140**, pp. 187–212, 10.1016/j.revpalbo.2006.04.003.

Walter, R.C., Aronson, J.L., 1993, Age and source of the Sidi Hakoma tuff, Hadar formation, Ethiopia. *Journal of Human Evolution*, **25**, pp. 229–240, 10.1006/jhev.1993.1046.

White, F., 1983, The vegetation of Africa., Courvoisier S.A. ed. Unesco, Paris, France.

White, T.D., Asfaw, B., Beyene, Y., Haile-Selassie, Y., Lovejoy, C.O., Suwa, G., WoldeGabriel, G., 2009, *Ardipithecus ramidus* and the Paleobiology of Early Hominids. *Science*, **326**, pp. 64–64, 75–86, 10.1126/science.1175802

WoldeGabriel, G., Ambrose, S.H., Barboni, D., Bonnefille, R., Bremond, L., Currie, B., DeGusta, D., Hart, W.K., Murray, A.M., Renne, P.R., 2009, The geological, isotopic, botanical, invertebrate, and lower vertebrate surroundings of Ardipithecus ramidus. *Science*, **326**, pp. 65–65e5, 10.1126/science.1175817.

Wynn, J.G., Alemseged, Z., Bobe, R., Geraads, D., Reed, D., Roman, D.C., 2006, Geological and palaeontological context of a Pliocene juvenile hominin at Dikika, Ethiopia. *Nature*, **443**, pp. 332–336, 10.1038/nature05048

CHAPTER 11

Ecosystem change and human-environment interactions of Arabia

Sarah J. Ivory

Department of Geosciences and the Earth and Environmental Systems Institute (EESI), Penn State University, University Park, USA

Michèle Dinies[1]

Institute of Geographical Sciences, Freie Universität, Berlin, Germany

Anne-Marie Lézine

Laboratoire d'Océanographie et du Climat, Expérimentation et Approche numérique/IPSL, Sorbonne University, CNRS-IRD-MNHN, Paris, France

ABSTRACT: The Arabian Peninsula hosts some of the most extreme environments in the world. Arabian ecosystems are largely semi-arid to hyper-arid and yet are often highly biodiverse and unique owing to a location at a biogeographic nexus. There is an urgent need to better understand ecosystem resilience to biotic and abiotic disturbance; however, the modern vegetation across much of the sub-continent is still poorly understood. Palaeoecological records in Arabia have the potential to fill in gaps in our fundamental understanding of vegetation responses to climate, disturbance, and human modification; however, to do this, data must be accessible and easy to find. To this end, within the framework of the relaunch of the African Pollen Database, here we review existing palaeoecological datasets from the Arabian Peninsula, inventory those which are available, and synthesize results from these records. Due to the dearth of pre-Holocene information, this synthesis focuses on reconstructing vegetation from the Holocene Humid Period to today with emphasis on the impact of aridification and changing human livelihoods and culture on landscapes. Finally, as Arabia is perhaps the least well-studied region within the African Pollen Database, we offer some suggestions about fruitful directions for future palaeoecological research in this area.

11.1 INTRODUCTION

The Arabian Peninsula hosts some of the world's most extreme environments, including landscape gradients from montane woodland to cloud forest to active dune fields over small spatial scales, vast gravel deserts, and the world's largest sand sea, the Rub al'Khali. Its flora is complex biogeographically as well as ecologically very valuable (Kuerschner 1998). Due to its location

---

[1] Other affiliation: *German Archaeological Institute (DAI), Scientific Department of the Head Office, Berlin, Germany*

DOI 10.1201/9781003162766-11

near Asia, Europe, and Africa, the flora of the Arabian Peninsula includes a high percentage of plants with affinities to neighboring regions co-occurring in unique formations, with many plants that only grow in these harsh climates (Ghazanfar 1992).. Further, the deserts of Arabia are quite diverse, with the southern half situated within the Horn of Africa Biodiversity hotspot (Mittermeier *et al.* 2011).

The impact of climate change in Arabia is likely to put these fragile ecosystems at greater risk (Huang *et al.* 2017). Projections of climate that suggest that the "dry get drier" could further reduce moisture in a landscape already with a strong moisture deficit (Dahinden *et al.* 2017; Pachauri *et al.* 2014; Pausata *et al.* 2020). Further, the addition of increased rainfall variability could also intensify drought periods. Although many plants in the region are already adapted to arid and hyper-arid conditions, limited areas of grassland and woodland could undergo significant changes.

Further, despite climate being the main determinant of vegetation, land cover throughout the region is rapidly changing due to human activities (Alquarashi *et al.* 2016; Galletti *et al.* 2016). Rapid development following recent events, like the discovery of oil reserves in the Gulf States, has resulted in increased pressure on ecosystems as nations industrialize (Albalawi et al. 2018). Additionally, in much of Arabia, millions of people still live pastoral or agropastoral lifestyles and depend on ecosystem services for food, medicine, and fodder for animals (Ball *et al.* 2020). Freshwater availability is another important issue that can also be directly tied to vegetation. Ecosystem change associated with tree cover loss in wadis and southern Arabia is known to result in decreased surface water, declining water tables, and less spring discharge (Friesen *et al.* 2018; Hildebrandt *et al.* 2007). Thus, between the dual threats of climate and land-use change, it is important to better understand ecosystem changes in order to help develop sustainable management practices and validate models for an understudied region.

To this end, information is needed to better understand the relationship of Arabian vegetation to climate change on centennial timescales. Palaeoecological records, primarily from fossil pollen, provide key insights into the natural history and origins of the modern ecosystems in Arabia, feedbacks between climate and vegetation, ecosystem resilience, anthropogenic influences, and disturbance. In other regions, dense networks of these palynological records even allow for spatially explicit mapping of past vegetation (e.g. Dawson *et al.* 2016; Huntley 1990; Webb and McAndrews 1976; Williams and Jackson 2007). In contrast, Arabia is very data poor. This results form a few linked problems: 1) due to aridity, there is a lack of traditional palaeoenvironmental archives like lakes that have existed continuously through multiple climatic cycles, 2) access to regions that have potential for palaeoenvironmental work has become increasingly difficult due to evolving global politics, 3) existing pollen data from the region were housed within the African Pollen Database (APD), which lapsed in 2007 due to lack of funding, making accessibility of data also difficult. While these first two issues are a challenge to address, this paper, in conjunction with others in this special issue, emerge from an effort to renew and relaunch the APD (Ivory *et al.* 2020). This comes at a very critical time for using renewed, accessible data to synthesize the state of the art of palaeoecology in Arabia. Therefore, in this paper, we seek to: 1) review the state of current APD data holdings and those which have been generated since 2007, 2) synthesize these results from within the region in order to evaluate changes in vegetation since the Late Pleistocene, 3) provide recommendations for future work.

11.2 MODERN SETTING

The Arabian Peninsula (Figure 1) is situated in a complex geographical and geological setting between Africa, Asia, and the Mediterranean. Covering some 2.7 million km^2, this sub-continent spans approximately 15° latitude from the tropics to the subtropics (Kuerschner 1998). Because of active tectonism, Arabia is bordered on most sides by mountains, including the Zagros Mountains to the north, Hajar Mountains to the east, the Hijaz, Asir and Yemen Highlands to the west, and Dhofar Mountains and Hadramawt escarpments to the south bordering the Arabian Sea. In the continental interior, due to significant rain shadows from these highlands, Arabia hosts extensive arid and hyper-arid environments including three sand seas (Rub al'Khali, Nafud, Wahiba Sands; Figure 1).

The climate of Arabia involves both northern and tropical atmospheric circulation systems. Rainfall is generally low (Figure 2), with most areas receiving less than 100mm/yr. This includes the hyper-arid interior and northwest coast (<50mm/yr), a large area from the Rub al'Khali (with 50–50mmyr), and *c*. 100mm/yr along the southern coasts. Notable exceptions to this are semi-arid regions in northern and southern Omani/Yemen mountains receiving *c*. 250mm/yr, and the southwestern highlands in Yemen receiving <400mm/yr (Almazroui *et al*. 2012; Fisher and Membery 1998). In the south, the summer rainfall is related to the yearly migration of the Intertropical Convergence Zone (ITCZ), which is located just south of the Oman/Yemen coasts in the summer months in its most northerly current position. However, despite the proximity of tropical convergence, southwesterly winds in June-July-August result in upwelling of cold waters off of the southern Yemen/Oman coast which limits evaporation of moisture into air masses. In this region, cool, damp winds encounter steep, coastal escarpments and form dense seasonal fogs in summertime (Fisher and Membery 1998; Hildebrandt *et al*. 2007). Thus, instead of rainfall, a portion of annual moisture received in the coastal mountains of Yemen and Oman occurs as dew or horizontal precipitation (Hildebrandt *et al*. 2007). Further, as a result of the trapping of moisture along the coasts, there is a strong rainshadow and many inland areas receive no reliable moisture at all. In the north, beyond the limit of summer moisture, northwesterly winds occur in the winter months. These winds are associated with mid-latitude westerly circulation. Rainfall typically occurs due to cyclonic systems that develop over the Mediterranean Sea. These systems bring gentle rainfall to the Near East and may also be funneled up along the Zagros Mountains to the Persian Gulf delivering moisture as far south as the United Arab Emirates (UAE) and northern Oman.

The vegetation in Arabia owes its character to tectonics, geology, relief, biogeographic histories of different adjacent regions, climate change, and human presence and cultural evolution (Figure 1; Kuerschner 1998). Large parts of the peninsula maintained strong floral affinities with Africa before separation of the continents due to rifting around 15 myrs ago (Kuerschner 1998). Since that time, evolutionary isolation from the tropical vegetation of Africa coupled with new biogeographic connections with Europe and Asia and drying during the later Cenozoic have strongly shaped vegetation structure and community composition (Patzelt 2015). Three main phytogeographical units are found: the Saharo-Sindian/Arabian regional zone, the Somali-Masai regional center of endemism, and an Afromontane phytogeographic unit similar to the high mountains of East Africa. These occur along with intruders in the north and southeast of Mediterranean and Irano-Turanian origin.

Beyond the coasts, the modern continental interior of the Arabian Peninsula is dominated by hyper-arid to semi-arid ecosystems dominated by plants with Saharo-Arabian affinities (Figure 1). These regions have sparse vegetation with extremely low ground cover in diffuse or contracted patterns on slopes or in depressions. A few scattered trees with tropical affinities such as *Maeura crassifolia, Boscia arabica,* and *Acacia tortilis* also occur. To the east, species composition changes to reflect the proximity of Asia in the Gulf of Oman desert and semi-desert, where tree communities commonly include *Prosopis cineraria*.

In the inland gravel and rock deserts, shrublands with *Haloxylon salicornicum* cover most of North Arabia, while shrublands with *Rhanterium epapposum* are most common in the eastern part of the Arabian Peninsula. Both formations constitute highly appreciated grazing resources. Shrublands with *Calligonum comosum, Artemisia jordanica,* and numerous annuals characterizes the Great Nafud and Dahna sands. In the hyper-arid Rub'al Khali, annuals are uncommon, and

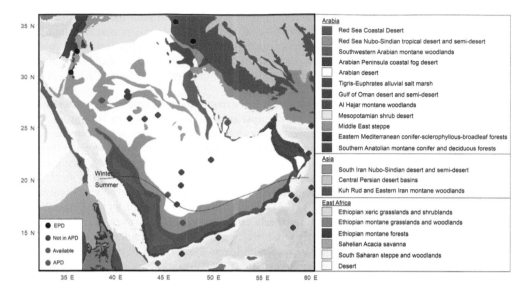

Figure 1. Map of terrestrial ecoregions of Arabia and adjacent areas (Olson *et al.* 2001). Dots indicate sites with datasets relevant to the African Pollen Database (APD), including those already in the APD data holdings (green), those available after 2007 (yellow), and those not within the APD (red). Black dots indicate datasets in the European Pollen Database (EPD). Dotted line indicates the boundary between dominant summer and winter rainfall.

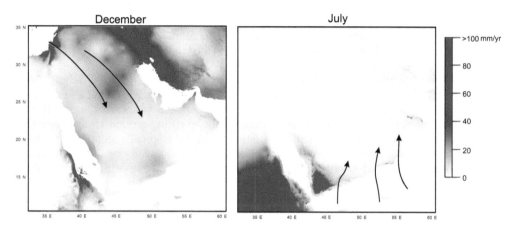

Figure 2. Map of rainfall throughout Arabia during December (left) and July (right). Average monthly precipitation at 2.5' resolution comes from Worldclim2 (Fick and Hijmans 2017). Arrows indicate direction of prevailing seasonal winds.

very open shrublands cover thousands of square kilometers, dominated by the endemic species *Cornulaca arabica* associated with *Calligonum crinitum, Cyperus conglomeratus,* and *Tribulus arabicus*.

Trees are mainly restricted to vegetation on mountain slopes, wadis, and coastal plains. Within the wadis, different *Acacia* species are accompanied by Mediterranean and Irano-Turanian taxa in the northern, northwestern, and southeastern regions, such as *Nerium oleander*. Sudanian taxa such as the toothbrush tree (*Salvadora persica*) are common in wadi communities in the southern part of Arabia.

In the northernmost mountains, Mediterranean taxa play an important role. Steppic forests with pistachio (*Pistacia khinjuk, P. atlantica*) and juniper (*Juniperus phoenicea*) trees are found in the Hijaz Mountains. In the summit zone, oak-woodlands (with *Quercus calliprinos*) replace the juniper-pistachio woodlands.

In contrast, the southern and eastern mountains (Asir mountains, Yemeni highlands, the Hadhramaut Plateau and Dhofar Mountains) show a strong Afromontane affinity. Typical representatives are olive trees (*Olea europaea* subsp. *cuspidata*) in the western and southern Arabian mountains and juniper (*Juniperus excelsa*) at highest altitudes (Kuerschner 1998). Tree heathers (*Erica arborea)* in Yemen, the southwestern Asir mountains, and the southernmost Hijaz mountains are another example of the close floristic connections between East-Africa (Baierle 1993; Deil and al Gifri 1998; Ghazanfar 1992; Gil-Lopez *et al.* 2017; Kuerschner 1998; Mandaville 1998).

Other tropical African elements are representatives of the Somali-Masai *Acacia-Commiphora* phytogeographic zone along the Oman and Yemen coasts (Ghazanfar 1992; Mittermeier *et al.* 2011; Patzelt 2015). These include dense stands of seasonal woodlands and forests dominated by *Terminalia* (formerly *Anogeissus*; including the endemics *T. dhofarica* and *T. bentii*; Ghazanfar 1992). Some of these restricted patches owe their existence to seasonal summer fogs and increased rainfall along the south facing seaward slopes where the rainfall may exceed 250 mm/yr. However, due to several factors, including human modification from overgrazing and wood harvesting, expansive coastal deserts including savanna grasslands and rocky xeromorphic scrubland replace tropical deciduous woodland and forest today even at elevation in many locations (Ball *et al.* 2020).

Mangroves are also found at stream outlets leading to coastal estuaries (kwars or khors) in some locations along the southern coasts of Oman and Yemen (Deil 1998). In contrast with the freshwater dominated mangroves common in southwestern Asia, with species like *Rhizophora mucronata*, these Arabian mangroves are dominated by halophytic trees such as *Avicennia marina*. These environments have also become relatively degraded by human land-use and are intensively grazed by animals such as camels during the dry season.

11.3 STATE OF RECORDS IN THE AFRICAN POLLEN DATABASE

While Arabian palaeoecological records are quite few, their representation as a percentage of those in existence in data repositories such as the APD is also exceptionally low in comparison to other regions of Africa (Table 1). For this review, we consider only Arabian records south of the Levant for consideration, as Levantine and Near Eastern records have been contributed commonly to the European Pollen Database due to their Mediterranean affinity. The original data holdings of the APD dating to 2007 includes only 7 datasets. Within the framework of the relaunch of the APD, work has begun to inventory all Arabian records and encourage new data contributions. As a result, Table 1 shows that, of the 40 publications including pollen datasets, 9 new datasets have been contributed to date (total 16 Arabian records in APD). Several observations can be made of the emerging datasets produced within the last 15 years, including new archives previously unrepresented in Arabia (*hyrax* middens, archeological contexts, estuary sediment cores) which have resulted in an influx of new information to help answer ecological questions.

11.4 VEGETATION HISTORY

11.4.1 Pre-Holocene

Little palaeoecological information exists about the Cenozoic flora of Arabia prior to the Late Pleistocene, and none which falls under the convention of the African Pollen Database. The little

Table 1. Inventory of all palynological records relevant to the African Pollen Database (APD).

Type	Site Name	Latitude (°N)	Longitude (°E)	Country/ Ocean	Reference
APD	Suwayh [Drill 2]	22.09	59.67	Oman	Lézine et al. (2002)
APD	Suwayh [Drill 1]	22.09	59.67	Oman	Lézine et al. (2002)
APD	Khor [F.B.]	25.50	51.42	Qatar	Bonnefille and Riollet (1988)
APD	Khor M	25.50	51.42	Qatar	Bonnefille and Riollet (1988)
APD	Site 36	25.53	51.50	Qatar	Bonnefille and Riollet (1988)
APD	Mundafan	18.53	45.38	Saudi Arabia	El-Moslimany (1983)
APD	al-Hawa	15.87	46.88	Yemen	Lézine et al. (1998)
APD	MD92-1002	12.03	44.32	Indian Ocean	Fersi et al., (2016).
available	SO90-56KA	24.83	65.92	Indian Ocean	Ivory and Lézine (2009)
available	Filim	20.60	58.19	Oman	Lézine et al. (2017)
available	Khawr Al Balid 1	17.00	54.34	Oman	Hoorn and Cremaschi (2004)
available	Khawr Al Balid 3	17.00	54.33	Oman	Hoorn and Cremaschi (2004)
available	Khawr Rawri 2	17.03	54.40	Oman	Hoorn and Cremaschi (2004)
available	Khawr Rawri 4	17.03	54.41	Oman	Hoorn and Cremaschi (2004)
available	Kwar al Jaramah	22.50	59.77	Oman	Lézine (2009)
available	Sumharam	17.04	54.43	Oman	Lippi et al. (2011)
available	Tayma	27.63	38.55	Saudi Arabia	Dinies et al. (2015)
Not in APD	MD 77 202	19.22	60.67	Indian Ocean	Van Campo (1983)
Not in APD	RC27-14	15.42	58.10	Indian Ocean	Overpeck et al. (1996)
Not in APD	RC27-28	18.50	57.98	Indian Ocean	Overpeck et al. (1996)
Not in APD	Hole 721B	16.68	59.86	Indian Ocean	Van Campo (1991)
Not in APD	Hole 723 A/B	18.05	58.44	Indian Ocean	Van Campo (1991)

Continued

Table 1. Continued

Type	Site Name	Latitude (°N)	Longitude (°E)	Country/ Ocean	Reference
Not in APD	MD 76 135	14.45	50.53	Indian Ocean	Van Campo (1983)
Not in APD	MD 76 136	12.88	46.80	Indian Ocean	Van Campo (1983)
Not in APD	Maqta oasis	22.77	58.96	Oman	Urban and Buerkert (2009)
Not in APD	an Nafud [profil 2]	25.85	41.43	Saudi Arabia	Schulz and Whitney (1986)
Not in APD	an Nafud [profil 3]	27.90	41.32	Saudi Arabia	Schulz and Whitney (1986)
Not in APD	an Nafud [profil 4]	28.02	41.15	Saudi Arabia	Schulz and Whitney (1986)
Not in APD	an Nafud [profil 5]	28.10	41.37	Saudi Arabia	Schulz and Whitney (1986)
Not in APD	an Nafud [profil 6]	28.45	41.22	Saudi Arabia	Schulz and Whitney (1986)
Not in APD	an Nafud [profil 7]	28.45	41.22	Saudi Arabia	Schulz and Whitney (1986)
Not in APD	As-Sirr	26.18	44.33	Saudi Arabia	Schulz and Whitney (1986)
Not in APD	Djebel Asmar	25.80	43.02	Saudi Arabia	Schulz and Whitney (1986)
Not in APD	Jebel Ghiran	20.66	46.73	Saudi Arabia	El-Moslimany (1983)
Not in APD	Rub'al-Khali	19.43	46.69	Saudi Arabia	El-Moslimany (1983)
Not in APD	Rub'al-Khali [AS-3-1]	21.88	49.72	Saudi Arabia	El-Moslimany (1983)
Not in APD	Rub'al-Khali [XX-B-5]	17.59	46.30	Saudi Arabia	El-Moslimany (1983)
Not in APD	Awafi	25.72	56.10	UAE	Parker *et al.* (2004)
Not in APD	Mleiha	25.13	55.86	UAE	Garcia Anton and Sainz Ollero (1999)
Not in APD	OS73	25.14	60.83	Gulf of Oman	Miller *et al.* (2016)
Not in APD	Sumharam	17.04	54.43	Oman	Bellini *et al.* (2020)

that can be inferred is based on the ancient biogeographic connection of Arabia with Eastern Africa. Prior to the inception of the East Africa Rift system, the Arabian and African plates were joined, forming a continuous stretch of land from what is now Ethiopia into southwestern Arabia (Yemen, Oman, and Saudi Arabia; Guba and Glennie 1998). This connection with Africa in the Eocene and Oligocene that sets the legacy of the modern floristic affinities with the African continent (Kuerschner 1998). By *c*. 30 Ma ago, the Arabian plate separated from Africa which likely had important consequences for the climate and flora of the region. Due to the relative biogeographic isolation of Arabia, endemic vascular plants emerged (Kuerschner 1998). During the last 10 myrs, global cooling led to increased aridity and the rise of C_4 tropical savanna and open desert ecosystems. Pollen and carbon isotopes on leaf waxes from DSDP Site 231 in the Gulf of Aden suggest a transition to aridity by 10 Ma and from C_3 woodland to C_4 savanna by *c*. 3.5 Ma (Bonnefille 2010; Feakins 2013). Aridification progressed at the beginning of the Quaternary. Decreases in sea levels related to glacial-interglacial fluctuations in the Red Sea and Persian Gulf led to decreased biogeographic barriers with Africa and Asia, facilitating plant migrations when sea levels were low (Kuerschner 1998). This likely resulted in the inclusion of many elements of Irano-Sindian elements in plant communities of western and northern Arabia.

Over the late Quaternary, marine sediment cores from the Arabian Sea and Gulf of Aden along with very sparse terrestrial records provide information about climate and vegetation. Geomorphological studies of palaeolakes and dune mobilization coupled with observation speleothem growth phases suggest repeated oscillations of wetting and drying over at least the last 400 kyrs (Figure 3; Burns *et al.* 2001; McClure 1984; Preusser *et al.* 2002; Rosenberg *et al.* 2011; Stokes and Bray 2005). This includes the development of inland lakes and aquifer recharge, such as have been observed during MIS5e in the Near East and Saudi Arabia (Issar and Bruins 1983; Petit-Maire et al. 2010; Rosenberg *et al.* 2011). $\delta^{18}O$ measured on benthic and planktic foraminifera from marine sediment cores MD77 202 and MD76 136 suggest that climate at least over the last 160 kyrs was strongly related to glacial-interglacial cycles (Van Campo *et al.* 1982). In tropical Africa, palaeoclimate and palaeoecology are also mechanistically connected to eccentricity-modulated insolation as well as teleconnections with northern latitude glaciation; however, the record in Arabia is not long enough to resolve the impact of changes in tropical insolation on climate.

Regardless, pollen analysis on marine sediment cores MD77 202 and MD76 136 highlight enhanced aridity during glacial periods such as MIS 6 and MIS2 characterized by high abundances of xerophytic herbs like Amaranthaceae (Figure 3). Despite relative wetness during interglacial periods, pollen assemblages appear to still contain many indicators that flora on the sub-continent has an important arid component. This includes high abundances of Asteraceae and *Tribulus*, but includes some more tropical elements, such as *Acacia* and higher abundances of grasses. This implies that coastal forests and mangroves may have expanded during interglacial periods (Fersi *et al.* 2016; Van Campo *et al.* 1982). In northern Arabia, drill cores from the Dead Sea provide insight into the environments from *c*. 147-89 ka. These show pronounced aridity through much of the penultimate glacial-interglacial transition with an abrupt transition to grass and woodland during the maximum interglacial from 124-115 ka (Chen and Litt 2018).

11.4.2 Holocene

Throughout the northern hemisphere tropics, the Pleistocene-Holocene transition was accompanied by an increase in summer insolation (Figure 4). This resulted in enhanced land-sea temperature contrasts which intensified northern tropical monsoon systems and is often referred to as the Holocene Humid Period (Mayewski *et al.* 2004). In North Africa, palaeoclimate and paleoecological research has focused on evidence of extensive closed-basin lake systems like Lake Mega-Chad and the resultant northward migration of savanna and tropical woodland ecosystems to form the "Green Sahara" (Hély and Lézine 2014; Hoelzmann *et al.* 1998; Watrin *et al.*

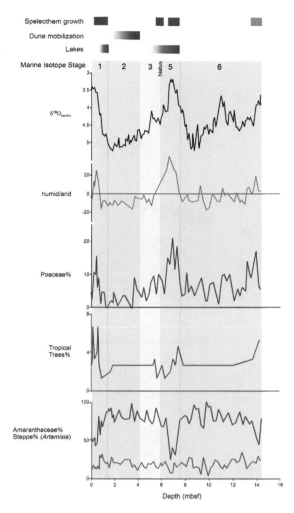

Figure 3. Late Quaternary climate and vegetation of Arabia, including (from top to bottom) indication of periods of speleothem growth (Burns *et al.* 2001), dune mobilization (Stokes and Bray 2005), closed basin lake deposits in Rub al-Khali (McClure 1984), geochemical and pollen data from marine sediment core MD77 202 including $\delta^{18}O_{benthic}$, humid-arid index, and pollen abundances (Van Campo 1983).

2009). In Arabia, however, although changes in atmospheric circulation and teleconnections with the high latitudes resulted in increased rainfall during this time, the pattern is quite complex and varies regionally, particularly the response of the vegetation to enhanced rainfall.

11.4.2.1 Palaeohydrology
Palaeoclimatic studies of cave and lacustrine deposits demonstrate changes hydrology consistent with altered atmospheric circulation involving both a more northerly ITCZ and intensified Mediterranean moisture source during the early and middle Holocene from at least 9-6 ka across much of the Arabian Peninsula (Dinies *et al.* 2015; Enzel *et al.* 2015; Fleitmann *et al.* 2007, 2011; Lézine *et al.* 1998, 2007, 2009, 2014; Parker *et al.* 2004, 2006; Schulz and Whitney 1987). In southern Arabia, $\delta^{18}O$ from speleothems collected across a latitudinal gradient from northern Oman (Hoti Cave, 22°N) to southern Oman (Qunf Cave, 17°N) show a depletion in

oxygen isotopes attributed to the "amount effect" as well as indications of a southerly moisture source (Figure 4; Fleitmann *et al.* 2007, 2011). This implies increased tropical summer moisture associated with a northward migration of the ITCZ between 10-9 ka (Fleitmann *et al.* 2007, 2011;

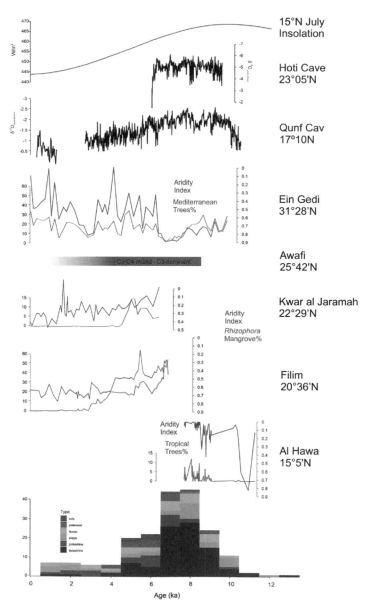

Figure 4. Holocene Humid Period climate and vegetation from Arabia including (from top to bottom) July solar insolation at 15N (Berger and Loutre 1991), speleothem $\delta^{18}O$ from Hoti Cave (northern Oman) and Qunf Cave (southern Oman; Fleitmann *et al.* 2007), aridity index and percent Mediterranean trees (*Quercus/Olea/Juniperus*) from Ein Gedi (Israel; Litt *et al.* 2012), vegetation at Awafi (UAE; Parker *et al.* 2004), aridity index and freshwater mangrove pollen (*Rhizophora*) from Kwar al Jaramah and Filim (Oman; Lézine *et al.* 2017), aridity index and percent tropical trees (*Acacia, Dodonaea viscosa, Podocarpus, Ziziphus, Salvadora persica*) from Al-Hawa (Yemen; Lézine *et al.* 1998), and frequency diagram of numbers of waterlain deposits (Lézine *et al.* 2014).

Neff *et al.* 2001). Numerous studies link monsoon intensification to high summer insolation in the northern hemisphere at the early to mid-Holocene.

However, monsoon intensity based on marine core records responded non-linearly to radiative forcing with significantly centennial and millennial-scale variability (Gupta *et al.* 2003; Sirocko *et al.* 1993). This work suggests both direct and indirect forcing of the monsoon with variability mechanistically linked to teleconnections with both temperatures in the North Atlantic as well as changes in albedo over the Asian landmass. Although no quantitative estimates of rainfall increase are associated with those studies, palaeoclimate model simulations by Lézine *et al.* (2017) imply that large rainfall anomalies occurred in the south between 14-18°N at least as late as 6 ka with increased summer moisture limited mostly to southern Oman and Yemen. Only marginal rainfall increase occurred as far north as 20-24°N.

In addition to changes in tropical circulation, winter Mediterranean rainfall systems varied over the Holocene. Speleothem $\delta^{18}O$ from caves in the Eastern Mediterranean, including Soreq Cave in Israel, lake level fluctuations, and the timing of sapropel deposition suggest that the winter rainfall zone underwent several transitions. Two major wetter phases were observed: an early Holocene Humid period (*c.* 10-7 ka) and a wetter phase at the end of the Middle Holocene, from *c.* 65-3 ka (Bar-Matthews *et al.* 1999; Bar-Matthews and Ayalon 2011; Litt *et al.* 2012; Migowski *et al.* 2004; Roberts *et al.* 2011; Robinson *et al.* 2006). This suggests an early phase of higher moisture associated with Mediterranean cyclonic systems coeval with early Holocene monsoon intensification in southern Arabia. However, unlike tropical Arabia, where indications of moisture persisted along the coast until 6-5 ka, the northern winter rainfall zone underwent a period of aridity during the mid-Holocene followed by renewed wetting after southern Arabia was already dry.

Higher rainfall across southern Arabia along coastal escarpments and highlands during the early Holocene resulted in increased streamflow into closed basins that comprise the continental interior. Waterlain deposits like lakes, ponds, and marshes are found that date to between 10-6 ka in southern Arabia (Figure 4; Enzel *et al.* 2015; Lézine *et al.* 2010). This includes palustrine deposits and extensive shallow endorheic lake systems in the Ramlat as-Sabatayn, Dhamar highlands in Yemen, and eastern fringes of the Rub al'Khali due to increased drainage from the landward sides of coastal escarpments (Davies 2006; Lézine *et al.* 1998; Parker *et al.* 2004). Wadis, or dry river valleys, which today rarely have flowing water, show evidence of fluvial activity during the early and mid-Holocene, including fine-grained deposits indicative of the development of permanent meandering streams in places like Wadi Sana, Yemen and Dhofar, Oman (Anderson 2007; Berger *et al.* 2012; Enzel *et al.* 2015; Sander 2006). Further, other geological evidence exists of groundwater recharge and higher water table from tufa and travertine deposits across Oman and Yemen which also date to the early Holocene (Cremaschi *et al.* 2015; Sander 2006). Carbonate-rich facies indicated by lithostratgraphic studies of the palaeolake deposits suggests that their development is due to enhanced runoff from adjacent highlands which trapped moisture rather than from direct rainfall onto lake surfaces (Lézine *et al.* 2007; Parker *et al.* 2004).

11.4.2.2 Southern Arabian palaeoecology

In southern Arabia, pollen, phytoliths, and macrobotanical remains show that the early Holocene was a period of major landscape transformation resulting in a reduction of arid and hyper-arid environments and expansion of diverse, savanna and woodland ecosystems due to increased moisture. As far north as 27°N, which today is largely outside of the summer southwest monsoon zone, phytolith, carbon isotopes, and pollen analysis from a palaeolake deposit at Awafi demonstrate the development of a C_3 Pooid grassland with higher abundances trees like *Acacia, Tamarix, Prosopis,* and palms occurred from 9-6 ka (Parker *et al.* 2004). At Suwayh, along the northern Omani coast, increased abundances of grasses and some tropical trees is in agreement with denser vegetation with many taxa with Somali-Masai affinities (Lézine *et al.* 2002). To the southwest, in Wadi Sana in Yemen, pollen and macrobotanical evidence from hyrax middens and alluvial sediments demonstrates the development of woodland with a dense grassy understory along a permanent stream until at least 6-5 ka (McCorriston *et al.* 2002). This woodland included a keystone tree taxon from coastal seasonal woodlands (*Terminalia dhofarica*), which during the mid-Holocene, appears to have had a more continuous range across coastal highlands, but also may have extended further inland along waterways (Oberprieler *et al.* 2009).

However, despite the expansion of tropical savanna and woodland in some locations in southern Arabia during the early and mid-Holocene, there are also many indications of persistent aridity. This suggests strong heterogeneity in local moisture and therefore strong local controls on vegetation. For example, pollen analysis also demonstrates the stability of many xerophytic taxa throughout this interval. Although pollen analysis on the palaeolake sediments suggest that the lake margin vegetation was dominated by local aquatic plants (*Typha*) due to the presence of water, the terrestrial flora around the lake was characterized by Arabian desert plants like Cyperaceae, *Calligonum*, and Amaranthaceae, common in the modern flora (Lézine *et al.* 1998, 2007). This is true as well in pollen records from coastal estuaries, which integrate pollen from a large regional watershed. For example, at Filim and Kwar al Jaramah along the Oman coast, desert taxa (*Artemisia*, Cyperaceae) increase in abundances after 6-5 ka but are a dominant component of the pollen spectra even prior to this (Lézine *et al.* 2017). These regionally integrating records do however support the idea of localized changes in vegetation structure and composition in distant headwaters. For example, at Al-Hawa, long distant transport of pollen from Afromontane and tropical woodland trees (*Podocarpus, Erica, Juniperus*) that are not observed today in modern pollen rain occurred in small abundances until the desiccation of the palaeolake around 7.2 ka (Lézine *et al.* 1998, 2007). This supports that idea that tropical African elements occurred more abundantly in the mountains that drain into the Ramlat as-Sabatayn and that these mountains likely acted as strong topographic barriers to rainfall even during times of enhanced monsoon intensity. This suggests that in contrast to widespread expansion of wetter biomes in Northern Africa, Arabian ecosystems even during the Holocene Humid Period were always on the margins of desert, perhaps due in part to the complex topography.

An additional impact of increased freshwater flux from heavier rainfall in southern Arabia during the Holocene Humid Period was the presence of *Rhizophora* mangrove vegetation throughout coastal estuaries. Today, mangroves are found only in very limited patches along the southern Arabian coast (Deil 1998). Further, species composition of modern mangroves is closely related to water chemistry, particularly salinity, within an estuary. For example, most mangroves remaining in arid and semi-arid regions are typically dominated by the halophytic taxon *Avicennia marina*. While this tree is also present in the early to mid-Holocene, increased abundances of mangrove pollen taxa that do not tolerate high salinity, such as *Rhizophora*, can only occur in the presence of increased freshwater flux from the land. *Rhizophora* pollen occurs in high abundances in sediments before 4 ka along the southern Arabian and east Asian coasts at Kwar al Jaramah, Suwayh, and Filim as well as being present in Arabian Sea marine sediment cores (Lézine *et al.* 2017). This suggests that increased freshwater runoff into coastal estuaries altered their water chemistry, allowing taxa like *Rhizophora* which do not tolerate very saline conditions to thrive during the mid-Holocene.

11.4.2.3 Northern Arabian palaeoecology

In northern Arabia, the record of early to mid-Holocene vegetation is quite different from southern Arabia. In particular, there is a marked difference in the timing of the maximum in Holocene moisture latitudinally from subtropical northern to tropical southern Arabia. Based on compilations of hydrological records from southern Arabia, Lézine *et al.* (2014) demonstrated that the Holocene Humid Period maximum was coeval at tropical sites and occurred from 10-6.5 ka. In contrast, in the subtropics, $\delta^{18}O$ measurements on lacustrine sediments suggest increased winter moisture from 9-8 ka, with a marked arid interval from 8-6 ka despite continued winter rain influence (Bar-Matthews *et al.* 1999; Litt *et al.* 2012; Roberts *et al.* 2011; Stevens *et al.* 2006). Lithostratigraphic observations and pollen analysis from Lake Mirabad in Iran suggest warm and dry conditions at this time favored a more arid, open vegetation characterized by low abundances of tree pollen in comparison to today (Van Zeist and Bottema 1977). Further south along the Dead Sea, lake level fluctuations indicate an early Holocene Humid Period (10-8.6 ka), increased aridity during 8-5.6 ka, and another phase of raised lake level indicating increased humidity during 5.6-3.5 ka (Migowski *et al.* 2006). The early Holocene vegetation transitions based on the core from the shoreline at Ein Gedi demonstrates low proportions of Mediterranean tree pollen such as deciduous oak, and scattered pistachio and olive pollen probably originating from the adjacent mountains. Amaranthaceae, together with Poaceae, *Artemisia, Ephedra,* and Asteraceae are the main constituents, associated to the Saharo-Arabian and Irano-Turanian biomes, indicating increased winter temperatures (Litt *et al.* 2012). A milder winter during early Holocene is in line with the increase of pistachio during this period in the Mediterranean (Rossignol-Strick 1999). Further south into Saudi Arabia, pollen from the Tayma record show regionally wetter conditions associated with increased Mediterranean woodlands (*Quercus, Pistachia*) in the Hijaz mountains in the west and expanded grasslands during the early Holocene only until 8 ka (Dinies et al. 2015, 2016). Other shallow lake sediments from the early Holocene corroborate an increase in humidity in this region (Crassard *et al.* 2013, Guagnin *et al.* 2017, Petraglia *et al.* 2020). At about 8 ka, a distinct vegetation shift at Tayma occurred, characterized by the retreat of grasslands and the re-expansion of desert dwarf-shrubs (i.e. *Haloxylon* type). This indicates the onset of arid conditions lasting at least until 4000 yrs BP, when the shallow-water lake dried out (Engel et al. 2012). Four pollen records (2–4 samples each) from diatomites in interdune depressions in the Nafud sand desert date to 9.5–5.8 ka and are similar to modern pollen spectra, pointing to persistent arid conditions throughout the Holocene (Schulz and Whitney 1987).

In the northern subtropics and into the Near East, winter moisture associated with enhanced westerly cyclonic activity increased at 6.5 ka. This is in agreement with palaeoecological work from Israel and Northern Jordan showing the expansion of Mediterranean forests with deciduous and evergreen oaks and olives (wild and/or cultivated) to lower elevations than present. At this same period, high lake levels at Dead Sea occurred between 6.3-3.3 ka (Litt *et al.* 2012). Indications of enhanced winter cyclonic moisture at this time extend into tropical Arabia, where Parker et al. (2004) observed a relatively stable period of mixed C_3-C_4 grassland and steppe at Awafi long after summer moisture had retreated from this region. However, there are no indications for wetter conditions in the more southern part of northwestern Arabia, where dwarf shrub steppe dominated until 4.2 ka.

11.4.2.4 Characteristics of the Arabian Holocene Humid Period
Modeling and empirical studies have focused on determining if the termination of the Holocene Humid Period in the tropics and the retreat of the ITCZ was abrupt or gradual. Within Arabia, in particular, a contrast is observed in the apparent pace of hydrological and ecological change recorded in marine versus terrestrial settings (Lézine *et al.* 2014). Marine sediment cores from the Indian Ocean and Arabian Sea capture dust sourced from deserts of Arabia and East Africa during periods of dry or windy climate (Sirocko *et al.* 1993). Throughout the early Holocene, abrupt transitions associated with centennial-millennial scale variability in monsoon intensity are observed (Gupta *et al.* 2003). Further, at the end of the Holocene Humid Period, dust influxes rose in several of these cores, interpreted as abrupt reduction of moisture in the region and increased dust transport (Sirocko *et al.* 1993). However, few terrestrial archives of both climate and palaeoenvironmental information capture this abrupt variability or termination. Speleothem $\delta^{18}O$ from southern Arabia exhibit a time transgressive reduction in rainfall (Fleitmann *et al.* 2007). At Hoti Cave in Northern Oman, a relatively rapid enrichment in $\delta^{18}O$ occurs at *c.* 6 ka leading to a hiatus marking the initial southward retreat of the ITCZ and a decrease in summertime moisture around 23°N. Further south at Qunf Cave (17°N) along the Arabian Sea, $\delta^{18}O$ instead followed a gradual enrichment beginning just after 6 ka and continuing until a depositional hiatus at around 2.5 ka. This gradual decline in rainfall is in agreement with palaeoclimate simulations across similar latitudinal transect, which suggest more northerly sites saw a more abrupt, earlier

decline in rainfall than southern sites (Lézine *et al.* 2017). Further evidence of a progressive decline in moisture comes from Lézine *et al.* (2014) who synthesized ages of waterlain deposits as indications of changes in rainfall. This analysis also suggests a gradual disappearance of surface water until *c.* 6 ka, when most areas dried up. Most palaeoecological records from southern Arabia also agree that ecosystem change followed a similar gradual transition toward semi-arid and arid ecosystems which resemble modern biomes, particularly after 6 ka. At Awafi, this time marked the transition back to from C_3 to mixed C_3–C_4 savanna (Parker *et al.* 2004). Macrofossils and pollen from hyrax middens from Wadi Sana, Yemen, show a transition from gallery woodland to semi-desert with an increase in *Artemisia* and Amaranthaceae (McCorriston *et al.* 2002). Along the coasts in estuaries, declining freshwater inputs result in a relatively gradual decline in freshwater mangrove taxon, *Rhizophora,* in favor of saltwater mangroves (Lézine *et al.* 2017).

Increases in rainfall and decreased rainfall seasonality throughout the Arabian Peninsula during the early to Mid-Holocene resulted in the development of ecosystems whose composition and dynamics were not analogous to those found today. In fact, most pollen spectra prior to 6 ka are difficult to compare directly with pollen samples representing modern vegetation (El-Moslimany 1990; Lézine *et al.* 2010). We illustrate some examples of these in Figure 5 which shows affinities of pollen taxa to modern phytogeographic regions. In this figure, most palaeoecological sites presented here have affinities based on pollen assemblages to multiple modern vegetation types. Here we will demonstrate those primary and secondary affinities to modern vegetation for no analog pollen spectra.

For example, in the Near East at Wadi Faynan, increases in arboreal taxa from at least 8-6 ka suggest that SW Jordan hosted woodlands along waterways and expanded montane forest communities (Hunt *et al.* 2004). Although many of these still occur either at the site or at least within the region (*Juniperus, Quercus*) today, *Ulmus* is not recorded in modern samples in that region. This is true for southern Arabia as well. For example, in Wadi Sana, in the modern stony desert, occasional trees such as *Maerua* and *Acacia* are observed. In the mid-Holocene, in addition to higher pollen abundances of these common modern arboreal taxa, higher abundances of trees with tropical affinities only were found at the southernmost coast, such as *Terminalia*, *Olea*, and *Dodonea viscosa* were also an important part of the arboreal community (McCorriston *et al.* 2002; Ivory *et al.*, 2021). However, it is important to note, that as discussed above, much of southern as well as northern Arabia remained arid or semi-arid even at the peak in the monsoon, thus we suggest that these changes in vegetation composition are likely linked to the presence of increased surface water amidst persistent climatic aridity.

Finally, the increase in biomass during the early to Mid-Holocene also appears to have resulted in a change in the dynamics of wildfires in some location in Arabia. Today through-out Arabia, wildfire is extremely uncommon due to lack of fuel and continuity of fuel on the landscape because of extreme aridity (Cao *et al.* 2015). Charcoal and microcharcoal datasets are rarer even than other palaeoecological information; however, increased charcoal deposition in palaeolake deposits, hyrax middens, and even the appearance of burnt surfaces in fine-grained stream deposits is a testament to the increase in fuel locally, linked to higher rainfall seasonality (Kimiaie and McCorriston 2014; McCorriston *et al.* 2002). Further, there are indications of increased inputs of charred plant material during this time coeval with increased land use and oasis cultivation, suggesting a relationship between fire and human presence (Dinies *et al.* 2016).

11.5 COUPLED HUMAN AND NATURAL SYSTEMS

Modern humans may have been present in Arabia for much of the Late Pleistocene. Arabian geography in fact likely played a strong role in early human migration (Armitage *et al.* 2011; Petraglia 2011). Environmental change, mainly hydrological variations related to glacial-interglacial cycles and monsoon intensity, is thought to have contributed to the movement of early humans out of

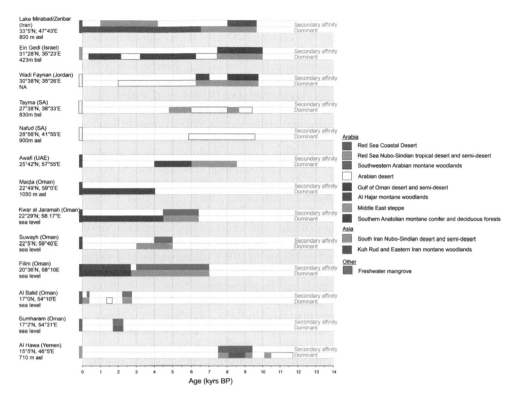

Figure 5. Holocene vegetation affinities in relation to ecoregion species composition as presented in Figure 1. For each site, there are two rows which represent the primary vegetation affinity (lower row) as represented by the pollen assemblages and the secondary affinity (upper row) in the case that pollen assemblage composition differed substantially from modern assemblages for numerous reasons (long distance transport, no modern analog, presence of significant aquatic vegetation).

Africa and continuous human presence (deMenocal and Stringer 2016; Tierney *et al.* 2017). Increasing evidence exists of earlier lacustrine phases in central Arabia which could have supported a more resource-rich landscape into which early hunter-gathers may have migrated (Parker 2010). Further, low sea levels during glacials, particularly along the Red Sea as well as the gulfs to the east, would have facilitated migration of early humans as they may have for plants (Lambeck 1996; Lokier *et al.* 2015). Evidence from mitochondrial DNA implies populations present in Arabia at least as early as 60 ka (Fernandes *et al.* 2015); however, direct archeological evidence of this time is scarce. For example, lithic piles are commonly observed throughout Arabia, but limitations in dating complicates an assessment beyond morphological attribution of assemblages to at least Early and Middle Palaeolithic (Groucutt and Petraglia 2012; Rose and Usik 2010). Little palaeoecological evidence exists beyond marine sediment cores to provide an environmental context of these migrations or potential footprint of these early hunters on the landscape. Evidence was preserved only from the Neolithic and the onset of the Holocene Humid Period to suggest continuous human presence throughout much of the sub-continent.

During Late Pleistocene and Holocene, human-environment interactions in Arabia began to shift through accelerated technological advancements. Notably, the cultivation and domestication processes of annual plants started in the Fertile Crescent at this time. Numerous studies point to a long-lasting, multi-phased, and multi-centered domestication process of cereals and pulses, connected to multiple triggers including climatic changes as well as cultural and demographic developments (e.g. Fuller *et al.* 2012; Hillman *et al.* 2001; Willcox 2012; Zeder 2011). These innovations are at the origin of far-reaching landscape transformations resulting in highly anthropogenic ecosystems such as gardens and fields. Perennial fruit trees such as olive and grapes were also domesticated and propagated from 9-6 ka (e.g. Zohary 1982). However, a much earlier start of these processes cannot be excluded for fruit trees, as indicated by the supposed early beginnings of domestication of figs near 11.4 ka (Kislev *et al.* 2006).

The spread of these Near Eastern domesticates into the Arabian Peninsula, the introduction of cultivars from other domestication centers such as African crops in Southern Arabia (e.g. millet; Tengberg 2003), and autochthonous agricultural developments on the Arabian Peninsula are not well known. With the exception of the southern mountains, modern precipitation is too low to practice durable dry-land farming which requires at least 250 mm/yr rainfall (e.g. Thalen 1979). Buried early-to-mid Holocene soils in Yemen show the potential for local agriculture during this wetter period, however these conditions have been lacking since Bronze Age (Pietsch *et al.* 2010). Arable farming thus was restricted to hydrologically favorable sites in most parts of the Arabian Peninsula, such as natural oases and wadis or depending on the development of irrigation. Terracing slopes or wadi beds are other basic features connected to gardening and oasis cultivation, (e.g. Avni *et al.* 2012; Buerkert and Schlecht 2010), often combined with run-off water harvesting installations at least in the past. The complex nature of sedimentary deposits from terraces and wadi beds is challenging for palaeoecological investigations. For example, these terrace features have fluvial accumulation resulting in discontinuous sedimentation, re-deposition, and often missing preservation of organic material. However, interdisciplinary research approaches indicate the potential in deciphering the fluctuating multiple uses, plant cultivation, and grazing during the past (Beckers *et al.* 2013; Charbonnier *et al.* 2017; Meister *et al.* 2016; Moraetis *et al.* 2020; Purdue *et al.* 2019) These basic landscape features necessitated adapted agricultural systems such as oasis agriculture, characterized today by it's a combination of cultivation of perennial fruit trees and annual crops.

Neolithic archeological assemblages in the Nafud, the northern sand desert of Saudi Arabia, show a strong Levantine influence (Crassard *et al.* 2013), while southern Arabian archeological findings point to the persistence of regional agricultural developments. Archeo-zoological records of domesticated sheep, goats, and cattle since the 8ka point to connections to the Near East as well. Herding (and hunting) is thought to have been a major subsistence strategy since at least the Neolithic (*c.* 7-6 ka). Somewhat more favorable environmental conditions during the early to mid-Holocene, i.e. denser and more diverse vegetation, facilitated the spread of this new livelihood (Crassard and Drechsler 2013; Uerpmann *et al.* 2010). Its impact on natural ecosystems on a regional level is quite difficult to estimate. In southern Jordan, vegetation disturbance due to grazing is supposed to be small as compared to the impacts in later times by industrial activities such as ore production (Hunt *et al.* 2007). In southern Arabia in particular, decades of research have focused on the observation that southern Arabian populations seem to have adopted agriculture relatively late in comparison with the Gulf region (within the last 3-2 ka; Harrower *et al.* 2010; McCorriston 2013; McCorriston and Johnson 1998). Instead, Late Stone Age and Bronze Age people in southern Arabia retained pastoral livelihoods through much of the Holocene. Throughout Yemen and Oman, evidence of ephemeral camps exist in the highlands as well as in the inland draining wadis and around palaeolake shorelines, suggesting mobile lifestyles centred around animal-husbandry of cattle and goats by 8 ka, and eventually camels by 3.5ka (Lézine *et al.* 2010; Martin *et al.* 2009; McCorriston 2013).

However, despite the adoption of agriculture in southern Arabia, there is still some equivocal palaeoecological evidence for human environmental modification of the early to mid-Holocene landscape by nomadic pastoralists. In Wadi Sana, Yemen, evidence of burned surfaces and hearths contained in terrace sediments are coeval with evidence of dated occupation by pastoralists (Kimiae and McCorriston 2014; McCorriston *et al.* 2002). This suggests that the practice of controlled burning of the landscape to deter woody encroachment and promote tender grassy shoot production for animal fodder, a practice common in modern pastoralist groups, may have been used at this time (Kimiae and McCorriston 2014). Hearths and settlement excavations from southern Oman and Yemen attest to wood harvesting, particularly taxa known for use as fodder, construction materials, or cooking fuel including *Acacia, Terminalia,* and *Ziziphus.* This evidence of fire and landscape modification is also visible in pollen and macrobotanical remains preserved in hyrax middens near archeological sites in Yemen (McCorriston *et al.* 2002; Ivory *et al.*, 2021). Further, following the initial decline in monsoon moisture around 6 ka, construction of simple check dams and irrigation features is observed along wadis in Oman and Yemen (Harrower 2008). This is also contemporaneous with the increase in construction of falaj systems in northern Oman to provide an additional source of freshwater in the lowlands. These water management structures are thought perhaps to mitigate increased moisture variability and decreased surface runoff associated with declining monsoon intensity prior to abandonment of the region around 5 ka (Harrower 2008).

The cultivation of crops however is unambiguous evidence of anthropogenic changes. Although pollen of *Cerealia* is present in moderate abundances as early as 10 ka in sediment cores (Hunt *et al.* 2004; Litt *et al.* 2012), these very few early Holocene pollen records do not offer clear clues due to the challenge in reliably distinguishing wild grasses from domesticated cereals on the Arabian Peninsula. However, since the 5 ka, the cultivation of cereals and other annuals is attested by seeds and fruits for different regions on Arabian Peninsula, indicating connectivity with the Near East (Boivin and Fuller 2009; Bouchaud *et al.* 2016). In the Gulf region, records of *Phoenix dactylifera* predate the cereals, and so far, the oldest date palm kernels are found on the Dalma islands dating to *c.* 8 ka (Beech and Shepherd 1994), with otherwise marine food remains dominating. Since 5 ka, the date palm seemed to have played an important role in the arable agriculture of the gulf region, while in other regions of the Arabian Peninsula, this now dominant tree was still missing (e.g. Schiettecatte *et al.* 2013) and oasis agriculture relied on other fruit trees such as grapes (Dinies *et al.* 2016).

The crop evidence since *c.* 5-4 ka agree with the emergence of large oasis settlements and increased connectivity, including maritime trade (e.g. Boivin and Fuller 2009; Hausleiter and Eichmann 2018). The storied "incense road" may be seen as part of this connectivity. Ongoing residue analyses in the oasis of Tayma, NW Arabia, point to likely local resin sources (*Pistacia, Commiphora*) during late mid-Holocene, while *Boswellia* became important during the last 3000-2000 years. The commercial harvest and trade of Boswellia resin (luban, frankincense) resulted in the construction of seaports such as Sumharam and Al Baleed in southern Oman to facilitate trade of frankincense along Arabian Sea trade routes (Hoorn and Cremaschi 2004; Lippi *et al.* 2011). Here too, the impact of oasis agriculture on the landscape on a regional scale is hard to estimate due to the very few palaeoecological records and probably varied depending on the local practices. Detailed archeobotanical analyses of a NW Arabian oasis indicate the persistence of desert vegetation without major shifts, pointing to exploitation of available natural resources, but no over-use (Bouchaud 2011). The modern heavily overused landscape situation seems to differ from that prevailing even just a few centuries ago, when grazing (and wood/shrub cutting) was restricted because of a missing permanent water supply. This today has become obsolete because of deep wells and mobile water tanks guaranteeing accessibility everywhere (Hunt *et al.* 2007; Thalen 1979).

11.6 CONCLUSIONS

The vegetation of Arabia is strongly imprinted by biogeographic and palaeoecological processes that unfolded over millions of years beginning with the separation from Africa during the Cenozoic. Although little empirical palaeoecological data exist, new efforts to relaunch the African Pollen Database including Arabia will galvanize new research efforts as well as syntheses by conforming to FAIR data standards that make existing data easier to find and work with (Ivory *et al.* 2020; Wilkinson *et al.* 2016). This review attempts to summarize some of this work and inventory data holdings currently within the APD.

Palaeoecological information from Arabia make it clear that while ecologically diverse and endemic, Arabian ecosystems have existed on the edge of extreme aridity for at least hundreds of thousands of years. We recognize the strong need for more palaeoecological data which represent pre-Holocene time periods in order to better disentangle the influence of climate events on the vegetation. Although deep, long-lived lakes are rare in the region, analysis of pre-Holocene palaeolake deposits as well as drilling campaigns in estuarine sites could help fill in some of this gap. Further, despite the highest data density within the early Holocene, these data are still very sparse. Improving spatial and temporal resolution of this period should also be considered a priority. This may be achieved by continued work on classical palaeoecological archives (palaeolakes, oases), complementing archeological investigations with archeobotanical research, and the development of novel palaeoecological archives adapted to arid regions such as hyrax middens, termite mounds, and fine-grained fluvial deposits.

The antiquity of the human legacy on the environment is not well known. The human imprint on the landscape in Arabia locally is strong through crop cultivation in oases or wadis but seems to be reversible and without indication of over-use in vast desert and semi-desert regions used as rangelands. The density of palaeoecological information associated with archeological investigations do not lend themselves to a very clear picture of human modification of landscapes through many interventions. This region displays evidence of the some of the earliest plant cultivation, many of which leave telltale signals in pollen and macrobotanical records. Further, early evidence of water management and burning connected with pastoralism have acted as a filter on species composition of local vegetation and ecosystems for thousands of years. Deeper connections between the palaeoecological record, archeological data, and anthropological theory through continued empirical data generation and combined empirical and modeling approaches could prove extremely helpful for bettering understanding feedbacks between changing human demography, culture, and the natural environment.

Finally, we observe that Arabian landscapes, particularly semi-desert and desert ecosystems today are extremely fragile; however, they show an overall strong resilience throughout the Holocene as observed through the lens of the palaeoecological record. However, while some land-use practices have been observed by pollen data to degrade resource-rich landscapes such as intensive agriculture, other forms of traditional management alternately enriched landscapes, such as water management for agriculture and pastoralism. Palaeoecology in Arabia provides an important perspective and framework for lessons in sustainable management of dryland ecosystems under changing climate in the future.

ACKNOWLEDGMENTS

This work contributes to the ACCEDE ANR-NSF Belmont Forum project (18 BELM 0001 05 and NSF-1929563) as well as NSF-CNH 1617185. Thanks are due to the African Pollen Database for data access. SJI is funded by Penn State University, AML is funded by CNRS, MD is funded by the DFG priority program SPP Entangled Africa.

REFERENCES

Abuelgasim, A., Ross, W., Gopal, S. and Woodcock, C., 1999, Change detection using adaptive fuzzy neural networks: Environmental damage assessment after the Gulf War. *Remote Sensing of Environment*, **70**, pp. 208–223, 10.1016/S0034-4257(99)00039-5.

Albalawi, E., Dewan, A. and Corner, R., 2018, Spatio-temporal analysis of land use and land cover changes in arid region of Saudi Arabia. *International Journal of GEOMATE*, **14**, pp. 73–81, 10.21660/2018.44.3708.

Almazroui, M., Nazrul Islam, M., Athar, H., Jones, P. D. and Rahman, M. A., 2012, Recent climate change in the Arabian Peninsula: annual rainfall and temperature analysis of Saudi Arabia for 1978–2009. *International Journal of Climatology*, **32**, pp. 953–966, 10.1002/joc.3446.

Alqurashi, A. F., Kumar, L. and Sinha, P., 2016, Urban land cover change modelling using time-series satellite images: A case study of urban growth in five cities of Saudi Arabia. *Remote Sensing*, **8**, article: 838, 10.3390/rs8100838.

Anderson, J., 2007, *Climatic and structural controls on the geomorphology of Wadi Sana, highland Southern Yemen*, PhD thesis, University of South Florida.

Armitage, S., Jasim, S., Marks, A., Parker, A., Usik, V. and Uerpmann, H., 2011, The southern route "out of Africa": evidence for an early expansion of modern humans into Arabia. *Science*, **331**, pp. 453–456, 10.1126/science.1199113.

Avni, Y., N. Porat, N., Avni, G., 2012, Pre-farming environment and OSL chronology in the Negev Highlands, Israel. *Journal of Arid Environments*, **86**, pp. 12 27, 10.1016/j.jaridenv.2012.01.002.

Baierle, H., 1993, Vegetation und Flora im südwestlichen Jordanien. *Dissertationes botanicae*, **200**, I-VIII.

Ball, L., MacMillan, D., Tzanopoulos, J., Spalton, A., Al Hikmani, H. and Moritz, M., 2020, Contemporary Pastoralism in the Dhofar Mountains of Oman. *Human Ecology*, **48**, pp. 267–277, 10.1007/s10745-020-00153-5.

Bar-Matthews, M. and Ayalon, A., 2011, Mid-Holocene climate variations revealed by high-resolution speleothem records from Soreq Cave, Israel and their correlation with cultural changes. *The Holocene*, **21**, pp. 163–171, 10.1177/095683610384165.

Bar-Matthews, M., Ayalon, A., Kaufman, A. and Wasserburg, G. J., 1999, The Eastern Mediterranean paleoclimate as a reflection of regional events: Soreq cave, Israel. *Earth and Planetary Science Letters*, **166**, pp. 85–95, 10.1016/S0012-821X(98)00275-1.

Beckers, B., Schütt, B., Tsukamoto, S. and Frechen, M. (2013): Age determination of Petra's engineered landscape optically stimulated luminescence (OSL) and radiocarbon ages of runoff terrace systems in the Eastern Highlands of Jordan. *Journal of Archaeological Science* **40**, pp. 333–348, 10.1016/j.jas.2012.06.041.

Beech, M. and Shepherd, E., 1994, Archaeobotanical evidence for early date consumption. *Antiquity*, **71**, pp. 288-99, 10.1017/S0003598X00052765.

Berger, A. and Loutre, M.-F., 1991, Insolation values for the climate of the last 10 million years. *Quaternary Science Reviews*, **10**, pp. 297–317, 10.1016/0277-3791(91)90033-Q.

Berger, J. F., Bravard, J. P., Purdue, L., Benoist, A., Mouton, M. and Braemer, F., 2012, Rivers of the Hadramawt watershed (Yemen) during the Holocene: Clues of late functioning. *Quaternary International*, **266**, pp. 142–161, 10.1016/j.quaint.2011.10.037.

Boivin, N. and Fuller, D., 2009, Shell middens, ships and seeds: Exploring coastal subsistence, maritime trade and the dispersal of domesticates in and around the ancient Arabian Peninsula. *Journal of World Prehistory*, **22**, pp. 113–180, 10.1007/s10963-009-9018-2.

Bonnefille, R., 2010, Cenozoic vegetation, climate changes and hominid evolution in tropical Africa. *Global and Planetary Change*, **72**, pp. 390–411, 10.1016/j.gloplacha.2010. 01.015.

Bouchaud, C., 2011, *Paysages et pratiques d'exploitation des ressources végétales en milieux semi-aride et aride dans le sud du Proche-Orient : Approche archéobotanique des périodes antique et islamique (IVe siècle av. J.-C. – XVIe siècle ap. J.-C.)*. Thèse, Paris 1, pp. 399.

Bouchaud, C., Dabrowski, V. and Tengberg, M., 2016, État des lieux de la recherche archéobotanique en péninsule Arabique. *Actualités des recherches archéologiques en Arabie. Routes de l'Orient*, **2**, pp. 21–37.

Buerkert, A. and Schlecht, E., 2010, *Oases of Oman*: Millenia old livelihood systems at the cossroads. Al Roya Press & Publishing House, Muscat, Oman, pp. 145.

Burns, S., Fleitmann, D., Matter, A., Neff, U. and Mangini, A., 2001, Speleothem evidence from Oman for continental pluvial events during interglacial periods. *Geology*, **29**, pp. 623–626, 10.1130/0091-7613(2001)029#<0623:SEFOFC#>2.0.CO;2.

Cao, X., Meng, Y. and Chen, J., 2015, Mapping grassland wildfire risk of the world. In *World Atlas of Natural Disaster Risk*, Berlin, Heidelberg, Springer, pp. 277–283.

Charbonnier, J., Purdue, L., Calastrenc, C., Regagnon, E., Sagory, T. and Benoist, A., 2017, Ancient agricultural landscapes in Southeast Arabia: Approach and first results of an archaeological, geo-archaeological, and spatial study of the Masāfī Palm Grove, Emirate of Fujairah. Proceedings of Water and Life in Arabia Conference, 14th - 16th December, Dec 2014, Abu Dhabi, United Arab Emirates. halshs-01792812.

Chen, C. and Litt, T., 2018, Dead Sea pollen provides new insights into the paleoenvironment of the southern Levant during MIS 6–5. *Quaternary Science Reviews*, **188**, pp. 15–27, 10.1016/j.quascirev.2018.03.029.

Crassard, R. and Drechsler, P., 2013, Towards new paradigms: multiple pathways for the Arabian Neolithic. *Arabian Archaeology and Epigraphy*, **24**, pp. 3–8, 10.1111/aae.12021.

Crassard, R., Petraglia, M., Drake, N., Breeze, P., Gratuze, B., Alsharekh, A. and Robin, C. J., 2013, Middle Palaeolithic and Neolithic occupations around Mundafan palaeolake, Saudi Arabia: implications for climate change and human dispersals. *PLoS One*, **8**, article: e69665.

Cremaschi, M., Zerboni, A., Charpentier, V., Crassard, R., Isola, I., Regattieri, E. and Zanchetta, G., 2015, Early–Middle Holocene environmental changes and pre-Neolithic human occupations as recorded in the cavities of Jebel Qara (Dhofar, southern Sultanate of Oman). *Quaternary International*, **382**, pp. 264–276, 10.1016/j.quaint.2014.12.058.

Dahinden, F., Fischer, E. M. and Knutti, R., 2017, Future local climate unlike currently observed anywhere. *Environmental Research Letters*, **12**, article: 084004, 10.1088/1748-9326/aa75d7.

Davies, C., 2006, Holocene paleoclimates of southern Arabia from lacustrine deposits of the Dhamar highlands, Yemen. *Quaternary Research*, **66**, pp. 454–464, 10.1016/j.yqres.2006.05.007.

Dawson, A., Paciorek, C., McLachlan, J., Goring, S., Williams, J. and Jackson, S., 2016, Quantifying pollen-vegetation relationships to reconstruct ancient forests using 19th-century forest composition and pollen data. *Quaternary Science Reviews*, **137**, pp. 156–175, 10.1016/j.quascirev.2016.01.012.

Deil, U., 1998, Coastal and sabkha vegetation. In *Vegetation of the Arabian Peninsula*, edited by Ghazanfar, S.A. and Fisher, M., (Dordrecht: Springer), pp. 209–228.

Deil, U., and al Gifri, A., 1998, Montane and wadi vegetation In *Vegetation of the Arabian Peninsula*, edited by Ghazanfar, S.A. and Fisher, M., (Dordrecht: Springer), pp. 125–174.

deMenocal, P. and Stringer, C., 2016, Human migration: Climate and the peopling of the world. *Nature*, **538**, pp. 49–50, 10.1038/nature19471.

Dinies, M., Plessen, B., Neef, R. and Kürschner, H., 2015, When the desert was green: Grassland expansion during the early Holocene in northwestern Arabia. *Quaternary International*, **382**, pp. 293–302, 10.1016/j.quaint.2015.03.007.

Dinies, M., Neef, R., Plessen, B., and Kürschner, H., 2016, Holocene vegetation, climate, land use and plant cultivation in the Tayma region, northwestern Arabia. *The Archaeology of North Arabia: Oases and Landscapes*, edited by Luciani, M., OREA – Oriental and European Archaeology, **4**, pp. 57–78.

El-Moslimany, A., 1990, Ecological significance of common nonarboreal pollen: examples from drylands of the Middle East. *Review of Palaeobotany and Palynology*, **64**, pp. 343–350, 10.1016/0034-6667(90)90150-H.

Engel, M., Brückner, H., Pint, A., Wellbrock, K., Ginau, A., Voss, P. and Frenzel, P., 2012, The early Holocene humid period in NW Saudi Arabia–Sediments, microfossils and palaeo-hydrological modelling. *Quaternary International*, **266**, pp. 131–141, 10.1016/j.quaint.2011. 04.028.

Enzel, Y., Kushnir, Y. and Quade, J., 2015, The middle Holocene climatic records from Arabia: Reassessing lacustrine environments, shift of ITCZ in Arabian Sea, and impacts of the southwest Indian and African monsoons. *Global and Planetary Change*, **129**, pp. 69–91, 10.1016/j.gloplacha.2015.03.004.

Feakins, S., 2013, Pollen-corrected leaf wax D/H reconstructions of northeast African hydrological changes during the late Miocene. *Palaeogeography, Palaeoclimatology, Palaeoecology*, **374**, pp. 62–71, 10.1016/j.palaeo.2013.01.004.

Fernandes, V., Triska, P., Pereira, J. B., Alshamali, F., Rito, T., Machado, A. and Richards, M. B., 2015, Genetic stratigraphy of key demographic events in Arabia. *PloS one*, **10**, article: e0118625, 10.1371/journal.pone.0118625.

Fersi, W., Lézine, A. M., and Bassinot, F., 2016, Hydro-climate changes over southwestern Arabia and the Horn of Africa during the last glacial–interglacial transition: A pollen record from the Gulf of Aden. *Review of Palaeobotany and Palynology*, **233**, pp. 176–185, 10.1016/j.revpalbo.2016.04.002.

Fick, S.E. and Hijmans, R.J., 2017, WorldClim 2: new 1-km spatial resolution climate surfaces for global land areas. *International Journal of Climatology*, **37**, pp. 4302–4315, 10.1002/joc.5086.

Fisher, M. and Membery, D., 1998, Climate. In *Vegetation of the Arabian Peninsula*, edited by Ghazanfar, S.A. and Fisher, M., (Dordrecht: Springer), pp. 5–38.

Fleitmann, D., Burns, S., Mangini, A., Mudelsee, M., Kramers, J., Villa, I. and Neff, U., 2007, Holocene ITCZ and Indian monsoon dynamics recorded in stalagmites from Oman and Yemen (Socotra). *Quaternary Science Reviews* **26**, pp. 170–188, 10.1016/j.quascirev.2006.04.012.

Fleitmann, D., Burns, S., Pekala, M., Mangini, A., Al-Subbary, A., Al-Aowah, M. and Matter, A., 2011, Holocene and Pleistocene pluvial periods in Yemen, southern Arabia. *Quaternary Science Reviews*, **30**, pp. 783–787, 10.1016/j.quascirev.2011.01.004.

Friesen, J., Zink, M., Bawain, A. and Müller, T., 2018, Hydrometeorology of the Dhofar cloud forest and its implications for groundwater recharge. *Journal of Hydrology: Regional Studies*, **16**, pp. 54–66, 10.1016/j.ejrh.2028.03.002.

Fuller, D., Asouti, E. and Purugganan, M., 2012, Cultivation as slow evolutionary entanglement: comparative data on rate and sequence of domestication. *Vegetation History and Archaeobotany*, **21**, pp. 131–145, 10.1007/s00334-011-0329-8.

Galletti, C., Turner, B. and Myint, S., 2016, Land changes and their drivers in the cloud forest and coastal zone of Dhofar, Oman, between 1988 and 2013. *Regional Environmental Change*, **16**, pp. 2141–2153, 10.1007/s10113-016-0942-2.

Ghazanfar, S., 1992, Quantitative and biogeographic analysis of the flora of the Sultanate of Oman. *Global Ecology and Biogeography Letters*, **2**, pp. 189–195, 10.2307/2997660.

Gil-López, M., Segarra-Moragues, J. and Ojeda, F., 2017, Influence of habitat patchiness on diversity patterns of a habitat specialist plant community. *Journal of Vegetation Science*, **28**, pp. 436–444, 10.1111/jvs.12488.

Groucutt, H. and Petraglia, M., 2012, The prehistory of the Arabian peninsula: deserts, dispersals, and demography. *Evolutionary Anthropology: Issues, News, and Reviews*, **21**, pp. 113–125, 10.1002/evan.21308.

Guagnin, M., Shipton, C., Martin, L. and Petraglia, M., 2017, The Neolithic site of Jebel Oraf 2, northern Saudi Arabia: First report of a directly dated site with faunal remains. *Archaeological Research in Asia*, **9**, pp. 63–67, 10.1016/j.ara.2017.02.001.

Guba, I., and Glennie, K., 1998, Geology and geomorphology. In *Vegetation of the Arabian Peninsula* edited by Ghazanfar, S.A. and Fisher, M., (Dordrecht: Springer), pp. 39–62.

Gupta, A., Anderson, D. and Overpeck, J., 2003, Abrupt changes in the Asian southwest monsoon during the Holocene and their links to the North Atlantic Ocean. *Nature*, **421**, pp. 354–357, 10.1038/nature01340.

Harrower, M., 2008, Mapping and dating incipient irrigation in Wadi Sana, Hadramawt (Yemen). *Proceedings of the Seminar for Arabian Studies*, **38**, pp. 187–201.

Harrower, M., McCorriston, J. and D'Andrea, A., 2010, General/specific, local/global: comparing the beginnings of agriculture in the Horn of Africa (Ethiopia/Eritrea) and southwest Arabia (Yemen). *American Antiquity*, pp. 452–472, 10.7183/0002-7316.75.3.452.

Hausleiter, A. and Eichmann, R., 2018, The archaeological exploration of the oasis of Tayma. In *Taymā' 1. Archaeological Exploration, Palaeoenvironment, Cultural Contacts*, edited by Hausleiter, A., Eichmann, R. and al-Najem, M. (Oxford: Archaeopress Publishing LTD), pp. 3–58.

Hély, C., and Lézine, A.-M., 2014, Holocene changes in African vegetation: Tradeoff between climate and water availability. *Climate of the Past*, **10**, pp. 681–686, 10.5194/cp-10-681-2014.

Hildebrandt, A., Al Aufi, M., Amerjeed, M., Shammas, M., and Eltahir, E., 2007, Ecohydrology of a seasonal cloud forest in Dhofar: 1. Field experiment. *Water Resources Research*, **43**. W10411, 10.1029/2006WR005261.

Hillman, G., Hedges, R., Moore, A., Colledge, S., and Pettitt, P., 2001, New evidence of Lateglacial cereal cultivation at Abu Hureyra on the Euphrates. *The Holocene*, **11**, pp. 383–393, 10.1191/095968301678302823.

Hoelzmann, P., Jolly, D., Harrison, S., Laarif, F., Bonnefille, R. and Pachur, H., 1998, Mid-Holocene land-surface conditions in northern Africa and the Arabian Peninsula: A data set for the analysis of biogeophysical feedbacks in the climate system. *Global biogeochemical cycles*, **12**, pp. 35–51, 10.1029/97GB02733.

Hoorn, C. and Cremaschi, M., 2004, Late Holocene palaeoenvironmental history of Khawr Rawri and Khawr Al Balid (Dhofar, Sultanate of Oman). *Palaeogeography, Palaeoclimatology, Palaeoecology*, **213**, pp. 1–36, 10.1016/j.palaeo.2004.03.014.

Huang, J., Li, Y., Fu, C., Chen, F., Fu, Q., Dai, A. and Zhang, L., 2017, Dryland climate change: Recent progress and challenges. *Reviews of Geophysics*, **55**, pp. 719–778, 10.1002/2016RG000550.

Hunt, C., Elrishi, H., Gilbertson, D., Grattan, J., McLaren, S., Pyatt, F. and Barker, G. W., 2004, Early-holocene environments in the Wadi Faynan, Jordan. *The Holocene*, **14**, pp. 921–930, 10.1191/0959-683604hl769rp.

Hunt, C., Gilbertson, D. and El-Rishi, H., 2007, An 8000-year history of landscape, climate, and copper exploitation in the Middle East: The Wadi Faynan and the Wadi Dana National Reserve in southern Jordan. *Journal of Archaeological Science*, **34**, pp. 1306–1338, 10.1016/j.jas.2006.10.022.

Huntley, B., 1990, European vegetation history: Palaeovegetation maps from pollen data-13 000 yr BP to present. *Journal of Quaternary Science*, **5**, pp. 103–122, 10.1002/jqs.3390050203.

Issar, A. S. and Bruins, H. J., 1983, Special climatological conditions in the deserts of Sinai and the Negev during the latest Pleistocene. *Palaeogeography, Palaeoclimatology, Palaeoecology*, **43**(1–2), pp. 63–72, 10.1016/0031-0182(83)90048-2.

Ivory, S., Lézine, A.-M., Grimm, E. and Williams, J.W., 2020, Relaunching the African pollen database: abrupt change in climate and ecosystems. PAGES, **51**, pp. 26–35.

Ivory. S., Cole, K., Anderson, R.S., Anderson, A. and McCorriston, M., 2021, Human landscape modification and expansion of tropical woodland in southern Arabia during the mid-Holocene from rock hyrax (Procavia capensis) middens, *Journal of Biogeography*, 10.1111/jbi.14226.

Kimiaie, M. and McCorriston, J., 2014, Climate, human palaeoecology and the use of fuel in Wadi Sana, Southern Yemen. *Vegetation History and Archaeobotany*, **23**, pp. 33–40, 10.1007/s00334-013-0394-2.

Kislev, M., Hartmann, A. and Bar-Yosef, O., 2006, Early domesticated fig in the Jordan valley. *Science*, **312**, pp. 1372–1374, 10.1126/science.1125910.

Kuerschner, H., 1998, Biogeography and Introduction to Vegetation. In *Vegetation of the Arabian Peninsula*, edited by Ghazanfar, S.A. and Fisher, M., (Dordrecht: Springer), pp. 63–98.

Lambeck, K., 1996, Shoreline reconstructions for the Persian Gulf since the last glacial maximum. *Earth and Planetary Science Letters*, **142**, pp. 43–57, 10.1016/0012-821X(96)00069-6.

Lézine, A.-M., 2009, Timing of vegetation changes at the end of the Holocene Humid Period in desert areas at the northern edge of the Atlantic and Indian monsoon systems. *Comptes Rendus Geosciences*, **341**, pp. 750–759, 10.1016/j.crte.2009.01.001.

Lézine, A.-M., Saliège, J., Robert, C., Wertz, F., and Inizan, M. L., 1998, Holocene lakes from Ramlat as-Sab'atayn (Yemen) illustrate the impact of monsoon activity in southern Arabia. *Quaternary Research*, **50**, pp. 290–299, 10.1006/qres.1998.1996.

Lézine, A.-M., Saliège, J., Mathieu, R., Tagliatela, T., Mery, S., Charpentier, V. and Cleuziou, S., 2002, Mangroves of Oman during the late Holocene: climatic implications and impact on human settlements. *Journal of Vegetation History and Archaeobotany* **11**, pp. 221–232, 10.1007/s003340200025.

Lézine, A.-M., Tiercelin, J.J., Robert, C., Saliège, J.F., Cleuziou, S., Inizan, M.L. and Braemer, F., 2007, Centennial to millennial-scale variability of the Indian monsoon during the early Holocene from a sediment, pollen and isotope record from the desert of Yemen. *Palaeogeography, Palaeoclimatology, Palaeoecology*, **243**, pp. 235–249, 10.1016/j.palaeo.2006.05.019.

Lézine, A.-M., Robert, C., Cleuziou, S., Inizan, M., Braemer, F., Saliège, J.F. and Charpentier, V., 2010, Climate change and human occupation in the Southern Arabian lowlands during the last deglaciation and the Holocene. *Global and Planetary Change*, **72**, pp. 412–428, 10.1016/j.gloplacha.2010.01.016.

Lézine, A.-M., Bassinot, F. and Peterschmitt, J.Y., 2014, Orbitally-induced changes of the Atlantic and Indian monsoons over the past 20,000 years: New insights based on the comparison of continental and marine records. *Bulletin de la Société Géologique de France*, **185**, pp. 3–12, 10.2113/gssgfbull.185.1.3.

Lézine, A.-M., Ivory, S., Braconnot, P. and Marti, O., 2017, Timing of the southward retreat of the ITCZ at the end of the Holocene Humid Period in Southern Arabia: Data-model comparison. *Quaternary Science Reviews*, **164**, pp. 68–76, 10.1016/j.quascirev.2017.03.019.

Lippi, M., Bellini, C., Benvenuti, M. and Fedi, M., 2011, Palaeoenvironmental signals in ancient urban setting: the heavy rainfall record in Sumhuram, a pre-Islamic archaeological site of Dhofar (S Oman). *The Holocene*, **21**, pp. 951–965, 10.1177/0959683611400203.

Litt, T., Ohlwein, C., Neumann, F. H., Hense, A. and Stein, M., 2012, Holocene climate variability in the Levant from the Dead Sea pollen record. *Quaternary Science Reviews*, **49**, pp. 95–105, 10.1016/j.quascirev.2012.06.012.

Lokier, S., Bateman, M., Larkin, N., Rye, P. and Stewart, J., 2015, Late Quaternary sea-level changes of the Persian Gulf. *Quaternary Research*, **84**, pp. 69–81, 10.1016/j.yqres.2015.04.007.

Mandaville, J., 1998, Vegetation of the sands. In *Vegetation of the Arabian Peninsula*, edited by Ghazanfar, S.A. and Fisher, M., (Dordrecht: Springer), pp. 191–282.

Martin, L., McCorriston, J. and Crassard, R., 2009, Early Arabian pastoralism at Manayzah in Wādī Ṣanā, Ḥaḍramawt. In *Proceedings of the Seminar for Arabian Studies*. Archaeopress, pp. 271–282.

Mayewski, P., Rohling, E., Stager, J., Karlén, W., Maasch, K., Meeker, L. and Lee-Thorp, J., 2004, Holocene climate variability. *Quaternary Research*, **62**, pp. 243–255, 10.1016/j.yqres.2004.07.001.

McClure, H., 1984, *Late Quaternary palaeoenvironments of the Rub'Al Khali*; PhD, University College London, UK.

McCorriston, J., 2013, The Neolithic in Arabia: a view from the south. *Arabian Archaeology and Epigraphy*, **24**, pp. 68–72, 10.1111/aae.12012.

McCorriston, J., and Johnson, Z., 1998, Agriculture and animal husbandry at Ziyadid Zabid, Yemen. *Proceedings of the Seminar for Arabian Studies*, **28**, pp. 175–188.

McCorriston, J., Walter, D., Oches, E., and Cole, K., 2002, Holocene paleoecology and prehistory in highland southern Arabia. *Paléorient*, **28**(1), pp. 61–88, 10.3406/paleo.2002.4739.

Meister, J., Krause, J., Müller-Neuhof, B., Portillo, M., Reimann, T. and Schütt, B., 2016, Desert agricultural systems at EBA Jawa (Jordan): Integrating archaeological and paleoenvironmental records. *Quaternary International* **434**, pp. 33–50, 10.1016/j.quaint.2015.12.086.

Migowski, C., Agnon, A., Bookman, R., Negendank, J. F. and Stein, M., 2004, Recurrence pattern of Holocene earthquakes along the Dead Sea transform revealed by varve-counting and radiocarbon dating of lacustrine sediments. *Earth and Planetary Science Letters*, **222**, pp. 301–314, 10.1016/j.epsl.2004.02.015.

Mittermeier, R., Turner, W., Larsen, F., Brooks, T. and Gascon, C., 2011, Global biodiversity conservation: the critical role of hotspots. In *Biodiversity hotspots*, (Berlin and Heidelberg: Springer), pp. 3–22.

Moraetis, D., Al Kindi, S., Al Saadib, S., Al Shaibani, A., Pavlopoulos, K., Scharf, A., Mattern, F., Harrower, M. and Pracejus, B., 2020, Terrace agriculture in a mountainous arid environment – A study of soil quality and regolith provenance: Jabal Akhdar (Oman). *Geoderma* **363**, pp. 1141–1152, 10.1016/j.geoderma.2019.114152.

Neff, U., Burns, S. J., Mangini, A., Mudelsee, M., Fleitmann, D. and Matter, A., 2001, Strong coherence between solar variability and the monsoon in Oman between 9 and 6 kyr ago. *Nature*, **411**, pp. 290–293, 10.1038/35077048.

Nellessen, T., 2013, *The Consciousness of Water: Narrative Flows, Environmental Change, and the Voice of Yemen*. PhD thesis, University of Arkansas, USA.

Oberprieler, C., Meister, J., Schneider, C. and Kilian, N., 2009, Genetic structure of Anogeissus dhofarica (Combretaceae) populations endemic to the monsoonal fog oases of the southern Arabian Peninsula. *Biological Journal of the Linnean Society*, **97**, pp. 40–51, 10.1111/j.1095-8312.2008.01173.x.

Olson, D.M., Dinerstein, E., Wikramanayake, E.D., Burgess, N.D., Powell, G.V.N., Underwood, E.C., D'Amico, J.A., Itoua, I., Strand, H.E., Morrison, J.C., Loucks, C.J., Allnutt, T.F., Ricketts, T.H., Kura, Y., Lamoreux, J.F., Wettengel, W.W., Hedao, P. and Kassem, K.R., 2001, Terrestrial ecoregions of the world: a new map of life on Earth. *Bioscience*, **51**, pp. 933–938, 10.1641/0006-3568(2001)051[0933:TEOTWA]2.0.CO;2.

Pachauri, R.K., Allen, M.R., Barros, V.R., Broome, J., Cramer, W., Christ, R., Church, J.A., Clarke, L., Dahe, Q., Dasgupta, P. and Dubash, N.K., 2014, *Climate change 2014: synthesis report. Contribution of Working Groups I, II and III to the fifth assessment report of the Intergovernmental Panel on Climate Change* (p. 151).

Parker, A., Eckersley, L., Smith, M., Goudie, A., Stokes, S., Ward, S. and Hodson, M., 2004, Holocene vegetation dynamics in the northeastern Rub'al-Khali desert, Arabian Peninsula: a phytolith, pollen and carbon isotope study. *Journal of Quaternary Science*, **19**, pp. 665–676.

Parker, A., 2010, Pleistocene climate change in Arabia: developing a framework for hominin dispersal over the last 350 ka. In *The evolution of human populations in Arabia*, edited by Petraglia, M.D. and Rose, J.I., (Dordrecht: Springer), pp. 39–49.

Parker, A., Goudie, A., Stokes, S., White, K., Hodson, M., Manning, M. and Kennet, D.. 2006, A record of Holocene climate change from lake geochemical analyses in southeastern Arabia. *Quaternary Research*, **66**, pp. 465–476, 10.1016/j.yqres.2006.07.001.

Patzelt, A., 2015, Synopsis of the flora and vegetation of Oman, with special emphasis on patterns of plant endemism. *Abhandlungen der Braunschweigischen Wissenschaftlichen Gesellschaft*, **282**, pp. 317.

Pausata, F., Zanchettin, D., Karamperidou, C., Caballero, R., and Battisti, D., 2020, ITCZ shift and extratropical teleconnections drive ENSO response to volcanic eruptions. *Science Advances*, **6**, eaaz5006, 10.1126/sciadv.aaz5006.

Petit-Maire, N., Carbonel, P., Reyss, J. L., Sanlaville, P., Abed, A., Bourrouilh, R. and Yasin, S., 2010, A vast Eemian palaeolake in Southern Jordan (29 N). *Global and Planetary Change*, **72**, pp. 368–373, 10.1016/j.gloplacha.2010.01.012.

Petraglia, M., 2011, Trailblazers across Arabia. *Nature*, **470**, pp. 50–51, 10.1038/470050a.

Petraglia, M., Groucutt, H., Guagnin, M., Breeze, P. and Boivin, N., 2020, Human responses to climate and ecosystem change in ancient Arabia. *Proceedings of the National Academy of Sciences*, **117**, pp. 8263–8270, 10.1073/pnas.1920211117.

Pietsch, D., Kühn, P., Scholten, T., Brunner, U., Hitgen, H. and Gerlach, I., 2010, Holocene soils and sediments around Ma'rib Oasis, Yemen: Further Sabaean treasures? *The Holocene*, **20**, pp. 785–799, 10.1177/0959683610362814.

Preusser, F., Radies, D. and Matter, A., 2002, A 160,000-year record of dune development and atmospheric circulation in Southern Arabia. *Science*, **296**, pp. 2018–2020, 10.1126/science.1069875.

Purdue, L., Charbonnier, J., Régagnon, E., Calastrenc, C., Sagory, T., Virmoux, C., Crépy, M., Costa, S. and Benoist, A., 2019, Geoarchaeology of Holocene oasis formation, hydro-agricultural management and climate change in Masafi, southeast Arabia (UAE). *Quaternary Research*, **92**, pp. 109–132, 10.1017/qua.2018.142.

Roberts, A., Rohling, E., Grant, K., Larrasoaña, J. and Liu, Q., 2011, Atmospheric dust variability from Arabia and China over the last 500,000 years. *Quaternary Science Reviews*, **30**, pp. 3537–3541, 10.1016/j.quascirev.2011.09.007.

Robinson, S., Black, S., Sellwood, B. and Valdes, P., 2006, A review of palaeoclimates and palaeoenvironments in the Levant and Eastern Mediterranean from 25,000 to 5000 years BP: setting the environmental background for the evolution of human civilisation. *Quaternary Science Reviews*, **25**, pp. 1517–1541, 10.1016/j.quascirev.2006.02.006.

Rose, J. and Usik, V., 2010, The "Upper Paleolithic" of South Arabia. In *The evolution of human populations in Arabia*, (Dordrecht: Springer), pp. 169–185.

Rosenberg, T., Preusser, F., Fleitmann, D., Schwalb, A., Penkman, K., Schmid, T. and Matter, A., 2011, Humid periods in southern Arabia: windows of opportunity for modern human dispersal. *Geology*, **39**, pp. 1115–1118, 10.1130/G32281.1.

Rossignol-Strick, M., 1999, The Holocene climatic optimum and pollen records of sapropel 1 in the eastern Mediterranean, 9000–6000 BP. *Quaternary Science Reviews*, **18**, pp. 515–530, 10.1016/S0277-3791(98)00093-6.

Sander, K., 2006, Holocene climate and hydrologic changes recorded in Tufa and Lacustrine deposits in Southern Yemen. PhD diss., University of South Florida, USA.

Schiettecatte, J., 2013, Le palmier dattier (*Phoenix dactylifera* L.) dans l'Arabie méridionale préislamique. *Revue d'ethnoécologie*, 4, 1356, 10.4000/ethnoecologie.

Schiettecatte, J., Chabrol, A., and Fouache, É., 2013, Landscape and settlement process in al-Kharj oasis (province of Riyadh). In The Archaeology of North *Arabia, Oases and Landscapes*, edited by Luciani, M., (Vienna: Verlag der Österreichischen Akademie der Wissenschaften), pp. 257–280.

Schulz, E., and Whitney, J., 1987, Upper Pleistocene and Holocene lakes in the an Nafud, Saudi Arabia. *Paleolimnology*, **4**, pp. 175–190, 10.1007/BF00026660.

Scotese, C. and Golonka, J., 1997, *Paleogeographic atlas* (pp. 1-45). Arlington: PALEOMAP Project, University of Texas at Arlington.

Sirocko, F., Sarnthein, M., Erlenkeuser, H., Lange, H., Arnold, M. and Duplessy, J., 1993, Century-scale events in monsoonal climate over the past 24,000 years. *Nature*, **364**, pp. 322–324, 10.1038/364322a0.

Stevens, L., Ito, E., Schwalb, A. and Wright Jr, H., 2006, Timing of atmospheric precipitation in the Zagros Mountains inferred from a multi-proxy record from Lake Mirabad, Iran. *Quaternary Research*, **66**, pp. 494–500, 10.1016/j.yqres.2006.06.008.

Stokes, S. and Bray, H., 2005, Late Pleistocene eolian history of the Liwa region, Arabian Peninsula. *Geological Society of America Bulletin*, **117**, pp. 1466–1480, 10.1130/B25554.1.

Tengberg, M., 2003, Archaeobotany in the Oman peninsula and the role of Eastern Arabia in the spread of African crops. *Food, fuel and fields: Progress in African archaeobotany*, pp.229–238.

Thalen, D., 1979, *Ecology and utilization of desert shrub rangelands in Iraq*, (The Hague: Dr. W. Junk. Publishers), 428 p.

Tierney, J., deMenocal, P. and Zander, P., 2017, A climatic context for the out-of-Africa migration. *Geology*, **45**, pp. 1023-1026, 10.1130/G39457.1.

Uerpmann, H., Potts, D. and Uerpmann, M., 2010, Holocene (re-) occupation of eastern Arabia. In *The evolution of human populations in Arabia*, edited by Petraglia, M.D., Rose, J.I. (Dordrecht, Springer), pp. 205–214.

Van Campo, E., Duplessy, J. and Rossignol-Strick, M., 1982, Climatic conditions deduced from a 150-kyr oxygen isotope–pollen record from the Arabian Sea. *Nature*, **296**, pp. 56–59, 10.1038/296056a0.

Van Zeist, W., and Bottema, S., 1977, Palynological investigations in western Iran. *Palaeohistoria*, **19**, pp. 19–85.

Watrin, J., Lézine, A.M. and Hély, C., 2009, Plant migration and plant communities at the time of the "green Sahara". *Comptes Rendus Geoscience*, **341**, pp. 656–670, 10.1016/j.crte.2009.06.007.

Webb III, T. and McAndrews, J., 1976, Corresponding patterns of contemporary pollen and vegetation in central North America. *Geological Society of America Memoir*, **145**, pp. 267–299, 10.1130/MEM145-p267.

Willcox, G., 2012, The beginnings of cereal cultivation and domestication in Southwest Asia. *A companion to the archaeology of the ancient Near East*, **1**, pp. 163–180.

Williams, J.W. and Jackson, S., 2007, Novel climates, no-analog communities, and ecological surprises. *Frontiers in Ecology and the Environment*, **5**, pp. 475–482, 10.1890/070037.

Wilkinson, M., Dumontier, M., Aalbersberg, I., Appleton, G., Axton, M., Baak, A. and Bouwman, J., 2016, The FAIR Guiding Principles for scientific data management and stewardship. *Scientific data*, **3**, pp. 1–9, 10.1038/sdata.2016.18.

Zeder, M., 2011, The origins of agriculture in the Near East. *Current Anthropology*, **52**, pp. S221–S235, 10.1086/659307.

Zohary, M., 1982, *Vegetation of Israel and Adjacent Areas*. (Wiesbaden: Dr. Ludwig Reichert Verlag).

CHAPTER 12

The challenge of pollen-based quantitative reconstruction of Holocene plant cover in tropical regions: A pilot study in Cameroon

Marie-José Gaillard[1] & Esther Githumbi[1]

Department of Biology and Environmental Science, Linnaeus University, Kalmar, Sweden

Gaston Achoundong

National Herbarium, IRAD, Yaoundé, Cameroon

Anne-Marie Lézine

Laboratoire d'Océanographie et du Climat, Expérimentation et Approche numérique/IPSL, Sorbonne University, CNRS-IRD-MNHN, Paris, France

Christelle Hély

Institut des Sciences de l'Evolution de Montpellier, Université de Montpellier and Ecole Pratique des Hautes Etudes, Université PSL, Montpellier, France

Judicaël Lebamba

Département de Biologie, Université des Sciences et Techniques de Masuku, Franceville, Gabon

Laurent Marquer[2]

Institute of Botany, University of Innsbruck, Innsbruck, Austria

Florence Mazier

Environmental Geography Laboratory, GEODE UMR-CNRS 5602, Université de Toulouse Jean Jaurès, Toulouse, France

---

[1]Other affiliation: *Department of Physical Geography and Ecosystem Science, Lund University, Lund, Sweden*
[2]Other affiliation: *Research Group for Terrestrial Palaeoclimates, Max Planck Institute for Chemistry, Mainz, Germany*

DOI 10.1201/9781003162766-12

Furong Li

School of Ecology, Sun Yat-sen University, Guangzhou, China

Shinya Sugita

Institute of Ecology, Tallinn University, Estonia

ABSTRACT: Quantitative pollen-based reconstruction of Holocene plant abundance has not been attempted in the tropics so far. Tropical vegetation is characterized by a large number of entomophilous species and pollen-vegetation modelling for such plants is widely considered too problematic. However, there is a pressing need for quantitative reconstructions of Holocene plant cover in tropical regions to better assess climate- and human-induced environmental crises over time. Of particular note is the environmental crisis at the end of the 'African Humid Period' (*c.* 3 thousand years before present [ka BP]) which remains unquantified. Here, we present results from a pilot study in the Cameroon highlands including estimates of relative pollen productivity (RPP) for major taxa of the Afromontane forests and grasslands using the Extended R-Value (ERV) model, and the first reconstructions of past regional and local Holocene plant cover using the Landscape Reconstruction Algorithm (LRA). RPP estimates were obtained for 12 taxa. We found that *Celtis, Alchornea, Syzygium, Macaranga/Mallotus, Nuxia* type, and *Schefflera* have RPPs larger than Poaceae, and Moraceae, Combretaceae/Melastomataceae, *Prunus africana/Rubus pinnatus*, Cyperaceae, and *Podocarpus* lower RPPs than Poaceae. The LRA was applied to three Holocene pollen records from Cameroon (Bambili, Mbi, Mbalang) and one from Nigeria (Tilla). The results confirm the abruptness of the forest collapse in Cameroon at 3 ka BP, and suggest that it was faster and of larger magnitude than earlier anticipated. The LRA reconstruction also indicates that the regional landscape in the mountain region was more open than uncorrected pollen percentages alone suggest. Moreover, it supports the earlier interpretation of a dense forest at Bambili 11–3.5 ka BP and the local occurrence of Afromontane forest taxa at Mbalang in the Mid-Holocene.

12.1 INTRODUCTION: THE CHALLENGE

A way to anticipate the impact of current and future anthropogenic climate change on the environment, in particular in relation to the ongoing '6th extinction' (Barnosky *et al.* 2011), is to study past crises as accurately as possible. Tropical Africa experienced a major climate-driven environmental crisis at the end of the Holocene (Lézine *et al.*, this volume, and references therein). This crisis affected all natural tropical African ecosystems. Near the equator this resulted in the collapse of Afromontane forests and the expansion of grasslands and woodland grasslands. While in the Sahara and Sahel tropical species disappeared. This crisis is well documented by numerous pollen records acquired over the last decades in western Africa (e.g. Hély *et al.* 2014; Vincens *et al.* 1999). However, the resulting land-cover change remains difficult to quantify because of the wide variety of pollen productivity and dispersal mechanisms of tropical plant species (Lézine *et al.* this volume; Bouimethraan *et al.*, this volume); which implies unexpected and variable representation of the surrounding vegetation by pollen in studies of modern pollen-vegetation relationships (e.g. Gajewski *et al.* 2002; Julier *et al.* 2019). The Intergovernmental Panel on Climate Change (IPCC) pointed out that proxy-based climate reconstructions in Africa were too limited to support regional climate change assessments (Masson-Delmotte *et al.* 2013). In this context pollen-based quantification of past land-cover change is essential for climate modelling and the understanding of past land cover-climate interactions (e.g. Gaillard *et al.* 2010) as they can be used to evaluate Dynamic Vegetation Model simulations (e.g. Lu *et al.* 2018) and Anthropogenic Land Cover Change scenarios (Kaplan *et al.* 2017; Li *et al.* 2020), and help to better evaluate carbon storage over time and its impact on climate.

Quantitative pollen-based reconstruction of Holocene plant abundance have not been attempted in the tropics so far. However, some studies have estimated Relative Pollen Productivity (RPP) of major taxa in tropical and subtropical regions of the world (e.g. Jiang *et al.* 2020; Wan *et al.* 2020), including two studies in South Africa (Duffin and Bunting 2008; Hill *et al.*, this volume) and one study in Namibia (Tabares *et al.* 2021). Here we present results from a pilot study in Cameroon testing methods and models previously used and validated in the northern Hemisphere, i.e. the Extended R-Value (ERV) model (Parsons and Prentice 1981; Prentice and Parsons 1983) and the Landscape Reconstruction Algorithm (LRA; Sugita *et al.* 2007a and 2007b). The aim was to obtain first estimates of RPPs for major plant taxa and reconstruct their cover over the Holocene at both regional and local spatial scale on the Adamawa Plateau and in the montane/sub-montane regions of Cameroon (Figure 1). Reconstruction of the relationship between the cover of woodland and open land over the 3 ka BP Holocene environmental crisis in Cameroon (e.g. Lézine *et al.* 2013) is of particular interest and represents a first attempt to quantify the crisis effect on the vegetation cover. We present major results of the study to illustrate the potential of using the ERV model and the LRA in tropical areas, and to highlight major methodological issues and possible improvements.

12.2 MATERIALS AND METHODS

12.2.1 Study Region And Study Sites

The 30 sites selected for collection of modern soil pollen and related vegetation surveys for the RPP study are distributed at altitudes between 432 and 2990 m asl. in the montane/submontane and Guineo-Congolian sectors, except one sample at the southern border of the Sudano-Guinean sector (Figure 1, Table S1). A stratified random distribution of sites was chosen to ensure a good representation of the major taxa occurring in the Holocene vegetation history of the study region. However, site distribution was also influenced by logistical issues related to the number and season of the project's field expeditions and other limitations due to local circumstances.

The Holocene pollen sites used for the LRA reconstructions fall within the geographical range of the modern sites and are located at different elevations in the volcanic mountain range of Cameroon, at 2000–2200 m asl. in the Afromontane forest sector (Bambili, Lézine *et al.* 2013; and Mbi, Lézine *et al.* 2021), and at 1110–1260 m asl. in the Guineo-Sudanian wooded grasslands (Mbalang, Vincens *et al.* 2010) (Figures 1, 2; Table S3). An additional pollen site (Tilla, Salzmann *et al.* 2002) from the Sudanian wooded grassland at 690 m asl. in Nigeria was selected for the REVEALS reconstruction of regional vegetation needed to apply the LOVE model for reconstruction of the local vegetation around Mbalang (Figure 1; Table S3). For further details on the choice of sites, please see Supporting Online Material (SOM) section 12.2.4. Henceforth, all dates are given in calibrated ^{14}C kyrs before present (CE 1950) (abbreviated ka BP).

12.2.2 Pollen And Vegetation Data

The collection of soil samples and vegetation surveys for the RPP study were performed in February 2010, October 2012, and April 2015. The pollen-vegetation dataset includes data from 4 sites in semi-deciduous forests, 10 sites in mountain forests, and 16 sites in savannahs (Figure 1;

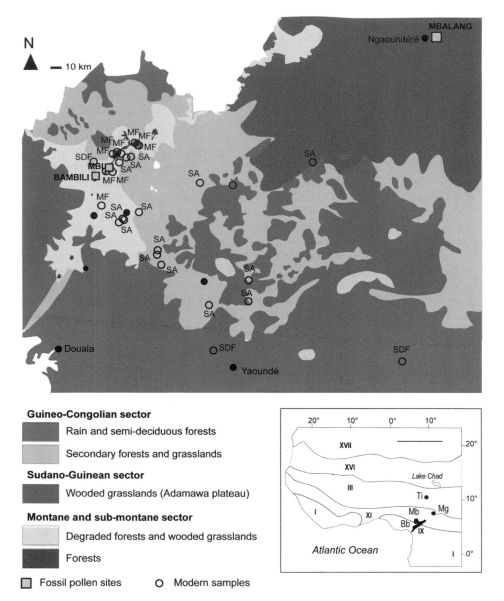

Figure 1. Location of sites for the study of relative pollen productivities (surface soil samples and vegetation surveys) and fossil pollen records used for the application of the Landscape Reconstruction Algorithm. Simplified vegetation map of Cameroon (from Letouzey (1985) and White (1983), modified). Small map, lower right corner: I. Guineo-Congolian centre of endemism; XI. Guineo-Congolian/Sudanian transition zone; III. Sudanian centre of endemism; XVI. Sahelian transition zone. XVII. Saharan transition zone. Abbreviations: modern samples: MF montane forest; SA savannah; SDF semi-deciduous forest; fossil pollen sites: Bb Bambili; Mb Mbi; Mg Mbalang; Ti Tilla.

Table S1). For further details on the methods used for collection of the pollen samples and vegetation surveys, pollen analysis, vegetation data handling, and on data access please see Supplementary Online Material, SOM-12.1.

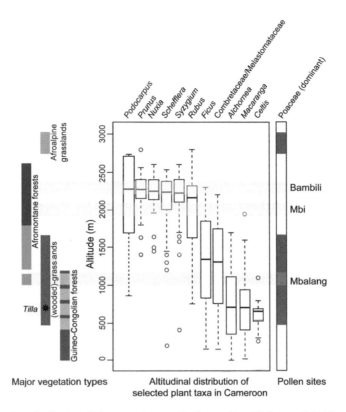

Figure 2. Altitudinal distribution of plant taxa discussed in the text (from Verlhac *et al.* 2018; modified). The altitudinal distribution of the main vegetation types (from Letouzey 1985) is shown on the left, while the altitudinal occurrence of Poaceae as a dominant taxon in wooded and open grasslands is indicated on the right. The colours refer to those in Figure 1 (except for Poaceae) and Figures 3, 5 and 6. The altitudinal location of pollen sites used for the LRA reconstructions is shown on the right, and the Nigerian site (Tilla) on the left. * indicates that the site Tilla belongs to another grassland system than the Cameroon sites (Figure 1).

12.2.3 The Erv Model and Erv Analysis

Detailed overviews of the underlying theory and assumptions of the ERV model and its developments can be found in Gaillard *et al.* (2008) and Bunting *et al.* (2013). Details on models and methods used for calculation of RPPs, and the computer programme for the ERV analysis are provided in SOM-12.1.4.

   In this study, we used the three ERV sub-models developed by Parsons and Prentice (1981), Prentice and Parsons (1983), and Sugita (1994). The application of the ERV sub-models requires the following inputs:

- Pollen data: pollen % of taxon i at site k. In this study the maximum number of taxa analysed is 26 and k is a soil sample from a 1 m² area, for 30 sites (SOM-12.1, Table S2)
- Radius of the sedimentary basin: distance in m. In this study this is the radius of the area from which soil was collected, i.e. 0.5 m.
- Reference taxon: the RPP of the reference taxon is set to 1, and the RPP of all other taxa included in the analysis are related to that reference taxon. In this study Poaceae is used as the reference taxon.

- Vegetation data: Non-distance-weighted vegetation abundance of taxon i around site k. In this study percentage cover is used, from 0.5 to 3000 m in 5 m-concentric rings working out from the sampling location.
- Z_{max}: distance within which most pollen comes from. In this study estimated to be 200 km (see SOM-12.1.4).

The ERV model also requires that vegetation data is distance-weighted. We used three alternative models, 1/d (Prentice and Webb 1986), 'Prentice's model' for bogs (uses a Gaussian plume diffusion model (P-GPM) and assumes pollen deposition at one point; Prentice (1985)), and the Lagrangian Stochastic Model (LSM) (Kuparinen 2006; Kuparinen *et al*. 2007). P-GPM and LSM describe dispersion and deposition of pollen in different ways (e.g. Theuerkauf *et al*. 2013). For the application of these two models, additional inputs are needed:

- Fall speed of pollen (FSP) (m/s) of each taxon used in the analysis; FSP was calculated for the 26 taxa that were used in the ERV 'start run' (Table S2; see SOM-12.1 for details on the ERV analysis strategy and methods to calculate fall speed of pollen).
- Wind speed (m/s), in this study 3 m/s.; this wind speed is widely used in the RPP studies performed in the Northern Hemisphere; it is also close to the modern average 10-m wind speed over land in Europe and China (e.g. Grassi *et al*. 2015), and the global average 10-m wind speed over land (3.28 m/s (class 1), Archer and Jacobson 2005).

The ERV analysis estimates the RPPs and their standard errors (SEs) using the maximum likelihood method (Stuart and Ord 1994) and quantifies the best 'linear fit' between pollen percentages and distance-weighted plant abundances using the likelihood function score (LFS, Prentice and Parsons 1983) or log likelihood (log L; e.g. Li *et al* 2017) over distance (see SOM-12.1 for details). Beyond the distance at which the best linear fit is achieved, the curve of LFSs and log Ls reach an asymptote, i.e. do not decrease, respectively increase with distance.

The main outputs of one ERV analysis using the three ERV sub-models and the three distance-weighting models, i.e. nine runs of ERV analysis in total are:

- Vegetation dispersion with distance (a measure of vegetation stationarity, see SOM-12.1.3) (Figure 3A),
- Mean plant composition with distance for each taxon used in the analysis (Figure 3A),
- Adjusted pollen proportions (submodels 1 and 2) or relative pollen loading (submodel 3) and vegetation proportions (1, 2) or absolute vegetation abundance (3) as calculated by the maximum likelihood method; empirical modern pollen data are expressed in percentages, which implies a non-linear pollen-vegetation relationship that would not have any algebraic solution for the RPPs and the 'background pollen' in the ERV model equation. The pollen-vegetation relationship needs therefore to be linearized (Figure 4) (see SOM-12.1.4 for details),
- LFS and log L with distance (Figure 3A),
- RSAP (*sensu* Sugita 1994) calculated with the moving-window regression approach developed by S. Sugita (*in* Gaillard *et al*. 2008); RSAP is Z_{RSAP} in the ERV model equation, i.e. the distance beyond which the 'background pollen' (second term in the ERV model's equation, ω_i) is constant between sites within a region characterized by stationary vegetation; it is also the distance at which the linearity of the pollen-vegetation relationship is best and does not improve with distance (see SOM-12.1.4 for details); stationary vegetation implies that any area (e.g. 10 km × 10 km) has the same species composition as any other such area within the study region,
- RPP and their standard errors (SEs) with distance for each taxon used in the analysis (Figure 3B).

Details on the ERV analysis strategy are provided in SOM-12.1.4. The first ERV analysis run used the 26 taxa with both pollen and vegetation data in ≥ 5 sites (Table S2). Based on plots

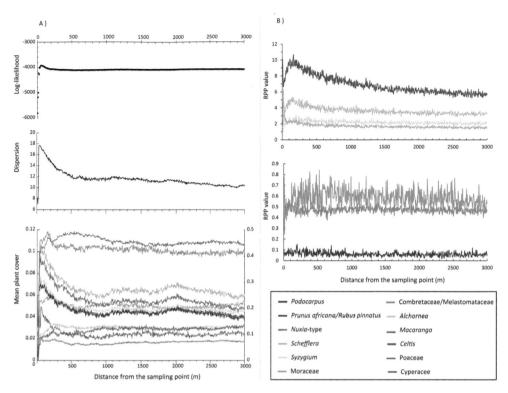

Figure 3. Examples of results from the ERV analysis of pollen and vegetation data from 30 sites in Cameroon (see Figure 1 for site distribution). This analysis used 12 taxa, ERV submodel 2 and the Prentice's Gaussian Plume Model for distance weighting of vegetation (see Methods and SOM-12.1.4 for details). (A) Log likelihood, vegetation dispersion, and mean taxa composition (in proportion) over distance from the pollen sample points; note that the scale for Poaceae is on the right of the diagram. (B) Relative Pollen Productivities (RPPs) with distance from the pollen sample points for seven taxa. The colours relate to those in Figures 1, 2, 5 and 6; the five shades of red (dark brownish red to light pink) relate to the red colour in Figures 1 and 2 and to the five shades of red in Figures 5 and 6; the three shades of green relate to Figures 1 and 2 two shades of green together and to the three shade of green in Figures 5 and 6.

of adjusted pollen percentages and vegetation proportions (Figure 4), the taxa with the best linear relationships were selected for the next runs of ERV analysis. In this paper, we present the results of the ERV analysis that produced the curve of log L closest to the theoretically correct curve (increases and reaches an asymptote at a certain distance; SOM-12.1.4), which was obtained with 12 taxa, ERV sub-model 2, and P-GPM for distance weighting of vegetation (Figure 3A; Table 1A).

12.2.4 Landscape Reconstruction Algorithm (LRA)

Details on the LRA and its assumptions are found in SOM-12.2.3. LRA comprises two models, REVEALS (REgional VEgetation Abundance from Large Sites) (Sugita 2007a) and LOVE (LOcal VEgetation Estimates) (Sugita 2007b). Both models reduce biases due to inter-taxonomic differences in pollen productivity and basin size. REVEALS estimates the mean regional vegetation composition of plant taxa in proportion (dimensionless) for a large region (minimum 100 km × 100 km; Hellman *et al.* 2008) using pollen records from large lakes. REVEALS assumes that no plants are growing on the deposition basin (Sugita 2007a). Therefore, model

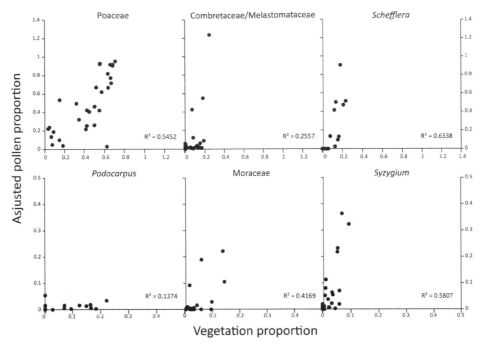

Figure 4. Examples of ERV-corrected pollen-vegetation relationships for six taxa from the ERV analysis using the full pollen-vegetation dataset (30 sites), 12 taxa, ERV submodel 2 and the Prentice's Gaussian Plume Model for distance weighting of vegetation (see Methods and SOM-12.1 for details). The coefficient of determination (R^2) is provided as a measure of linearity, the higher the value the more linear the relationship.

estimates from pollen assemblages in large bogs will be biased by pollen coming from the surface of the bog. LOVE estimates 'background pollen' using REVEALS estimates of regional vegetation, and subtracts 'background pollen' to derive local vegetation cover within the RSAP (see ERV model above and SOM-12.1). The RSAP represents the smallest spatial scale of the LOVE reconstruction. Both REVEALS and LOVE provide estimates of plant abundance/cover with their SEs. Note that the plant abundance estimated by LOVE is distance weighted.

The application of the LRA requires a minimum of two pollen records from the same vegetation/landscape type, one from a large lake for a REVEALS-based reconstruction of the regional plant cover, and one from a small lake for a LOVE reconstruction of the local plant cover. However, calculation of the RSAP of small sites requires pollen records from several sites (see SOM-12.2.3 for further explanations). Due to the small number of pollen records appropriate for a LRA in the study region, we made the following assumptions:

(1) Pollen of Cyperaceae are primarily of local origin, growing on the swamp (Mbi) or at lake shores (Bambili, Tilla, Mbalang), and Cyperaceae are the dominant plants in these biotopes,
(2) The RSAP of the small sites is 2 km (SOM-12.2.3),
(3) The pollen record (Cyperaceae excluded) from Mbi and Tilla represent the mean regional vegetation cover around Bambili and Mbalang, respectively.

Possible departures from these assumptions are described in SOM-12.2.3. A local origin of most pollen from Cyperaceae is supported by the analysis of > 1000 modern pollen samples from Africa, Madagascar and Arabia (Gajewski *et al.* 2002) suggesting that Cyperaceae pollen is related primarily to humid regions and habitats in Africa.

Table 1. Habitus, pollination, Fall Speed of Pollen (FSP) and Relative Pollen Productivities (RRP) with their standard errors (SE) for 12 major taxa in: (A) Cameroon (this study), (B) from other studies in South Africa (Hill *et al.*, this volume); Namibia (Tabares *et al.* 2021); Bolivia (Whitney *et al.* 2018); Brazil (Piraquive Bermúdez *et al.* 2019); and (C) from other studies in Subtropical China: left (Fang *et al.* 2020), centre (Jiang *et al.* 2020), right (Chen *et al.* 2019); Tropical China (Wan *et al.* 2020). ˆ FSP from Jiang *et al.* (2020). Reference taxon (Poaceae=1): * indicates that Poaceae is the reference taxon used in the analysis. ** indicates that other taxa were used as reference and RPP values were converted to Poaceae=1. Abbreviations: H herb, S shrub, T tree; Pollinat Pollination; A anemophilous, E entomophilous; (A) and (E) indicate that the one type of pollination is less common than the other.

(A) Cameroon (this study)

Taxa	Habitus	Pollinat.	FSP (m/s)	RPP	SE
Alchornea	S/T	E/A	0.012	4.00	0.188
Celtis	T	E/A	0.010	7.00	0.270
Combretaceae/Melastomataceae	T/H	E	0.012	0.50	0.039
Cyperaceae	H	A/(E)	0.022	0.23	0.045
Macaranga	T	E/(A)	0.013	1.70	0.054
Moraceae	T	A/E	0.006	0.60	0.070
Nuxia type	T	E	0.003	1.40	0.397
Poaceae	H	A	0.035	1.00*	0.000
Podocarpus	T	A	0.017	0.06	0.018
Prunus africana/Rubus pinnatus (Rosaceae)	T	E	0.013 (0.021)ˆ	0.40	0.037
Schefflera	T	E	0.009	1.80	0.164
Syzygium	T	E	0.003	2.20	0.107

(B) Other studies:	S Africa	Namibia		Bolivia	Brazil			
Taxa	RPP	RPP	SE	RPP	RPP	SE	RPP	SE
Alchornea								
Celtis								
Comb./Mel.		0.217	0.014					
Cyperaceae					0.48	0.11	0.63	0.33
Macaranga								
Moraceae				0,25				
Nuxia type								
Poaceae	**1.00***	**1.00***	**0.00**	1.00**	1.00**	0.00	1.00**	0.00
Podocarpus	6.50							
P. africana/R. pinnatus (Rosaceae)								
Schefflera								
Syzygium								

(Continued)

Table 1. Continued

(C) Other studies:	Subtropical China						Tropical China	
Taxa	RPP	SE	RPP	SE	RPP	SE	RPP	SE
Alchornea								
Celtis								
Comb./Mel.								
Cyperaceae								
Macaranga							2.21	0.08
Moraceae					0.50	0.35	6.52	0.08
Nuxia type								
Poaceae	**1.00***	**0.00**	1.00**	0.00	1.00**	0.00	**1.00***	**0.00**
Podocarpus								
P. africana/R. pinnatus								
(Rosaceae)	0.41	0.28	0.84	0.045	1.00	0.45		
Schefflera								
Syzygium								

Relying on the three assumptions above, we used the pollen records from Mbi and Tilla (large lake sites) for REVEALS reconstructions of the regional plant abundance, and from Bambili and Mbalang (small lake sites) for the LOVE reconstructions of the local plant abundance around each lake. We excluded Cyperaceae from the pollen records and used the 11 taxa for which we have RPP estimates (Table 1A). In addition, we tested alternative RPP values for *Podocarpus*, i.e. the mean RPPs of *Pinus* (6.38) and *Picea* (2.62) in Europe, to evaluate their effect on the reconstructions and discuss them in the light of the existing knowledge on the dispersal and deposition of *Podocarpus* pollen. Only one additional RPP value is available for *Podocarpus* to date, i.e. 6.54 (related to Poaceae) from a study in South Africa (Duthie 2015; Hill *et al.*, this volume) (Table 1B). *Pinus*, *Picea* and *Podocarpus* all belong to coniferous plants for which large pollen productivity is common. Note also that Hill *et al.*'s value is very close to *Pinus* mean RPP in Europe. The REVEALS and LOVE reconstructions were performed for continuous 500-year time windows over the last 12 ka years, except for Mbalang (last 6 ka years). Z_{max} was set to 200 km and wind speed to 3 m/s (see 12.1.4, above, and SOM-12.1.4). We used the computer programme LRA.REVEALS.v6.2.4.exe and LRA.LOVE.v6.2.3.exe (Sugita, unpublished) to calculate the REVEALS estimates of regional plant cover at Mbi and Tilla and LOVE estimates of local plant cover within 2 km-radius areas around Bambili and Mbalang. See SOM-12.2.3 for details on access to the computer programmes.

12.3 RESULTS

12.3.1 ERV analysis and RPP estimates

Vegetation dispersion and mean plant cover over distance from the pollen sample points (Figure 3A) based on the collected vegetation dataset (SOM-12.1.3) indicate that the ERV model's assumption of stationary vegetation is violated in the study region. Both vegetation dispersion and mean plant cover increase with distance until maximum values at *c.* 50–100 m and then decrease until *c.* 600 m instead of reaching an asymptote. The values of mean plant cover for *Alchornea, Macaranga, Celtis*, Moraceae and Poaceae indicate that these taxa have a more stable composition already from *c*. 200 m. The log likelihood curve shows a similar pattern as vegetation dispersion, although with a much less prominent 'overshoot' of the curve at 50 m. From 600 m, the values of mean plant cover remain relatively constant for all plant taxa. This distance also corresponds to the RSAP (585 m) as calculated from the log L curve (Figure 3A) with the moving-window regression method. An attempt to analyse forest and savannah sites separately to decrease vegetation dispersion did not improve the results; instead, the log L curves diverged from the theoretically correct curve even more (results not shown).

Most curves of RPPs plotted against distance from the pollen sample point deviate from the expected behaviour (increase to reach an asymptote at a certain distance), but the RPP values are relatively constant from *c*. 1500 m for all taxa except *Celtis*. The RPPs of *Podocarpus* exhibit more or less constant values over 3 km. The RPPs of Combretaceae/Melastomataceae and *Schefflera* reach maximum values at very short distances, decrease abruptly, and maintain relatively constant values from a few metres. The RPPs of *Alchornea* and *Celtis* increase to maximum values at *c*. 200 m and then decrease gradually until 1500 m and 3000 m, respectively. The RPPs of Moraceae and *Syzygium* increase until 200 m where they reach an asymptote, although the values of Moraceae are very unstable. Because RPPs still decrease after the calculated RSAP distance (585 m), we selected the final RPP values at the distance for which they stabilize for all taxa, i.e. 1500 m (Table 1A).

Figure 4 presents examples of ERV model-corrected pollen-vegetation relationships for six of the 12 taxa. *Schefflera*, Poaceae and *Syzygium* exhibit the most linear relationships with coefficients of determination (R^2) of 0.63, 0.54 and 0.58, respectively. Moraceae and Combretaceae/Melastomataceae have weaker linear relationships with R^2 values of 0.42 and 0.26, respectively. A R^2 value of 0.14 indicates a very weak linearity for *Podocarpus*, Combretaceae/Melastomataceae, *Schefflera*, and *Syzygium* are strongly overrepresented, Poaceae is slightly overrepresented, and *Podocarpus* strongly underrepresented by pollen. The pollen-vegetation relationship of Moraceae clearly indicates either overrepresentation or underrepresentation by pollen depending on the site.

12.3.2 Landscape Reconstruction Algorithm (LRA): Regional and local plant cover

The REVEALS reconstructions represent regional plant abundance in an area of *c*. 100 km × 100 km around Mbi and Tilla, and the LOVE reconstructions the local plant abundance in an area of minimum 2 km radius around Bambili and Mbalang (see Methods and SOM-12.2.4). The REVEALS and LOVE estimates with their SEs for the 11 taxa are provided for Mbi and Bambili in Figures S2–S5) and highlight the effect of the RPPs' SEs on the reconstructions. *Podocarpus*, *Nuxia*-type, Moraceae, and *Prunus africana/Rubus pinnatus* have the largest SEs in relation to their RPP value (Table 1A). These taxa also exhibit the largest SEs of LRA reconstructions. When RPP 0.06 is used for *Podocarpus*, SEs are very large for *Podocarpus* and relatively large for *Nuxia*-type (for both REVEALS and LOVE plant cover) (Figure S2). SEs are also quite large for Moraceae and *Prunus africana/Rubus pinnatus* (for LOVE plant cover) in the Early Holocene (Figures S2 and S3). All other taxa exhibit small SEs over most of the Holocene. There are generally few values not different from zero (i.e. SD ≥ REVEALS/LOVE estimate) and they occur mainly for very low values of plant cover, except for *Podocarpus* (RPP 0.06) that has also very large SEs for high LOVE plant cover over most of the Holocene (Figure S2).

12.3.2.1 Mountain woodlands: Mbi and Bambili (Figure 5; Figures S2–S5)
The pollen sum of the 11 taxa used in the LRA reconstructions represent 61–81% (mean 69.4%) (Mbi) and 59–76% (mean 65.5%) (Bambili) of the total pollen sum of terrestrial plant taxa (SOM-12.2.1, Table S3A). The REVEALS and LOVE reconstructions are strongly influenced by the RPP used for *Podocarpus*. The low RPP (0.06) results in a very strong dominance of *Podocarpus* 9.5–0.5 ka BP in the regional vegetation (Mbi) and 10.5–2 ka BP locally around

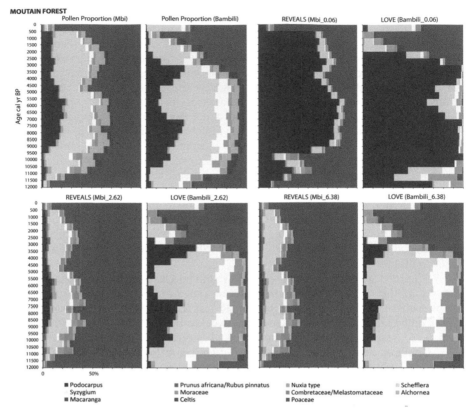

Figure 5. REVEALS-based regional and LOVE-based local plant cover using the pollen records from Mbi and Bambili, respectively, and three alternative RPPs for *Podocarpus* (0.06, 2.62, and 6.38). Plant cover in percentage of an area of 100 km × 100 km for Mbi, and of minimum 2 km radius for Bambili. Upper panel: pollen percentages for Mbi and Bambili, REVEALS (Mbi)- and LOVE (Bambili)-based plant cover using 0.06 for *Podocarpus*; Lower panel: REVEALS (Mbi)- and LOVE (Bambili)-based plant cover using RPP 2.62 (left) and 6.38 (right) for *Podocarpus*. The colours relate to those in Figures 1, 2 and 3; the five shades of red correspond to carmine red and three shades of green to the two shades of green together in Figures 1 and 2.

Bambili. The other nine tree taxa have a reduced abundance compared to their pollen %, except Moraceae. Regional woodland cover (Mbi) is larger than total tree pollen %, and local woodland cover at Bambili is 100% 12–3 ka BP while percentages of total tree pollen increase from *c.* 50% (12 ka BP) to 97–98% (8–4.5 ka BP) to decrease back to *c.* 55% at 3 ka BP.

When higher RPPs (2.62 and 6.38) are used for *Podocarpus*, regional openland (Poaceae) around Mbi represents > 50% of the total vegetation cover over the entire Holocene, while pollen % exceed 50% only 12–11 ka BP, 5–4.5 ka BP, and 1 ka BP to present. In contrast, local vegetation cover at Bambili is 100% wooded 11–3.5 ka BP, and the decrease of woodland from 3 ka BP is much larger and more abrupt than the decrease in total tree pollen %. Woodland cover drops from 100% to 29% (RPP 2.62) or 22% (RPP 6.38) between 3.5 and 3 ka BP, whereas pollen % decrease from 98% to 55% between 4.5 and 3 ka BP. While *Podocarpus* maintains pollen % > 10% at Bambili 7–4.5 ka BP, its reconstructed local cover around the site is < 10% and not different from zero 6–5.5 ka BP (Figure S2). The major effect of using RPP 2.62 rather than 6.38 on the reconstruction of regional vegetation is a larger cover (*c.* double) of *Podocarpus* throughout the Holocene at the expense of Poaceae, while the cover of other trees does not change significantly. The largest differences in composition of tree taxa between pollen %

and regional plant cover (Mbi) is mainly a larger cover of Combretaceae/Melastomataceae in relation to *Alchornea, Macaranga,* and *Celtis.* In terms of the local tree composition around Bambili, the major difference between pollen percentage and plant cover is a larger cover of *Schefflera* at the expense of *Podocarpus* 10.5–3.5 ka BP, and a larger cover of Moraceae and *Macaranga* at the expense of *Alchornea* and *Celtis* (except for the last 50 years). In recent time, the cover of Combretaceae/Melastomataceae is larger than earlier and *Schefflera* dominates the tree cover.

12.3.2.2 The Sudanian and Guineo-Congolian wooded grasslands: Tilla and Mbalang

The pollen sum of the 11 taxa used in the LRA reconstructions represent 68.1–95.8% (mean 82.7%) (Tilla) and 48.7–94% (mean 72.2%) (Mbalang) of the total pollen sum of terrestrial plant taxa (SOM-12.2.1, Table S1A) (Figure 6). The REVEALS reconstructions of regional plant cover around Tilla are characterized by a very large cover of openland (Poaceae) throughout the Holocene whatever RPP is used. Moreover, the cover of Poaceae is larger than their pollen %.

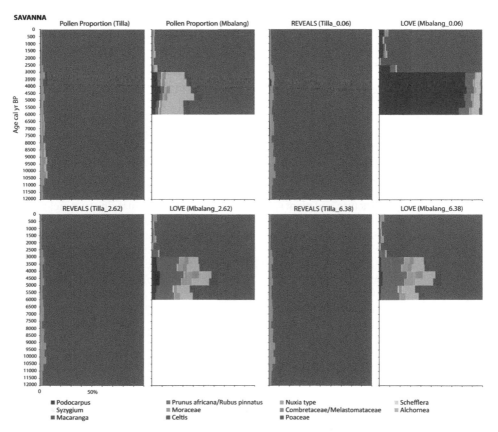

Figure 6. REVEALS-based regional and LOVE-based local plant cover using the pollen records from Tilla and Mbalang, respectively, and three alternative RPPs for *Podocarpus* (0.06, 2.62, and 6.38). Plant cover in percentage of an area of 100 km × 100 km for Tilla, and an area of minimum 2 km radius for Mbalang. Upper panel: pollen percentages for Tilla and Mbalang, REVEALS (Tilla)- and LOVE (Mbalang)-based plant cover using 0.06 for *Podocarpus*; Lower panel: REVEALS (Tilla)- and LOVE (Mbalang)-based plant cover using RPP 2.62 (left) and 6.38 (right) for *Podocarpus.* The colours relate to those in Figures 1 and 2; the five shades of red correspond to carmine red and three shades of green to the two shades of green together in Figures 1 and 2.

There is no significant difference between the three reconstructions except a slightly larger cover of *Podocarpus* when RPP 0.06 is used. The total cover of trees is smaller than tree pollen %, and Combretaceae/Melastomataceae are dominant over the other tree taxa, which is not the case in pollen %.

The LOVE reconstructions for Mbalang exhibit a total cover of tree of 100% with RPP 0.06 for *Podocarpus* over the period 6–3 ka BP, and of 40–60% with RPPs 2.62 and 6.38. With RPP 0.06, *Podocarpus* is dominant, while all other trees have a cover smaller or much smaller than their pollen percentage, except *Prunus africana/Rubus pinnatus* and Moraceae that have a higher, respectively slightly higher cover than their pollen %. With RPPs 2.62 and 6.38, Poaceae has a cover slightly smaller than its pollen %, especially 6–3 ka BP. *Prunus africana/Rubus pinnatus* has a larger cover than its pollen % over the entire period 6–0 ka BP, as well as Moraceae and *Macaranga* 6–3 ka BP, and Combretaceae/ Melastomataceae 4.5–3 ka BP. *Podocarpus*, Combretaceae/ Melastomataceae, *Celtis* and *Alchornea* have a smaller cover than their pollen %. LOVE plant cover and pollen of *Syzygium* and *Schefflera* are similar in proportions. *Podocarpus* has a larger cover in the reconstruction with RPP 2.62 than in the one with RPP 6.38. The decrease in total trees at 3 ka BP is sharp in both cover and pollen %, but the decrease in cover (48 to 9%) is larger than in pollen % (30 to 7%).

12.4 DISCUSSION

12.4.1 Relative Pollen Productivity (RPP)

12.4.1.1 Effects of violation of model assumptions and other factors

The effects on RPPs of deviations from the ERV model assumptions and other factors have been discussed earlier in syntheses of RPP values in Europe (Broström *et al.* 2008; Mazier *et al.* 2012), temperate China (Li *et al.* 2017) and the Northern hemisphere (Wieczorek and Herzschuh 2020). The number of factors affecting RPPs and/or causing between-study variability in RPP values for a taxon are numerous. They span from between-study region differences in climate, land use, vegetation type/structure and plant species (including their pollination strategies), between-study variation in methods used to collect the pollen-vegetation dataset (including methods in field and laboratory, and data handling), to differences between models and/or model implementation (including dissimilarities in parameter setting, parameter values, and calculation procedures). Because of the relatively low number of RPP values (response variables) compared to the number of factors (explanatory variables), statistical quantification of the effect of these factors on RPP estimates has never been attempted so far. In this pilot study, the major factors that need to be considered with care are the violation of the ERV model's assumptions in terms of stationarity of the vegetation in the study area, random distribution of sites, mode of dispersal and deposition of pollen, and taphonomy and preservation of pollen in soil samples.

Vegetation dispersion and mean plant composition over distance (Figure 3A) indicate that vegetation is not stationary in the study region, and site distribution did not follow a true random scheme, but a stratified random one. Both factors may explain the shape of the log L curves obtained from the ERV runs (example in Figure 3A) (e.g. Broström *et al.* 2005; Li *et al.* 2017). Moreover, the difference in vegetation data collection within and beyond 100 m (also seen in the difference in mean plant composition within and beyond 100 m, Figure 3A) may also cause the 'overshoot' of log L values at short distances (< 100 m) before the curve decreases and/or plans out. The shape of the log L curve implies that the definition of the RSAP is uncertain. Based on the empirical and simulation study of Broström *et al.* (2005) on the effect of non-random site distribution on likelihood function scores and the identification of the RSAP, we assume that the large RSAP, i.e. the distance at which all curves of RPP with distance stabilize (1500 m), is most realistic.

Eight of the twelve plant taxa involved in the analysis are entomophilous, of which three have been described as partly anemophilous (Table 1A). The assumption of the ERV model that all pollen is transported by wind is therefore violated. However, these taxa are common in lake sediments and peat deposits of tropical Africa, which suggests that part of the pollen produced is transported by wind. Therefore, an estimate of the relationship between vegetation and pollen transported to palaeoecological archives such as lakes and bogs remains useful. Nevertheless, such estimate will not be comparable with true pollen productivity in the plant stamina, but can be used as a measure of the proportion of pollen transported and deposited in soils, moss pollsters, lakes and bogs. The use of soil samples instead of moss pollsters in regions where mosses are uncommon and/or poor in pollen (e.g. Wan *et al.* 2020) may also imply biases of the pollen assemblages compared to mosses/lake sediment/peat. Gajewski *et al.* (2002) analysed a database of all surface pollen data available in Africa and concluded that percentages of the three most abundant pollen taxa did not show any bias among sampling methods. However many studies have indicated biases of pollen assemblages in soil samples compared to, e.g. pollen traps due to poor preservation conditions (e.g. Zhang *et al.* 2015). We therefore assume that the most common pollen types in soils are those that are well preserved due to their species/taxon-specific wall structure, and that their amount corresponds to the pollen transported and deposited in the soil. To our knowledge, little is known on the number of years of pollen deposition in soil samples (e.g. Havinga 1984). Thus, we assume that they represent some years of pollen deposition. One of the basic assumptions of pollen analysis as a tool to reconstruct past vegetation change is that pollen productivity is a constant, although pollen-trap studies have shown that inter-annual variation occurs in the absence of changes in plant cover (e.g. Abraham *et al.* 2021; Gosling *et al.* 2005; Haselhorst *et al.* 2013; Julier *et al.* 2019). These known variations are considered as 'noise' in terms of the pollen productivity measure needed to interpret sedimentary pollen records (e.g. Fang *et al.* 2020). Individual pollen assemblages from lake sediments or peat used? to be from multiple years-samples, and each quantitative reconstruction of plant cover using the LRA (REVEALS and LOVE) seldom represents less than 100 years, and generally covers 100 to 500 years (e.g. Cui *et al.* 2014; Trondman *et al.* 2015).

Fall Speed of Pollen (FSP) is an important parameter in pollen-vegetation modelling. Between-study differences in FSP for a given taxon will influence the RPP results (e.g. Bunting *et al.* 2004). However, there is little difference in the estimated FSPs of pollen taxa between this study and other studies in the tropics and subtropics, except for *Podocarpus* (0.0175 m/s in this study; 0.074 in Hill *et al.*, this volume). Fall speed for Poaceae varies between studies from 0.0133 m/s in tropical China (Wan *et al.* 2020) to 0.050 m/s in South Africa (Hill *et al.*, this volume). However, the FSP estimate from the study in Namibia (0.031 m/s, Tabares *et al.* 2021) is close to the FSP we are using (0.035 m/s; Broström *et al.* 2004).

12.4.1.2 Comparison with RPPs from other studies in the tropics

RPP values are available from other studies in the tropics for six taxa included in our study (number of values in brackets, our study excluded): Combretaceae/Melastomataceae (1), Cyperaceae (2), *Macaranga* (1), *Moraceae* (3), *Podocarpus* (1) and Rosaceae (3; for comparison with the RPP of *Prunus africana/Rubus pinnatus*) (Table 1B). RPPs for each of these taxa are surprisingly similar given that studies are from geographically very different regions and often use different methods. The comparison confirms RPPs smaller than that of Poaceae for Combretaceae/Melastomataceae, Cyperaceae, and Rosaceae, and RPPs larger than that of Poaceae for *Macaranga*. However, Moraceae exhibit contradicting results with RPPs that are either larger (one value in tropical China) or smaller (two values) than the RPP of Poaceae. Similarly, *Podocarpus* has a much smaller RPP than Poaceae in our study, but much larger RPP than Poaceae in South Africa (see Hill et al., this volume). Earlier studies have also reported that both Moraceae and *Podocarpus* can exhibit high and low pollen quantities in surface samples within only a few meters (e.g. Gosling *et al.* 2005; Maley *et al.* 1990; Vincens 1984; Wan *et al.* 2020). Moreover, results indicated that *Podocarpus* pollen % in surface soils were often high within the forest (up to 40% in Vincens (1984) or 60% in Maley *et al.* (1990)), and decreased abruptly over the forest edge (to *c.* 10%) and beyond, to < 5% already 10 m from the edge (Maley *et al.* 1990) or to *c.* 5% and 1% 10 km and 50 km from the edge, respectively (Vincens 1984). Lézine (unpublished) also reported that *Podocarpus* pollen percentage in soil samples within the *Podocarpus* forests of Cameroon can vary between very low and high values within a few meters. The very low RPP of *Podocarpus* obtained in our study is probably due in part to the low proportions of the taxon in the studied forests (*c.* 8–23%, max. 29%, in distance-weighted plant abundance within 100 m of the studied sites) with corresponding pollen percentages of 0.2–2.5%, maximum 8% for 29% plant cover. Such 'fuzzy' pollen-vegetation relationships need to be further investigated. New, carefully designed RPP studies are necessary before we can conclude on this issue and the relevance of quantitative reconstruction for this type of taxa. Moreover, the pollen-vegetation relationship of *Podocarpus* can be biased by long-distance pollen transport, as reported in earlier studies (e.g. Maley *et al.* 1990; Vincens 1984; Wan *et al.* 2020).

Earlier studies of modern and fossil pollen spectra and related vegetation provide some hints on whether the RPP we obtained for the remaining six taxa roughly agree with those results in terms of representation of vegetation by pollen. Pollen of several entomophilous tropical taxa have been found in large numbers, e.g. *Alchornea* (Elenga *et al.* 2000; Reynaud-Farrera 1995; Vincens *et al.* 1997, 2000; Watrin *et al.* 2007), and *Celtis* and *Syzygium* (e.g. Verlhac *et al.* 2018; Watrin *et al.* 2007). Lézine *et al.* (2013) suggested that *Schefflera* might be overrepresented in the pollen record from Bambili due to high pollen productivity and long-distance transport. Our results suggest that these four taxa all have RPPs larger (*Schefflera*, *Syzygium*) or much larger (*Alchornea*, *Celtis*) than Poaceae (Table 1A).

12.4.2 LRA reconstructions of Holocene plant cover at regional and local spatial scales

Although the 11 pollen taxa used in the LRA reconstructions generally represent > 60–70% of the total pollen sum of terrestrial pollen types in the four pollen records (Table S3A), it is essential to keep in mind that we are missing in many cases a third of the pollen record. Major taxa for which we would need RPP estimates to achieve more complete reconstructions of the plant cover are: *Aerva*-type, *Antidesma*-type, *A. venosum*, *Ficus*, *Gnidia*, *Justicia*, *Maesa*, and *Rapanea* (mountain woodlands), and *Cussonia*, *Hymenocardia*, *Ixora*, *Olea*, *Tetrorchidium*, and *Uapaca* (wooded grasslands). One RPP value (19.19 relative to Poaceae) is available for *Justicia* (Navya 2020). Based on a large database of surface pollen samples in Africa, Gajewski *et al.* (2002) concluded that *Olea* pollen (*O. capensis*-type) represent the presence and abundance of the plant correctly. *Uapaca* was found to be either over-represented or under-represented by pollen (Julier *et al.* 2019). The information we have so far on the pollen-vegetation relationships of the taxa mentioned above is still too poor to anticipate the effect they would have on the LRA reconstructions. They could influence (in terms of plant cover) the timing, rate and magnitude of change in the woodland/openland relationship, and the taxa composition in woodlands over time.

We consider that the LRA reconstructions using the lowest RPP (0.06) for *Podocarpus* are unrealistic. The low RPP results in an extreme dominance of *Podocarpus* over all other tree taxa, and over Poaceae for most of the Holocene (see also discussion on RPPs above). We therefore focus on the reconstructions performed with the alternative RPPs 2.62 and 6.38, i.e. pollen productivities higher than that of Poaceae. These RPP values are also in a comparable range of values as common entomophilous tree taxa in African pollen records such as *Alchornea*, *Celtis*, *Schefflera*, and *Syzygium*.

12.4.2.1 Mountain woodlands: Mbi and Bambili

The assumed minimum size of the landscape represented by REVEALS estimates of plant cover is 100 km × 100 km (see Methods), which corresponds to an area around the site Mbi covering today montane and submontane vegetation (*c*. 55%), Guineo-Congolian rain and semi-deciduous forests (*c*. 20%), and secondary forests and grassland (25%), i.e. vegetation types from altitudes between *c*. 1000 and 3000 m asl. (Figures 1 and 2).

The REVEALS reconstructions suggest a clear dominance of grasses in the regional vegetation over the entire Holocene. The composition of trees is significantly different in plant cover compared to the pollen percentages, with a generally smaller cover than pollen percentage of *Schefflera*, *Podocarpus* (high altitude taxa) and *Alchornea* (low altitude taxon), and larger cover than pollen percentage of Moraceae and Combretaceae/Melastomataceae (mid-altitude taxa) and Poaceae (mainly mid- and low altitude) (see Figure 2 for the altitudinal distribution of taxa). Thus, the REVEALS reconstruction suggests that the regional landscape was much more open than suggested by the pollen percentages, and the high altitude and low altitude trees less abundant in relation to mid altitude trees. However, the REVEALS reconstruction does not alter the overall history and timing of changes in woodland abundance documented by the pollen percentages.

The LOVE reconstructions represent plant cover over a 2 km-radius area (assumed RSAP, Methods and SOM-12.1.4) around the lake of Bambili, which includes essentially afro-alpine grasslands and montane forests. The reconstructions suggest that the vegetation at Bambili was 100% forested 11–3.5 ka BP with a stronger dominance of the high altitude taxa (including Moraceae; *Ficus* in Figure 2) over the low altitude taxa than pollen percentages indicate. The sudden decreases in cover of *Podocarpus* at 7 ka and 3 ka BP are more pronounced than the declines in percentages, and the general collapse of the mountain forest more rapid. The landscape openness (cover of Poaceae) 3–0.5 ka BP is significantly larger than suggested by the pollen percentages.

12.4.2.2 The Sudanian and Guineo-Congolian wooded grasslands: Tilla and Mbalang

The REVEALS reconstructions of Tilla represent the regional plant cover in the Sudanian dry forests with grasslands and shrublands at low altitudes (Figures 1 and 2) and confirm that tree cover was very small all through the Holocene in contrast to Sahel's more tree-rich vegetation over early and Mid-Holocene (Salzmann *et al.* 2002). However, Combretaceae/Melastomataceae might have been more abundant in relation to *Alchornea*, *Celtis*, and *Macaranga* than pollen percentages suggest.

The assumed RSAP of Lake Mbalang (2 km) lies within the Sudano-Guinean wooded grasslands. The LOVE reconstruction suggests a 10–15% higher cover of tree taxa in the vegetation compared to the pollen percentages 6–3 ka BP and confirms a local presence of *Podocarpus* and *Prunus africana/Rubus pinnatus*. However, the larger abundance of trees compared to pollen percentages might be a bias due to the pollen record from Tilla representing only the very open regional vegetation North of Mbalang and not the areas West and South of the lake that might have been more wooded. Therefore, the LRA might have substracted too large amounts of Poaceae (background pollen inferred from the REVEALS-based regional vegetation of Tilla). If this is the case, pollen percentage of Poaceae might be closer to the actual vegetation openness at Mbalang than its LOVE-estimated cover indicates. A pollen record from a large lake closer to Mbalang would be more appropriate for a LOVE reconstruction.

12.5 CONCLUSIONS

The results from this first attempt at estimating RPP for major taxa of the Holocene vegetation of Cameroon confirm the insights gained earlier from traditional studies of the relationship between pollen and vegetation. *Alchornea, Celtis, Macaranga, Schefflera*, and *Syzygium* have been found to be overrepresented by pollen (Elenga *et al.* 2000; Lézine *et al.* 2013; Reynaud-Farrera 1995; Verlhac *et al.* 2018; Vincens *et al.* 1997, 2000; Watrin *et al.* 2007) and are also the taxa for which the estimated RPP are larger than that of Poaceae. Of the taxa for which RPP values are available from other studies, Combretaceae, Cyperaceae, *Macaranga* and *Prunus africana/Rubus pinnatus* exhibit relatively small between-study differences in RPPs. Existing RPP values for both Moraceae and *Podocarpus* are either lower or higher than that of Poaceae, which corroborates earlier observations on the representation of these taxa in modern pollen samples compared to the plant abundance in the vegetation (e.g. Gosling *et al.* 2005; Maley *et al.* 1990; Vincens 1984; Wan *et al.* 2020).

The major uncertainties of our RPP study are due to the strong effect of non-stationary vegetation in the study region on the calculation of RPPs using the maximum likelihood approach. It implies that the RSAP of the pollen samples (here surface soil) is difficult to identify, and therefore the best RPP values are uncertain. Nevertheless, whatever RSAP we would choose (either a few hundred meters or 3 kilometres), ranking of the RPP values would remain the same and, therefore, the correction of the pollen percentages identical. However, the relative cover/abundance of taxa would probably be influenced. In regions where it might be difficult to find large enough areas of stationary vegetation to estimate RPP of forest taxa in particular, one might test the robustness of the obtained RPPs using other models and calculation approaches that might be less sensitive to non-stationary vegetation (Fang *et al.* 2020; Tabares *et al.* 2021). In future work, the use of combined satellite and drone data might help to decrease the amount of field work and increase the consistency of vegetation datasets over large areas around each pollen sample site. The latter would probably improve the performance of models for estimation of RPPs.

The LRA reconstructions in this study provide a first tentative quantification of changes in the Afromontane woodlands and wooded grasslands of Cameroon over the Holocene. The results suggest that the regional landscape in the mountain region was more open than pollen percentages suggest, and that the collapse of the high altitude forests around Bambili at 3 ka BP was faster and of larger magnitude than indicated by the pollen record. The latter will need to be confirmed by further studies involving a larger number of major taxa characteristic of the Holocene vegetation history of Cameroon. The reconstructions also suggest that the woodlands' taxa composition, at both regional and local scales, is different once the pollen percentages are corrected using the RPP dataset and RPPs 2.62 or 6.38 for *Podocarpus*. The LRA reconstruction for Mbalang confirms the local occurrence of mountain woodland around the lake in Mid-Holocene. The correction of the pollen percentages indicate a larger cover of *Prunus africana/Rubus pinnatus* and Moraceae than their pollen percentages. *Podocarpus* and Moraceae are the most problematic taxa to model with the RPPs currently available. The reconstructions using high RPPs for *Podocarpus* are the most probable so far. It is not possible, based on this study, to assess the reliability of the available RPPs for Moraceae.

ACKNOWLEDGEMENTS

This research was supported by the National Research Agency (ANR), France, through the C3A (ANR-09-PEXT-001) and VULPES (ANR-15-MASC-0003) projects and by the Labex L-IPSL. We gratefully acknowledge funds from the CNRS (A.-M. Lézine), and Linnaeus University (Faculty of Health and Life Sciences, Kalmar, Sweden) and the Swedish Strategical Research Area 'ModElling the Regional and Global Ecosystem' (MERGE; http://www.merge.lu.se/) (M.-J. Gaillard, E. Githumbi, F. Li). This study is also a contribution to the Past Global Change (PAGES) project and its working group LandCover 6k (http://pastglobalchanges.org/landcover6k), which in turn received support from the Swiss National Science Foundation, the Swiss Academy of Sciences, the US National Science Foundation, and the Chinese Academy of Sciences. We thank

U. Salzmann (University of Northumbria, Newcastle) and A. Vincens (CEREGE, France) for sharing their pollen data (Tilla and Mbalang, respectively). All pollen records are archived in the African Pollen Database (APD).

REFERENCES

Abraham, V., Hicks, S., Svobodová-Svitavská, H., Bozilova, E., Panajiotidis, S., Filipova-Marinova, M., Eldegard Jensen, C., Tonkov, S., Pidek, I. A., Swieta-Musznicka, J., Zimny, M., Kvavadze, E., Filbrandt-Czaja, A., Hättestrand, M., Karlıoglu Kılıç, N., Kosenko, J., Nosova, M., Severova, E., Volkova, O., Hallsdóttir, M., Kalnina, L., Noryskiewicz, A. M., Noryskiewicz, B., Pardoe, H., Christodoulou, A., Koff, T., Fontana, S. L., Alenius, T., Isaksson, E., Seppä, H., Veski, S., Pedziszewska, A., Weiser, M. and Giesecke, T., 2021, Patterns in recent and Holocene pollen influxes across Europe – the Pollen Monitoring Programme Database as a tool for vegetation reconstruction. *Biogeosciences*, **18**(15), 4511–4534. 10.5194/bg-18-4511-2021.

Archer, C.L. and Jacobson, M.Z., 2005, Evaluation of global wind power. *Journal of Geophysical Research*, **110**, article: D12110, 10.1029/2004JD005462.

Barnosky, A., Matzke, N., Tomiya, S., Wogan, G. O. U., Swartz, B., Quental, T. B., Marshall, C., McGuire, J. L., Lindsey, E. L., Maguire, K. C., Mersey, B. and Ferrer, E. A., 2011, Has the Earth's sixth mass extinction already arrived? *Nature*, **471**, pp. 51–57, 10.1038/nature09678.

Bouimetarhan, I., Dupont, L., Reddad, H., Baqloul, A., Lézine, A.-M. and APD contributors this volume, Vegetation history and climate change of Africa: A synthesis of deep-sea pollen records. Vegetation response to millennial- and orbital-scale climate changes: A view from the Ocean. *Palaeoecology of Africa*, **35**, chapter: 22, 10.1201/9781003162766-22.

Broström, A., Sugita, S. and Gaillard M.-J., 2004, Pollen productivity estimates for the reconstruction of past vegetation cover in the cultural landscape of southern Sweden. *The Holocene*, **14**, pp. 368–381, 10.1191/0959683604hl713rp.

Broström, A., Sugita, S., Gaillard, M.-J. and Pilesjö, P., 2005, Estimating the spatial scale of pollen dispersal in the cultural landscape of southern Sweden. *The Holocene*, **15**, pp. 252–262, 10.1191/0959683605hl790rp.

Broström, A., Nielsen, A.B., Gaillard, M.-J., Hjelle, K.L., Mazier, F., Binney, H., Bunting, M.J., Fyfe, R., Duffin, K., Meltsov, V., Poska, A., Räsänen, S., Soepboer, W., von Stedingk, H., Suutari, H. and Sugita, S., 2008, Pollen productivity estimates of key European plant taxa for quantitative reconstruction of past vegetation — a review. *Vegetation History and Archaeobotany*, **17**, 461–478, 10.1007/s00334-008-0148-8.

Bunting MJ, Gaillard MJ, Sugita S, Middleton, R. and Broström, A., 2004, Vegetation structure and pollen source area. *The Holocene*, **14**, pp. 651–660, 10.1191/0959683604hl744rp.

Bunting, M.J., Armitage, R., Binney, H.A. and Waller, M., 2005, Estimates of 'relative pollen productivity' and 'relevant source area of pollen' for major tree taxa in two Norfolk (UK) woodlands. *The Holocene*, **15**, pp. 459–465, 10.1191/0959683605hl821rr.

Bunting, M. J., Farrell, M., Broström, A., Hjelle, K., Mazier, F., Middleton, R., Nielsen, A. B., Rushton, E., Shaw, H. and Twiddle, C. L., 2013, Palynological perspectives on vegetation survey: A critical step for model-based reconstruction of Quaternary land cover. *Quaternary Science Reviews*, **82**, pp. 41–55, 10.1016/j.quascirev.2013.10.006.

Chen, H., Xu, Q., Zhang, S., Sun, Y., Wang, M. and Zhou, Z., 2019, Relative pollen productivity estimates of subtropical evergreen and deciduous braodleaved mixed forest in Ta-pieh Mountains. *Quaternary Sciences*, **39**, pp. 469–482, 10.11928/j.issn.1001-7410.2019.02.19.

Cui, Q.Y., Gaillard, M.-J., Lemdahl, G., Stenberg, L., Sugita, S. and Zernova, G., 2014, Historical land-use and landscape change in southern Sweden and implications for present and future biodiversity. *Ecology and Evolution*, **4**, pp. 555–3, 10.1002/ece3.1198,570.

Duffin, K.I. and Bunting, M.J., 2008, Relative pollen productivity and fall speed estimates for southern African savanna taxa. *Vegetation History and Archaeobotany*, **17**, pp. 507–525, 10.1007/s00334-007-0101-2.

Duthie, T. J., 2015, Relative pollen productivity estimates (PPE) and relevant source area of pollen (RSAP) for key taxa from vegetation communities in Cathedral Peak, KwaZulu-Natal Drakensburg. *MSc Thesis, Univ. of KwaZulu-Natal*.

Elenga, H., Namur, C.D, Vincens, A., Roux, M. and Schwartz, D., 2000, Use of plots to define pollen–vegetation relationships in densely forested ecosystems of Tropical Africa. *Review of Palaeobotany and Palynology*, **112**, pp. 79–96, 10.1016/S0034-6667(00)00036-1.

Fang, Y., Ma, C. and Bunting, M J., 2019, Novel methods of estimating relative pollen productivity: A key parameter for reconstruction of past land cover from pollen records. *Progress in Physical Geography*, **46**, pp. 731–753, 10.1177/0309133319861808.

Gaillard, M.-J., Sugita, S., Bunting, M.J., Middleton, R., Broström, A., Caseldine, C., Giesecke, T., Hellman, S. E. V., Hicks, S., Hjelle, K., Langdon, C., Nielsen, A. B., Poska, A., von Stedingk, H., Veski, S. and POLLANDCAL members, 2008, The use of modelling and simulation approach in reconstructing past landscapes from fossil pollen data: A review and results from the POLLANDCAL network. *Vegetation History and Archaeobotany*, **17**, pp. 419–443, 10.1007/s00334-008-0169-3.

Gaillard, M.-J., Sugita, S., Mazier, F., Trondman, A.-K., Broström, A., Hickler, T., Kjellström, E., Punes, P., Lemmen, C., Olofsson, J., Smith, B., Strandberg, G., Kokfelt, Ulla., Miller, P., Poska, A., Rundgren, M. and Barnekow, L., 2010, Holocene land-cover reconstructions for studies on landcover-climate feedbacks. *Climate of the Past*, **6**, pp. 483–499, 10.5194/cp-6-483-2010.

Gajewski, K., Lézine, A.-M., Vincens, A., Delestan, A., Sawada, M. and APD members, 2002, Modern climate-vegetation-pollen relations in Africa and adjacent areas. *Quaternary Science Reviews*, **21**, pp. 1611–1631, 10.1016/S0277-3791(01)00152-4.

Gosling, W.D., Mayle, F.E., Tate, N.J. and Killeen, T., 2005, Modern pollen-rain characteristics of tall terra firme moist evergreen forest, southern Amazonia. *Quaternary Research*, **64**, pp. 284–297, 10.1016/j.yqres.2005.08.008.

Grassi, S., Veronesi, F., Schenkel, R., Peier, C., Neukom, J., Volkwein, S., Raubal, M. and Hurni, N., 2015, Mapping of the global wind energy potential using open source GIS data. 2nd International Conference on Energy and Environment: bringing together Engineering and Economics. Guimarães, Portugal 18–19 June, 2015.

Haselhorst, D.S., Moreno, J.E. and Punyasena, S.W., 2013, Variability within the 10-year pollen rain of a seasonal neotropical forest and its implications for paleoenvironmental and phenological research. *PLoS One*, **8**, article: e53485, 10.1371/journal.pone.0053485.

Havinga, A.J., 1984, A 20-year experimental investigation into the differential corrosion susceptibility of pollen and spores in various soil types. *Pollen et Spores*, **26**, pp. 541–558.

Hellman, S., Gaillard, M.-J., Broström, A. and Sugita, S., 2008, The REVEALS model, a new tool to estimate past regional plant abundance from pollen data in large lakes: validation in southern Sweden.*Journal of Quaternary Science*, **23**, pp. 21–42, 10.1002/jqs.1126.

Hellman, S., Bunting, M. J. and Gaillard, M. J., 2009a, Relevant Source Area of Pollen in patchy cultural landscapes and signals of anthropogenic landscape disturbance in the pollen record: a simulation approach. *Review of Palaeobotany and Palynology*, **153**, pp. 245–258, 10.1016/j.revpalbo.2008.08.006.

Hellman, S., Gaillard, M.-J., Bunting, J. M. and Mazier, F., 2009b, Estimating the relevant source area of pollen in the past cultural landscapes of southern Sweden - a forward modelling approach. *Review of Palaeobotany and Palynology*, **153**, pp. 259–271, 10.1016/j.revpalbo.2008.08.008.

Hély, C., Lézine, A.-M. and APD contributors, 2014, Holocene changes in African vegetation: tradeoff between climate and water availability. *Climate of the Past*, **10**, pp. 681–686, 10.5194/cp-10-681-2014.

Hill, T. R., Duthie, T.J., and Bunting, M.J., this volume, Relevant source area of pollen and pollen productivity estimates from KwaZulu-Natal. *Palaeoecology of Africa*, **35**, chapter: 16, 10.1201/9781003162766-16.

Jiang, F., Xu, Q., Zhang, S., Li, F., Zhang, K., Wang, M., Shen, W., Sun, Y. and Zhou, Z., 2020, Relative pollen productivities of the major plant taxa of subtropical evergreen–deciduous mixed woodland in China. *Journal of Quaternary Science*, **35**, pp. 526–538, 10.1002/jqs.3197.

Julier, A.C.M., Jardine, P. E., Adu-Bredu, S., Coe, A. L., Fraser, W. T., Lomax, B. H., Malhi, Y., Moore, S. and Gosling, W. D., 2019, Variability in modern pollen rain from moist and wet tropical forest plots in Ghana, West Africa. *Grana*, **58**, pp. 45–62, 10.1080/00173134.2018. 1510027.

Kaplan, J.O., Krumhardt, K.M., Gaillard, M.-J., Sugita, S., Trondman, A.-K., Fyfe, R., Marquer, L., Mazier, F. and Nielsen, A.B., 2017, Constraining the deforestation history of Europe: evaluation of historical land use scenarios with pollen-based land cover reconstructions. *Land*, **6**, article: 91, 10.3390/land6040091.

Kuparinen, A., 2006, Mechanistic models for wind dispersal. *Trends in Plant Science*, **11**, pp. 296–301, 10.1016/j.tplants.2006.04.006.

Kuparinen, A., Markkanen, T., Riikonen, H. and Vesala, T., 2007, Modeling air-mediated dispersal of spores, pollen and seeds in forested areas. *Ecological Modelling*, **208**, pp. 177–188, 10.1016/j.ecolmodel.2007.05.023.

Letouzey, R., 1985, Notice de la carte phytogeographique du Cameroun au 1: 500,000, (Toulouse: Institut de la carte internationale de la végétation).

Lézine, A.-M., Assi-Kaudjhis, C., Roche, E., Vincens, A. and Achoundong, G., 2013, Towards an understanding of West African montane forest response to climate change. *Journal of Biogeography*, **40**, pp. 183–196, 10.1111/j.1365-2699.2012.02770.x.

Lézine, A.-M., Lemonnier, K., Waller, M.P., Bouimetarhan, I. and Dupont, L., this volume, Changes in the West African landscape at the end of the African Humid Period *Palaeoecology of Africa*, **35**, chapter: 6, 10.1201/9781003162766-6.

Lézine, A-M., Izumi, K. and Achoundong, G., 2021, Mbi Crater (Cameroon) illustrates the relations between mountain and lowland forests over the past 15,000 years in western equatorial Africa. *Quaternary International*. DOI: 10.1016/j.quaint.2020.12.014.

Li, F.-R., Gaillard, M.-J., Sugita, S., Mazier, F., Xu, Q.-H., Zhou, Z.-Z., Zhang, Y., Li, Y. and Laffly, D., 2017, Relative pollen productivity estimates for major plant taxa of cultural landscapes in central eastern China. *Vegetation History and Archaeobotany*, **26**, pp. 587–605, 10.1007/s00334-017-0636-9.

Li, F., Gaillard, M.-J., Cao, X., Herzschuh, U., Sugita, S., Tarasov, P.-E., Wagner, M., Xu, Q., Ni, J., Wang, W., Zhao, Y., An, C., Beusen, A.H.W., Chen, F., Feng, Z., Klein Goldewijk, C.G.M., Huang, X., Li, Y., Li, Y., Liu, H., Sun, A., Yao, Y., Zheng, Z. and Jia, X., 2020, Towards quantification of Holocene anthropogenic land-cover change in temperate China: A review in the light of pollen-based REVEALS reconstructions of regional plant cover, *Earth-Science Reviews*, **203**, article 103119, 10.1016/j.earscirev.2020.103119.

Lu, Z., Miller, P.A., Zhang, Q., Zhang, Q., Warlind, D., Nieradzik, L., Sjolte, J. and Smith, B., 2018, Dynamic vegetation simulations of the mid-Holocene green Sahara. *Geophysical Research Letters* **45**, pp. 8294–8303, 10.1029/2018GL079195.

Maley, J. Caballe, G. and Sita, P., 1990, Etude d'un peuplement résiduel à basse altitude de *Podocarpus latifolius* sur le flanc congolais du massif du Chaillu. Implications paléoclimatique et biogéographiques. Etude de la pluie pollinique actuelle. In : Paysages Quaternaires de l'Afrique Centrale Atlantique, edited by Lanfranchi R. and Schwartz D., *Paysages quaternaires de l'Afrique centrale atlantique*, pp. 336-352 (Paris : ORSTOM).

Masson-Delmotte, V., Schulz, M., Abe-Ouchi, A., Beer, J., Ganopolski, A., Gonzaalez Rouco, J.F., Jansen, E., Lambeck, K., Luterbacher, J., Naish, T., Osborn, T., Otto-Bliesner, B., Quinn,

T., Ramesh, R., Rojas, M., Shao, X. and Timmermann, A., 2013, Information from paleoclimate archives. In *Climate Change 2013: the Physical Science Basis. Contribution of Working Group I to the Fifth Assessment Report of the Intergovernmental Panel on Climate Change*, edited by Stocker, T.F., Qin, D., Plattner, G.-K., Tignor, M., Allen, S.K., Boschung, J., Nauels, A., Xia, Y., Bex, V. and Midgley, P.M. pp. 383–464 (Cambridge University Press).

Mazier, F., Gaillard, M. J., Kuneš, P., Sugita, S., Trondman, A. K. and Broström, A., 2012, Testing the effect of site selection and parameter setting on REVEALS-model estimates of plant abundance using the Czech Quaternary Palynological Database. *Review of Palaeobotany and Palynology*, **187**, pp. 38–49, 10.1016/j.revpalbo.2012.07.017.

Navya R., 2020, *Pollen based estimates of Holocene vegetation in southern India: An LRA (Land-cover Reconstruction Algorithm) approach "*, PhD Thesis, Pondicherry University, Puducherry, India, pp. 1–268.

Parsons, R.W. and Prentice, I.C., 1981, Statistical approaches to R-values and the pollen-vegetation relationship. *Review of Palaeobotany and Palynology*, **32**, pp. 127–152, 10.1016/0034-6667(81)90001-4.

Piraquive Bermúdez, D., Theuerkauf, M. and Giesecke, T., 2019, Towards quantifying Holocene changes in forest cover in the Araucaria forest grassland mosaic of southern Brazil. Abstract. 20th Congress of the International Union for Quaternary Research (INQUA), Dublin 25–31 July 2019, https://app.oxfordabstracts.com/events/574/program-app/submission/92213.

Prentice, I.C., 1985, Pollen representation, source area, and basin size – toward a unified theory of pollen analysis. *Quaternary Research*, **23**, pp. 76–86, 10.1016/0033-5894(85)90073-0.

Prentice, I.C. and Parsons, R.W., 1983, Maximum likelihood linear calibration of pollen spectra in terms of forest composition, *Biometrics*, **39**, pp. 1051–1057, 10.2307/2531338.

Prentice, I.C. and Webb III, T., 1986, Pollen percentages, tree abundances and the Fagerlind effect. *Journal of Quaternary Science*, **1**, pp. 35–43, 10.1002/jqs.3390010105.

Reynaud-Farrera, I., 1995, *Histoire des paléoenvironnements forestiers du sud Cameroun à partir d'analyses palynologiques et statistiques de dépôts holocènes et actuels*. PhD Thesis, University of Montpellier, France.

Salzmann, U., Hoelzmann, P. and Morczinek, I., 2002, Late Quaternary climate and vegetation of the Sudanian zone of northeast Nigeria. *Quaternary Research*, **58**, pp. 73–83, 10.1006/qres.2002.2356.

Sugita, S., 1994, Pollen representation of vegetation in Quaternary sediment: theory and method in patchy vegetation. *Ecology*, **82**, pp. 881–897, 10.2307/2261452.

Sugita, S., 2007a, Theory of quantitative reconstruction of vegetation I: pollen from large sites REVEALS regional vegetation composition. *The Holocene*, **17**, pp. 229–241, 10.1177/0959683607075837.

Sugita, S., 2007b, Theory of quantitative reconstruction of vegetation II: all you need is LOVE. *The Holocene*, **17**, pp. 243–257, 10.1177/0959683607075838.

Tabares, X., Ratzmann, G., Kruse, S., Theuerkauf, M., Mapani, B. and Herzschuh, U., 2021, Relative pollen productivity estimates of savanna taxa from southern Africa and their application to reconstruct shrub encroachment during the last century. *The Holocene*.

Theuerkauf, M., Kuparinen, A. and Joosten, H., 2013, Pollen productivity estimates strongly depend on assumed pollen dispersal. *The Holocene*, **23**, pp. 14–24, 10.1177/0959683612450194.

Trondman, A. K., Gaillard, M.-J., Mazier, F., Sugita, S., Fyfe, R., Nielsen, A. B., Twiddle, C., Barratt, P., Birks, H. J. B., Bjune, A. E., Bjorkman, L., Broström, Anna. Caseldine, C., David, R., Dodson, J., Doerfler, W., Fischer, E., van Geel, B., Giesecke, T., Hultberg, T., Kalnina, L., Kangur, M., van der Knaap, P., Koff, T., Kunes, P., Lageras, P., Latalowa, M., Lechterbeck, J., Leroyer, C., Leydet, M., Lindbladh, M., Marquer, L., Mitchell, F. J. G., Odgaard, B. V., Peglar, S. M., Persson, T., Poska, A., Roesch, M., Seppa, H., Veski, S. and Wick, L., 2015, Pollen-based quantitative reconstructions of Holocene regional vegetation cover (plant-functional types and land-cover types) in Europe suitable for climate modelling. *Global Change Biology*, **21**, pp. 676–697.

Verlhac, L., Izumi K., Lézine, A.-M., Lemonnier, K., Buchet, G., Achoundong, G. and Tchiengué, B., 2018, Altitudinal distribution of pollen, plants and biomes in the Cameroon highlands. *Review of Palaeobotany and Palynology*, **259**, pp. 21–28, 10.1016/j.revpalbo.2018.09.011.

Vincens, A., 1984, Environnement végétal et sédimentation pollinique lacustre actuelle dans le Bassin du Lac Turkana (Kenya). *Revue de Paléobiologie*, volume spécial, **1**, pp. 235–242.

Vincens, A., Ssemmanda, I., Roux, M. and Jolly, D., 1997, Study of the modern pollen rain in western Uganda with a numerical approach. *Review of Palaeobotany and Palynology*, **96**, pp. 145–168, 10.1016/S0034-6667(96)00022-X.

Vincens, A., Schwartz, D., Elenga, H., Reynaud-Farrera, I., Alexandre, A., Bertaux, J., Mariotti, A., M., Meunier, J.D., Nguetsop, F., Servant, M., Servant-Vildary, S. and Wirrmann, D., 1999, Forest response to climate changes in Atlantic Equatorial Africa during the last 4000 years BP and inheritance on the modern landscapes. *Journal of Biogeography*, **26**, pp. 879–885, 10.1046/j.1365-2699.1999.00333.x.

Vincens, A., Dubois, M.A., Guillet, B., Achoundong, G., Buchet, G., Kamgang Kabeyene Beyala, V., de Namur, C. and Riera, B., 2000, Pollen-rain–vegetation relationships along a forest– savanna transect in southeastern Cameroon. *Review of Palaeobotany and Palynology*, **110**, pp. 191–208, 10.1016/S0034-6667(00)00009-9.

Vincens, A., Lézine, A.M., Buchet, G., Lewden, D. and Le Thomas, A., 2007, African pollen database inventory of tree and shrub pollen types. *Review of Palaeobotany and Palynology*, **145**, pp. 135–141, 10.1016/j.revpalbo.2006.09.004.

Vincens, A., Buchet, G., Servant, M. and ECOFIT Mbalang collaborators, 2010, Vegetation response to the "African Humid Period" termination in Central Cameroon (7°N) - new pollen insight from Lake Mbalang. *Climate of the Past*, **6**, pp. 281–294, 10.5194/cp-6-281-2010.

Wan, Q., Zhang, Y., Huang, K., Sun, Q., Zhang, X., Gaillard, M.-J., Xu, Q., Li, F. and Zheng, Z., 2020, Evaluating quantitative pollen representation of vegetation in the tropics: A case study on the Hainan Island, tropical China. *Ecological Indicators*, **114**, article: 106297, 10.1016/j.ecolind.2020.106297.

Watrin, J., Lézine, A.-M., Gajewski, K. and Vincens, A., 2007, Pollen plant–climate relationships in sub-Saharan Africa. *Journal of Biogeography*, **34**, pp. 489–499, 10.1111/j.1365-2699.2006.01626.x.

White, F., 1983, *The vegetation of Africa*. (Paris: UNESCO).

Whitney, B.S., Smallman, T.L., Mitchard, E.T., Carson, J.F., Mayle, F.E., and Bunting, M.J., 2018, Constraining pollen-based estimates of forest cover in the Amazon: a simulation approach. *The Holocene*, **29**, pp. 262–269, 10.1177/0959683618810394.

Wieczorek, X. and Herzschuh, U., 2020, Compilation of relative pollen productivity (RPP) estimates and taxonomically harmonised RPP datasets for single continents and Northern Hemisphere extratropics. *Earth System Science. Data*, **12**, 3515–3528, 10.5194/essd-12-3515-2020.

Zhang, S.-R., Xu, Q.-H., Gaillard, M.-J., Cao, X.-Y., Li, J., Zhang, L., Li, Y., Tian, F., Zhou, L., Lin, F., and Yang, X., 2016, Characteristic pollen source area and vertical pollen dispersal and deposition in a mixed coniferous and deciduous broad-leaved woodland in the Changbai mountains, northeast China. *Vegetetation History and Archaeobotany*, **25**, pp. 29–43, 10.1007/s00334-015-0532-0.

CHAPTER 13

A Holocene pollen record from Mboandong, a crater lake in lowland Cameroon

Keith Richards[1]

KrA Stratigraphic, Deganwy, Conwy, North Wales, United Kingdom

ABSTRACT: Mboandong is a small crater lake situated approximately 30 km to the north of Mount Cameroon. A core 13 m in length was collected from the lake centre in December 1981 with the objective of studying Holocene vegetation change in the lowlands of West Cameroon. Radiocarbon dating indicates that the base on the core is close to 7000 years old. Volcanic eruptions of Mount Cameroon between *c*. 7000 and 6000 years ago are indicated by ash deposits which are overlain by diatomite. Otherwise the lake sediments consist of fine-grained organic mud, with no obvious sedimentary breaks or any sign of the lake having dried out. A fairly rapid sedimentation rate of at least 1 m per 500 years is indicated. Predominantly forested conditions existed throughout much of the Holocene, possibly of the Congo type (everwet) rainforest prior to *c*. 2700 cal yr BP. At this time man was probably encouraging fruits of the '*atili*' tree *Canarium schweinfurthii* as an oil-rich food. A major increase in Poaceae pollen after *c*. 2500 cal yr BP is most probably linked to a drying climate, as has now been inferred at numerous sites in tropical West Africa. An increase in pollen from regrowth and semi-deciduous trees possibly indicates a change to an Atlantic type (more seasonal) rainforest after this time. This climate change probably led to subsequent forest clearance, perhaps linked to the southwards migration of the Bantu people and spread of iron technology. Increased numbers of *Elaeis guineensis* pollen recorded after 1700 cal yr BP are indicative of cultivation. The oil palm appears to have replaced the *atili* tree as an oil-rich food source.

13.1 INTRODUCTION

The fact that major changes in climate and vegetation occurred in Africa during the Quaternary is now well established, with evidence from a large number of sites from around the continent (e.g. Dupont *et al.* 2000; Rossignol-Strick and Duzer 1979; Lézine *et al.* 2013b, 2019; Maley and Brenac 1998; Miller and Gosling 2014; Vincens *et al.* 1999; Sowunmi 1981; Talbot *et al.* 1984). Hamilton (1976), in a review of botanical and zoological evidence, however, had proposed that the forested areas of present-day Cameroon were part of a refugium where forest had remained relatively unaffected by periods of climatic drying during the Quaternary. A primary aim of this study was to collect a core from this suggested refugium region of Cameroon in order to determine the extent, if any, of vegetation change by means of pollen analysis (Figure 1). Mboandong was chosen for a number of reasons. A variety of lakes in Cameroon have been described, for example by Gèze (1943), Green (1972) and Corbet *et al.* (1973). Of these lakes, Barombi Mbo was too

---

[1] Other affiliation: Department of Geography and Planning, University of Liverpool, Liverpool, United Kingdom

DOI 10.1201/9781003162766-13

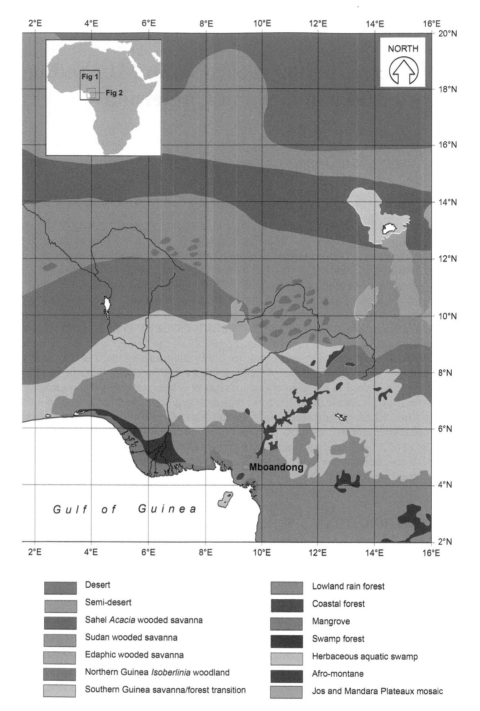

Figure 1. Simplified map of West African vegetation (based on Keay 1959; White 1983) showing the Mboandong study site location.

deep for coring at the time and Barombi Kotto was excluded as its shores and island are inhabited and it was a known centre for schistosomiasis (Awo *et al*. 2018, 2020; Duke and Moore 1971; Zahra 1953). Mboandong is only 5 m deep (Green 1972), relatively undisturbed by man and accessible by road. The base of the core has been dated within the range of 6480 to 7020 cal yr BP and therefore provides a record of more or less the last 7000 years. 'Mbo' is the local name for 'lake' hence there is no need to add the prefix of 'lake' to Mboandong (Corbet *et al*. 1973).

13.2 REGIONAL SETTING

Mboandong (4.4500°N, 9.2689°E) is situated in a shallow crater at about 120 m altitude on the low-lying plain between Mount Cameroon, the Roumpi Hills and Mount Koupé. The area consists mainly of basaltic rocks and evidence for widespread volcanic activity is provided by the hundred or so historic eruption points (Figure 2), several of which now contain crater lakes. Geological information from Mboandong is lacking but the lake may be of the same approximate age as nearby Barombi Kotto, a larger crater lake, created by explosive action during a geologically recent, presumably Quaternary, eruptive phase (Gèze 1943). Mboandong lies in the rain shadow of Mount Cameroon. Annual rainfall is the region of 2000 to 2600 mm per year (Hawkins and Brunt 1965); this compares with rainfall typically of 8500 to 10,300 mm per year to the west of the mountain, in the Atlantic coastal region. The wettest months tend to be July, August and September and the driest December, January and February. Even so, Richards (1963) noted that this region of Cameroon, which falls within the 'sous climat bas Camerounian' of Aubréville (1949), is characterised by the absence of a well-defined and prolonged dry season.

A detailed study of the vegetation of Cameroon was carried out by Letouzey (1968) and the main vegetation types are shown in Figure 1. The western boundary of the Southern Bakundu Forest Reserve is only 2 km from Mboandong. In a survey of vegetation carried out in 1948, Richards (1963) described the Bakundu forest as 'essentially similar' to other African Mixed Rain Forests, though not uniform in structure or floristic composition. Patches of high forest with emergent trees occurred in a mosaic, together with gaps with less tall trees and climbing plants. The exceptionally high floristic diversity of this area prompted Richards (1963) to consider the region a likely refugium. Since the 1948 study of Bakundu, timber extraction has increased substantially and this has also been the case for the area around Mboandong. Figure 3 shows the vegetation in the immediate vicinity of the lake before 1976, but large tracts of forest have been entirely removed and plantations extended since that date. In December 1981, the lake was surrounded by a narrow band of rainforest including both primary and secondary forest elements. Most distinctive were *Musanga cecropioides*, *Raphia hookeri* and *Ceiba pentandra* (Baker *et al*. 1986). A few patches of relatively undisturbed rainforest remained on the eastern side of the lake. A small area had been cleared on the north western shore for cocoa plantations. Away from the lake, large areas of land had been cleared for cultivation including oil palm, banana, plantain, cocoa, rubber and cassava.

13.3 METHODS

A single core 13 m in length was collected from the approximate centre of the lake, using a modified Livingstone mud sampler (Livingstone 1955; Walker 1964) operated from a raft. Aluminium casing was used to prevent the coring rods from bending and to ensure that each section of the core was collected from precisely the same location. A continuous sediment core was obtained except for the interval 12.50-12.00 m which was lost. The core was extruded and described in the field, before being sealed in polythene sheet to prevent contamination and

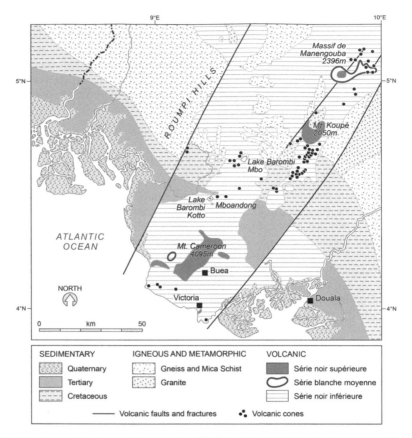

Figure 2. Geological map of South-west Cameroon (modified after Gèze 1943) showing key geographical features and study sites. Note: the many additional volcanic cones formed in recent years in the Mount Cameroon vicinity (e.g. Wantim *et al.* 2013) are not shown.

oxidation during transportation. Twenty-seven samples were prepared for pollen analysis using KOH (10%), cold HCl (20%) and warm HF (40%) following the methods described by Moore and Webb (1978). One Stockmarr *Lycopodium* tablet (batch 201890) was added to 30 g dry weight of sample. Samples were sieved using ultrasound to remove particles less than 5 μm in diameter and subject to brief acetolysis (using a 9:1 mix of acetic anhydride and sulphuric acid) and oxidation (using sodium chlorate and concentrated HCl). Residues were mounted in silicone oil. A minimum of 400 pollen and spores were counted for each sample, with the exception of 10.89 m where recovery of pollen was particularly low. All laboratory preparations were carried out at Hull University. Microscopy was carried out in 1983 and 1984 at Hull University and Duke University. Selected grains were photographed from ancient and modern reference material to ensure consistency in identification (Supporting Online Material [SOM] Plates I-V). Three bulk mud samples were radiocarbon dated and results recalibrated by Beta Analytic in October 2020 using the High Probability Density Range method (Bronk Ramsey 2009). The results of the pollen analysis were constrained by CONISS cluster analysis (Grimm 1987). The core was kept in cold storage at Hull University but unfortunately deteriorated and was disposed of around the year 2010 (Jane Bunting, personal communication, 2020).

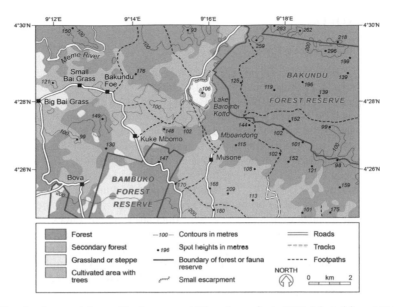

Figure 3. Map showing vegetation and land-use around Mboandong prior to 1976. Adapted from 1:50,000 Map of Cameroon (Buea-Douala NB-32-IV sheets 1c and 1d), Centre Géographique National, Yaoundé.

Table 1. Radiocarbon ages from Mboandong bulk mud samples. Ages were calibrated by Beta Analytic Inc. in October 2020 using the INTCAL13 database and High Probability Density Range Method (HPD). No adjustments for Reservoir Effect were made.

Depth (m)	Reference	^{14}C age	δ^{13} C (‰)	cal yrs BP (95%)
4.37 – 4.48	Beta-11198	2380 ± 70	−26.01	2310 to 2720 cal BP
8.37 – 8.48	Beta-11197	4520 ± 90	−33.29	4870 to 5330 cal BP
12.93 – 12.98	Harwell-5028	5930 ± 110	−32.30	6480 to 7020 cal BP

13.4 SEDIMENT DESCRIPTION AND RADIOCARBON DATING

Between 13 m and 10 m the core exhibits several distinct lithological changes. At least two bands of volcanic ash occur (12.71 to 12.50 m, 12.00 to 11.80 m and 11.64 to 11.00 m). Samples from these intervals were prepared for pollen analysis but recovery was too low to give a useable count. The upper ash layer is directly overlain by a single bed of diatomite (11.00 m to 10.05 m). Above 10.05 m the core consists of uniform fine-grained organic mud. Results of radiocarbon dating are shown in Table 1. The base of the core produced a conventional ^{14}C age of 5930±110 BP which is within the range (95% confidence) of 6480 to 7020 cal yr BP. The radiocarbon dates appear to be fairly consistent and give no indication of any serious errors. The δ^{13}C values of −26.01 to −32.30‰ suggest no significant old carbon contamination. Calcareous deposits of Cretaceous age do occur in the south-west province of Cameroon but the water catchment of Mboandong is more or less restricted to the overlying volcanic strata. Inwashing of old carbon from calcareous rock is therefore considered unlikely, especially as there are no rivers inflowing to the lake. No traces of modern rooted vegetation or other sources of possible modern carbon contamination were observed.

13.5 PALYNOLOGY

13.5.1 General observations

The results of the pollen analysis are given in Figure 4. Assigned names have been amended to follow the format of the African Pollen Database (Vincens *et al.* 2007). The pollen taxa have been grouped in 'ecological groups'. However, as pollen determination has often only been possible to generic or even family level, this is problematic when a recorded pollen type could have originated from more than one life form or vegetation type. The groupings in Figure 4, therefore, represent the life forms and vegetation types from which the recorded pollen types are most probably derived, although this is an over-simplification. In making this zonation, reference has been made, in particular, to Hutchinson and Dalziel (1954–1968), Hall and Swaine (1981), Letouzey (1968, 1979), Nielson (1965) and Sowunmi (1981). Quite high percentages of the pollen sum are taken up by *Alchornea* and Urticaceae undiff. but no real ecological significance can be attributed to these widely fluctuating occurrences. *Alchornea* pollen is very probably derived mostly from *Alchornea cordifolia*, a common understory rainforest tree, which occurs also in swamp forest, montane forest and savanna. In Africa, Urticaceae are mainly forest herbs, some of which are very high pollen producers (Hamilton 1982). The pollen of this group is very small, often less than 10 μm in diameter, and may therefore have been subject to losses during decantation and ultrasonic sieving, although all samples were processed using identical methods. A complete list of pollen and spore taxa recorded is provided in SOM (Appendix I).

13.5.2 Pollen in relation to local vegetation

Due to time constraints, it was not possible to obtain information on modern pollen rain from the Mboandong locality. However, the near-surface sample from the lake core (0.14 m) probably gives a reasonable indication of modern pollen deposition at the site, and can be compared with floral data from the nearby Bakundu Forest Reserve. The floral data from Bakundu (Richards 1963) included all tree taxa over 30 cm girth, listed as number of boles in 15 size classes ranging from 30 cm to over 4.0 m. These data were re-calculated and compared to the pollen data from Mboandong (Richards 1987), grouped by family rather than genera or species, except where a finer taxonomic division could be detected in the pollen record. Percentage girth data from Bakundu were plotted against the percentage of total tree pollen at 0.14 m in the lake core. Considerable over representation in the pollen sample was shown by *Elaeis guineensis*, *Alchornea* and Urticaceae undiff. These taxa are part of the 'local' pollen element and were not recorded in the 'regional' tree data from Bakundu. The pollen to vegetation relationship can be summarised as follows: 1) Taxa which feature strongly in the pollen record but have limited or no representation at Bakundu. These include a) Trees associated with semi-deciduous forest, secondary forest or understory conditions such as *Ceiba pentandra*, *Macaranga*, *Lophira*, *Bosqueia*, *Musanga*, *Antidesma* and *Celtis*; b) Trees associated with swampy conditions such as *Uapaca*, *Raphia* and Palmae. These taxa are most probably derived predominantly from the forest edge, swamp forest or disturbed forest and therefore represent the local pollen element, mainly from around the study site. Rhizophoraceae pollen is most probably sourced from *Anopyxis klaineana*, a rainforest tree (a mangrove pollen source is unlikely). 2) Taxa which are represented in both the pollen record and the vegetation survey, and are therefore likely to be good indicators of the regional pollen component. These include: Anacardiaceae, Combretaceae, Euphorbiaceae e.g. *Phyllanthus*, Rubiaceae, *Pycnanthus*, *Coulea*, *Vitex* and Sapotaceae. 3) Tree taxa recorded in

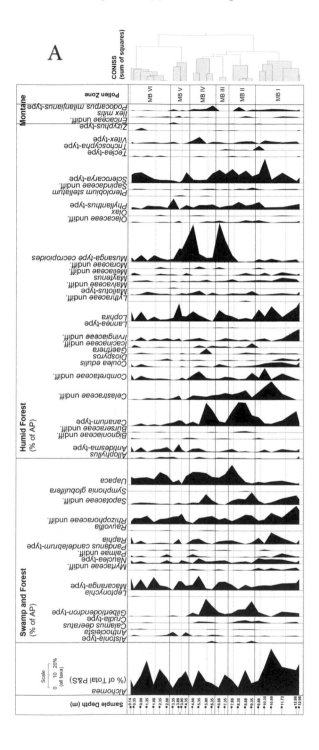

(Continued)

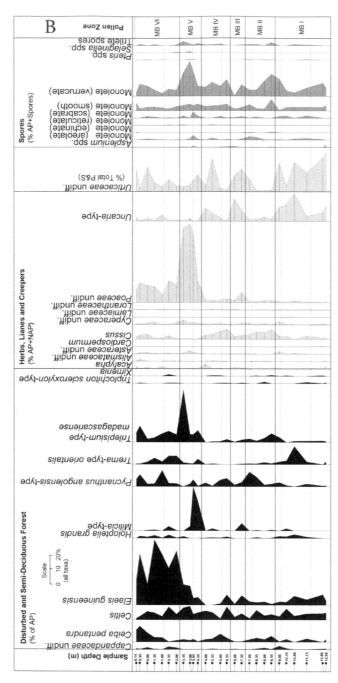

Figures 4A and 4B. Diagrams showing the pollen and spores recorded in the Mboandong study. Tree and shrub pollen are shown as a percentage of Arboreal Pollen (AP). Non Arboreal Pollen (NAP) from herbs and lianes are calculated as a percentage of AP+NAP. Spores are calculated as a percentage of AP+Spores. *Alchornea*, Urticaceae undiff. and undetermined pollen are excluded from all sums.

the Bakundu forest that show little or no representation in the pollen record, such as Olacaceae, *Diospyros*, Irvingiaceae, *Cola*, Annonaceae, Flacourtiaceae, Sterculiaceae, Clusiaceae, Lecythidaceae, *Alstonia*, Violaceae and Lauraceae. These taxa may be entomophilous, have low pollen production and/or poor dispersal, have poor preservational properties or may have pollen that is not readily distinguishable. Most of the trees and shrubs represented in the pollen study were also identified in an ecological survey of the riparian vegetation around Lake Barombi Kotto, situated around 2 km to the north of Mboandong (Awo 2018).

13.5.3 Pollen diagram zonation

In this section values for trees and shrubs are percentages of Arboreal Pollen (AP) whereas values for non-arboreal pollen (NAP) are percentages of AP+NAP. Values for spores are percentages of AP+Spores. Pollen of *Alchornea*, Urticaceae undiff. and undetermined taxa are excluded from all sums. Values given for the 'ecological groups' are calculated in the same way. Zones are assigned on the basis of CONISS clusters (Figure 4). Age ranges shown are estimates based on extrapolation of the radiocarbon ages (Table 1).

13.5.3.1 Zone MB I: 12.98 -9.88 m (6800–5600 BP)
This interval is characterised by the presence of common *Alchornea* (11-36%) and Urticaceae undiff. (up to 26%). Swamp forest elements are represented more or less consistently at between 28% and 36%, with a slightly lower abundance of 19% at 10.35 m. Important components in this interval are Rhizophoraceae undiff. (3-14%), *Macaranga*-type (3-8%), *Gilbertiodendron*-type (up to 4%) and Sapotaceae undiff. (up to 7%). Lowland forest elements are co-dominant in this interval with abundances of between 43% and 61%, with taxa including *Canarium*-type (up to 11%), *Lophira* (up to 15%), Irvingiaceae undiff. (up to 9%) and Olacaceae undiff. (less than 1%). Minor peaks occur in Meliaceae undiff. and *Coulea edulis*. Semi-deciduous forest elements are consistently represented at between 14% and 25%. Montane elements are consistent but rare with *Podocarpus milanjianus*-type occurring at 2% or less. Other taxa of note are *Uncaria*-type (up to 19%) which accounts for a significant proportion of NAP (21% at 10.89 m), and Celastraceae undiff. (up to 14%). Fern spores account for between 7% and 14% of the sum. Of the disturbance indicators, Poaceae pollen is rare (less than 1%) as are *Elaeis guineensis* (2-7%), *Trema*-type *orientalis* (up to 12%) and *Musanga*-type *cecropioides* (up to 3%).

13.5.3.2 Zone MB II: 9.35–7.88 m (5400–4700 BP)
This interval is characterised by a predominance of pollen most probably from lowland forest and swamp arboreal sources. Frequent *Canarium*-type pollen (up to 18%) occurs together with common *Gilbertiodendron*-type (up to 12%) and Sapotaceae undiff. (up to 9%). Pollen from likely swamp elements is common, occurring fairly consistently at 28% to 37%. Important components of this group are *Uapaca* which reaches 15% at 7.88 m and Rhizophoraceae undiff. (up to 7%). The lowland forest elements are dominant in this interval, ranging between 42 and 56%, and include *Maytenus* (up to 2%), *Antidesma*-type (up to 2%) and Olacaceae undiff. (up to 2%) and many other rare but consistent taxa. Other important pollen types present include two types thought likely to originate from the Anacardiaceae family: *Sclerocarya*-type is consistently common (5-11%), and *Trichoscypha*-type occurs rarely (less than 1%) but is more or less restricted to this interval of the core, as is Icacinaceae undiff. (3% at 9.35m). Pollen of *Cissus* (up to 7%), derived from a climber in the Vitaceae family, and fern spores (up to 20%) are consistent in this interval. Semi-deciduous forest elements are less common but occur consistently at between 15% and 22% and include *Pycnanthus angolensis*-type (up to 10%). Poaceae, *Elaeis guineensis* and *Musanga*-type *cecropioides* pollen are present in very low numbers.

13.5.3.3 Zone MB III: 7.35–6.88 m (4400–4100 BP)
Elaeis guineensis pollen ranges from 1% or less to 6% at 7.35 m. A slight increase in Poaceae pollen (7%) occurs in the same sample, which is also marked by the first major increase in *Musanga* comp. (16%).

13.5.3.4 Zone MB IV: 6.35–4.88 m (3700–2800 BP)
This interval is characterised by a common presence of pollen of *Canarium*-type (up to 18%), *Gilbertiodendron*-type (up to 13%) as well as a strong input from the Urticaceae, Rubiaceae and Euphorbiaceae families. A slight increase in *Podocarpus milanjianus*-type occurs. Poaceae and *Elaeis guineensis* pollen occur but are not frequent (maximum of 3% and 7% respectively). Monolete fern spores are consistently present (6-11%).

13.5.3.5 Zone MB V: 4.35–3.35 m (2500–1900 BP)
This interval is characterised by the overwhelming abundance of grass pollen (family Poaceae) which shows a gradual increase towards the top of the interval where it accounts for 50 to 53% of the pollen sum. Consistent *Elaeis guineensis* (up to 14%) is also present together with other disturbed forest types such as *Musanga*-type *cecropioides* (up to 17%). This interval is also marked by an increase in fern spores (up to 32%). Lowland rainforest elements are also reduced (24-40%) and are characterised by a large number of rare taxa such as *Lophira* (up to 13%) and *Antidesma*-type (up to 6%). Also of interest is the presence of pollen types associated with disturbed or semi-deciduous forest (up to 62%), notably *Trilepisium*-type *madagascariense* [=*Bosqueia angolensis*] (up to 36%), which occurs with *Milicia*-type [=*Chlorophora*] (up to 30%), *Trema*-type *orientalis* (up to 4%) and *Celtis* (up to 9%).

13.5.3.6 Zone MB VI: 2.88–0.14m (1700–100 BP)
This interval is characterised by the presence of abundant *Elaeis guineensis* (up to 44%), and by common Poaceae (up to 15%) but with lower proportions of grass pollen than in Zone MB V. Pollen of understory and/or secondary forest trees, such as *Alchornea* (up to 26%), is well represented. Lowland rainforest elements occur fairly consistently (18-40%) as do swamp forest elements (13-27%). Also notably common is tree pollen from disturbed or semi-deciduous vegetation (up to 69%), including *Trilepisium*-type *madagascariense* (up to 11%), *Pycnanthus angolensis*-type (up to 11%) and *Celtis* (up to 9%). Pollen of *Ceiba pentandra*, another semi-deciduous tree, occurs commonly (up to 11%) towards the top of this interval.

13.6 INTERPRETATION

13.6.1 The pollen and sedimentary record

Zone MB I (12.98-9.88 m; 6800–5600 BP) contains a mixture of sediment types including at least two volcanic ash deposits and thin layers of organic mud, which are overlain by a 1 m thick deposit of diatomite (11.00 m to 10.05 m). In all likelihood, the diatomite has formed in response to an increased availability of silica, provided by volcanic ash fall, as described by Wallace *et al.* (2006). It is likely that the ash originates from nearby Mount Cameroon which has been active during the Late Quaternary: at least 19 eruptive events of the mountain have been documented during the last 2500 years (Global Volcanism Program 2013). No confirmed eruption dates are known from the earlier part of the Holocene. On the basis of the radiocarbon age obtained at the base of the core, the volcanic ash in the core would have been deposited between approximately 7000 and 6000 years ago. This is several thousand years after the ash layers recorded in core at Barombi Mbo (Maley *et al.* 1991). Interpretation of the vegetation is problematic due to the very high numbers of *Alchornea* and Urticaceae pollen. The vegetation close to the site probably consisted of mixed primary and secondary rainforest with some swamp elements. The very slight increase of *Trema*-type *orientalis* and *Uncaria*-type pollen at 10.89 m, immediately above the uppermost of the volcanic ash bands, may be due to localised volcanic disturbance. Very low levels of *Musanga*-type *cecropioides* and Poaceae pollen argue against widespread forest disturbance at this time.

Zone MB II (9.35-7.88 m; 5400-4700 BP) is characterised primarily by humid forest and swamp tree taxa, whereas Zone MB III (7.35-6.88 m; 4400-4100 BP) probably marks the first significant forest disturbance around the study site, after 4400 BP, on the basis of frequent *Musanga*-type *cecropioides* pollen. A mechanism for disturbance cannot be established with certainty, but man cannot be ruled out. The overall pollen assemblages are consistent with predominantly lowland humid rainforest with swamp and regrowth components. Zone MB IV (6.35-4.88 m; 3700-2800 BP) contains mostly tree pollen indicating closed forest. The co-occurrence of Caesalpiniaceae (*Gilbertiodendron*-type), *Canarium*-type and *Podocarpus milanjianus*-type in this interval at Mboandong is explained by Giresse *et al.* (2020) as 'cloud forest effect' due to increased condensation associated with cooling temperatures. This may also account for the relatively high numbers of fern spores recorded.

The main feature of Zone MB V (4.35-3.35 m; 2500-1900 BP) is the major increase in pollen from Poaceae and semi-deciduous tree taxa such as *Trilepisium*-type *madagascariense* and *Milicia*-type. Swamp elements and lowland rainforest elements are reduced, whereas fern spores are frequent. The overall assemblages from this interval conclusively show a change from humid rainforest to more open vegetation, characterised mainly by Poaceae and semi-deciduous trees. The uppermost interval Zone MB VI (2.88-0.14 m; 1700-100 BP) shows a marked decrease in Poaceae pollen offset by significantly increased *Elaeis guineensis*. Most tree pollen taxa are represented although it is not possible to determine if these arboreal taxa (e.g. Rhizophoraceae undiff., *Macaranga*-type and *Uapaca*) are derived from the immediate vicinity of the lake or from further afield. Reduced percentages of lowland and swamp forest taxa are balanced by an increase in semi-deciduous elements. The overall assemblage points to forest disturbance with likely cultivation of oil palm, *Elaeis guineensis*.

13.6.2 Vegetation representation

13.6.2.1 Swamp pollen
A fairly consistent input of pollen from swamp sources suggests that a fringing element of swamp forest has persisted around the lake for much of the last 7000 years. Scattered *Raphia* palms overhang the lake at the present time although, from the low levels of *Raphia* pollen recorded, it is unlikely that extensive '*raphiales*' swamp communities existed by the lake. Rhizophoraceae pollen is probably derived predominantly from trees such as *Anopyxis klaineana* which are known to occur in freshwater conditions, rather than as a result of long-distance transport from mangrove.

13.6.2.2 Montane pollen
Podocarpus pollen is an important indicator of Afromontane vegetation (Lézine *et al.* 2013a). *Podocarpus milanjianus* [=*P. latifolius*, Migliore *et al.* 2020] grows on several West African mountains but it is apparently absent from Mount Cameroon (Letouzey 1968). Low levels of this easily dispersed pollen type occur in the Mboandong core, but had *Podocarpus* occurred on Mount Cameroon at any time during the last 7000 years, the amount of *Podocarpus* pollen recorded in the lake is likely to have been greater. The small amount of *Podocarpus* pollen present in the lake core may be derived from the Roumpi Hills, some 15 km to the north or from peaks such as Mount Koupé which supports well-developed montane vegetation (Baker 1986).

13.6.2.3 Dry forest pollen

Trees such as *Triplochiton scleroxylon* and *Celtis*, which are characteristic of semi-deciduous forest, may also occur in disturbed areas within humid forest. *Trilepisium*-type *madagascariense* pollen, common in the upper regions of the core in Zones MB-V and MB-VI (after 2500 cal yr BP), is most probably derived from *Bosqueia angolensis*, a tree characteristic of semi-deciduous forest patches within the humid rainforest zone. Pollen of *Milicia*-type is probably derived from the semi-deciduous tree *Chlorophora excelsa* (Moraceae), although the pollen is also morphologically similar to *Antiaris toxicaria*. *Milicia*-type pollen is notably rare in the core with the exception being a brief peak at around 4 m (*c*. 2500-2300 BP). This may be related to a short-lived abundance of these trees at around the time when the first signs of major forest disturbance are visible, or a response to a drying climate. The slight increase in *Ceiba pentandra* pollen in the uppermost few samples in the core may be due to species protection because of its useful 'kapok'. The consistent presence of *Ceiba pentandra* pollen appears to confirm the view of Baker (1965) who considered the species native to West Africa; Hutchinson and Dalziel (1954–1968) and others believed it to be an ancient introduction from tropical America.

13.7 DISCUSSION

13.7.1 Floristic affinities of forest communities

The pollen assemblages indicate that predominantly humid rainforest has been present in the Mboandong region for the last 7000 years, although it is unclear whether the rainforest represented is of the Atlantic or Congo type, as defined by Letouzey (1968). Two important tree pollen taxa well represented in the lower intervals of the core are *Gilbertiodendron*-type and Sapotaceae undiff. The pollen attributed to *Gilbertiodendron*-type is fairly large and coarsely striate but several other genera within the Caesalpiniaceae produce similar pollen. These include *Berlinia*, *Didelotia*, *Microberlinia* and others which are all genera characteristic of the Atlantic type forest, although none of these trees were found during the floristic survey of Bakundu by Richards (1963). *Gilbertiodendron*-type pollen occurs only very rarely in the upper reaches of the core. If, therefore, forest of the Atlantic type has persisted in the Mboandong region for the last 6000 or more years, the marked decline in *Gilbertiodendron*-type pollen after *c*. 3000 BP must be linked either to the direct felling of these leguminous trees or the overwhelming dominance of pollen such as Poaceae and *Elaeis guineensis*.

Another possibility is that the abundance of *Gilbertiodendron*-type pollen in the lower regions of the core is from *Gilbertiodendron dewevrei*. This species is characteristic of Congo type forests and is now mainly restricted to the Dja region of south-east Cameroon, (Letouzey 1968). It is possible, therefore, that the former distribution of this forest type may have extended further westwards than at present. The abundance of pollen from the Sapotaceae family shows an almost identical pattern of distribution in the core to *Gilbertiodendron* comp. Both are more common in the basal parts of the core, with peaks at 8.35 m and 5.90 m. It is possible that the Sapotaceae pollen in the basal parts of the core could be derived from *Baillonella toxisperma*, a large forest emergent which shows similar relict distribution patterns to *Gilbertiodendron dewevrei* in Cameroon (Letouzey 1968). *Baillonella toxisperma* was the only Sapotaceae species of note recorded at Bakundu (Richards 1963), occurring as a single, very large individual tree. It is certainly possible that forest of Congo floristic affinities could have occurred in the vicinity of Mboandong, particularly between *c*. 5400 and 2500 BP. The 'relict' characteristics evident both at Bakundu (Richards 1963) and in the present distributions shown by *Gilbertiodendron dewevrei* and *Baillonella toxisperma* (Letouzey 1968) are certainly consistent with a more extensive distribution of Congo type forest in the past.

Letouzey (1968) also lists *Alchornea floribunda* and *Trichoscypha congoensis* [= *T. acuminata*] as characteristic species of the Congo type forest, although they also occur in Atlantic type forest. *Alchornea* pollen is common throughout the Mboandong core and is therefore of little interpretative value. Pollen attributed to *Trichoscypha*-type is present in the core but only in low quantities in Zones MB I, II and III where *Gilbertiodendron* comp. and Sapotaceae are common. Its presence, therefore, is consistent with, but not indicative of, a Congo type forest. Similarly, pollen of *Cissus* and Celastraceae undiff. are also common in Zones MB I, II, III and IV. These are also taxa considered characteristic of Congo forest by Letouzey (1968) but they are not exclusive to it. In view of the large degree of species overlap between the Congo and Atlantic forest types, the distinction between the two forest types could at least partly be due to the higher degree of forest disturbance in the western regions of the country, as noted by Letouzey (1968). The boundary between Zones MB-IV and MB-V which sees a sharp decline in the abundance of *Gilbertiodendron*-type and Sapotaceae pollen also coincides with increased evidence of forest disturbance. Additionally, the Atlantic forest occurs in the region with a more seasonal (i.e. periodically dry) climate whereas the Congo forest occurs in the more equatorial (i.e. everwet) regions. Hence a climatic control may partly account for the present distributions and species composition of Atlantic and Congo type forests.

13.7.2 Climate change or man-made change?

The cause of the vegetation change evident at Mboandong after 2500 cal yr BP is the subject of ongoing debate. Increasing aridity is invoked by Maley (2002) and Lebamba *et al.* (2012) as the primary cause of the 'late Holocene rainforest crisis' between 2500 and 2000 years BP. Bostoen *et al.* (2013, 2015) provided supporting evidence based on linguistics. Garcin *et al.* (2018), in a study from Lake Barombi, however, considered that the 'late Holocene rainforest crisis' after 2600 cal yr BP was a consequence of human activity, and resulted from the Bantu expansion, although this view was countered by Clist *et al.* (2018). The balance of pollen-based evidence from multiple sites presented by Giresse *et al.* (2020), however, does support a drying climate as the initial cause of the change from forest-dominated to grass-dominated vegetation around 2500 years ago. This is a change of emphasis from that previously reported by Richards (1986, 1987) but is made on the basis of the wealth of new information from the region that has become available since that time.

The significant increase of *Elaeis guineensis* pollen (up to 23%) after *c.* 1700 BP suggests that cultivation of oil palm was taking place in the Mboandong region. Low levels of this pollen type in the lower reaches of the core may represent the natural presence in forest clearings. The relationship between the oil palm (*Elaeis guineensis*) and the 'atili tree' (*Canarium schweinfurthii*) is also relevant (Lézine *et al.* 2013b; Maley and Chepstow-Lusty 2001; Sowunmi 1999). At Mboandong, based on pollen evidence, *Canarium* appears to have been more important in the past than at the present time, as *Canarium*-type pollen accounts for 13% of the pollen sum at 8.35 m, around 5000 years ago, but is absent from the uppermost levels of the core, where *Elaeis* is an important element. Based on comparisons with reference pollen, it is almost certain that the pollen recorded in the core is from *Canarium schweinfurthii*. It is therefore highly likely that *Canarium* was a significant oil-food source in Cameroon, very probably an example of 'vegeculture' which has subsequently declined with *Elaeis* cultivation within the last 2000 years.

13.8 CONCLUSIONS

Early Holocene volcanic activity in the region, presumably of Mount Cameroon, is indicated by the deposits of volcanic ash in the Mboandong core deposited between *c.* 7000 and 6000 years ago. Silica from the ash enabled a bloom of diatoms, resulting in a 1 m thick diatomite deposited after *c*. 6000 cal yr BP. Since this time, the Mboandong lake sediments consist of fine-grained organic mud, but with no evidence of any hiatus or other significant stratigraphic change. No signs of the lake having dried out within the last 7000 years were observed. From extrapolation of the radiocarbon ages, a fairly rapid sedimentation rate of at least 1 m per 500 years is indicated, equivalent to *c*. 2 mm per year.

The pollen record indicates that predominantly forested conditions have prevailed in the vicinity of the study site throughout much of the Holocene. Between *c*. 7000 and 2500 cal yr BP, the dominant forest elements were derived from humid rainforest and swamp forest, with important components derived from the Caesalpiniaceae and Sapotaceae families. There is a possible affinity with Congo type (everwet) rainforest although this is speculative. During this time, man was probably encouraging the tree *Canarium schweinfurthii* as an oil-rich food source. After *c*. 2500 cal yr BP increased proportions of regrowth and semi-deciduous trees are represented in the pollen record, possibly indicating a change to an Atlantic type (more seasonal) rainforest, as occurs in undisturbed vegetation in West Cameroon at the present time. Pollen from the oil palm *Elaeis guineensis* occurs throughout the Mboandong core, initially in low numbers, suggesting that oil palm at first grew naturally as secondary forest element. Significantly increased numbers of *Elaeis* pollen recorded after 1700 cal yr BP are indicative of cultivation. The oil palm *Elaeis guineensis* appears to have replaced the *atili* tree *Canarium schweinfurthii* as a source of oil-rich food.

A major increase in Poaceae pollen at *c*. 2500 cal yr BP is most probably linked to a drying climate, as has now been inferred at numerous sites in tropical West Africa. This climate change probably led to subsequent forest clearance, shown by an increase in pollen from semi-deciduous trees such as *Bosqueia angolensis*, which probably exploited openings within the forest canopy. This time-line is broadly consistent with dates for iron working in West Africa, which suggests that iron technology and forest clearance were related events, but which followed on from a climatic shift. It is quite conceivable that a significant drying within northern and western Africa within the last few thousand years prompted movement of people southwards to more humid latitudes. It is probable that iron technology was also disseminated at this time, perhaps linked to the migration of the Bantu people. This regional picture from West Africa is supported by the pollen data from Mboandong and suggests that a changing climate first affected the regional vegetation around 2500 years ago, followed by forest clearance and crop cultivation within the last 2000 years.

ACKNOWLEDGEMENTS

This work was carried out as an MSc project in the early 1980's at the Department of Geography, University of Hull under the guidance of the late John R. Flenley and at Duke University, assisted by the late Daniel A. Livingstone. This paper is dedicated to the fond memory of both JRF and DAL whose support and encouragement made this study possible. Partial funding was received from The Nuffield Foundation. Margaret Adebisi Sowunmi (University of Ibadan) provided assistance with some pollen identification and Miranda Awo (University of Buea) provided up-to-date information on the vegetation, plant-pollen relationships and limnology of the study region. Ron Hatfield of Beta Analytic Inc. assisted with calibration of radiocarbon ages. Figures 2 and 3 were re-drawn by Alison Davies of The Mapping Company, UK, from originals drawn at Hull University, Department of Geography. The core was collected by members of the 1981-82 Hull University Cameroon Expedition: Richard Baker, Stephen Compton, Sean Edwards, David Newsome, Keith Richards, Carrie Rimes, Howard Smith and Ollie Smith.

REFERENCES

Aubréville, A., 1949, *Climats. Fôrets et desertification de l'Afrique tropicale.* Paris : Sociéte d'Editions Géographiques, Maritimes et Coloniales, Paris.

Awo M.E., 2018, *Water quality, riparian vegetation and algal ecological assessment of Lake Barombi Kotto, Cameroon.* PhD Thesis, University of Buea, Cameroon.

Awo, M.E., Tabot, P.T., Goodenough, N. and Ambo, F.B., 2018, Spatiotemporal variation of phytoplankton community structure in the crater lake Barombi Kotto, Cameroon. *International Journal of Current Research in Biosciences and Plant Biology*, **5**(2), pp. 36–55, 10.20546/ijcrbp.2018.502.005.

Awo, M.E., Fonge, B.A., Tabot, P.T. and Akoachere, J.T.K., 2020, Water quality of the volcanic crater lake, Lake Barombi Kotto, in Cameroon. *African Journal of Aquatic Sciences*, **45**(4), pp. 401–411, 10.2989/16085914.2020.1737799

Baker, H.G., 1965, The evolution of the cultivated Kapok tree; a probable west African product. In *Ecology and Economic Development in Tropical Africa* edited by Brokensha, D., pp.185–216..Institute of International Studies, University of California.

Baker, R.G.E, 1986, Introduction: The Biogeography of Cameroon. *University of Hull Department of Geography Miscellaneous Series*, **30**, pp. 1–13.

Baker, R.G.E., Richards, K., and Rimes, C.A., 1986, The Hull University Cameroun Expedition: 1981-82 Final Report. *University of Hull Department of Geography Miscellaneous Series*, **30**, pp. 1–109.

Bostoen, K., Clist, B., Doumenge, C., Grollemund, R., Hombert J.-M., Muluwa, J.K. and Maley, J. 2015, Middle to Late Holocene paleoclimate change and the early Bantu expansion in the rain forests of western central Africa. *Current Anthropology*, **56**(3), pp. 354–384, 10.1086/681436.

Bostoen, K., Grollemund, R., Hombert J.-M. and Muluwa, J.K., 2013, Climate-induced vegetation dynamics and the Bantu Expansion: Evidence from Bantu names for pioneer trees *(Elaeis guineensis, Canarium schweinfurthii, and Musanga cecropioides). Comptes Rendus Geoscience*, **345**, pp. 336–349, 10.1016/j.crte.2013.03.005.

Bronk Ramsey, C., 2009, Bayesian analysis of radiocarbon dates. *Radiocarbon* **51**(1), pp. 337–360, 10.1017/S0033822200033865.

Clist, B., Bostoen, K., de Maret, P., Eggert, M.K., Höhn, A., Mindzié, C.M., Neumann, K. and Seidensticker, D., 2018, Did human activity really trigger the late Holocene rainforest crisis in Central Africa? *Proceedings of the National Academy of Sciences*, **115**(21), pp. E4733–E4734, 10.1073/pnas.1805247115.

Corbet, S.A., Green, J., Griffith, J. and Betney, E., 1973, Ecological studies on crater lakes in west Cameroon: Lakes Kotto and Mboandong. *Journal of Zoology*, **170**, pp. 309–324, 10.1111/j.1469-7998.1973.tb01380.x.

Duke, B.C.L. and Moore, P.J., 1971, The control of *Schistosoma haematobium* in west Cameroon. *Transactions of the Royal Society of Tropical Medicine and Hygiene*, **65**, pp. 841–843, 10.1016/0035-9203(71)90106-4.

Dupont, L.M., Jahns, S., Marret, F. and Ning, S. 2000, Vegetation change in equatorial West Africa: time-slices for the last 150ka. *Palaeogeography, Palaeoclimatology, Palaeoecology*, **155**, pp. 95–122, 10.1016/S0031-0182(99)00095-4.

Garcin, Y., Deschamps, P., Ménot, G., de Saulieu, G., Schefuß, E., Sebag, D., Dupont, L.M., Oslisly, R., Brademann, B., Mbusnum, K.G., Onana, J.-M., Akon, A.A., Epp, L.S., Tjallingii, R., Strecker, M.R., Brauer, A. and Sachse, D., 2018, Early anthropogenic impact on Western Central African rainforests 2,600 y ago. *Proceedings of the National Academy of Sciences*, 115(13), pp. 3261–3266, 10.1073/pnas.1715336115.

Gèze, B., 1943, Géographie physique et géologie du Cameroun occidental. *Mémoires du Muséum Nationale d'Histoire Naturelle Paris*, **17**, pp. 1–272.

Giresse, P., Maley, J. and Chepstow-Lusty, A., 2020, Understanding the 2500 yr BP rainforest crisis in West and Central Africa in the framework of the Late Holocene: pluridisciplinary analysis and multi-archive reconstruction. *Global and Planetary Change*, **192**, article: 103257, 10.1016/j.gloplacha.2020.103257.

Global Volcanism Program, 2013, Cameroon (224010) in Volcanoes of the World, v. 4.9.1 (17 Sep 2020) Smithsonian Institution. Downloaded 09 Oct 2020. https://volcano.si.edu/volcano.cfm?vn=224010

Green, J., 1972, Ecological studies on crater lakes in west Cameroun: zooplankton of Barombi Mbo, Mboandong, Lake Kotto and Lake Soden. *Journal of Zoology*, **166**, pp. 283–301, 10.1111/j.1469-7998.1972.tb03099.x.

Grimm, E.C., 1987, CONISS: a FORTRAN 77 program for stratigraphically constrained cluster analysis by the method of incremental sum of squares. *Computers and Geosciences*, **13**(1), pp. 13–35, 10.1016/0098-3004(87)90022-7.

Hall, J.B. and Swaine, M.D., 1981, *Distribution and Ecology of Vascular Plants in a Tropical Rain Forest: Forest Vegetation in Ghana*. Geobotany Series 1 (The Hague and London: Junk).

Hamilton, A.C., 1976, The significance of patterns of distribution shown by forest plants and animals in tropical Africa for the reconstruction of Upper Pleistocene palaeoenvironments: a review. *Palaeoecology of Africa*, **9**, pp. 63–97.

Hamilton, A.C., 1982, *Environmental History of East Africa. A Study of the Quaternary*, (London: Academic Press).

Hawkins, P. and Brunt, M., 1965, *The Soils and Ecology of West Cameroon*, Volumes 1-2, Food and Agriculture Organization of the United Nations, report 2083, Rome, pp. 1–516.

Hutchinson, J. and Dalziel, J.M., 1954–1968, *Flora of West Tropical Africa*, Edition 1, (London: Crown Agents for Overseas Governments).

Keay, R.W.J., 1959, *An Outline of Nigerian Vegetation*, (Lagos: Federal Government Printer) pp. 1–34.

Lebamba, J., Vincens, A. and Maley, J., 2012, Pollen, vegetation change and climate at Lake Barombi Mbo (Cameroon) during the last ca. 33,000 cal. yr BP: a numerical approach. *Climate of the Past*, **8**, pp. 59–78, 10.5194/cp-8-59-2012.

Letouzey, R., 1968, *Etude Phytogéographique du Cameroun*, Encyclopédie Biologique, volume LXIX, Paris.

Letouzey, R., 1979, Végétation. In *Atlas de la République Unie du Cameroon*, edited by Laclavère, G. (Paris: Editions Jeune Afrique).

Lézine, A.-M., Assi-Kaudjhis, C., Roche, E., Vincens, A. and Achoundong, G., 2013a, Towards an understanding of West African montane forest response to climate change. *Journal of Biogeography*, **40**(1), pp. 183–196, 10.1111/j.1365-2699.2012.02770.x.

Lézine, A.-M., Holl, A.F.C., Lebamba, J., Vincens, A., Assi-Khaudjis, C., Février, L. and Sultan, E., 2013b, Temporal relationship between Holocene human occupation and vegetation change along the northwestern margin of the Central African rainforest. *Comptes Rendus Geoscience*, 345(7–8), pp. 327–335, 10.1016/j.crte.2013.03.001.

Lézine, A.-M., Izumi, K., Kageyama, M. and Achoundong, G., 2019, A 90,000-year record of Afromontane forest responses to climate change. *Science* **363**, pp. 177–181, 10.1126/science.aav6821.

Livingstone, D.A., 1955, A light-weight piston sampler for lake deposits. *Ecology*, **36**, pp. 137–139, 10.2307/1931439.

Maley, J., 2002, A catastrophic destruction of African forests about 2,500 years ago still exerts a major influence on present vegetation. *Institute of Development Studies Bulletin*, **33**(1), pp. 13–30, 10.1111/j.1759-5436.2002.tb00003.x.

Maley, J. and Brenac, P., 1998, Vegetation dynamics, palaeoenvironments and climatic changes in the forests of West Cameroon during the last 28,000 years. *Review of Palaeobotany and Palynology*, **99**, pp. 157–88, 10.1016/S0034-6667(97)00047-X.

Maley, J., Livingstone, D.A., Giresse, P., Brenac, P., Kling, G., Stager, C., Thouveny, N., Kelts, K., Haag, M., Fournier, M., Bandet, Y., Williamson, D. and Zogning, A., 1991, West Cameroon Quaternary lacustrine deposits: preliminary results. *Journal of African Earth Sciences*, **12** (1–2), pp. 147–157, 10.1016/0899-5362(91)90065-7.

Maley, J. and Chepstow-Lusty, A., 2001, *Elaeis guineensis* Jacq. (oil palm) fluctuations in central Africa during the late Holocene: climate or human driving forces for this pioneering species? *Vegetation History and Archaeobotany*, **10**, pp. 117–20, 10.1007/PL00006920.

Migliore, J., Lézine, A.-M. and Hardy, O.J., 2020, The recent colonization history of the most widespread *Podocarpus* tree species in Afromontane forests. *Annals of Botany*, **126**, pp. 73–83, 10.1093/aob/mcaa049.

Miller, C.S. and Gosling, W.D., 2014, Quaternary forest associations in lowland tropical West Africa. *Quaternary Science Reviews*, **84**, pp. 7–25, 10.1016/j.quascirev.2013.10.027.

Moore, P.D. and Webb, J.A., 1978, *An Illustrated Guide to Pollen Analysis*, (London: Hodder and Stoughton), pp. 1–133.

Nielson, M.S., 1965, *Introduction to the Flowering Plants of West Africa*, (London: University of London Press), pp. 1–246.

Richards, K., 1986, Preliminary results of pollen analysis of a 6,000 year core from Mboandong, a crater lake in Cameroun. *University of Hull Department of Geography Miscellaneous Series*, **30**, pp. 14–28.

Richards, K., 1987, *A palynological study of the Late Quaternary vegetational history of Mboandong, a lowland lake in Cameroun*. MS. Thesis, University of Hull.

Richards, P.W., 1963, Ecological notes on West African vegetation, II. Lowland forest of the Southern Bakundu Forest Reserve. *Journal of Ecology*, **51**, pp. 123–149, 10.2307/2257510.

Rossignol-Strick, M. and Duzer, D., 1979, West African vegetation and climatic change since 22,500 BP from deep-sea cores palynology. *Pollen et Spores* **21**, pp. 105–134.

Sowunmi, M.A., 1981, Aspects of the late Quaternary vegetational changes in West Africa. *Journal of Biogeography*, **8**, pp. 457–474, 10.2307/2844565.

Sowunmi, M.A., 1999, The significance of the oil palm (*Elaeis guineensis* Jacq.) in the late Holocene environments of west and west central Africa: a further consideration. *Vegetation History and Archaeobotany*, **8**, pp. 199–210, 10.1007/BF02342720.

Talbot, M.R., Livingstone, D.A., Palmer, P.G., Maley, J., Melack, J.M., Delibrias, G. and Guilliksen, S., 1984, Preliminary results from sediment cores from Lake Bosumtwi, Ghana. *Palaeoecology of Africa*, **16**, pp. 173–192.

Vincens, A., Lézine, A.M., Buchet, G., Lewden, D. and Le Thomas, A., 2007, African pollen database inventory of tree and shrub pollen types. *Review of Palaeobotany and Palynology*, **145**(1-2), pp. 135–141, 10.1016/j.revpalbo.2006.09.004.

Vincens, A., Schwartz, D., Elenga, H., Reynaud-Farrera, I., Servant, M. and Wirrmann, D., 1999, Forest response to climate changes in Atlantic Equatorial Africa during the last 4000 years BP and inheritance on the modern landscapes. *Journal of Biogeography*, **26**, pp. 879–85, 10.1046/j.1365-2699.1999.00333.x.

Walker, D., 1964, A modified Vallentyne mud sampler. *Ecology*, **45**, pp. 642–644, 10.2307/1936118.

Wallace, A. R., Frank, D. G. and Founie, A., 2006, Freshwater diatomite deposits in the western United States. *US Department of the Interior, US Geological Survey Fact Sheet* 2006–3044.

Wantim, W.N., Kervyn, M., Ernst, G.G.J, Del Marmol, M.-A., Suh, C.E. and Jacobs, P., 2013, Morpho-Structure of the 1982 Lava Flow Field at Mount Cameroon Volcano, West-Central Africa. *International Journal of Geosciences*, **4**, pp. 564–583, 10.4236/ijg.2013.43052.

White, F., 1983, *The vegetation of Africa*, (Paris: UNESCO).

Zahra, A., 1953, Some notes on the incidence of *Schistosomiasis* in the southern Cameroons. *West African Medical Journal*, **2**(1), pp. 26–29.

CHAPTER 14

Future directions of palaeoecological research in the hyper-diverse Cape Floristic Region: The role of palynological studies

Lynne J. Quick

African Centre for Coastal Palaeoscience, Nelson Mandela University, Port Elizabeth, South Africa

ABSTRACT: The Cape Floristic Region (CFR) is a key focus area within southern Africa due to its botanical importance in terms of high levels of biodiversity as well as its rich cultural and archaeological heritage. The area is sensitive to cycles of regional and global environmental change, and palynological records obtained from the region can potentially provide valuable information regarding past vegetation dynamics and climate variability. Prior to the last decade, few high resolution palaeoenvironmental records were recovered from the CFR, and therefore its Late Quaternary environmental history was previously poorly understood. Significant progress was made over recent years and a considerable body of new palynological (as well as palaeoclimatological) evidence emerged. These new records provide greater insight into the nature and timing of past vegetation shifts and improve our understanding of how different subregions of the CFR have responded to past climate changes. They also highlight that there is a much higher degree of complexity, in terms of both vegetation and climate change, than previously thought. This paper provides a perspective on the progress made towards elucidating the palaeoecological history of the CFR, it highlights the importance of continuing and expanding upon the existing body of work and outlines current and future directions for palynological research in this hyper-diverse southwestern corner of southern Africa.

14.1 INTRODUCTION

The southwestern Cape of South Africa has long been recognized as a unique region particularly in terms of the distinctiveness of its flora. Due to the extraordinarily high levels of biodiversity and plant endemism found within the region it has previously been recognized as one of the six global floral kingdoms (Marloth 1908; Takhtajan *et al.* 1986) and is currently world-renowned as a key biodiversity hotspot (Mittermeier *et al.* 2004; Myers *et al.* 2000). The Fynbos Biome (Figure 1) and Cape Floristic Region (CFR) are considered by many to be synonymous, however the 'biome' refers only to the two dominant vegetation groups (Fynbos and Renosterveld), whereas the 'region' refers to the general geographical area and includes other vegetation types in the Afrotemperate Forest, Nama Karoo, Succulent Karoo and Albany Thicket Biomes. The CFR has been the subject of decades of intensive botanical research due to its unique floristic, evolutionary and ecological characteristics as well as its conservation appeal (e.g. Allsopp *et al.* 2014; Bond and Goldblatt 1984; Campbell 1985; Cowling 1992; Cowling *et al.* 1995; Goldblatt and Manning 2000; Rebelo *et al.* 2006).

DOI 10.1201/9781003162766-14

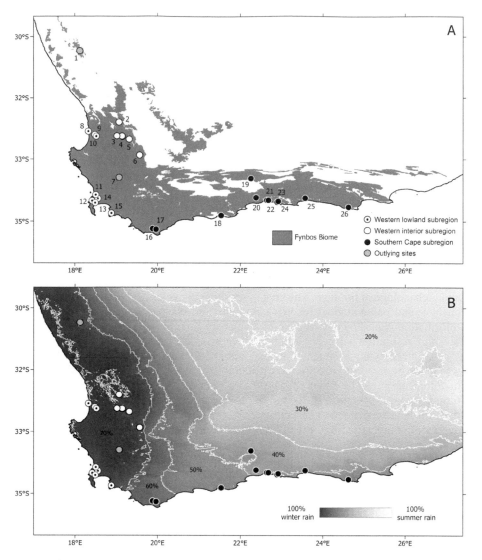

Figure 1. The distribution of fossil pollen records for the Cape Floristic Region (as demarcated by the Fynbos Biome) (A) and in relation to seasonality of rainfall (defined in terms of percentage of winter rainfall) (B). Records are grouped according to subregion and the type of archive is demarcated in brackets after the site name (w = wetland deposit; h = hyrax midden and c = cave deposit, hc =hyena coprolites). 1 – Groenkloof (w) (MacPherson *et al.* 2018), 2 – Pakhuis Pass (h) (Scott and Woodborne 2007b, a), 3 – De Rif (h) (Quick *et al.* 2011; Valsecchi *et al.* 2013), 4 – Driehoek and Sneeuberg vleis (w) (Meadows and Sugden 1991), 5 – Truitjes Kraal 4 (h) (Chase *et al.* 2015b; Meadows *et al.* 2010), 6 – Katbakkies 1 (h) (Chase *et al.* 2015b; Meadows *et al.* 2010), 7 – Vankraal Spring (w) (Forbes *et al.* 2018), 8 – Elands Bay Cave (c) (Meadows and Baxter 1999), 9 – Grootdrift (w) (Meadows *et al.* 1996), 10 – Klaarfontein Springs (w) (Meadows and Baxter 2001), 11 – Rietvlei (w) (Schalke 1973), 12 - Cecilia Cave (c) (Meadows and Baxter 1999), 13 – Princessvlei (w) (Cordova *et al.* 2019; Neumann *et al.* 2011), 14 – Cape Flats (w) (Schalke 1973), 15 – Cape Hangklip (w) (Schalke 1973), 16 – Voelvlei (w) (Carr *et al.* 2006), 17 – Soetendalsvlei (w) (Carr *et al.* 2006), 18 – Rietvlei Still Bay (w) (Quick *et al.* 2015), 19 – Boomplaas (c) (Deacon *et al.* 1984), 20 – Norga (w) (Scholtz 1986), 21 – Eilandvlei (w) (Quick *et al.* 2018), 22 – Bo Langvlei (w) (du Plessis *et al.* 2020), 23 – Groenvlei (w) (Martin 1968), 24 – Vankervelsvlei (w) (Irving and Meadows 1997; Quick *et al.* 2016), 25 – Platbos 1 (w) (MacPherson *et al.* 2019), 26 – Oyster Bay (hc) (Carrion *et al.* 2000) [Data: South African National Biodiversity Institute (2006-). The Vegetation Map of South Africa, Lesotho and Swaziland, Mucina, L., Rutherford, M.C. and Powrie, L.W. (Editors), Online, http://bgis.sanbi.org/SpatialDataset/Detail/18, Version 2018].

The southwestern Cape is also important from a climatic perspective as it is a particularly dynamic region in terms of long-term climate change. Unlike most of South Africa which receives summer rainfall, the CFR mainly falls within the winter rainfall zone (WRZ) (*sensu* Chase and Meadows 2006), presently receiving most of its rainfall during the austral winter when the southern westerly storm track migrates northward (Taljaard 1996; Tyson 1986; Tyson and Preston-Whyte 2000). Whereas during the austral summer, the westerlies and the South Atlantic Anticyclone shift southward limiting the influence of both frontal systems and tropical moisture sources (Reason *et al.* 2006; Tyson and Preston-Whyte 2000), resulting in warm, dry summer conditions. This Mediterranean-type climatic setting has played an important part in the development of the CFR and is thought to have helped foster the region's botanical diversity (Bradshaw and Cowling 2014; Cowling and Lombard 2002; Cowling *et al.* 2015; Goldblatt 1978; Linder 2005; Linder *et al.* 1992). While long-term climatic stability during the Pliocene and Pleistocene is hypothesized to be one of the key factors for the diversification of the Cape flora (Verboom *et al.* 2014; Cowling and Lombard 2002; Cowling *et al.* 2015; Cowling *et al.* 1997; Linder 2005), little direct evidence is available. Furthermore, species diversity within the CFR is not homogeneous and at finer spatial and temporal scales the concept of climatic stability driving diversity and species richness becomes more complex (Cowling 1992; Cowling and Lombard 2002; Cowling et al. 1997).

Palaeoecological studies within the CFR have the potential to contribute important insight into biodiversity patterns and processes, providing long-term perspectives of climatic and ecosystem changes. The CFR has a rich archaeological history as it is the locality for speciation events that culminated in the evolution of anatomically and behaviourally modern humans (Brown *et al.* 2009; Brown *et al.* 2012; Henshilwood *et al.* 2001; Henshilwood *et al.* 2002; Marean 2010; Marean *et al.* 2007). Determining the environmental context for these events represents an essential tool to understand human evolution, behaviour, and dispersal patterns. Not only can palaeoecology provide important insights into long-term human evolutionary trends but palaeoecological studies within the CFR can also help contextualize more recent archaeological trends such as migration patterns of indigenous pastoralists (e.g. Deacon 1992; Sadr 1998, 2008)) and their possible impact on late Holocene environments and ecosystems (e.g. Meadows and Baxter 2001; Neumann *et al.* 2011).

In terms of palaeoecological data, palynology represents the principal method available to determine the responses of vegetation to environmental change over various spatial and temporal scales (Bennett and Willis 2001; Birks *et al.* 2016; Faegri and Iversen 1989). Despite being a key focus area within southern Africa (as outlined above) no systematic study of the palynology of the Cape Flora has been conducted, considerable efforts have however been made towards the generation of individual fossil pollen records.

This paper presents a brief synopsis of the current state of the palaeoecological record of the CFR, with a focus on the pollen records that have been produced over the last decade. The role that palynology can play in advancing palaeoecological research is further explored by outlining current initiatives and future directions for palynological research within the CFR.

14.2 THE LATE QUATERNARY PALAEOECOLOGICAL HISTORY OF THE CAPE FLORISTIC REGION

As this is a short perspective piece, this section is not an exhaustive review of all the palynological records that fall within the CFR (although all pollen records are presented in Figure 1), but rather focuses on several of the most recently established records.

In general, the significant advances that were made in palaeoenvironmental/palaeoecological research could not have been achieved without the generation and evaluation of multiple lines of proxy evidence as well as the use of unique environmental archives. This is particularly the case for the CFR as organic sediments (which are traditionally the main source of fossil pollen) are relatively rare, given the semi-arid nature of the region. Records from rock hyrax (*Procavia capensis*) middens represent the single most significant contribution to a more nuanced and detailed understanding of the palaeoclimates of the CFR as they are remarkably sensitive to environmental changes and contain a great diversity of proxies (e.g. fossil pollen, stable isotopes, plant biomarkers) (Chase *et al.* 2012; Scott 1990). Hyrax middens are however largely preserved only within the more arid inland mountainous regions of the CFR (i.e. within mountain ranges associated with the Cape Fold Belt) while more conventional palaeoecological archives, such as coastal lakes and wetlands, are found along the west and southern Cape coasts. Records derived from these sites helped to provide a more balanced coverage across the CFR (Figure 1). While inherent differences exist between pollen assemblages derived from hyrax middens and those from wetland deposits (i.e. pollen spectra from wetland deposits often include high proportions of local hydrophilic taxa whereas pollen found in hyrax middens is likely a more direct representation of the terrestrial vegetation), each record is carefully considered in the context of its unique environmental, ecological, hydrological and geomorphological setting – all of which can influence pollen production, dispersal and representation. These factors are also taken into account when comparing records across the CFR.

That the existing spatial heterogeneity within the CFR is a key determining factor for the nature of any identified palaeoenvironmental changes is not a new observation (Chase and Meadows 2007; Cowling 1992; Cowling and Holmes 1992; Meadows and Baxter 1999), however, with more records generated, and the increasing availability of high-resolution data, we are now able to explore this theory to a fuller extent. To best illustrate how distinct subregions of the CFR have responded differently to palaeoenvironmental changes and to highlight the contribution of the most recent published pollen records, two key subregions – the southern Cape coast and the Cederberg mountains – are briefly considered.

14.2.1 Palaeoecology of the southern Cape: The significance of coastal environments

The southern Cape coastal plain forms the eastern lowland subregion of the CFR (Figure 1) and is influenced by atmospheric dynamics associated with both the winter and summer rainfall zones (the aseasonal or year-round rainfall zone; *sensu* Chase and Meadows 2007).

While several sediment cores have been extracted from the Vankervelsvlei wetland site (Irving 1998; Irving and Meadows 1997), limited reliable palaeoenvironmental evidence has emanated from these studies due to poor chronological control and inconsistent pollen preservation. However, with the addition of optically stimulated luminescence (OSL) as a dating technique, Quick et al. (2016) established a chronology for a sediment sequence spanning the last 140,000 years. The variability within the pollen assemblage from this record indicates a complex response to the transition from interglacial to glacial conditions. In general, the dominance of afrotemperate forest pollen from Marine Isotope Stage (MIS) 5d (*c.* 108–96 ka) alludes to enhanced summer rainfall under warmer conditions allowing for the development of more extensive forests in the area (Quick *et al.* 2016). In contrast, the pollen from the last glacial period was dominated by fynbos types indicating cooler temperatures and a more seasonal winter-dominated rainfall regime (Quick *et al.* 2016).

Due to the virtual absence of palaeoenvironmental data from the southern Cape covering MIS 5 to MIS 3, the Vankervelsvlei pollen data represents an important contribution, however as no pollen was preserved from *c.* 27–8.7 ka, the record is incomplete. This highlights a particularly frustrating trend of the Last Glacial Maximum (LGM; 26.5–19 cal ky BP (Clark et al. 2009)) period being absent from pollen sequences from the southern Cape coast. Either the pollen records recovered from the subregion do not extend as far back as the LGM (e.g. du Plessis *et al.* 2020; Martin 1968; Scholtz 1986) or as is the case of Vankervelsvlei and Rietvlei-Still Bay (Quick *et al.* 2015), there is limited to no pollen preserved. Supplementary proxy data from Vankervelsvlei and Rietvlei-Still Bay suggest that conditions were likely quite arid during the LGM (Quick *et al.* 2015; Quick *et al.* 2016), which could explain the lack of pollen preservation, and is supported by evidence of decreased moisture availability from other studies from the subregion (Chase and Meadows 2007, and references therein). Other factors that could potentially have contributed to decreased wetland productivity and a lack of pollen preservation within these coastal or near-coastal wetlands during the LGM are lower sea-levels (affecting groundwater recharge and regional hydrology) and enhanced aeolian activity. As pollen is found within the LGM sections of records from the western interior (e.g. the Pakhuis Pass and De Rif hyrax midden records (Quick *et al.* 2011; Scott and Woodborne 2007a, b; Valsecchi *et al.* 2013)) and the southwestern Cape coast within the modern WRZ (Pearly Beach record (Quick *et al.*, 2021), the lack of preservation seems to be confined to the year-round rainfall portion of the southern Cape (Southern Cape subregion, Figure 1) and therefore likely to be tied to subregion-scale changes in moisture availability. This theory can be tested with the acquisition of new sites from both the coast and the interior.

With an average sample resolution of 57 years, the pollen record from the coastal lake of Eilandvlei (Quick *et al.* 2018; Figure 1) represents the most continuous and highest resolution record of Holocene vegetation change for the southern Cape subregion. Similar to the Vankervelsvlei record, the Eilandvlei pollen assemblage primarily reflects shifts in the abundance of afrotemperate forest and fynbos taxa. For example, fynbos dominates the assemblage during the early Holocene, indicating generally cool, seasonal conditions, whereas increasing percentages of afrotemperate forest taxa from 3.5 cal ky BP to present day suggest that the late Holocene was characterized by a long-term trend of steadily increasing moisture availability (Quick *et al.* 2018). Comparisons of this high-resolution record with regional and extra-regional palaeoclimatic data highlight the complex nature of southern Cape climate dynamics (Quick *et al.* 2018). The similarities between the Eilandvlei record and records from more tropical regions of South Africa led to the proposal that summer rainfall is of great importance in terms of maintaining high moisture availability and that the Agulhas Current could be responsible for transmitting this signal of tropical variability along the coast (Chase and Quick 2018; Quick *et al.* 2018). These insights emphasize the importance of considering proxy evidence, including fossil pollen, from sites on the coast separately from those from the interior and caution against extrapolating results beyond the subregion scale.

14.2.2 Cederberg palaeoenvironments: Environmental stability *vs.* extreme palaeoclimatic changes

The Cederberg mountains represent the core of the western interior CFR subregion (Figure 1) and while free from the direct, localized influence of coastal dynamics, the palaeoenvironmental history of the Cederberg is no less complex than that of the southern Cape coast.

Several fossil pollen records from this subregion infer that extremely muted changes in vegetation composition took place during the last glacial and Holocene (Meadows *et al.* 2010; Meadows and Sugden 1991; Meadows and Sugden 1993; Quick *et al.* 2011). It has been suggested that mountain fynbos (the dominant vegetation type in the area) is highly resilient to changes in moisture availability, which could account for the low-amplitude variability within these palaeoecological records (Quick *et al.* 2011). However, this is contrasted by a growing body of data that indicates more dynamic patterns in vegetation and, especially, climate change (Chase *et al.* 2019; Chase *et al.* 2015a; Chase *et al.* 2011; Scott and Woodborne 2007a, b; Valsecchi *et al.* 2013). This evidence mainly consists of stable isotope records derived from hyrax middens situated within the central (Chase *et al.* 2011; Chase *et al.* 2015), northern (Chase *et al.* 2019) and eastern (Chase *et al.* 2015) sections of the Cederberg. These records indicate that substantial, rapid changes in hydroclimatic conditions occurred since the LGM, and that while the drivers of climate dynamics appear to be shared across the subregion, there are distinct local responses, resulting in a great degree of variability across short spatial gradients (Chase *et al.* 2019).

The discrepancies between the older pollen records and the more recent stable isotope records could be related to differences in the nature of the information obtained from these two proxies. Stable carbon and nitrogen isotope data from hyrax middens are excellent recorders of changes in moisture availability (Chase *et al.* 2012) whereas palynological data is more nuanced and it can be difficult to disentangle ecological and climatic drivers of vegetation change. A further possibility is that while species turnover may be high, because it is generally only possible to identify pollen to genus or family level, this variability is not captured within the pollen assemblages.

It is clear from the emerging complexity of the palaeoenvironmental dynamics within these two subregions that substantial variability can occur across relatively short distances and interpolations beyond site level cannot always be reliably applied to regional syntheses.

14.3 CURRENT AND FUTURE DIRECTIONS FOR PALAEOECOLOGICAL RESEARCH IN THE CAPE FLORISTIC REGION

We now have a more nuanced understanding of palaeoenvironmental change within the CFR, however new high-resolution data are still required to fully explore subregional variability and to provide a more comprehensive understanding of the climate dynamics, drivers and the impact of climate change on CFR environments. There is great potential for fossil pollen data to contribute to this research agenda and to provide long-term perspectives of ecological dynamics within the CFR. The following sections outline current and potential future research directions that align to these goals.

14.3.1 The influence of the Agulhas Current on coastal palaeoenvironments

To expand on the theory outlined in Chase and Quick (2018) that a strong dipole in hydroclimatic conditions exist between the coast and interior during the Holocene, and that the Agulhas Current acts as a vector for the propagation of tropical southeast African climate signals to the southern Cape, more data is required. For a clearer understanding of the spatiotemporal dynamics of this hypothesized relationship additional high-resolution records extending into the last glacial period are needed from both the interior and coastal subregions of the southern Cape as well as from offshore (i.e. sea-surface temperature records). To begin to address this gap, a new project (by the author and collaborators) is currently underway (2020–2023) that focuses on the generation of palaeoenvironmental data, particularly fossil pollen, from wetland sites situated on a west-east transect along the southern Cape coastline.

14.3.2 Exploring steep palaeoenvironmental gradients in the Cederberg mountains

As previously outlined, the findings from the high-resolution stable isotope data from the Cederberg suggest that considerable palaeoclimatic change occurred across the subregion during the last glacial and Holocene (Chase *et al.* 2019). To fully evaluate and contextualize these findings, more sites from transects across the subregion are required. A range of proxies (such as stable isotopes, pollen, plant biomarkers, phytoliths and microcharcoal) will be analysed. New pollen data would be especially useful as it is still unclear as to whether vegetation change was in fact relatively muted or if the pollen records currently in existence are not of high enough spatial or temporal resolution to adequately capture the variability evident in the stable isotope records.

14.3.3 Palaeoenvironments of the eastern Cape: The final frontier?

Encompassing the transition from fynbos to Albany/subtropical thicket as well as from fynbos to Succulent Karoo, the eastern Cape represents an important, understudied eastern boundary of the CFR. Palaeoecological data for the eastern Cape is extremely limited and no palynological research has been conducted within the Albany Thicket biome. Only one pollen record exists for the region and this record, derived from hyena coprolite samples from a Middle Stone Age archaeological site, has no chronology and the pollen content was not sufficient for a detailed palaeoenvironmental reconstruction to be performed (Carrion *et al.* 2000; Figure 1). There are two additional publications that include pollen data from sites situated within the Eastern Cape Province, however both of these sites fall outside the CFR: the Dunedin and Salisbury vlei pollen records from the Winterberg escarpment (Meadows and Meadows 1988) and the more recently published preliminary pollen results from ongoing archaeological research at Waterfall Bluff, eastern Pondoland (Esteban *et al.* 2020; Fisher *et al.* 2020).

There is great scope for the generation of new pollen records from both the montane (e.g. the Baviaanskloof mountains) and the coastal regions (e.g. between Tsitsikamma and Port Elizabeth). Stable isotope data from a hyrax midden site within the Baviaanskloof provide new evidence for a high degree of spatiotemporal climatic variability during the Holocene (Chase *et al.* 2020). The generation of pollen data from this midden is currently underway, this data will be able to shed light on how vegetation within this ecotonal subregion of the CFR may have responded to the documented changes in hydroclimatic variability.

14.3.4 Bridging the gap between palaeoecology and ecology

Despite the similarities in their names, ecology and palaeoecology have historically been almost entirely distinct fields of research. Both fields have contributed to this disassociation: While ecologists generally recognize the need for long-term perspectives of vegetation change, they have been sceptical of the robustness of palaeoecological research methods and the inferences that are drawn from proxy records. Consequently, they have had reservations about the usefulness and reliability of palaeoecological data for addressing ecological questions. While this may once have been a legitimate concern, there have been vast methodological improvements to the generation of palaeoecological data which have resulted in much higher resolution, more accurately dated, proxy sequences (Birks *et al.* 2016). Conversely, due to the different spatial and temporal scales, palaeoecologists have often assumed that contemporary ecology is too far removed methodologically and theoretically from their field to be truly applicable. They have tended to focus on directly answering climate-related questions and in doing so have often overlooked the ecological associations of the data which has significant implications for palaeoclimatic reconstructions.

While several studies integrating ecological theory and palynological research have been conducted elsewhere in South Africa (e.g. Gillson and Duffin 2007; Gillson and Ekblom 2020; Gillson and Marchant 2014) until recently, this approach had not been attempted for the CFR. Drawing from the relatively muted changes within two late Holocene pollen records derived from sediment cores from biome boundaries (MacPherson *et al.* 2018; MacPherson *et al.* 2019), Gillson *et al.* (2020) present new hypotheses exploring possible ecological mechanisms to explain the apparent resilience of fynbos to environmental change. An assessment of how the ecosystem resilience and resistance hypotheses outlined in Gillson *et al.* (2020) align to the greater shifts in fynbos-forest dynamics documented in other palaeoecological records from ecotonal regions of the CFR (e.g. du Plessis *et al.* 2020; Quick *et al.* 2018; 2016) remains to be undertaken. However these studies as well as Forbes *et al.* (2018) highlight the potential for applying a more analytical ecological lens to CFR palynological records.

Fire is a major driving force in most Mediterranean-type ecosystems including fynbos (Cowling 1992). Without fire, fynbos communities become senescent and other plant types can invade into the biome (Cowling *et al.* 1992; Mucina and Rutherford 2006). Information on past fire regimes is therefore particularly important in order to understand changes at the plant community level as well as shifts in biome boundaries, especially boundaries shared with fire-intolerant communities such as forest and thicket (Mucina and Rutherford 2006). Microcharcoal abundances (which are routinely counted simultaneously with pollen) can be used to reconstruct local and regional fire regimes (Cowling *et al.* 1997). Determining fire intensities from charcoal found in sedimentary sequences is now possible as a result of the development of a new method (Gosling *et al.* 2019), this method should be explored within the context of the CFR.

By providing a better understanding of ecological responses to natural and anthropogenic impacts, palaeoecological data can directly contribute to conservation management and restoration practices, however this will rely on the establishment of effective multidisciplinary working groups (Gillson, this volume; Manzano *et al.* 2020).

14.3.5 New methods and resources

The CREST (Climate REconstruction SofTware) method developed by Chevalier *et al.* (2014) has been very successful in producing quantitative climate reconstructions for southern Africa using contemporary climate data, modern plant distributions and fossil pollen records (Chase *et al.* 2015a; Chase *et al.* 2015b; Chevalier and Chase 2015; Chase *et al.* 2017; Chevalier *et al.* 2021; Cordova *et al.* 2017; Lim *et al.* 2016). A major advantage of this method is that a large modern pollen database (which does not currently exist for the CFR) is not required. The steep environmental gradients and coastal settings of many of the CFR pollen records have proven to introduce a great degree of complexity to the application of CREST within this region. However, there certainly still is scope for its application particularly with the addition of a new resource:

A 'by-product' of the CREST method has given rise to the 'Atlas of southern African pollen taxa: Distributions and Climatic Affinities' (Chevalier *et al.* this volume) which will greatly improve, refine and standardize qualitative interpretations of fossil pollen data as well as quantitative reconstructions (such as CREST). This reference atlas includes a series of diagnostic tools and results based on statistical analyses of modern distributions of common southern African pollen taxa in relation to key climate variables. This new resource reduces the subjectivity often inherent in more qualitative interpretations and presents more accurate determination of absolute and relative climate sensitivities of southern African pollen types, including a large proportion of CFR taxa.

The recent digitization of CFR pollen reference material (originally from the Palaeoecological Laboratory in the Department of Environmental and Geographical Science, University of Cape Town) and the establishment of a new open access, searchable electronic database (http://pollen.mandela.ac.za) represents a new primary palynological resource. While the resurrection of the African Pollen Database (APD, Vincens *et al.* 2007) and its integration into the Neotoma Palaeoecology Database (Goring *et al.* 2015), will ensure that previously published and new pollen datasets are archived within an open access online platform (Ivory *et al.* 2020). If the majority of fossil pollen records for the CFR were included in the new APD - Neotoma database, new forms of spatiotemporal analyses could be conducted which would provide greater insight into ecosystem-climate linkages across the region.

14.4 CONCLUSIONS

While our understanding of paleoenvironmental change within the CFR has been greatly improved by the establishment of recent high-resolution palaeoclimatic and palaeoecological records, fundamental questions still remain relating to the full extent of the subregional variability of both climate and vegetation change, and how these subregions are associated/disassociated with each other. Further insight into the long-term environmental dynamics within the CRF will be possible if the spatial and temporal coverage of sites are expanded and by employing the new methods and resources now available.

In the face of global climate change and ever-increasing ecological uncertainty as a result of unprecedented climatic and societal changes, there has never been a more urgent need for scientific collaboration and the establishment of multidisciplinary studies. The wide sphere of applicability of pollen analysis to a multitude of fields (e.g. vegetation modelling, restoration ecology, conservation biology, aerobiology and archaeology) highlights the incredible potential, both current and future, of palaeoecology's role within multidisciplinary studies. Hopefully in the coming years palaeoecological research within the CFR will continue to advance and contribute to a greater understanding of the region's unique biodiversity and climatic history.

ACKNOWLEDGEMENTS

While this paper is a personal perspective, the conceptual frameworks, ideas and insights are the culmination of collaboration particularly with Michael Meadows and Brian Chase, who originally set me on this research path. Two anonymous reviewers are thanked for their insightful comments and suggestions on an earlier version of this paper.

REFERENCES

Allsopp, N., Colville, J.F. and Verboom, G.A., 2014, Fynbos: ecology, evolution, and conservation of a megadiverse region. Oxford University Press, USA.

Bennett, K.D. and Willis, K.J., 2001, Pollen. In *Tracking Environmental Change Using Lake Sediments. Volume 3: Terrestrial, Algal, and Siliceous Indicators*, edited by Smol, J.P., Birks, H.J.B., Last, W.M., (Dordrecht: Kluwer Academic Publishers), pp. 5–26.

Birks, H.J.B. and Birks, H.H. and Ammann, B., 2016, The fourth dimension of vegetation. *Science* **354**, pp. 412–413, 10.1126/science.aai8737.

Bond, P. and Goldblatt, P., 1984, Plants of the Cape flora: A descriptive catalogue, In *Journal of South African Botany*, Supplementary Volume, edited by Eloff, J.N., pp. 1–455.

Bradshaw, P.L. and Cowling, R.M., 2014. Landscapes, rock types, and climate of the Greater Cape Floristic Region. In *Fynbos: Ecology, Evolution, and Conservation of a Megadiverse Region*, edited by Allsopp, N., Colville, J.F., Verboom, G.A. (Oxford: Oxford University Press) pp. 26–46.

Brown, K.S., Marean, C.W., Herries, A.I.R., Jacobs, Z., Tribolo, C., Braun, D., Roberts, D.L., Meyer, M.C. and Bernatchez, J., 2009, Fire as an engineering tool of early modern humans. *Science*, **325**, pp. 859–862, 10.1126/science.1175028.

Brown, K.S., Marean, C.W., Jacobs, Z., Schoville, B.J., Oestmo, S., Fisher, E.C., Bernatchez, J., Karkanas, P. and Matthews, T., 2012, An early and enduring advanced technology originating 71,000 years ago in South Africa. *Nature* **491**, pp. 590-593, 10.1038/nature11660.

Campbell, B.M., 1985, *A Classification of the Mountain Vegetation of the Fynbos Biome*. (Pretoria: Botanical Research Institute, Deptartment of Agriculture and Water Supply).

Carr, A.S., Thomas, D.S.G., Bateman, M.D., Meadows, M.E. and Chase, B., 2006, Late Quaternary palaeoenvironments of the winter-rainfall zone of southern Africa: Palynological and sedimentological evidence from the Agulhas Plain. *Palaeogeography, Palaeoclimatology, Palaeoecology*, **239**, pp. 147–165, 10.1016/j.palaeo.2006.01.014.

Carrion, J.S., Brink, J.S., Scott, L. and Binneman, J.N.F., 2000, Palynology and palaeoenvironment of Pleistocene hyaena coprolites from an open-air site at Oyster Bay, Eastern Cape coast, South Africa. *South African Journal of Science*, **96**, pp. 449–453.

Chase, B., Boom, A., Carr, A., Chevalier, M., Quick, L., Verboom, A. and Reimer, P., 2019, Extreme hydroclimate gradients within the western Cape Floristic region of South Africa since the Last Glacial Maximum. *Quaternary Science Reviews*, **219**, pp. 297–307, 10.1016/j.quascirev.2019.07.006.

Chase, B.M. and Meadows, M.E., 2007, Late Quaternary dynamics of southern Africa's winter rainfall zone. *Earth-Science Reviews*, **84**, pp. 103–138, 10.1016/j.earscirev.2007.06.002.

Chase, B.M., Quick, L.J., Meadows, M.E., Scott, L., Thomas, D.S.G. and Reimer, P.J., 2011, Late-glacial interhemispheric climate dynamics revealed in South African hyrax middens. *Geology*, **39**, pp. 19–22, 10.1130/G31129.1.

Chase, B.M., Scott, L., Meadows, M.E., Gil-Romera, G., Boom, A., Carr, A.S., Reimer, P.J., Truc, L., Valsecchi, V. and Quick, L.J., 2012, Rock hyrax middens: A palaeoenvironmental archive for southern African drylands. *Quaternary Science Reviews*, **56**, pp. 107–125, 10.1016/j.quascirev.2012.08.018.

Chase, B.M., Boom, A., Carr, A.S., Carré, M., Chevalier, M., Meadows, M.E., Pedro, J.B., Stager, J.C. and Reimer, P.J., 2015a, Evolving southwest African response to abrupt deglacial North Atlantic climate change events. *Quaternary Science Reviews*, **121**, pp. 132–136, 10.1016/j.quascirev.2015.05.023.

Chase, B.M., Lim, S., Chevalier, M., Boom, A., Carr, A.S., Meadows, M.E. and Reimer, P.J., 2015b, Influence of tropical easterlies in southern Africa's winter rainfall zone during the Holocene. *Quaternary Science Reviews*, **107**, pp. 138–148, 10.1016/j.quascirev.2014.10.011.

Chase, B.M., Chevalier, M., Boom, A. and Carr, A.S., 2017, The dynamic relationship between temperate and tropical circulation systems across South Africa since the last glacial maximum. *Quaternary Science Reviews*, **174**, pp. 54–62, 10.1016/j.quascirev.2017.08.011.

Chase, B.M. and Quick, L.J., 2018, Influence of Agulhas forcing of Holocene climate change in South Africa's southern Cape. *Quaternary Research*, **90**, pp. 303–309, 10.1017/qua.2018.57.

Chase, B.M., Boom, A., Carr, A.S., Quick, L.J. and Reimer, P.J., 2020, High-resolution record of Holocene climate change dynamics from southern Africa's temperate-tropical boundary, Baviaanskloof, South Africa. *Palaeogeography, Palaeoclimatology, Palaeoecology*, **539**, article: 109518, 10.1016/j.palaeo.2019.109518.

Chevalier, M., Cheddadi, R. and Chase, B.M., 2014, CREST (Climate REconstruction SofTware): a probability density function (PDF)-based quantitative climate reconstruction method. *Climate of the Past*, **10**, pp. 2081–2098, 10.5194/cp-10-2081-2014.

Chevalier, M. and Chase, B.M., 2015. Southeast African records reveal a coherent shift from high- to low-latitude forcing mechanisms along the east African margin across last glacial–interglacial transition. *Quaternary Science Reviews* **125**, pp. 117–130, 10.1016/j.quascirev.2015.07.009.

Chevalier, M., Chase, B.M., Quick, L.J., Dupont, L.M., and Johnson, T.C., 2021, Temperature change in subtropical southeastern Africa during the past 790,000 yr. *Geology*, **49**(1), pp. 71–75, 10.1130/G47841.1.

Chevalier, M., Chase, B.M., Quick, L.J. and Scott, L., this volume, An atlas of southern African pollen types and their climatic affinities. *Palaeoecology of Africa*, **35**, chapter 15, 10.1201/9781003162766-15.

Clark, P.U., Dyke, A.S., Shakun, J.D., Carlson, A.E., Clark, J., Wohlfarth, B., Mitrovica, J.X., Hostetler, S.W. and McCabe, A.M., 2009, The last glacial maximum. *Science*, **325**, pp. 710–714, 10.1126/science.1172873.

Cordova, C.E., Scott, L., Chase, B.M. and Chevalier, M., 2017, Late Pleistocene-Holocene vegetation and climate change in the Middle Kalahari, Lake Ngami, Botswana. *Quaternary Science Reviews*, **171**, pp. 199–215, 10.1016/j.quascirev.2017.06.036.

Cordova, C.E., Kirsten, K.L., Scott, L., Meadows, M. and Lücke, A., 2019, Multi-proxy evidence of late-Holocene paleoenvironmental change at Princessvlei, South Africa: The effects of fire, herbivores, and humans. *Quaternary Science Reviews*, **221**, article: 105896, 10.1016/j.quascirev.2019.105896.

Cowling, R.M., 1992, T*he Ecology of Fynbos: Nutrients, Fire, and Diversity*. (Cape Town: Oxford University Press).

Cowling, R.M. and Holmes, P.M., 1992, Endemism and speciation in a lowland flora from the Cape Floristic Region. *Biological Journal of the Linnean Society*, **47**, pp. 367–383, 10.1111/j.1095-8312.1992.tb00675.x.

Cowling, R.M., Richardson, D.M., Paterson-Jones, C., 1995, *Fynbos: South Africa's Unique Floral Kingdom*. (Vlaeberg: Fernwood Press).

Cowling, R.M., Richardson, D.M., Schulze, R.E., Hoffman, M.T., Midgley, J.J. and Hilton-Taylor, C., 1997. Species diversity at the regional scale. In *Vegetation of Southern Africa*, edited by Cowling, R.M., Richardson, D.M., Pierce, S.M, (Cambridge: Cambridge University Press), pp. 447–273.

Cowling, R.M. and Lombard, A.T., 2002, Heterogeneity, speciation/extinction history and climate: explaining regional plant diversity patterns in the Cape Floristic Region. *Diversity and Distributions* **8**, pp. 163-179, 10.1046/j.1472-4642.2002.00143.x.

Cowling, R.M., Potts, A.J., Bradshaw, P.L., Colville, J., Arianoutsou, M., Ferrier, S., Forest, F., Fyllas, N.M., Hopper, S.D. and Ojeda, F., 2015, Variation in plant diversity in mediterranean-climate ecosystems: The role of climatic and topographical stability. *Journal of Biogeography*, **42**, pp. 552–564, 10.1111/jbi.12429.

Deacon, H.J., Deacon, J., Scholtz, A., Thackeray, J.F. and Brink, J.S., 1984, Correlation of palaeoenvironmental data from the Late Pleistocene and Holocene deposits at Boomplaas Cave, southern Cape. In *Late Cainozoic Palaeoclimates of the Southern Hemisphere*, edited by Vogel, J.C., (Rotterdam: Balkema), pp. 339–360.

Deacon, H.J., 1992, Human settlement. In *The Ecology of Fynbos: Nutrients, Fire and Diversity*, edited by Cowling, R.M. (Cape Town: Oxford University Press), pp. 260–270.

du Plessis, N., Chase, B.M., Quick, L.J., Haberzettl, T., Kasper, T. and Meadows, M.E., 2020, Vegetation and climate change during the Medieval Climate Anomaly and the Little Ice Age on the southern Cape coast of South Africa: Pollen evidence from Bo Langvlei. *The Holocene*, **30**(12), pp. 1716–1727, 10.1177/0959683620950444.

Esteban, I., Bamford, M.K., House, A., Miller, C.S., Neumann, F.H., Schefuß, E., Pargeter, J., Cawthra, H.C. and Fisher, E.C., 2020. Coastal palaeoenvironments and hunter-gatherer plant-use at Waterfall Bluff rock shelter in Mpondoland (South Africa) from MIS 3 to the Early Holocene. *Quaternary Science Reviews*, **250**, article: 106664, 10.1016/j.quascirev.2020. 106664.

Faegri, K. and Iversen, J., 1989, *Textbook of Pollen Analysis*, 4th Edition, (Chichester: John Wiley & Sons).

Fisher, E.C., Cawthra, H.C., Esteban, I., Jerardino, A., Neumann, F.H., Oertle, A., Pargeter, J., Saktura, R.B., Szabó, K. and Winkler, S., 2020, Coastal occupation and foraging during the last glacial maximum and early Holocene at Waterfall Bluff, eastern Pondoland, South Africa. *Quaternary Research*, **97**, pp. 1–41, 10.1017/qua.2020.26.

Forbes, C.J., Gillson, L. and Hoffman, M.T., 2018, Shifting baselines in a changing world: Identifying management targets in endangered heathlands of the Cape Floristic Region, South Africa. *Anthropocene* **22**, pp. 81–93, 10.1016/j.ancene.2018.05.001.

Gillson, L. and Duffin, K., 2007, Thresholds of potential concern as benchmarks in the management of African savannahs. *Philosophical Transactions of the Royal Society B: Biological Sciences*, **362**, pp. 309–319, 10.1098/rstb.2006.1988.

Gillson, L. and Marchant, R., 2014, From myopia to clarity: sharpening the focus of ecosystem management through the lens of palaeoecology. *Trends in Ecology & Evolution*, **29**, pp. 317–325, 10.1016/j.tree.2014.03.010.

Gillson, L. and Ekblom, A., 2020, Using palaeoecology to explore the resilience of southern African savannas. *Koedoe*, **62**, pp. 1–12, 10.4102/koedoe.v62i1.1576.

Gillson, L., MacPherson, A.J. and Hoffman, M.T., 2020, Contrasting mechanisms of resilience at mesic and semi-arid boundaries of fynbos, a mega-diverse heathland of South Africa. *Ecological Complexity*, **42**, article: 100827, 10.1016/j.ecocom.2020.100827.

Gillson, L., this volume, The role of palaeoecology in conserving African ecosystems. *Palaeoecology of Africa*, **35**, chapter: 24, 10.1201/9781003162766-24.

Goldblatt, P., 1978, An analysis of the flora of southern Africa: its characteristics, relationships, and origins. *Annals of the Missouri Botanical Garden* **65**, pp. 369–436, 10.2307/2398858.

Goldblatt, P. and Manning, J., 2000, *Cape Plants. A Conspectus of the Cape Flora of South Africa*. (Pretoria and St. Louis: National Botanical Institute and Missouri Botanical Gardens).

Goring, S., Dawson, A., Simpson, G., Ram, K., Graham, R., Grimm, E. and Williams, J., 2015, Neotoma: A programmatic interface to the Neotoma Paleoecological Database. *Open Quaternary* 1(1), article 2, 10.5334/oq.ab.

Gosling, W.D., Cornelissen, H. and McMichael, C.N.H., 2019, Reconstructing past fire temperatures from ancient charcoal material. *Palaeogeography, Palaeoclimatology, Palaeoecology*, **520**, pp. 128–137, 10.1016/j.palaeo.2019.01.029.

Henshilwood, C.S., d'Errico, F., Marean, C.W., Milo, R.G. and Yates, R., 2001, An early bone tool industry from the Middle Stone Age at Blombos Cave, South Africa: implications for the origins of modern human behaviour, symbolism and language. *Journal of Human Evolution*, **41**, pp. 631–678, 10.1006/jhev.2001.0515.

Henshilwood, C.S., d'Errico, F., Yates, R., Jacobs, Z., Tribolo, C., Duller, G.A.T., Mercier, N., Sealy, J.C., Valladas, H., Watts, I. and Wintle, A.G., 2002, Emergence of modern human behavior: Middle stone age engravings from South Africa. *Science*, **295**, pp. 1278–1280, 10.1126/science.1067575.

Irving, S.J.E. and Meadows, M.E., 1997, Radiocarbon chronology and organic matter accumulation at Vankervelsvlei, near Knysna, South Africa. *South African Geographical Journal*, **79**, pp. 101–105, 10.1080/03736245.1997.9713630.

Irving, S.J.E., 1998. Late Quaternary palaeoenvironments at Vankervelsvlei, near Knysna, South Africa, Department of Environmental and Geographical Sciences. University of Cape Town, Cape Town.

Ivory, S., Lézine, A.-M., Grimm, E. and Williams, J.W., 2020, Relaunching the African Pollen Database: Abrupt change in climate and ecosystems, *Past Global Changes Magazine*, **28**(1), pp. 26.

Lim, S., Chase, B.M., Chevalier, M. and Reimer, P.J., 2016, 50,000 years of vegetation and climate change in the southern Namib Desert, Pella, South Africa. *Palaeogeography, Palaeoclimatology, Palaeoecology*, **451**, pp. 197–209, 10.1016/j.palaeo.2016.03.001.

Linder, H.P., 2005. Evolution of diversity: The Cape flora. *Trends in Plant Science*, **10**, pp. 536–541, 10.1016/j.tplants.2005.09.006.

Linder, H.P., Meadows, M. and Cowling, R.M., 1992, History of the Cape Flora. In *The Ecology of Fynbos: nutrients, fire and diversity*, edited by Cowling, R.M. (Cape Town: Oxford University Press), pp. 113–134.

MacPherson, A., Gillson, L. and Hoffman, M., 2018, Climatic buffering and anthropogenic degradation of a Mediterranean-type shrubland refugium at its semi-arid boundary, South Africa. *The Holocene*, **28**, pp. 651–666, 10.1177/0959683617735582.

MacPherson, A.J., Gillson, L. and Hoffman, M.T., 2019, Between-and within-biome resistance and resilience at the fynbos-forest ecotone, South Africa. *The Holocene*, **29**, 1801–1816, 10.1177/0959683619862046.

Manzano, S., Julier, A.C.M., Dirk, C.J., Razafimanantsoa, A.H.I., Samuels, I., Petersen, H., Gell, P., Hoffman, M.T. and Gillson, L., 2020, Using the past to manage the future: the role of palaeoecological and long-term data in ecological restoration. *Restoration Ecology*, **28**, pp. 1335–1342, 10.1111/rec.13285.

Marean, C.W., Bar-Matthews, M., Bernatchez, J., Fisher, E., Goldberg, P., Herries, A.I.R., Jacobs, Z., Jerardino, A., Karkanas, P., Minichillo, T., Nilssen, P.J., Thompson, E., Watts, I., and Williams, H.M., 2007, Early human use of marine resources and pigment in South Africa during the Middle Pleistocene. *Nature*, **449**, pp. 905–908, 10.1038/nature06204.

Marean, C.W., 2010, Pinnacle Point Cave 13B (Western Cape Province, South Africa) in context: The Cape Floral kingdom, shellfish, and modern human origins. *Journal of Human Evolution*, **59**, pp. 425–443, 10.1016/j.jhevol.2010.07.011.

Marloth, R., 1908. Das Kapland: Insonderheit das Reich der Kapflora, das Waldgebiet und die Karroo, pflanzengeographisch dargestellt. (Jena: G. Fischer).

Martin, A.R.H., 1968. Pollen analysis of Groenvlei lake sediments, Knysna (South Africa). *Review of Palaeobotany and Palynology* 7, pp. 107–144, 10.1016/0034-6667(68)90029-8.

Meadows, M.E. and Baxter, A.J., 1999, Late Quaternary palaeoenvironments of the southwestern Cape, South Africa: A regional synthesis. *Quaternary International*, **57–8**, pp. 193–206.

Meadows, M.E. and Baxter, A.J., 2001, Holocene vegetation history and palaeoenvironments at Klaarfontein Springs, Western Cape, South Africa. *The Holocene* **11**, pp. 699–706, 10.1191/09596830195726.

Meadows, M.E., and Meadows, K.F., 1988, Late Quaternary vegetation history of the Winterberg Mountains, Eastern Cape, South Africa. *South African Journal of Science*, **84**, pp. 253–259.

Meadows, M.E. and Sugden, J.M., 1991, A vegetation history of the last 14,000 years on the Cederberg, southwestern Cape Province. *South African Journal of Science*, **87**, 34–43.

Meadows, M.E. and Sugden, J.M., 1993, The late Quaternary palaeoecology of a floristic kingdom: the southwestern Cape, South Africa. *Palaeogeography, Palaeoclimatology, Palaeoecology*, **101**, pp. 271–281, 10.1016/0031-0182(93)90019-F.

Meadows, M.E., Baxter, A.J. and Parkington, J., 1996, Late Holocene environments at Verlorenvlei, Western Cape Province, South Africa. *Quaternary International*, **33**, pp. 81–95, 10.1016/1040-6182(95)00092-5.

Meadows, M.E., Seliane, M. and Chase, B.M., 2010, Holocene palaeoenvironments of the Cederberg and Swartruggens mountains, Western Cape, South Africa: pollen and stable isotope evidence from hyrax dung middens. *Journal of Arid Environments*, **74**, pp. 786–793.

Mittermeier, R.A., Robles Gil, P., Hoffman, M., Pilgrim, J., Brooks, T., Goettsch Mittermeier, C., Lamoreux, J. and da Fonseca, G.A.B., 2004, *Hotspots Revisited*. (Mexico City: CEMEX, Agrupación Sierra Madre, S.C.).

Myers, N., Mittermeier, R.A., Mittermeier, C.G., da Fonseca, G.A.B. and Kent, J., 2000, Biodiversity hotspots for conservation priorities. *Nature*, **403**, pp. 853–858, 10.1038/35002501.

Neumann, F.H., Scott, L., Bamford, M.K., 2011, Climate change and human disturbance of fynbos vegetation during the late Holocene at Princess Vlei, Western Cape, South Africa. *The Holocene*, **21**, pp. 1137–1149, 10.1177/0959683611400461.

Quick, L.J., Chase, B.M., Meadows, M.E., Scott, L. and Reimer, P.J., 2011, A 19.5 kyr vegetation history from the central Cederberg Mountains, South Africa: Palynological evidence from rock hyrax middens. *Palaeogeography, Palaeoclimatology, Palaeoecology* 309, pp. 253–270, 10.1016/j.palaeo.2011.06.008.

Quick, L.J., Carr, A.S., Meadows, M.E., Boom, A., Bateman, M.D., Roberts, D.L., Reimer, P.J. and Chase, B.M., 2015, A Late Pleistocene–Holocene multi-proxy record of palaeoenvironmental change from Still Bay, southern Cape Coast, South Africa. *Journal of Quaternary Science*, **30**, pp. 870–885, 10.1002/jqs.2825.

Quick, L.J., Meadows, M.E., Bateman, M.D., Kirsten, K.L., Mäusbacher, R., Haberzettl, T. and Chase, B.M., 2016, Vegetation and climate dynamics during the last glacial period in the fynbos-afrotemperate forest ecotone, southern Cape, South Africa. *Quaternary International*, **404**, pp. 136–149, 10.1016/j.quaint.2015.08.027.

Quick, L.J., Chase, B.M., Wündsch, M., Kirsten, K.L., Chevalier, M., Mäusbacher, R., Meadows, M.E. and Haberzettl, T., 2018, A high-resolution record of Holocene climate and vegetation dynamics from the southern Cape coast of South Africa: pollen and microcharcoal evidence from Eilandvlei. *Journal of Quaternary Science*, **33**, pp. 487–500, 10.1002/jqs.3028.

Quick, L.J., Chase, B.M., Carr, A.S., Chevalier, M., Grober, B.A. and Meadows, M.E., 2021, A 25,000-year record of climate and vegetation change from the southwestern Cape coast, South Africa. *Quaternary Research*, 10.1017/qua.2021.31.

Reason, C.J.C., Landman, W. and Tennant, W., 2006, Seasonal to decadal prediction of southern African climate and its links with variability of the Atlantic ocean. *Bulletin of the American Meteorological Society*, **87**, pp. 941–955, 10.1175/BAMS-87-7-941.

Rebelo, A.G., Boucher, C., Helme, N., Mucina, L. and Rutherford, M.C., 2006, Fynbos Biome. In *The vegetation of South Africa, Lesotho and Swaziland*, edited by Mucina, L., Rutherford, M.C. (Preroria: South African National Biodiversity Institute), pp. 221–299.

Sadr, K., 1998, The first herders at the Cape of Good Hope. *African Archaeological Review* **15**, pp. 101-132, 10.1023/A:1022158701778.

Sadr, K., 2008, Invisible herders? The archaeology of Khoekhoe pastoralists. *Southern African Humanities*, **20**, pp. 179–203.

Schalke, H.J.W.G., 1973, The Upper Quaternary of the Cape Flats area. *Scripta Geologica*, **15**, pp. 1–57.

Scholtz, A., 1986, Palynological and Palaeobotanical Studies in the Southern Cape. University of Stellenbosch, Stellenbosch, South Africa.

Scott, L., 1990. Hyrax (*Procaviidae*) and dassie rat (*Petromuridae*) middens in palaeoenvironmental studies in Africa. In *Packrat Middens: The Last 40,000 Years of Biotic Change,* edited by Betancourt, J.L., van Devender, T.R., Martin, P.S. (Tucson: University of Arizona Press), pp. 408–427.

Scott, L. and Woodborne, S., 2007a, Vegetation history inferred from pollen in Late Quaternary faecal deposits (hyraceum) in the Cape winter-rain region and its bearing on past climates in South Africa. *Quaternary Science Reviews*, **26**, pp. 941–953, 10.1016/j.quascirev.2006.12.012.

Scott, L. and Woodborne, S., 2007b, Pollen analysis and dating of late Quaternary faecal deposits (hyraceum) in the Cederberg, Western Cape, South Africa. *Review of Palaeobotany and Palynology*, **144**, pp. 123–134, 10.1016/j.revpalbo.2006.07.004.

Takhtajan, A., Crovello, T.J. and Cronquist, A., 1986, Floristic Regions of the World. (Berkeley: University of California Press).

Taljaard, J., 1996, *Atmospheric Circulation Systems, Synoptic Climatology and Weather Phenomena of South Africa*. South African Weather Service Technical Paper 32. (Pretoria: South African Weather Service).

Tyson, P.D., 1986, *Climatic Change and Variability in Southern Africa*. (Cape Town: Oxford University Press).

Tyson, P.D. and Preston-Whyte, R.A., 2000, The Weather and Climate of Southern Africa. Oxford University Press, Cape Town.

Valsecchi, V., Chase, B.M., Slingsby, J.A., Carr, A.S., Quick, L.J., Meadows, M.E., Cheddadi, R. and Reimer, P.J., 2013, A high resolution 15,600-year pollen and microcharcoal record from the Cederberg Mountains, South Africa. *Palaeogeography, Palaeoclimatology, Palaeoecology*, **387**, pp. 6–16, 10.1016/j.palaeo.2013.07.009.

Vincens, A., Lézine, A.-M., Buchet, G., Lewden, D. and Le Thomas, A., 2007, African pollen database inventory of tree and shrub pollen types. *Review of Palaeobotany and Palynology*, **145**, pp. 135–141, 10.1016/j.revpalbo.2006.09.004.

Verboom, G.A., Linder, H.P., Forest, F., Hoffmann, V., Bergh, N.G. and Cowling, R.M., 2014, Cenozoic assembly of the Greater Cape flora. In *Fynbos: Ecology, Evolution, and Conservation of a Megadiverse Region*, edited by Allsopp, N., Colville, J.F., and Verboom, G.A., (Oxford: Oxford University Press), pp. 93–118.

CHAPTER 15

An atlas of southern African pollen types and their climatic affinities

Manuel Chevalier[1]

Institute of Earth Surface Dynamics, Geopolis, University of Lausanne, Switzerland

Brian M. Chase[2]

Institut des Sciences de l'Evolution-Montpellier (ISEM), University of Montpellier, Centre National de la Recherche Scientifique (CNRS), EPHE, IRD, Montpellier, France.

Lynne J. Quick

African Centre for Coastal Palaeoscience, Nelson Mandela University, Port Elizabeth, South Africa

Louis Scott

Department of Plant Sciences, University of the Free State, Bloemfontein, South Africa

ABSTRACT: Interpretations of fossil pollen data are often limited to broad, qualitative assessments of past climatic and environmental conditions (e.g. colder *vs.* warmer, wetter *vs.* drier, open *vs.* closed landscape). These assessments can be particularly imprecise in regions such as southern Africa, where botanical biodiversity is high, and there exists an associated uncertainty regarding the climatic/environmental sensitivities of the plants contributing to a given pollen type. This atlas addresses this limitation by characterising the climate sensitivities of the 140 pollen morphotypes most often recorded in Late Quaternary palaeoecology studies in southern Africa, relying on their parent plant distributions as one of the basic factors that determine their presence. The atlas is designed as a suite of graphical diagnostic tools and photographs together with analyses of the modern geographical distribution of more than 22,000 plant species to identify their primary climatic sensitivities across southern Africa. Together, the elements included span the complete workflow from pollen identification through interpretation and climate reconstruction. The atlas can be accessed from https://doi.org/10.5281/zenodo.4013452.

[1] Other affiliation: *Institute of Geosciences, Sect. Meteorology, Rheinische Friedrich-Wilhelms-Universität Bonn, Bonn, Germany*
[2] Other affiliation: *Department of Environmental and Geographical Science, University of Cape Town, Rondebosch, South Africa*

DOI 10.1201/9781003162766-15

15.1 INTRODUCTION

Fossil pollen records have been a key source of palaeoenvironmental information in southern Africa for decades (Coetzee 1967; Finch and Hill 2008; Lim *et al.* 2016; Neumann *et al.* 2010, 2011, 2014; Quick *et al.* 2016, 2018; Scott and Nyakale 2002; Scott and Woodborne 2007; Scott *et al.* 2012; Scott 1982, 1989, 1999; Valsecchi *et al.* 2013; van Zinderen Bakker and Coetzee 1988; van Zinderen Bakker 1957, 1982; e.g. Bousman *et al.* 1988). While it is generally understood that a range of edaphic and ecological factors may influence plant distributions and vegetation composition, it is common for interpretations of pollen records to focus on plant climatic affinities and the reconstruction of past climate change. However, plant-climate relationships are often complex, and interpretive frameworks based on individuals' assessments of climate conditions where a given taxon occurs often lack the comprehensive consideration required to establish the relative influence of specific climatic parameters. Further, as pollen morphotypes (hereafter referred to as pollen types) commonly comprise more than one single species, defining the climatic affinity of a pollen type quickly becomes exceedingly complicated. These issues are amplified in regions with high botanical diversity, such as southern Africa, where tens or even hundreds of species, sometimes with distinct climate requirements, are subsumed within undifferentiated pollen types that can cover wide ranges of ecological and climatic conditions. In practice, robust statistical estimates of the specific climate affinities (both in terms of their tolerances and associated optima) are missing for the vast majority of the commonly observed pollen types, which often limits interpretations of fossil pollen data to relatively coarse-scale qualitative – and often subjective – climatological and/or environmental considerations.

The development of the CREST (Climate REconstruction SofTware) method (Chevalier *et al.* 2014) has enabled the production of the first quantitative climate reconstructions using fossil pollen records in southern Africa. CREST is based on the definition of independent, taxon-specific probability density functions (pdfs) to model a probabilistic relationship between pollen and climate from an ensemble of modern plant geolocalised occurrence data (hereafter 'plant distributions'). While more commonly used reconstruction techniques – such as modern analogues (Overpeck 1985; e.g. Imbrie and Kipp 1971), or regression techniques (ter Braak and Looman 1986; e.g. ter Braak and Juggins 1993) – rely on large collections of modern pollen samples to estimate pollen-climate relationships, such samples cannot be obtained for every environment or vegetation type of southern Africa. This limitation has strongly restricted their use in the region (Chevalier *et al.* 2014). In contrast, CREST uses widely-available plant distributions as calibration data (Chevalier 2019), and has proven efficient at reconstructing the different climates and vegetation types of the region (Chase *et al.* 2015a, 2015b, 2017; Chevalier and Chase 2016; Chevalier *et al.* 2021a; Cordova *et al.* 2017; Lim *et al.* 2016). As with all reconstruction techniques, the most reliable results are usually obtained when the method is run using a subset of taxa that are strongly influenced by the climatic parameter being reconstructed (Juggins *et al.* 2015; Kühl *et al.* 2002; Truc *et al.* 2013). The selection of the subset of taxa for climate reconstructions can be, however, subjective, as these selections are often based on the field experience of individuals, which could lead to inaccuracies because either the full distribution of the taxon is not considered, or the taxon is misattributed to certain conditions.

In this paper and accompanying supplementary atlas (Chevalier *et al.* 2021b), we present a series of diagnostic tools and results based on statistical analyses of the modern distributions of the parent plants of 140 southern African pollen types most commonly found in fossil records in relation to five key climate variables that capture the main climate gradients of the regions. This paper discusses how these diagnostics tools can be used to improve both palaeoecological interpretations and palaeoclimatological analyses derived from fossil pollen records. This atlas allows for a more accurate determination of the absolute and relative climate sensitivities of plants that produced the pollen types considered, which will facilitate both improved qualitative interpretations and the use of quantitative reconstruction methods. The user should be cognisant of the fact that this atlas does not account for the effect of *non-climatic* parameters, such as soil characteristics or fire, which may also be important determinants of some plant distributions. As such, the plant-climate relationships described in this atlas should always be balanced with considerations of the ecology of the studied taxa.

15.2 MATERIAL AND METHODS

15.2.1 Botanical Data

To estimate the climate responses of the different taxa, we used botanical data from the Global Biodiversity Information Facility (GBIF) database (GBIF.org 2020a, 2020b, 2020c, 2020d, 2020e, 2020f, 2020g, 2020h, 2020f; 'GBIF: Global Biodiversity Information Facility,' 2018) using the curated dataset of Chevalier (2019, 2018). These data are available as 'presence-only' observations (i.e. the absences and abundances are not documented) and their spatial resolution are homogenised at $0.25° \times 0.25°$ grid square (*c*. 900 km^2 at the latitudes of southern Africa). To build the atlas, we have used a total of 475,712 occurrence records, which correspond to 22,496 species grouped in 140 pollen types. The number of observed species in each grid cell varies between 1 and 2815, with a median of 17 species and a mean of 81 species.

The distribution of these data is spatially heterogenous (Figure 1A), with >75% of the records being concentrated in South Africa and its biodiversity hotspot (Rutherford *et al.* 2012; the Greater Cape Floristic Region; Myers *et al.* 2000). This sampling bias is also visible along

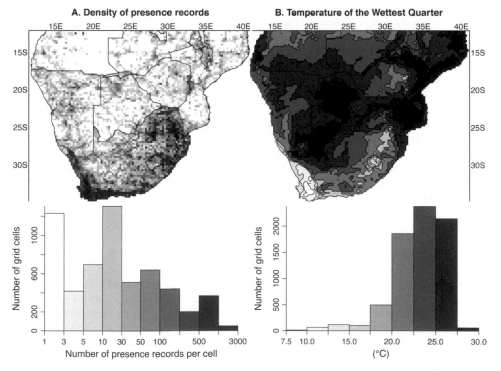

Figure 1. (A) Distribution of presence records (plant unique observations) across the study area (note the log scale for the colour gradient). This figure highlights the two biodiversity hotspots of the region, as well as a different degree of sampling across the study area. (B) TWetQ across the study area.

the South African border. The other countries composing our study area are less well represented, especially eastern Angola and western Zambia. However, the modern climate and vegetation gradients are less steep in those regions than they are in the southern part of the study area, where orography increases spatial variability. As such, it may be considered that the lower sampling density is sufficient to capture the full range of climate response of the studied taxa in these regions. It should also be remembered that some of the botanical data has been obtained from environments that have been influenced by human activity. While this could be considered as a factor that might undermine the natural plant-climate relationships that form the basis for this work, studies have shown that the plant-climate relationship remains robust as long as the full range of conditions experienced by the plant are sampled (Chevalier *et al.* 2014), and any anthropogenic bias is minimal.

Finally, it is important to note that a strict classification of plant species into pollen types is often not possible and our classification may contain related errors and/or regional biases. As plant and pollen taxonomies are being regularly improved, our classification is a compromise between the most recent changes and decades of palynological expertise. In general, the classification scheme adopted here maintains older names to remain consistent with published pollen records. For example, we use '*Acacia*-type', even as African *Acacia* are now *Vachellia* and *Senegalia*. The full list of species composing each pollen type in southern Africa is available at the end of the atlas (Chevalier *et al.* 2021b).

15.2.2 Climate Data

We used a subset of the Worldclim 2 climatology (Fick and Hijmans 2017) to generate the atlas. These climate data represent the mean climate between the years 1970 and 2000 and are available at several spatial resolutions. Here, we used the 5 arcmin resolution climatologies (*c.* 0.083°) and upscaled them to the resolution of the botanical data (Chevalier 2019). Among the wide range of annual and seasonal bioclimatic variables that are generally considered important elements in studying the eco-physiological tolerance of plants species (Guisan and Thuiller 2005; e.g. Elith *et al.* 2006), we identified five key variables based on the following criteria: (i) they have a direct impact on plants life cycle (i.e. variables also called 'direct gradients' by Guisan and Zimmermann 2000), which is critical to estimate reliable plant-climate relationships, (ii) they are important descriptors of southern African climates and also (iii) they are largely uncorrelated across the study area (Table 1). Together, the selected five variables reflect many important aspects of the regional climatic drivers and are expected to be major determinants of plant distributions. By extension, it is expected that these variables can be reconstructed from regional fossil pollen records (Chevalier *et al.* 2020). The five variables we selected are:

(1) the mean temperature of the wettest quarter (TWetQ [°C], Figure 1B), which we use here as a proxy for the temperature of the growing season,
(2) the minimum temperature of the coldest month (TColdM [°C], Figure 2A) used here as a proxy for cold/frost tolerance,
(3) the accumulated precipitation of the warmest Quarters (PWarmQ [mm], Figure 2B),
(4) the accumulated precipitation of the coldest Quarters (PColdQ [mm], Figure 2C), and
(5) an Aridity Index (Aridity [unitless], Figure 2D) also derived from Worldclim 2 as a ratio between potential evapotranspiration (PET) and mean annual precipitation (Trabucco and Zomer 2019; MAP; Trabucco *et al.* 2008). With this definition, environmental changes along the Aridity gradient are not linear and environmental changes are larger across specific sections of the gradient. We, therefore, calculated the square-root of the Aridity values to 'stretch' the spatially-abundant arid to sub-humid section of the southern African Aridity

Table 1. Spearman's correlation indices between the five selected climate descriptors. None of these correlations are significant at the 5% level (without even correcting for multiple testing), highlighting the statistical independence of our climate descriptors across the entire study area.

	TWetQ	TColdM	PWarmQ	PColdQ	Aridity
TWetQ	1				
TColdM	-0.017	1			
PWarmQ	0.004	0.014	1		
PColdQ	-0.006	0.019	-0.006	1	
Aridity	-0.002	0.011	0.003	-0.014	1

gradient and 'shrink' the most humid sections to increase the sensitivity of our modelling approach (e.g. Turner *et al*. 2020).

15.2.3 Pollen Photographs

Most of the pollen images are taken from the reference collection of African pollen types that was created by E.M. van Zinderen Bakker and co-workers at the University of the Free State, for a series entitled South African Pollen Grains and Spores, between 1953 and 1970 after extensive collection from several herbaria in Africa and Europe (Coetzee 1955; van Zinderen Bakker and Coetzee 1959; van Zinderen Bakker *et al*. 1970; van Zinderen Bakker 1953, 1956). Other photos were taken from the University of Cape Town reference collections (created by Jean Sugden and Sue De Villiers) and the African Pollen Database.

15.3 THE CREST METHOD

CREST uses probability density functions (pdfs) to estimate univariate statistical relationships between pollen and climate variables (Chevalier *et al*. 2014; Chevalier 2019). The process is based on two steps (Figure 3). First, each plant species is considered independently to calculate its pdfs, which can be interpreted as the realised niche of the species (Kearney 2006). Species pdfs are fully defined by two parameters, a mean and a standard deviation, which can be interpreted as the plant's climate preference (the optimum position on the climate gradient) and tolerance (the shape/width of the niche), respectively. To ensure a robust estimation of these two parameters, only species with a minimum of 20 grid cells in their distributions are considered in this study. Climate values are obtained by associating their occurrence data with climate gradients of interest, each being weighted according to the inverse of their abundance in the climate space (i.e. the histograms on Figures 1 and 2; Bray *et al*. 2006; Kühl *et al*. 2002). Following the recommendations of Chevalier *et al*. (2014), we imposed a symmetrical normal distribution for the species pdfs of TWetQ and TColdM and a right-skewed log-normal distribution (negative values have a probability of 0) for the species pdfs of PWarmQ, PColdQ and Aridity.

Finally, all the species pdfs of the species composing a pollen type are given a weight that corresponds to the square-root of the number of grid cells composing their distribution (i.e. species with larger distributions get a higher weight). A weighted average of their pdfs is then performed to integrate the full climatic spectrum occupied by the different species into one single response curve and meet the taxonomic resolution of the pollen type. No additional constraints are imposed to the shape of the resulting curve, which can be asymmetric (Figure 3) or multimodal

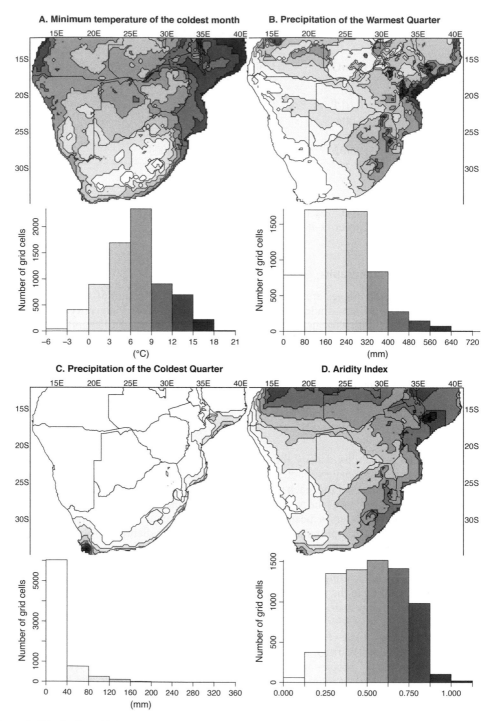

Figure 2. Distribution of (A) TColdM, (B) PColdQ, (C) PWarmQ and (D) Aridity Index values (low to high values for arid to humid conditions) across the study area.

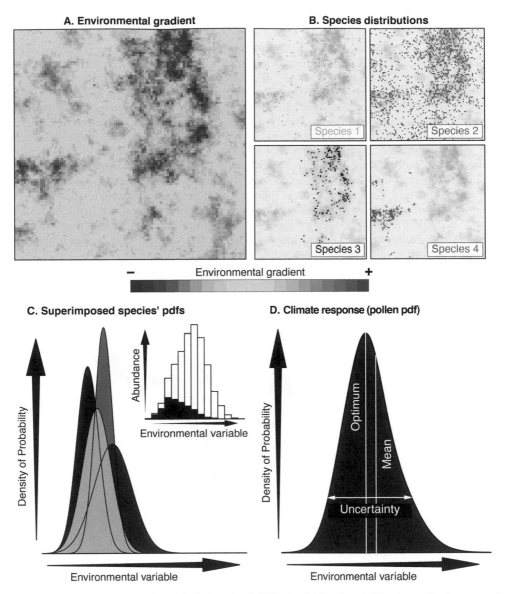

Figure 3. Conceptual representation of the fitting of probability density functions (pdfs) using randomly generated data. (A) Modern distribution of the environmental variable of interest. (B) Modern distribution of four species producing undifferentiated pollen grains in the environment. (C) Four curves representing four pdfs of the species represented in B. Inset histogram represents distribution of modern environment (white) that is occupied by at least one of the four species of interest (black), highlighting the preference for lower values. (D) Representation of statistics (optimum, mean, and width/uncertainty) that can be measured from each *pdf* to infer climate preferences and tolerances.

(i.e. exhibiting two or more different climate preferences) if necessary. In this study, the pdfs are only used to characterise the climatic sensitivities of various pollen types. For more details on how to use them to reconstruct climate, we refer to the original publications (Chevalier *et al*. 2014; Chevalier 2019).

15.4 IDENTIFICATION OF STRONGEST CLIMATIC DETERMINANTS

This section presents how the different diagrams and pdfs curves generated for each taxon in the atlas (a combination of Figures 3 to 6) can be used to evidence specific or relative climate sensitivities. The first two examples (Poaceae – Figure 4 – and Menispermaceae – Figure 5) provide simple diagnostic criteria to identify the taxa that cannot be used as clear indicators of climate in (semi-)quantitative frameworks, while the last one (*Artemisia* – Figure 6) uses the graphical diagnostics tools to identify the most influential climatic parameters of the plant distribution. However, it is important to clearly differentiate 'non-informative taxa' from 'taxa from which quantitative information cannot be extracted' and to stress that excluding pollen types, such as Poaceae, from quantitative interpretations does not mean they do not convey any valuable information. Integrated in an ecological/qualitative, rather than in a statistical/quantitative approach, they can provide critical indications of palaeoenvironmental changes, as shown by the sensitive, but biome-dependent, Asteraceae:Poaceae pollen ratio at Pella, South Africa (Lim *et al.* 2016) and as previously suggested in central South Africa (Coetzee 1967; Cooremans 1989). Non-informative taxa can also be used as a semi-independent, but complementary source of information to support climate reconstructions.

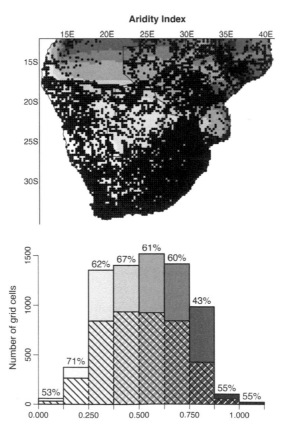

Figure 4. Distribution of Poaceae (black dots) compared to Aridity Index values across the study area; refer to Figure 2D to see the masked parts (top panel). Histogram that represents the proportion of the aridity space (grey scale) that is occupied (hatched) by at least one Poaceae species (bottom panel).

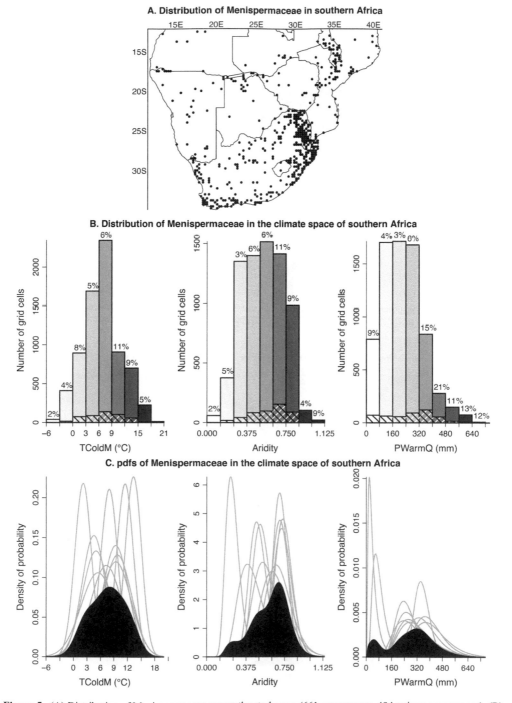

Figure 5. (A) Distribution of Menispermaceae across the study area (661 occurrences, 494 unique occurrences). (B) Histogram that represents the proportion of the climate space (grey scalee histogram) that is occupied (hatched histogram) by at least one Menispermaceae species. 5-10% of each class are occupied by at least one species. (C) pdfs fitted on each species (grey) highlighting the diversity of the individual responses and their broad combined pdf (black).

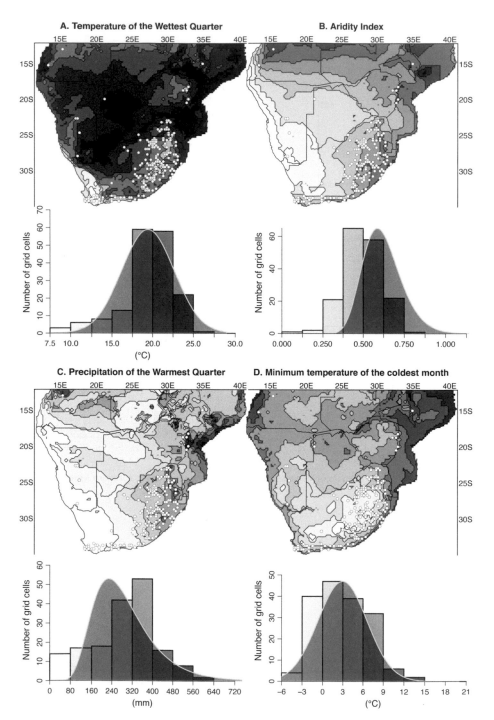

Figure 6. Distribution of *Artemisia* (white dots; 172 unique occurrences) plotted against four different climate gradients and summarised as histograms and respective pdf. The combination of TWetQ and Aridity (top row) allows inferring that *Artemisia* grows in cool and mesic conditions in southern Africa.

Finally, it should also be noted that any assessments of sensitivity from this study relate to the southern African context and the geographical and climatic spaces here defined (Figures 1 and 2), with a clear focus on the southern half of the study area due to distribution of plant occurrence data (Figure 1A). These results are also mainly valid at the spatial scale of the atlas and they could be superseded by local constraints (e.g. edaphic conditions). As such, extrapolating these results beyond this geographical range and spatial resolution should only be done carefully.

15.4.1 Identification of climatically 'uninformative' taxa

Some pollen types are not useful for quantitative climatic reconstructions because of the diversity of species producing pollen of indistinguishable morphologies. Best illustrated by Poaceae (1266 species grouped into 242 genera in our southern African botanical database; Figure 4), the diversity issue highlights one limitation of many statistical tools as there is no simple, unique link between the presence records and the associated climate values. Presumably present (almost) everywhere in southern Africa (Poaceae species have been observed in about 60% of all Aridity classes, see Figure 4), only their spatially-varying relative abundance in the modern environment (unavailable data) may contain reliable climatic information. In fact, when the number of species becomes too large, and the niches too diverse, the pdfs of such taxa are not flat - as with the pdfs of uninformative taxa - but rather exhibit a peak that reflects the distribution of the whole climate space rather than specific sensitivities (see Poaceae on pp. 234–235 of the atlas (Chevalier *et al.* 2021b), but also other taxa, such as Scrophulariaceae-type [pp. 272–273], Asteraceae [pp. 60–61] or Fabaceae [pp. 124–125]).

The degree of scattering of the different species pdfs along the climate gradient is another characteristic that may mask any potential climatic signal from being conveyed. As an example, Menispermaceae - composed of 30 species in southern Africa, but with only 10 having at least 20 grid cells in their distribution - can be found in a large range of distinct environments, as each species occupies a unique geographical and climatic niche (Figure 5). This means that Menispermaceae pollen cannot easily be interpreted in terms of climate (at least not for the five climate variables selected here) and, as such, cannot be directly used to reconstruct past climates without additional assumptions to deconvolve the different species signals. Such situations can be identified by observing the geographical distribution of the taxon (Figure 5A), but also by looking at the distribution of the taxon in the climate space (Figure 5B) and the shape of the pdfs, which are generally multimodal and/or associated with a high variance despite a limited number of composing species (Figure 5C).

15.4.2 Identification of climate sensitivities

Geographical distributions, species diversity, and the shapes of the pdfs can be reliable criteria for identifying taxa from which climatic information cannot be extracted from their distributions. However, identifying the most important climate sensitivities of the 'informative' taxa (i.e. the taxa from which specific climate sensitivities can be inferred from their distribution) is less direct, especially since plants are often sensitive to a range of climatic parameters. In practice, the distribution, availability and accessibility of different climates within any definite climate space generally reduce that multifactorial sensitivity aspect to one or two specific parameters that most strongly limit the plants' distributions.

Using the monospecific *Artemisia* pollen type (*Artemisia afra*) as a case study to illustrate the process of identification of climate sensitivities, the criteria aforementioned - especially the shapes of the pdfs - are unreliable indicators of climate sensitivities (Figure 6). By construction, the shape of monospecific pollen types is constrained and will appear smooth. In such situations, looking at the distribution of *Artemisia* in the southern African climate space, and more specifically where it grows and where it does not grow, provides additional valuable information to discriminate between constraining and non-constraining variables. Real absence data are not available here, but it has been suggested that when the modern presence record dataset is sufficiently extensive, pseudo-absence records can be inferred (Elith and Leathwick 2009; Elith *et al.* 2006; e.g. Birks *et al.* 2010). Superimposing plant distributions on climate layers thus sheds light on whether a variable is important or not. Maps in Figure 6 show that most records of *Artemisia* (*c.* 70% of the distribution) occur between TWetQ values of 17.5 to 25°C. *Artemisia* is rarely found in warmer or colder areas, and, therefore, TWetQ seems to apply a relatively strong constraint on the plant's distribution. Similarly, the distribution of Aridity Index values seems to also constrain the distribution of *Artemisia* which is mainly found at values between 0.25–0.75. In contrast, *Artemisia* is observed across the entire PWarmQ (23 to 616 mm) and PColdQ (1 to 340 mm) precipitation gradients. The false impression that *Artemisia* needs elevated summer rainfall is caused by the orographic effect of the higher altitude regions of eastern South Africa that locally correlates cool temperatures with high summer precipitation. Combined these maps suggest that *Artemisia* is more directly limited by a combination of average-to-cool temperatures of the growing season and mesic conditions rather than by precipitation amount specifically, at least in the southern African context.

The same approach of superposing the distributions with climate gradients can be used to highlight some specific sensitivities. For example, the distribution of *Podocarpus* is strongly correlated with high humidity values (Chevalier *et al.* 2021b, pp. 236–237), while *Zygophyllum's* distribution is concentrated on the other side of the aridity gradient in the most arid grid cells [pp. 312–313]. *Brachystegia* [pp. 66–67] lives in the hottest and most humid sections of our climate space, while *Burkea* [pp. 68–69] has a similar thermal niche but a preference for more mesic climates. These assessments are consistent with previous interpretations, but the data presented here clarifies and validates these interpretations, and perhaps more importantly, contributes to a common understanding of a broader range of plant-climate relationships and environmental sensitivities in southern Africa.

15.4.3 Relative comparisons of the climate preferences

The pdfs allow for an individual assessment of each pollen type for each variable. However, it can also be useful to assess relative sensitivities (Does taxon X prefer colder temperatures than taxon Y?) to better understand vegetation dynamics from a pollen diagram. To do so, the atlas also contains comparative figures of the variable-specific pdfs and also an analysis of their climate preferences by Principal Component Analysis (PCA) run on the five climate optima (Chevalier *et al.* 2021b, pp. 315–338). Here, we present the same analysis performed on the 40 most common pollen types in southern African records (Figures 7 and 8).

The first two axes represent >80% of the total variance and separate plant taxa according to different groups. The bottom left corner is occupied with taxa with preferences for low temperatures during the growing season, the top left quadrant shows a gradient of taxa that preferentially live in arid to semi-arid conditions, while the right half consists of taxa that prefer summer-rainfall conditions. The position of *Artemisia* near the centre of the PCA biplot supports our interpretation that this taxon is preferentially associated with average climate conditions in southern Africa. However, it should be noted that this PCA approach is only one element to consider and is not an absolute descriptor of relative tolerances. This PCA analysis is only based on climate optima and not the tolerances and of secondary optima of the pdfs. As shown in Figure 8, Geraniaceae has an arid climate optimum (*c.* 0.25), but its tolerance is large (*c.* 0.15 – 0.8), as distinct Geraniaceae species can grow across the entire aridity gradient of the region (Chevalier

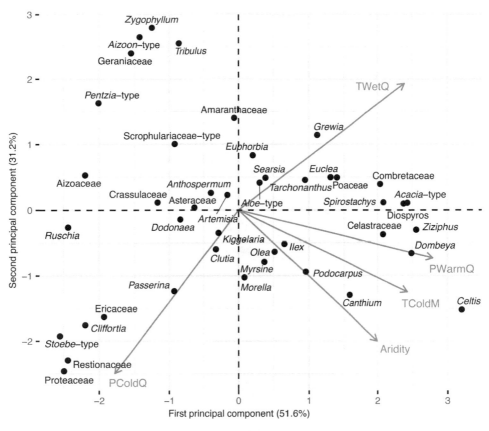

Figure 7. Scatterplot of the two first principal components of the PCA run with the five climate optima of the 40 most frequent pollen types in southern African records.

et al. 2021b, pp. 136–137). Scrophulariaceae-type (Chevalier *et al.* 2021b, pp. 272–273) as a similarly broad tolerance to the aridity gradient but with an optimum towards sub-humid values.

15.5 DISCUSSION AND CONCLUSIONS

The diagnostic tools presented in this paper are derived from the statistical analyses of modern plant distributions and provide a robust framework from which consistent interpretations of pollen data can be obtained. The three case studies were chosen to represent specific issues associated with the identification of climate sensitivities (high taxonomic diversity with Poaceae, scattering in the climate space with Menispermaceae and locally-correlated factors with *Artemisia*). While the first two are easily identified, confounding factors may be more difficult, and can lead to erroneous interpretations, especially when transposed back in time (Juggins 2013; Telford and Birks 2011). Confounding factors can only be detected when all the climatic and environmental aspects that matter to the plant life cycle are being considered. However, it should be kept in mind that since we only considered five climate variables, the full range of possibilities may not be always realised, and climatic factors important to certain taxa may not have been included in the atlas.

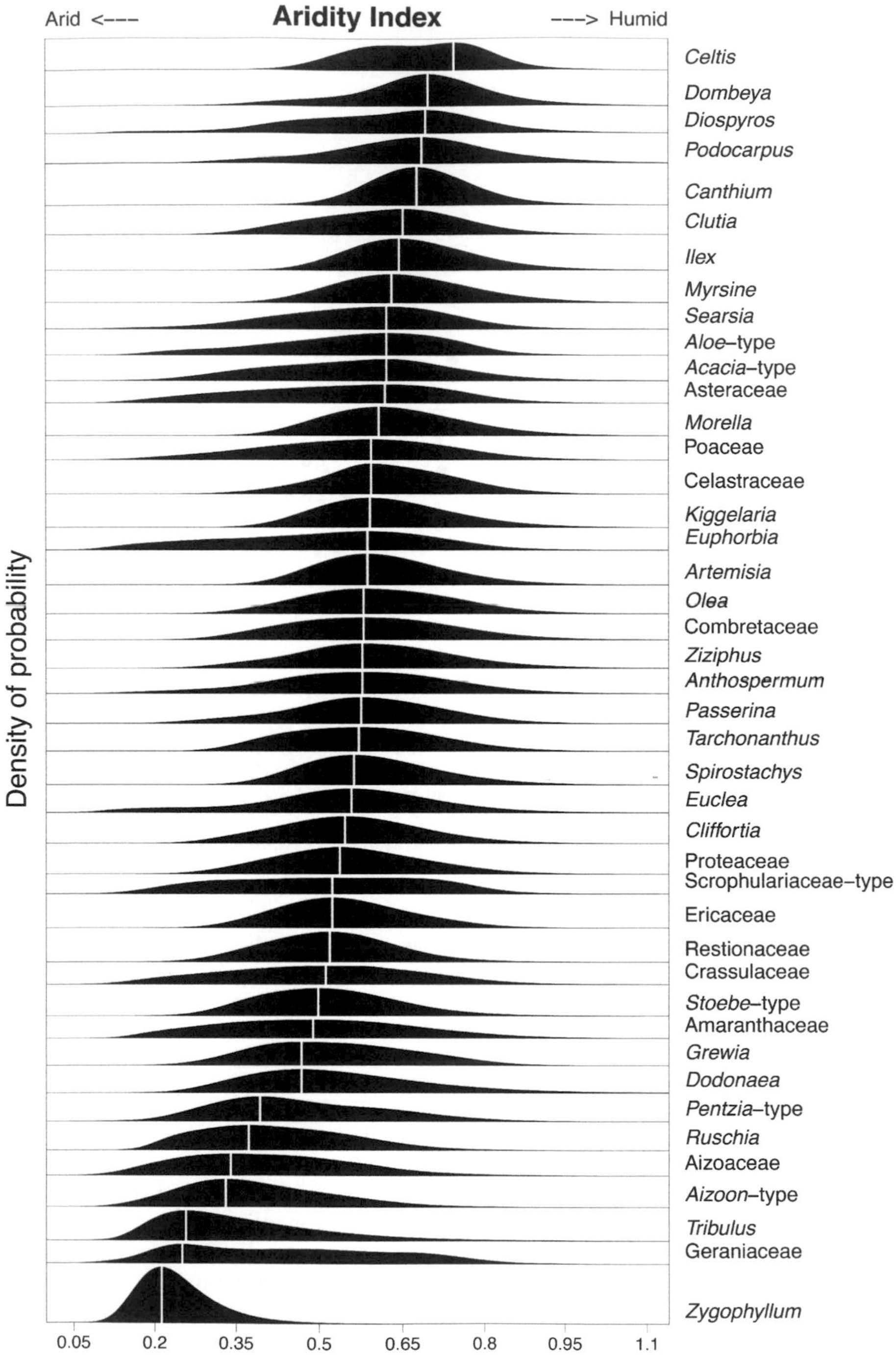

Figure 8. Aridity Index pdfs of the 40 most frequent taxa in southern African pollen records. Taxa are sorted from bottom (drier) to top (more humid) according to their climate optimum, not their tolerance.

Further, plant taxon distributions do not necessarily equal their pollen distributions precisely, as indicated by Hamilton (1972) for East Africa. Where available, knowledge of pollen distribution and preservation mechanisms could be used to adjust pollen-interpreted reconstructions. It is also important to remember that non-climatic factors such as fire or edaphic conditions are also important - even decisive - drivers of some plant distributions. Therefore, even if these diagnostic tools represent a valuable source of information regarding plant-climate relationships, they should always be considered in combination with ecological/botanical knowledge to ensure that essential aspects have not been ignored.

In addition, it is important to note that a climate sensitivity does not give any indication of the actual sign of the signal conveyed by the pollen (e.g. wetter or drier, warmer or colder). In fact, the sign of a signal is a relative concept that depends exclusively on the assemblage in which the taxon was observed. For instance, the relatively xeric *Pentzia*-type (Chevalier *et al.* 2021b, pp. 224–225) was on average an indicator of more humid environments in the Desert Biome at Pella, South Africa during the Holocene (Lim *et al.* 2016), while it was generally indicative of more arid environments in records from the more humid climates of eastern South Africa (Chevalier and Chase 2016). This relativity includes temporally-varying signatures relating to long-term changes in climate (e.g. *Tarchonanthus* [pp. 292–293] as a high and low precipitation indicator at Wonderkrater, South Africa (Chevalier and Chase 2015)).

Finally, while this paper is dedicated to the analysis of southern African pollen types, all the diagnostic tools presented are transferable and applicable to other regions of the World and/or different ecosystems as they are based on the *crestr* R package (https://github.com/mchevalier2/crestr) and can be generated using R scripts available at https://github.com/mchevalier2/Papers/tree/master/2021_Chevalier_etal_Pollen_Atlas_SA. This technique will help to formalise and standardise pollen-climate interpretations and facilitate and refine interpretations of regional fossil pollen data.

15.6 DATA AVAILABILITY

The 'Atlas of southern African pollen types and their climatic affinities' is available from https://doi.org/10.5281/zenodo.4013452.

ACKNOWLEDGEMENTS

This work was supported by the European Research Council (ERC) under the European Union's Seventh Framework Programme (FP7/2007-2013)/ERC Starting Grant 'HYRAX', grant agreement no. 258657. The authors also acknowledge the support of the South African National Botanical Institute (SANBI) in sharing their botanical data. Lloyd Rossouw, Alma Fuller and Frederick Scott for scanning and arranging negatives of the pollen reference collection of the Department of Plant Sciences University of the Free State in a database. LS was supported by the National Research foundation (South Africa) (NRF Grant no. 85903). Any opinion, finding, and conclusion or recommendation expressed in this material is that of the authors, and the NRF does not accept any liability in this regard.

REFERENCES

Birks, H.J.B., Heiri, O., Seppä, H. and Bjune, A.E., 2010, Strengths and weaknesses of quantitative climate reconstructions based on Late-Quaternary biological proxies. *The Open Ecology Journal*, **3**, pp. 68–110, 10.2174/1874213001003020068.

Bousman, C.B., Metcalfe, S.E., Partridge, T.C., Vogel, J.C. and Brink, J.S., 1988, Palaeoenvironmental implications of Late Pleistocene and Holocene valley fills in Blydefontein basin, Noupoort, CP, South Africa. *Palaeoecology of Africa*, **19**, pp. 43–67.

Bray, P.J., Blockley, S.P.E., Coope, G.R., Dadswell, L.F., Elias, S.A., Lowe, J.J. and Pollard, A.M., 2006, Refining mutual climatic range (MCR) quantitative estimates of palaeotemperature using ubiquity analysis. *Quaternary Science Reviews*, **25**, pp. 1865–1876, 10.1016/j.quascirev.2006.01.023.

Chase, B.M., Boom, A., Carr, A.S., Carré, M., Chevalier, M., Meadows, M.E., Pedro, J.B., Stager, J.C. and Reimer, P.J., 2015a, Evolving southwest African response to abrupt deglacial North Atlantic climate change events. *Quaternary Science Reviews*, **121**, pp. 132–136, 10.1016/j.quascirev.2015.05.023.

Chase, B.M., Lim, S., Chevalier, M., Boom, A., Carr, A.S., Meadows, M.E. and Reimer, P.J., 2015b, Influence of tropical easterlies in the southwestern Cape of Africa during the Holocene. *Quaternary Science Reviews*, **107**, pp. 138–148, 10.1016/j.quascirev.2014.10.011.

Chase, B.M., Chevalier, M., Boom, A. and Carr, A.S., 2017, The dynamic relationship between temperate and tropical circulation systems across South Africa since the last glacial maximum. *Quaternary Science Reviews*, **174**, pp. 54–62, 10.1016/j.quascirev.2017.08.011.

Chevalier, M., 2018, GBIF database for CREST. https://doi.org/10.6084/m9.figshare.6743207

Chevalier, M., 2019, Enabling possibilities to quantify past climate from fossil assemblages at a global scale. *Global and Planetary Change*, **175**, pp. 27–35, 10.1016/j.gloplacha.2019.01.016.

Chevalier, M. and Chase, B.M., 2015, Southeast African records reveal a coherent shift from high- to low-latitude forcing mechanisms along the east African margin across last glacial–interglacial transition. *Quaternary Science Reviews*, **125**, pp. 117–130, 10.1016/j.quascirev.2015.07.009.

Chevalier, M. and Chase, B.M., 2016, Determining the drivers of long-term aridity variability: a southern African case study. *Journal of Quaternary Science*, **31**, pp. 143–151, 10.1002/jqs.2850.

Chevalier, M., Chase, B.M., Quick, L.J., Dupont, L.M. and Johnson, T.C., 2021a, Temperature change in subtropical southeastern Africa during the past 790,000 yr. *Geology*, **49**, pp. 71–75, 10.1130/G47841.1.

Chevalier, M., Chase, B.M., Quick, L.J. and Scott, L., 2021b, Atlas of southern African pollen taxa (V1.1). https://doi.org/10.5281/zenodo.4013452

Chevalier, M., Cheddadi, R. and Chase, B.M., 2014, CREST (Climate REconstruction SofTware): a probability density function (PDF)-based quantitative climate reconstruction method. *Climate of the Past*, **10**, pp. 2081–2098, 10.5194/cp-10-2081-2014.

Chevalier, M., Davis, B.A.S., Heiri, O., Seppä, H., Chase, B.M., Gajewski, K., Lacourse, T., Telford, R.J., Finsinger, W., Guiot, J., Kühl, N., Maezumi, S.Y., Tipton, J.R., Carter, V.A., Brussel, T., Phelps, L.N., Dawson, A., Zanon, M., Vallé, F., Nolan, C., Mauri, A., de Vernal, A., Izumi, K., Holmström, L., Marsicek, J., Goring, S., Sommer, P.S., Chaput, M. and Kupriyanov, D., 2020, Pollen-based climate reconstruction techniques for late Quaternary studies. *Earth-Science Reviews*, **210**, article: 103384, 10.1016/j.earscirev.2020.103384.

Coetzee, J.A., 1955, The morphology of Acacia pollen. *South African Journal of Science*, **52**, pp. 23–27.

Coetzee, J.A., 1967, Pollen analytical studies in East and Southern Africa. *Palaeoecology of Africa and the Surrounding Islands*, **3**, pp. 1–146.

Cooremans, B., 1989, Pollen production in central southern Africa. *Pollen et Spores*, **31**, pp. 61–78.

Cordova, C.E., Scott, L., Chase, B.M. and Chevalier, M., 2017, Late Pleistocene-Holocene vegetation and climate change in the Middle Kalahari, Lake Ngami, Botswana. *Quaternary Science Reviews*, **171**, pp. 199–215, 10.1016/j.quascirev.2017.06.036.

Elith, J. and Leathwick, J.R., 2009, Species Distribution Models: Ecological Explanation and Prediction Across Space and Time. *Annual Review of Ecology, Evolution, and Systematics*, **40**, pp. 677–697, 10.1146/annurev.ecolsys.110308.120159.

Elith, J., Graham, C.H., Anderson, R.P., Dudík, M., Ferrier, S., Guisan, A., Hijmans, R.J., Huettmann, F., Leathwick, J.R., Lehmann, A., Li, J., Lohmann, L.G., Loiselle, B.A., Manion, G., Moritz, C., Nakamura, M., Nakazawa, Y., McC. M. Overton, J., Townsend Peterson, A., Phillips, S.J., Richardson, K., Scachetti-Pereira, R., Schapire, R.E., Soberón, J., Williams, S., Wisz, M.S. and Zimmermann, N.E., 2006, Novel methods improve prediction of species' distributions from occurrence data. *Ecography*, **29**, pp. 129–151, 10.1111/j.2006.0906-7590. 04596.x.

Fick, S.E. and Hijmans, R.J., 2017, WorldClim 2: new 1-km spatial resolution climate surfaces for global land areas. *International Journal of Climatology*, **37**, pp. 4302–4315, 10.1002/joc.5086.

Finch, J.M. and Hill, T.R., 2008, A late Quaternary pollen sequence from Mfabeni Peatland, South Africa: Reconstructing forest history in Maputaland. *Quaternary Research*, **70**, pp. 442–450.

GBIF: Global Biodiversity Information Facility, 2018.

GBIF.org, 2020a, Anthoccrotopsida occurrence data downloaded on September 24th, 2020. https://doi.org/10.15468/dl.t9zenf

GBIF.org, 2020b, Cycadopsidae occurrence data downloaded on September 24th, 2020. https://doi.org/10.15468/dl.sfjzxu

GBIF.org, 2020c, Liliopsida occurrence data downloaded on September 24th, 2020. https://doi.org/10.15468/dl.axv3yd

GBIF.org, 2020d, Polypodiopsida occurrence data downloaded on September 24th, 2020. https://doi.org/10.15468/dl.87tbp6

GBIF.org, 2020e, Gnetopsidae occurrence data downloaded on September 24th, 2020. https://doi.org/10.15468/dl.h2kjnc

GBIF.org, 2020f, Gingkoopsidae occurrence data downloaded on September 24th, 2020. https://doi.org/10.15468/dl.da9wz8

GBIF.org, 2020g, Pinopsidae occurrence data downloaded on September 24th, 2020. https://doi.org/10.15468/dl.x2r7pa

GBIF.org, 2020h, Magnoliopsida occurrence data downloaded on September 24th, 2020. https://doi.org/10.15468/dl.ra49dt

GBIF.org, 2020i, Lycopodiopsida occurrence data downloaded on September 24th, 2020. https://doi.org/10.15468/dl.ydhyhz

Guisan, A. and Thuiller, W., 2005, Predicting species distribution: offering more than simple habitat models. *Ecology Letters*, **8**, pp. 993–1009, 10.1111/j.1461-0248.2005. 00792.x.

Guisan, A. and Zimmermann, N.E., 2000, Predictive habitat distribution models in ecology. *Ecological Modelling*, **135**, pp. 147–186, 10.1016/S0304-3800(00)00354-9.

Hamilton, A.C., 1972, The interpretation of pollen diagrams from highland Uganda. *Palaeoecology of Africa*, **7**, pp. 45–149.

Imbrie, J. and Kipp, N.G., 1971, A new micropaleontological method for quantitative paleoclimatology: application to a Late Pleistocene Caribbean core. *The Late Cenozoic Glacial Ages.*, **3**, pp. 71–181.

Juggins, S., 2013, Quantitative reconstructions in palaeolimnology: New paradigm or sick science? *Quaternary Science Reviews*, **64**, pp. 20–32, 10.1016/j.quascirev.2012. 12.014.

Juggins, S., Simpson, G.L. and Telford, R.J., 2015, Taxon selection using statistical learning techniques to improve transfer function prediction. *The Holocene*, **25**, pp. 130–136, 10.1177/0959683614556388.

Kearney, M., 2006, Habitat, environment and niche: what are we modelling?. *OIKOS*, **115**, pp. 186–191, 10.1111/j.2006.0030-1299.14908.x.

Kühl, N., Gebhardt, C., Litt, T. and Hense, A., 2002, Probability Density Functions as Botanical-Climatological Transfer Functions for Climate Reconstruction. *Quaternary Research*, **58**, pp. 381–392, 10.1006/qres.2002.2380.

Lim, S., Chase, B.M., Chevalier, M. and Reimer, P.J., 2016, 50,000 years of climate in the Namib Desert, Pella, South Africa. *Palaeogeography, Palaeoclimatology, Palaeoecology*, **451**, pp. 197–209, 10.1016/j.palaeo.2016.03.001.

Myers, N., Mittermeier, R.A., Mittermeier, C.G., da Fonseca, G.A.B. and Kent, J., 2000, Biodiversity hotspots for conservation priorities. *Nature*, **403**, pp. 853–858, 10.1038/35002501.

Neumann, F.H., Scott, L., Bousman, C.B. and van As, L., 2010, A Holocene sequence of vegetation change at Lake Eteza, coastal KwaZulu-Natal, South Africa. *Review of Palaeobotany and Palynology*, **162**, pp. 39–53, 10.1016/j.revpalbo.2010.05.001.

Neumann, F.H., Scott, L. and Bamford, M.K., 2011, Climate change and human disturbance of Fynbos vegetation during the late Holocene at Princess Vlei, Western Cape, South Africa. *The Holocene*, **21**, pp. 1137–1149, 10.1177/0959683611400461.

Neumann, F.H., Botha, G.A. and Scott, L., 2014, 18,000 years of grassland evolution in the summer rainfall region of South Africa: evidence from Mahwaqa Mountain, KwaZulu-Natal. *Vegetation History and Archaeobotany*, **23**, pp. 665–681, 10.1007/s00334-014-0445-3.

Overpeck, J.T., 1985, A pollen study of a late Quaternary peat bog, south-central Adirondack Mountains, New York. *Geological Society of America Bulletin*, **96**, pp. 145–154, 10.1130/0016-7606(1985)96#<145:APSOAL#>2.0.CO;2.

Quick, L.J., Meadows, M.E., Bateman, M.D., Kirsten, K.L., Mäusbacher, R., Haberzettl, T. and Chase, B.M., 2016, Vegetation and climate dynamics during the last glacial period in the fynbos-afrotemperate forest ecotone, southern Cape, South Africa. *Quaternary International*, **404**, pp. 136–149, 10.1016/j.quaint.2015.08.027.

Quick, L.J., Chase, B.M., Wündsch, M., Kirsten, K.L., Chevalier, M., Mäusbacher, R., Meadows, M.E. and Haberzettl, T., 2018, A high-resolution record of Holocene climate and vegetation dynamics from the southern Cape coast of South Africa: pollen and microcharcoal evidence from Eilandvlei. *Journal of Quaternary Science*, **33**, pp. 487–500, 10.1002/jqs.3028.

Rutherford, M.C., Mucina, L. and Powrie, L., 2012, The South African National Vegetation Database: History, development, applications, problems and future. *South African Journal of Science*, **108**, pp. 1–8, 10.4102/sajs.v108i1/2.629.

Scott, L., 1982, Late Quaternary fossil pollen grains from the Transvaal, South Africa. *Review of Palaeobotany and Palynology*, **36**, pp. 241–278, 10.1016/0034-6667(82)90022-7.

Scott, L., 1989, Climatic conditions in Southern Africa since the last glacial maximum, inferred from pollen analysis. *Palaeogeography, Palaeoclimatology, Palaeoecology*, **70**, pp. 345–353, 10.1016/0031-0182(89)90112-0.

Scott, L., 1999, Vegetation history and climate in the Savanna biome South Africa since 190,000 ka: a comparison of pollen data from the Tswaing Crater (the Pretoria Saltpan) and Wonderkrater. *Quaternary International*, **58**, pp. 215–223, 10.1016/S1040-6182(98)00062-7.

Scott, L. and Nyakale, M., 2002, Pollen indications of Holocene palaeoenvironments at Florisbad spring in the central Free State, South Africa. *The Holocene*, **12**, pp. 497–503, 10.1191/0959683602hl563rr.

Scott, L. and Woodborne, S., 2007, Vegetation history inferred from pollen in Late Quaternary faecal deposits (hyraceum) in the Cape winter-rain region and its bearing on past climates in South Africa. *Quaternary Science Reviews*, **26**, pp. 941–953, 10.1016/j.quascirev.2006.12.012.

Scott, L., Neumann, F.H., Brook, G.A., Bousman, C.B., Norström, E. and Metwally, A.A.S.A.H., 2012, Terrestrial fossil-pollen evidence of climate change during the last 26 thousand years in Southern Africa. *Quaternary Science Reviews*, **32**, pp. 100–118, 10.1016/j.quascirev.2011.11.010.

Telford, R.J. and Birks, H.J.B., 2011, A novel method for assessing the statistical significance of quantitative reconstructions inferred from biotic assemblages. *Quaternary Science Reviews*, **30**, pp. 1272–1278, 10.1016/j.quascirev.2011.03.002.

ter Braak, C.J.F. and Looman, C.W.N., 1986, Weighted averaging, logistic regression and the Gaussian response model. *Vegetatio*, **65**, pp. 3–11, 10.1007/BF00032121.

ter Braak, C.J.F. and Juggins, S., 1993, Weighted averaging partial least squares regression (WA-PLS): an improved method for reconstructing environmental variables from species assemblages. *Hydrobiologia*, **269–270**, pp. 485–502, 10.1007/BF00028046.

Trabucco, A. and Zomer, R.J., 2019, Global Aridity Index and Potential Evapotranspiration (ET0) Climate Database v2.

Trabucco, A., Zomer, R.J., Bossio, D.A., van Straaten, O. and Verchot, L. V., 2008, Climate change mitigation through afforestation/reforestation: A global analysis of hydrologic impacts with four case studies. *Agriculture, Ecosystems & Environment*, **126**, pp. 81–97, 10.1016/j.agee.2008.01.015.

Truc, L., Chevalier, M., Favier, C., Cheddadi, R., Meadows, M.E., Scott, L., Carr, A.S., Smith, G.F. and Chase, B.M., 2013, Quantification of climate change for the last 20,000 years from Wonderkrater, South Africa: implications for the long-term dynamics of the Intertropical Convergence Zone. *Palaeogeography, Palaeoclimatology, Palaeoecology*, **386**, pp. 575–587, 10.1016/j.palaeo.2013.06.024.

Turner, M.G., Wei, D., Prentice, I.C. and Harrison, S.P., 2020, The impact of methodological decisions on climate reconstructions using WA-PLS. *Quaternary Research*, pp. 1–16, 10.1017/qua.2020.44.

Valsecchi, V., Chase, B.M., Slingsby, J.A., Carr, A.S., Quick, L.J., Meadows, M.E., Cheddadi, R. and Reimer, P.J., 2013, A high resolution 15,600-year pollen and microcharcoal record from the Cederberg Mountains, South Africa. *Palaeogeography, Palaeoclimatology, Palaeoecology*, **387**, pp. 6–16, 10.1016/j.palaeo.2013.07.009.

van Zinderen Bakker, E.M., 1953, *South African Pollen Grains and Spores*, Volume I, (Cape Town: AA Balkema).

van Zinderen Bakker, E.M., 1956, S*outh African Pollen Grains and Spores*, Volume II, (Cape Town: AA Balkema).

van Zinderen Bakker, E.M., 1957, A pollen analytical investigation of the Florisbad deposits (South Africa), in: Proceedings of the Third Pan African Congress on Prehistory. Livingstone, Chatto and Windus, London. pp. 56–67.

van Zinderen Bakker, E.M., 1982, Pollen analytical studies of the Wonderwerk Cave, South Africa. *Pollen et Spores*, **24**, pp. 235–250.

van Zinderen Bakker, E.M. and Coetzee, J.A., 1959, *South African Pollen Grains and Spores*, Volume III, (Cape Town: AA Balkema).

van Zinderen Bakker, E.M. and Coetzee, J.A., 1988, A review of late Quaternary pollen studies in east, central and southern Africa. *Review of Palaeobotany and Palynology*, **55**, pp. 155–174, 10.1016/0034-6667(88)90083-8.

van Zinderen Bakker, E.M., Welman, M. and Kuhn, L., 1970, S*outh African Pollen Grains and Spores*, Volume IV, (Cape Town: AA Balkema).

CHAPTER 16

Pollen productivity estimates from KwaZulu-Natal Drakensberg, South Africa

Trevor R. Hill & Tristan J. Duthie[1]

Discipline of Geography, School of Agriculture, Earth and Environmental Science, University of KwaZulu-Natal, Pietermaritzburg, South Africa

Jane Bunting

Department of Geography, Geology and Environment, University of Hull, Hull, United Kingdom

ABSTRACT: Pollen assemblages from sedimentary sequences reflect the landscapes surrounding the basin at the time of deposition, however the translation of pollen records into estimates or maps of past landcover is complex. Models of the pollen-vegetation relationship can help, but need to be calibrated first against modern data on pollen and vegetation for the pollen types of interest. This study presents initial estimates of Relative Pollen Productivity in the KwaZulu-Natal Drakensberg, South Africa, a palaeoecologically significant region of southern Africa, using the Extended R-Value approach. Modern pollen spectra from soil surface samples were extracted from three dominant vegetation communities, and vegetation data were collected around each sample point using a 3-tiered ring surveying approach. Fall speeds of key taxa were calculated using Stokes Law for spherical pollen grains and Falck's assumption for ellipsoidal grains. Pollen and vegetation data were subjected to Extended R-value analysis to calculate Relevant Source Area of Pollen (RSAP) and Relative Pollen Productivity estimates (RPP). RSAPs were found to be approximately 100–150 m in all three communities. Asteraceae and Proteaceae have low RPPs relative to Poaceae, the RPP of Ericaceae is comparable to Poaceae, and Pteridophyte spores and Rosaceae and Podocarpaceae pollen have significantly higher RPP values than Poaceae. These results imply that the cover of some important shrub taxa may well be underestimated in palaeoecological records from the area relative to forested communities. The number of modern pollen samples collected and analysed was small, and therefore these results can only be considered initial estimates, however they demonstrate that it is possible to use soil surface samples to obtain RSAP and PPE estimates in this important ecological and cultural region of southern Africa.

16.1 INTRODUCTION

A key aim of Quaternary palynology investigations is the quantitative reconstruction of past vegetation (Broström 2002; Sugita 2007a). A quantified understanding of the relationship between pollen assemblages retrieved at a site and the surrounding vegetation provides a sound basis for quantitative reconstruction of past environments (Jackson 1994). Plants have a range of growth

---

[1] Other affiliation: *University of Suwon, South Korea*

DOI 10.1201/9781003162766-16

forms and reproductive strategies, and these lead to differences in the amount of pollen produced by equivalent amounts of plant and to the dispersal potential of those pollen grains. This means pollen diagram interpretation is complicated (Jackson 1994; Prentice 1988); despite recent developments, much of the interpretation of pollen diagrams remains subjective and based on intuition. The challenge thus faced by contemporary palynologists has been to develop and supply the study of pollen analysis with improved tools for quantifying past vegetation and reconstruction techniques of palaeoenvironments (Räsänen *et al.* 2007).

There are three main approaches to the reconstruction of past land cover from pollen data: biomisation, modern analogue approaches and modelling the pollen-vegetation relationship. Biomisation approaches convert pollen data into plant functional types, then use ratios of these types to identify the biomes present in past landscapes (Jolly *et al.* 1998; Vincens *et al.* 2006), combining sites to map past land cover at biome level across regions and continents. Modern analogue approaches take spatially extensive datasets of modern pollen assemblages and associated environmental parameters (e.g. climate factors such as precipitation or mean annual temperature, or indices of landcover), develop transfer functions or other mathematical models such as the CREST model (e.g. Chevalier *et al.* 2014), then apply these models to pollen records to reconstruct the proxies of interest for the duration of the record (e.g. Chevalier *et al.* 2021). Both approaches assume that modern biomes/ecosystems, and therefore the pollen assemblages derived from them, are comparable with those present in past landscapes. As pollen is widely dispersed in the landscape from the original plants, pollen assemblages reflect both the local environment and the wider region, often referred to as a source of 'background' pollen. This 'background' component can make up 40–60% of the pollen influx recorded in a small lake or wetland (Sugita 1994), and whilst nature reserves and areas with lower human occupation may protect ecosystems which strongly resemble past conditions, the wider landscape has been heavily modified. The pollen-vegetation relationship modelling approach aims to address this concern by understanding how the pollen signal is formed, developing models of the relationship which are parameterised against modern data. The most widely used models of the pollen-vegetation relationship are linear, building on the R-value approach suggested by Davis (1963) (Andersen 1970; Davis 1963; Prentice 1985; Sugita 1993; reviewed by Jackson and Lyford 1999). This approach assumes that the pollen-vegetation relationship is constant in time, but is explicitly designed to reconstruct landcover options which have no modern analogues (Sugita 2007a 2007b; Bunting and Middleton 2009). This paper calibrates one such model for the main plant taxa found today in the KwaZulu-Natal Drakenberg, South Africa.

In the pollen-vegetation modelling approach, plants and their signal in a pollen assemblage are connected through models of pollen dispersal and deposition (Abraham and Kozáková 2012; Bunting 2008; Bunting and Hjelle 2010; Davis 2000). Using the modelling approach allows reconstruction of both abundance and spatial distribution of past vegetation (e.g. Broström 2002; Gaillard *et al.* 2008; Prentice 1985; Sugita 2007a, 2007b). The development of modern models of pollen dispersal and deposition has been reviewed by many authors and is not considered further here (e.g. Davis 2000; Li *et al.* 2018). The linear models used depend on estimates of Relative Pollen Productivity (RPP) and pollen fallspeed for the taxa of interest, and over the last 20 years, sets of these parameters have been developed for many regions of Europe and China (e.g. Abraham and Kozáková 2012; Broström *et al.* 2004; Bunting *et al.* 2005; Hjelle 1998; Li *et al.* 2018; Mazier *et al.* 2008; Mazier *et al.* 2012; Nielsen 2003; Räsänen *et al.* 2007; Soepboer *et al.* 2007; Sugita *et al.* 1999; von Stedingk *et al.* 2008). To date, there has been one such study published from Africa, in which Duffin and Bunting (2008) calculated RSAP and RPP estimates for southern Africa savanna taxa in the Savanna Biome of the Kruger National Park; a second such paper is included in this volume (Githumbi *et al.*).

The aim of this study was to obtain estimates of pollen productivity in the KwaZulu-Natal Drakensberg, to extend the range of taxa available for model use in South Africa and as a first step towards land cover reconstruction in the wider region, in order to strengthen the quality and

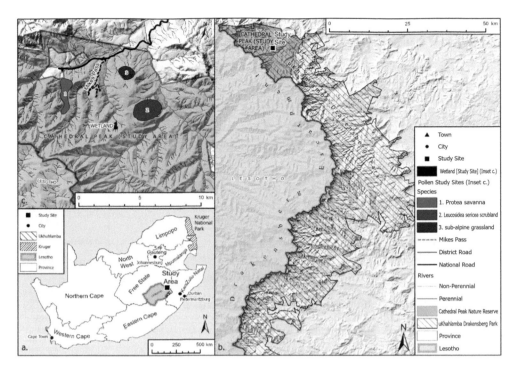

Figure 1. Location of Cathedral Peak study site in relation to: (a) orovinces and geographic features of South Africa, (b) the Drakensberg mountains in KwaZulu-Natal, and (c) kocal geographic features. Note: the wetlands highlighted in part c are those referred to in the study mentioned in the text by Lodder *et al.* (2018).

reliability of interpretations of long term palaeoecological records and improve the usefulness of palaeoecological data for allied disciplines such as conservation and archaeology.

16.2 MATERIALS AND METHODS

16.2.1 Study site

The study was carried out in the Cathedral Peak region of the KwaZulu-Natal Drakensberg (28°56′26″ S, 29°14′06″ E) (Figure 1). The Drakensberg mountain range is an ecologically important area of South Africa with a number of archaeological and palaeo-ecological records, and is therefore a significant area of study (Mitchell and Lane 2013). The vegetation shows a strong altitudinal progression from Drakensberg Foothill Moist Grassland to Northern Drakensberg Highland Grassland, then to uKhahlamba Basalt Grassland and finally, Drakensberg Afro-alpine Heathland at the highest altitudes (Mucina and Rutherford 2011).

Three vegetation communities were focused on for this study; *Themeda* dominated Highland Sourveld (Acocks, 1988) that will be referred to as *Themeda* grassland, *Protea* savanna and *Leucosidea sericea* scrubland. These three communities are the dominant vegetation communities represented in the region, contain between them the plants producing the major pollen types seen in palaeoecological-records, and occupy different altitudinal climate zones, therefore changes in the distribution of these communities are expected in response to environmental change (Hill 1996). The range of altitudes sampled varies between 1300–2000 m asl mean annual rainfall in this altitudinal range fluctuates between 700–1500 mm, and predominately falls in the summer months between October and March.

16.2.2 Methods

A suite of standard methods were applied to collect and analyse the data required (Figure 2).

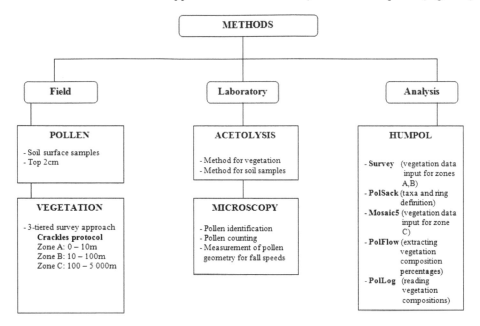

Figure 2. Schematic breakdown of methods used in this research: Field, Laboratory and Analysis.

16.2.2.1 Field methods

A total of five samples in each of the three vegetation communities were randomly located using ArcGIS 10.2, giving a total of 15 sample sites. Random positioning of sample locations is important wherever possible to obtain reliable RSAP and RPP estimates (Broström *et al.* 2005; Mazier *et al.* 2008). The central sampling point was located within a geo-rectified aerial photograph of the site, and locations in the field were found using a hand-held Garmin eTrex 10 GPS. At each sample site, the upper 2 cm of surface soil was collected following the methods described by Adam and Mehringer (1975) and Bunting *et al.* (2013) for collection of modern pollen surface samples. Vegetation surveys were carried out using a 3-tiered surveying approach, known as the Crackles vegetation protocol (Bunting *et al.* 2013), which was developed to standardise vegetation data collection for model calibration and thereby facilitate comparison of results between modelling studies. The vegetation surveying method is partitioned into three zones (A, B, C), with a comprehensive vegetation survey undertaken in zone A (0–10 m from the sample point), an intermediate vegetation survey in zone B (10–100 m), and GIS and remote sensing techniques used to extract broad-scale vegetation data from maps in zone C (100-1000+m).

16.2.2.2 Laboratory methods

Soil surface samples were processed following standard chemical extraction methods as described by Moore *et al.* (1991) and Duffin and Bunting (2008) at the University of KwaZulu-Natal (UKZN) Palaeoecology laboratory. Count size was determined by initially processing one soil surface sample from each of the vegetation communities, counting the pollen and recording the results with increasing sum size, and carrying out an 2-way Analysis of Variance (ANOVA) to assess if there was any significant difference in assemblages between sample counts of 250, 500 and 1 000 pollen grains in each community (Sokal and Rohlf 1969). In all three communities it was found that no statistically significant difference existed between the population means of pollen counts of 250, 500 and 1 000 at the level of significance ($\alpha = 0.05$), therefore remaining samples were counted to a sum of 250 grains (Hill 1996). All counting was conducted at x630 magnification using a Leica DM750 light microscope, with reference to the African Pollen database and UKZN Pollen repository specific to the region.

16.2.3 Data analysis

16.2.3.1 Estimating pollen fall speed

To use the taxon-specific Prentice-Sutton distance-weighting model to the vegetation data, estimates of the sedimentation velocity of the pollen types are required. An estimation method based on grain shape and size was used (e.g. Duffin and Bunting 2008). Pollen grains were categorised into two shape classes – spherical and ellipsoidal. Pollen grain geometries were measured using a Leica DM750 light microscope at ×400 magnification. Spherical grains were measured along one axis only, recording a diameter measurement, whereas ellipsoidal grains were measured along major and minor axes. For each taxon, a total of 50 random pollen grains were measured from pollen reference slides collected in the region, and an average diameter (spherical) and major and minor axis (ellipsoidal) measurement calculated. Fall speed was estimated using Stokes Law for spherical grains (Equation 1) and Falck's Assumption for ellipsoidal grains (Equation 2) (Gregory 1973).

$$v_s = \frac{2r^2 \cdot g(\rho_0 - \rho)}{9\mu} \quad \text{Equation 1. Stokes Law}$$

where:
v_s = spherical settling velocity (cm s^{-1})
r = pollen grain radius (cm)
g = acceleration due to gravity constant taken as 981 cm s^{-2}
ρ_0 = grain density (cm^{-3}) taken as 1 cm^{-3}
ρ = air density (cm^{-3}) taken as 1.27×10^{-3} cm^{-3}
μ = dynamic viscosity (cm^{-1} s^{-1}) taken as 1.8×10^{-4} cm^{-1} s^{-1}

$$v_e = v_s \sqrt[3]{\frac{a}{b}} \quad \text{Equation 2. Falck's Assumption}$$

where:
v_e = ellipsoidal settling velocity (cm s^{-1})
v_s = spherical settling velocity (cm s^{-1})
a = major axis (cm)
b = minor axis (cm)

16.2.3.2 Distance weighting vegetation data

Distance weighting of vegetation data is necessary before comparing vegetation with pollen assemblages formed at a specific location, so as to account for the empirical observation that plants situated closer to a sampling location contribute more pollen than those situated further away. This study used the Prentice-Sugita distance weighting term (Prentice 1985, 1988; Sugita 1993), which assumes aerial transport of pollen above the vegetation canopy is the mode of connection between plant and pollen assemblage. Vegetation data were collated into rings and distance weighted plant abundance (dwpa) calculated using the ring-based approach described in the Crackles vegetation protocol (Bunting et al. 2013).

The Prentice-Sugita-Sutton weighting term g_i for taxon i at distance z is calculated as:

$$g_i(z) = b_i \gamma z^{\gamma-1} e^{b_i z^{\gamma}} \quad \text{Equation 3}$$

where

$$b_i = \frac{4v_g}{nu\sqrt{\pi C_z}} \quad \text{Equation 3a}$$

and,

z = distance
γ = a coefficient of 0.125 (Prentice 1985)
v_g = approximated by v_s (fall speed, m sec^{-1})
C_z = the vertical diffusion coefficient (m$^{1/8}$)
n = a dimensionless turbulence parameter equal to 2
u = windspeed (m sec^{-1}), set equal to 3.

16.2.3.3 Extended R-value analysis (ERV)

Extended R-value analysis is an iterative computational approach to estimating values of relative pollen productivity from a bivariate dataset of modern pollen counts and distance-weighted vegetation abundance (Parsons and Prentice 1981; Prentice and Parsons 1983). By applying this analysis to vegetation data surveyed out to different distances from the sample point, it is possible to assess the spatial sensitivity of the pollen signal in the samples (by identifying the Relevant Source Area of Pollen – Sugita 1994), which provide the best estimates of relative pollen productivity (RPP – amount of pollen produced per unit vegetation, expressed as a ratio relative to a reference taxon) for the studied landscapes and can inform the spatial interpretation of past land cover around pollen sites with similar characteristics.

On examination of the pollen counts and vegetation data, it was clear that the species lists of the three communities were too different to allow all 15 samples to be analysed together, therefore data analysis was carried out separately for each community. It is generally recommended that the number of samples included in an analysis should be at least twice the number of pollen taxa considered (Bunting and Hjelle 2010), however this was not possible in this study. All pollen taxa recorded as present in both the pollen and vegetation data from all five samples in each community were included in the analysis, and ERV analyses were run using the HUMPOL (Hull Method of Pollen dispersal and deposition model) software suite on each of the vegetation communities separately to estimate RSAP (Relevant Source Area of Pollen, the distance beyond which adding additional vegetation survey data does not improve the fit between data and model) and to calculate RPP values for dominant taxa from each community. ERV analysis was run using sub-models 1 and 2 only, as these models use pollen and vegetation percentage data (as was collected in this research), whereas sub-model 3 was not used as it requires absolute vegetation data (Gaillard *et al.* 2008).

ERV model 1 assumes a taxon-specific background pollen component for each taxon i that is constant relative to the total influx of pollen at a site (Parsons and Prentice, 1981), whereas ERV model 2 assumes a taxon-specific background pollen component for each taxon i that is a constant relative to the total plant abundance around a site (Prentice and Parsons, 1983). ERV model 3 was proposed by Sugita (1994) and can be utilised if pollen proportion and absolute plant abundance data are available, which it is not for this study. Gaillard *et al.* (2008) suggest that similar results should be obtained from the different ERV models if the background pollen loading is low relative to the total pollen loading at a sampling site. Broström (2002) points out that there is no apparent motive for choosing ERV model 1 over 2 (or *vice versa*) a priori, and that both should be used simultaneously to see if they provide comparable model outputs (i.e. relative pollen productivity α_i and background pollen component z_i to deduce their robustness).

16.2.4 Application of RPP estimates to palaeoecology

One published pollen record is available from the study area, an unnamed small wetland within the management catchment called 'catchment six' (*c.* 0.1 ha – see Figure 1) (Lodder *et al.* 2018). The small size of the site and lack of other nearby sites means that full reconstruction of past landcover using the Landscape Reconstruction Algorithm (Sugita 2007a; b) is not appropriate, since this method relies on the availability of multiple pollen sites. The Multiple Scenario Approach (e.g. Bunting *et al.* 2018) could be applied to a single small site, however it is time-consuming and will be considered for future research. As a simple demonstration of how RPP values can be used to inform interpretation of pollen diagrams, therefore, we have undertaken a simple correction, taking no account of background pollen influx. Using count data for the taxa listed in Table 1, we summed data into a single sample for each pollen zone then calculated unweighted and weighted pollen percentages for comparison. Weightings were calculated as follows:

$$p_{iw} = \frac{n_i/R_i}{\sum\limits_{m}^{1} n_j/R_j} \qquad \text{Equation 4. Simple weighting}$$

where
   p_{iw} is the weighted proportion of taxon i
   n_i is the count of grains of taxon i
   R_i is the $RPP_{Poaceae}$ value for taxon i
   m is the number of taxa for which $RPP_{Poaceae}$ values are available.

Table 1. Palynological equivalent taxa identified from Cathedral Peak. Shown are the ten most prevalent taxa for all sample sites.

Vegetation Community	Sub-sample number	Anacardiaceae	Asteraceae	Cyperaceae	Ericaceae	*Pinus*	Poaceae	*Podocarpus*	Proteaceae	Pteridiophyte spores	*Leucosidea sericea*
Themeda	1	2.68	9.96	2.30	0.00	1.15	55.17	0.00	0.00	5.75	0.00
grassland	2	1.18	10.20	1.96	0.39	3.53	60.78	0.00	0.00	14.12	0.00
	3	0.00	8.95	2.72	0.00	1.95	62.26	1.17	0.00	14.79	0.00
	4	0.79	7.91	2.77	0.00	1.98	67.98	0.79	0.40	9.88	0.00
	5	0.78	5.08	4.30	0.39	3.52	73.05	0.78	0.00	5.47	0.00
Protea	1	0.00	6.25	0.00	0.39	5.08	57.81	2.34	9.38	11.33	0.00
savanna	2	0.79	3.56	9.49	0.40	1.19	54.94	1.19	6.32	8.30	0.00
	3	0.39	4.28	3.50	0.00	2.33	56.81	0.39	8.17	14.40	0.00
	4	1.20	3.19	2.79	0.40	1.20	41.83	0.80	5.58	36.25	0.00
	5	0.39	5.43	3.88	1.94	3.88	46.12	1.94	9.69	15.89	0.00
Leucosidea	1	0.79	15.75	3.54	0.79	2.36	47.24	0.00	0.00	11.42	10.63
sericea	2	0.00	18.33	3.19	1.99	1.59	54.98	0.00	0.00	9.56	2.79
scrubland	3	0.39	24.02	2.36	1.97	1.57	41.73	0.00	0.00	14.17	1.57
	4	0.40	14.68	3.97	1.59	0.79	44.84	0.00	0.00	25.79	2.38
	5	1.20	14.34	3.59	1.20	0.00	44.62	0.00	0.00	20.32	1.99

Percentage of pollen sum (%)

16.3 RESULTS

16.3.1 Vegetation data

Poaceae is the dominant vegetation cover type in all three studied vegetation communities, interspersed with shrubs and herbs (Asteraceae, Ericaceae, Euphorbiaceae, Fabaceae, Rubiaceae and Thymelaeaceae species). Bracken fern (*Pteridium aquilinum* (L.) Kuhn) is omnipresent throughout all vegetation communities. The *Themeda* grassland primarily consisted of Poaceae and small herbs and shrubs of Acanthaceae (*Barleria monticola*), Asteraceae (*Helichrysum aureonitens, Vernonia natalensis, Senecio coronatus*), Euphorbiaceae (*Acalypha punctata*), and Fabaceae (*Eriosema simulans*). Predominant woody vegetation cover was in the form of patchy indigenous *Podocarpus* type forest in all communities. Higher altitudes and valley areas along river courses are dominated by scrub vegetation which was particularly evident in the *Leucosidea* and *Protea* communities. Proteaceae (*Protea caffra, P. roupelliae*) was scattered throughout the *Protea* community, which is otherwise dominated by grasses and shrubs. Rosaceae (*Leucosidea sericea*) was pervasive throughout the *Leucosidea* community; vegetation in this landscape was dominated by grasses, ferns and *Leucosidea sericea*, a woody species that will invade un-burnt grassland.

16.3.2 Pollen data

Forty palynological equivalent taxa were identified from a total of 15 samples (Table 1 and 2). All samples had high proportions of Poaceae pollen (40–70%). The *Themeda* grassland community samples consistently contained Pteridophyte spores (5–15%), and herbaceous Asteraceae pollen (5–10%). Anacardiaceae, Cyperaceae and Pinaceae pollen grains were recorded in significant quantities. Anacardiaceae can be explained by the omnipresence of *Searsia dentata* shrubs throughout the *Themeda* grassland landscape and Cyperaceae reflects the presence of aquatic-type environments, explained by a nearby wetland and stream system close to all *Themeda* samples. The presence of exotic Pine is interesting considering the closest known Pinaceae plantation is approximately 7 km away, however isolated pine trees were observed invading nearby surrounding grasslands. Samples from *Protea* savanna communities contained Pteridophyte spores (10–40%), Proteaceae (5–10%) and Asteraceae (6%) pollen. The *Leucosidea* scrubland samples have substantial proportions of Asteraceae (10–25%) pollen and Pteridophyte spores (5–25%), and lower of Rosaceae (10%), Ericaceae (<2%) and Cyperaceae (5%) pollen values.

16.3.3 Data analysis

Taxa chosen for ERV analysis of samples from the *Themeda* grassland were Poaceae, Pteridophyta and Asteraceae, ERV analysis from the Protea savannah included Poaceae, Pteridophyta, Asteraceae, Proteaceae and Podocarpaceae, and ERV analysis from the *Leucosidea* shrubland included Poaceae, Pteridophyta, Ericaceae, and Rosaceae. Outcomes of ERV analysis (using both ERV model 1 and 2) show that all communities have a RSAP range of 100–150 m, suggesting that the local vegetation cover has a strong influence in the pollen spectra (Figure 3). Plots of likelihood function score reach an asymptote approximately 100–150 m from sample points (Figure 3). While RSAP values from sub-model 1 and 2 were similar, Relative Pollen Productivity values were markedly different, and the model which produced the most strongly clustered RPP values was chosen as the most appropriate one for each habitat (Broström 2002); this was ERV sub-model 1 in all three cases (Table 3). $RPP_{Poaceae}$ (pollen productivity relative to Poaceae) values were calculated by averaging all ERV analysis estimates from distances beyond 150 m.

Table 2. Shape, diameter and fall speeds of dominant taxa, Cathedral Peak.

Cathedral Peak taxa		Approx. shape	Avg. diameter (μm)	Fall speed ($m \cdot s^{-1}$)	Avg. fall speed ($m \cdot s^{-1}$)
Poaceae	*Themeda triandra*	Spherical	50.98	0.079	0.050
	Alloteopsis semialata		33.91	0.035	
	Digitaria flaccida		37.44	0.042	
	Panicum natalensis		30.78	0.029	
	Tristachya leucothrix		46.67	0.066	
Asteraceae	*Aster bakeranus*	Spherical	33.21	0.033	0.026
	Helichrysum aureonitens	Spherical	20.16	0.012	
	Vernonia type		32.62	0.032	
Cornaceae	*Curtisia dentata*	Spherical	16.89	0.009	0.009
Ebenaceaea	*Diospyros austro-africana*	Spherical	48.32	0.071	0.071
Ericaceae	*Erica drakensbergensis*	Spherical	33.24	0.033	0.077
	Erica cerinthoides		65.97	0.132	
	Erica straussianna		47.05	0.067	
Euphorbiaceae	*Acalypha punctata*	Spherical	15.48	0.007	0.007
Fabaceae	*Eriosema simulans*	Spherical	36.81	0.041	0.041
Geraniaceae	*Pelargonium* type	Spherical	91.77	0.255	0.255
Rubiaceae	*Pentanisia prunelloides*	Spherical	42.12	0.054	0.054
Thymelaeaceae	*Gnidia kraussiana*	Spherical	25.79	0.020	0.020
Pteridophyte spores	*Pteridium aquilinum*		38.18	0.044	0.044
Proteaceae	*Protea caffra*	Spherical	23.38	0.017	0.018
	Protea roupillae		24.45	0.018	
Anacardiaceae	*Searsia dentata*	Ellipsoid			
	x-axis		26.95	0.018	0.018
	y-axis		22.13		
Commelinaceae	*Commelina* type	Ellipsoid			
	x-axis		43.33	0.037	0.037
	y-axis		28.61		
Podocarpaceae	*Podocarpus latifolius*	Ellipsoid			
	x-axis		68.52	0.074	0.074
	y-axis		38.19		
Pinaceae	*Pinus patula*	Ellipsoid			
	x-axis		64.94	0.073	0.073
	y-axis		37.20		
Rosaceae	*Leucosidea sericea*	Ellipsoid			
	x-axis		29.51	0.016	0.016
	y-axis		17.98		
Cyperaceae	*Carex austro-africana*	Ellipsoid			
	x-axis		38.49	0.035	0.036
	y-axis		30.26		
	Fionia type	Ellipsoid			
	x-axis		37.17	0.037	
	y-axis		25.39		

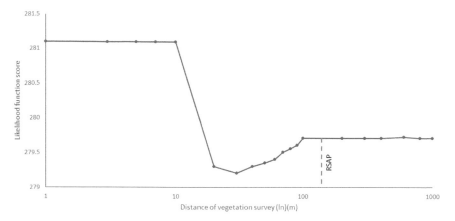

Figure 3. ERV analysis output for RSAP estimations. Illustrated are likelihood function score graphs for pollen-vegetation datasets run with Sutton's taxon specific distance-weighting using ERV model 1. RSAP is visually estimated from the graphs where the curve tends to an asymptote. While the maximum vegetation survey distance was 5000 m, a logarithmic scale was used to 1000 m from sample points as an asymptote had been reached by this distance.

Table 3. Calculated RPP estimates for taxa selected from sampled sites. All values indicated are unit-less ratios relative to a reference taxon, where RPP = 1. TT = *Themeda* grassland, PS = *Protea* savanna, LS = *Leucosidea* scrubland. *Taxon family names used in ERV analysis: Poa = Poaceae, Aster = Asteraceae, Pteri = Pteridophyte, Podo = Podocarpaceae, Prot = Proteaceae, Rosa = Rosaceae, Erica = Ericaceae.

Vegetation type	ERV model	PPE						
		Poa	Aster	Pterio	Podo	Prot	Rosa	Erica
Themeda	1	1	0.00015	4.60171	–	–	–	–
triandra	2	1	0.00026	3.14094	–	–	–	–
Protea savanna	1	1	0.00038	8.49329	6.54313	0.22356	–	–
	2	1	0.00057	9.51827	5.13538	0.00050	–	–
Leucosidea	1	1	–	25.71204	–	–	3.33815	0.80950
sericea	2	1	–	12988	–	–	7291	743

In the *Themeda* grassland, Asteraceae pollen (RPP$_{Poaceae}$ = 0.00026) producing plant species are substantially less productive in terms of pollen production than grasses, whereas Pteridophyte (RPP$_{Poaceae}$ = 3.14) are significantly higher producers. In the Protea savanna, Asteraceae pollen (RPP$_{Poaceae}$ = 0.00038) and Proteaceae pollen (RPP$_{Poaceae}$ = 0.22) producing plant species are less productive in terms of pollen production than grasses, whereas Pteridophyte (RPP$_{Poaceae}$ = 8.49) and Podocarpaceae pollen (RPP$_{Poaceae}$ = 6.54) producing species are significantly higher producers. RPP estimates obtained for the *Leucosidea* scrubland show Ericaceae pollen (RPP$_{Poaceae}$ = 0.81) producing plant species are less productive, whereas Pteridophyte (RPP$_{Poaceae}$ = 25.71) and Rosaceae (RPP$_{Poaceae}$ = 3.34) are higher palynomorph producers (Table 2).

Figure 4. Original and RPP-corrected pollen data for the four pollen zones from the Cathedral Peak Wetland pollen diagram (Lodder *et al.* 2018). (A) comparison of uncorrected and corrected values for 7 taxa (Table 1 other than Asteraceae – see text for details) – note variable x-axis scale. (B) Poaceae:Podocarpaceae ratio (an index of landscape openness).

16.3.4 Application of RPP values to palaeoecological data

Counts of the taxa listed in Table 1 made up 60–74% of the total pollen count – Cyperaceae (a local wetland taxon which dominates the pollen diagram; Lodder *et al.* 2018) of the four pollen zones from the Cathedral Peak Wetland. Figure 4 (a and b) shows the results of correction with Asteraceae excluded, since some Asteraceae may be local wetland vegetation; correction increases the apparent abundance of Poaceae, Ericaceae, Anacardaceae and Proteaceae, and reduces the apparent abundance of Podocarpaceae, Proteaceae and Rosaceae. An example of the importance of considering RPP when interpreting fossil pollen data is given by the ratio between Poaceae and Podocarpaceae pollen (Figure 4b). Podocarpaceae are forest taxa with well dispersed pollen, whilst Poaceae dominate grasslands and rarely flower under forest canopies, therefore this ratio provides an indication of the balance between open grassland and forested areas in the wider landscape. Without weighting, the ratio ranges from 11 to 49 between zones, whilst with the weighting included, values range from 44–196, giving a different impression of the role of forest in the wider landscape.

16.4 DISCUSSION

16.4.1 Vegetation data

The vegetation survey method and distance-weighting methods used allocates surveyor effort according to the relative importance of plants as potential pollen sources for the sample location, since plants closer to a sample site contribute more directly to the pollen signal than those further away. The dominant species recorded were as expected for the plant communities mapped in the area.

16.4.2 Pollen data

Pollen assemblages from all surface samples were dominated by Poaceae, reflecting the general character of the research area. Bracken fern was pervasive throughout all environments, and this was recorded by high spore counts.

16.4.3 Data analysis – estimates of RSAP and PPE

RSAP values are similar in all three communities, with estimates ranging between 100–150 m, suggesting that the variation in pollen signal between sample locations in each community is controlled by vegetation composition and distribution within a 100–150 m radius of the sample point (Sugita 1994).

Where a taxon appears in more than one community, the absolute RPP values obtained differ, however the rank of taxon RPP relative to Poaceae is consistent, suggesting that despite the small number of samples included, real features of the pollen vegetation relationship are being reliably detected by the analysis which can be used as the basis for future interpretation of pollen records.

Asteraceae is under-represented compared to Poaceae in both the *Themeda* grassland samples and the *Protea* savannah, Proteaceae are under-represented in *Themeda* grassland and Ericaeae are under-represented in *Leucosidea* scrubland. Pteridophyta are over-represented in all three habitats, whilst Podocarpaceae are over-represented in *Protea* savannah and Rosaceae are over-represented in *Leucosidea* scrubland.

Duffin and Bunting (2008) adopted the same approach to analyse modern pollen spectra from 34 surface sediment samples in the Kruger National Park and present RPP values. The RSAP for all sites was estimated at 700 m, which is considerably greater than the 150 m found in this study, however that is expected as Duffin and Bunting (2008) sampled small ponds rather than soil samples (e.g. Sugita 1994). RSAP differences could be partially attributed to the different sampling strategies in zones A and B of the vegetation surveys done between the two research projects or due to the differences in the Kruger National Park landscape and the topographical complexity of Cathedral Peak (Gaillard *et al.* 2008).

The taxa studied by Duffin and Bunting (2008), other than Poaceae, were different from those presented here. Combining the results (Table 4) and rounding RPP values, which is appropriate given the uncertainties associated with both studies, forms a useful basis for simulation studies and land cover reconstruction in advance of more detailed modern pollen studies in the region.

Results of analysis were considered more robust when using ERV model 1 rather than ERV model 2. In theory both models should be equivalent, however model 1 assumes that the taxon-specific background pollen (pollen travelling from beyond the RSAP) is constant relative to the total influx of pollen at the sample location, whereas model 2 assumes that the component is constant relative to the total vegetation abundance around the site. In landscapes with discontinuous vegetation cover, it can be argued that the model 1 assumption is more justifiable, which would accord with the results presented here.

16.4.4 Implications of findings for palaeoecological research in the region

The set of RPP values presented in Table 3 are an essential first step for the application of pollen-vegetation model-based reconstruction approaches in the region. For detailed landcover reconstruction, the next step is to compile all available pollen records and, where gaps exist, to seek out suitable sedimentary archives. One small wetland record in the KwaZulu-Natal Drakenberg area has produced a pollen record (Lodder *et al.* 2018) and a simple demonstration of how interpretation of the record can be altered by using RPP values to adjust the pollen counts taking into account differential pollen production presented in Figure 4.

Table 4. Pollen productivity and fall speed estimates for taxa in South Africa; RPP estimates rounded to powers of 2.

Taxon	RPP$_{Poaceae}$ (categorised)	Fallspeed estimate (ms^{-1})	Source
Acacia-type	0.5	0.096	Duffin and Bunting (2008)
Anacardiaceae	0.5	0.019	Duffin and Bunting (2008)
Asteraceae	0.01	0.026	This study
Colophospermum mopane	0.25	0.042	Duffin and Bunting (2008)
Combretaceae	0.5	0.014	Duffin and Bunting (2008)
Ericaceae	1.0	0.077	This study
Podocarpaceae	4.0	0.074	This study
Protea	0.25	0.017	This study
Pteridophyte spores	8.0	0.044	This study
Rosaceae	4.0	0.016	This study

16.4.5 Limitations and future direction

Soil pollen assemblages are considered to be relatively poor analogues for the pollen trapping properties of the sedimentary sequences palaeoecologists use for reconstruction compared to moss or lake sediment samples (e.g. Adam and Mehringer 1975; Li *et al.* 2018), but are often the only natural option in savannah and grassland environments. Future studies using a larger number of samples, or including pollen trap studies and surface sediment studies, would enable more robust conclusions to be drawn. However, the consistency of the results presented here, between the different communities, suggests that despite small numbers of samples the results do reflect meaningful properties of the studied ecosystem, allowing us to present initial values to support simulation studies and landcover reconstruction in the region. Balancing effort/cost with outcomes is a known and challenging issue in modern pollen studies, but the implications of being able to better translate pollen assemblages into past land cover demonstrate the incentive to continue to work.

16.5 CONCLUSIONS

Modern pollen and vegetation data from three dominant vegetation communities in the KwaZulu-Natal Drakensberg were collected and analysed, producing estimates of RSAP (100–150 m) and RPP for the dominant palynomorph-equivalent taxa in three communities, *Themeda* grassland, *Protea* savanna and *Leucosidea* scrubland. Interpretation of ERV results has shown that, while the model outputs are not 'perfect' as would be the case in an open, homogeneous and flat landscape, they are perfectly coherent and within the realm of expectations of results from a study with a relatively small amount of samples in a varied landscape. The findings of this study are therefore that developments made in pollen modelling have a very relevant place in South African pollen research and can significantly impact future work by strengthening the foundation from which we base our understanding – the interpretation of results.

The results, combined with the findings of Duffin and Bunting (2008), form a solid basis for the application of simulation and landcover reconstruction approaches in the KwaZulu-Natal Drakensberg area, which is an area possessing rich archaeological and palaeo-ecological archives. Further validation of these models in South African pollen research can only further serve to improve our confidence in pollen data, and it is the hope that this research has in the very least informed and sparked interest for future researchers in this new and significant aspect in pollen science.

ACKNOWLEDGEMENTS

To the University of KwaZulu-Natal Geography Department for the use of field equipment and laboratory space. Ezemvelo KZN Wildlife for agreeing to allow the research be conducted in Cathedral Peak Research Station. This research was supported by financial assistance from the National Research Foundation.

REFERENCES

Abraham, V. and Kozáková, R., 2012, Relative pollen productivity estimates in the modern agricultural landscape of Central Bohemia (Czech Republic). *Review of Palaeobotany and Palynology*, **179**, pp. 1–12, 10.1016/j.revpalbo.2012.04.004.

Adam, D.P. and Mehringer, P.J., 1975, Modern pollen surface samples – an analysis of subsamples. *Journal of Research of the U.S. Geological Survey*, **3(6)**, pp. 733–736.

Andersen ST (1970) The relative pollen productivity and pollen representation of northEuropean trees, and correction factors for tree pollen spectra. *Danmarks Geologiske Undersogelse*, **96**, pp.1–99.

Broström, A., 2002, Estimating source area of pollen and pollen productivity in the cultural landscapes of southern Sweden – developing a palynological tool for quantifying past plant cover. LUNDQUA Thesis 46. Department of Geology, Lund University, Lund.

Broström, A., Sugita, S. and Gaillard, M.-J., 2004, Pollen productivity estimates for the reconstruction of past vegetation cover in the cultural landscape of southern Sweden. *The Holocene*, **14(3)**, pp. 368–381, 10.1191/0959683604hl713rp.

Broström, A., Sugita, S. and Gaillard, M.-J., 2005, Estimating the spatial scale of pollen dispersal in the cultural landscape of southern Sweden. *The Holocene*, **15**, pp. 252–262, 10.1191/0959683605hl790rp.

Bunting, M.J., 2008, Pollen in wetlands: using simulations of pollendispersal and deposition to better interpret the pollen signal. *Biodiversity and Conservation*, **17**, pp. 2079–2096, 10.1007/s10531-007-9219-x.

Bunting, M.J., Armitage, R., Binney, H.A. and Waller, M., 2005, Estimates of "relative pollen productivity" and "relevant source area of pollen" for major tree taxa in two Norfolk (UK) woodlands. *The Holocene*, **15**, pp. 459–465, 10.1191/0959683605hl821rr.

Bunting, M.J., Farrell, M., Broström, A., Hjelle, K.L., Mazier, F., Middleton, R., Nielsen, A.B., Rushton, E., Shaw, H. and Twiddle, C.L., 2013, Palynological perspectives on vegetation survey: a critical step for model-based reconstruction of Quaternary land cover. *Quaternary Science Reviews*, **82**, pp. 41–55, 10.1016/j.quascirev.2013.10.006.

Bunting, M.J. and Hjelle, K.L., 2010, Effect of vegetation data collection strategies on estimates of relevant source area of pollen (RSAP) and relative pollen productivity estimates (relative PPE) for non-arboreal taxa. *Vegetation History and Archaeobotany*, **19**, pp. 365–374, 10.1007/s00334-010-0246-2.

Chevalier, M., Cheddadi, R., and Chase, B.M., 2014, CREST (Climate REconstruction SofTware): A probability density function (PDF)–based quantitative climate reconstruction method. *Climate of the Past*, **10**, pp. 2081–2098, 10.5194/cp-10-2081-2014.

Chevalier, M., Chase, B.M., Quick, L.J., Dupont, L.M. and Johnson, T.C., 2021. Temperature change in subtropical southeastern Africa during the past 790,000 yr. *Geology*, **49(1)**, pp.71–75, 10.1130/G47841.1.

Davis, M.B., 1963, On the theory of pollen analysis. *American Journal of Science*, **261**, pp.897–912, 10.2475/ajs.261.10.897.

Davis, M.B., 2000, Palynology after Y2K—Understanding the Source Area of Pollen in Sediments. *Annual Review of Earth and Planetary Sciences*, **28**, pp. 1–18, 10.1146/annurev.earth.28.1.1.

Duffin, K.I. and Bunting, M.J., 2008, Relative pollen productivity and fall speed estimates for southern African savanna taxa. *Vegetation History and Archaeobotany*, **17**, pp. 507–525, 10.1007/s00334-007-0101-2.

Gaillard, M-J., Sugita, S., Bunting, M.J., Middleton, R., Broström, A., Caseldine,C., Giesecke, T., Hellman, S.E.V., Hicks, S., Hjelle, K., Langdon, C., Nielsen, A.B., Poska, A., von Stedingk, H., Veski, S. and POLLANDCAL members, 2008, The use of modelling and simulation approach in reconstructing past landscapes from fossil pollen data: a review and results from the POLLANDCAL network. *Vegetation History and Archaeobotany*, **17**, pp. 419–443, 10.1007/s00334-008-0169-3.

Githumbi, E.N., Courtney, C.J. and Marchant, R., this volume sedimentological, palynological and charcoal analyses of the hydric palustrine sediments fromthe lielerai-kimana wetlands, kajiado, southern kenyA. *Palaeoecology of Africa*, **35**, chapter 8, 10.1201/9781003162766-8.

Gregory, P.H., 1973, *The microbiology of the Atmosphere*. 2nd ed. Leonard Hill, Aylesbury.

Hill, T., 1996, Description, classification and ordination of the dominant vegetation communities, Cathedral Peak, KwaZulu-Natal Drakensberg. *South African Journal of Botany*, **62**(5), pp. 263–269, 10.1016/S0254-6299(15)30655-4.

Hjelle, K., 1998, Herb pollen representation in surface moss samples from mown meadows and pastures in western Norway. *Vegetation History and Archaeobotany*, **7**, pp. 79–96, 10.1007/BF01373926.

Jackson, S.T., 1994, Pollen and spores in Quaternary lake sediments as sensors of vegetation composition: theoretical models and empirical evidence. In: *Sedimentation of Organic Particles*, edited by Traverse, A., (Cambridge: Cambridge University Press).

Jackson, S.T. and Lyford, M.E., 1999. Pollen dispersal models in Quaternary plant ecology: assumptions, parameters, and prescriptions. *The Botanical Review*, **65**(1), pp.39–75, 10.1007/BF02856557.

Jolly, D., Prentice, I.C., Bonnefille, R., Ballouche, A., Bengo, M., Brenac, P., Buchet, G., Burney, D., Cazet, J.P., Cheddadi, R. and Edorh, T., Elenga, H., Elmoutaki, S., Guiot, J., Laarif, F., Lamb, H., Lézine, A.-M., Maley, J., Mbenza M., Peyron, O., Reille, M., Reynaud-Farrera, I., Riollet, G., Ritchie, J.C., Roche, E., Scott, L., Ssemmanda, I., Straka, H., Umer, M., Van Campo, E., Vilimumbalo, S., Vincens, A.andWaller, M., 1998, Biome reconstruction from pollen and plant macrofossil data for Africa and the Arabian peninsula at 0 and 6000 years. *Journal of Biogeography*, **25**(6), pp.1007–1027, 10.1046/j.1365-2699.1998.00238.x.

Li. F., Gaillard, M.-J., Xu, Q., Bunting, M.J., Li, Y., Li, J., Mu, H., Lu, J., Zhang, P., Zhang, S., Cui, Q., Zhang, Y., and Shen, W., 2018, A review of relative pollen productivity estimates from temperate China for pollen-based quantitative reconstruction of past plant cover. *Frontiers in Plant Science*, **9**, article: 1214, 10.3389/fpls.2018.01214.

Mazier, F., Broström, A., Gaillard, M-J., Sugita, S., Vittoz, P. and Buttler, A., 2008, Pollen productivity estimates and relevant source area of pollen for selected plant taxa in a pasture woodland landscape of the Jura Mountains (Switzerland). *Vegetation History and Archaeobotany*, **17**, pp. 479–495, 10.1007/s00334-008-0143-0.

Mazier, F., Gaillard, M.J., Kuneš, P., Sugita, S., Trondman, A.K. and Broström, A., 2012, Testing the effect of site selection and parameter setting on REVEALS-model estimates of plant abundance using the Czech Quaternary Palynological Database. *Review of Palaeobotany and Palynology*, **187**, pp. 38–49, 10.1016/j.revpalbo.2012.07.017.

Mitchell, P.J. and Lane, P., 2013, *The Oxford Handbook of African Archaeology*. (New York: Oxford University Press).

Moore, P.D., Webb, J.A., Collinson, M.E., 1991, *Pollen Analysis*. 2nd edition, (Oxford: Blackwell Scientific Publications).

Mucina, L., Rutherford, M.C., 2011, *The Vegetation of South Africa, Lesotho and Swaziland. Strelitzia*, Volume 19, (Pretoria: South African National Biodiversity Institute).

Nielsen, A.B., 2003, Pollen-based quantitative estimation of land cover-relationships between pollen sedimentation in lakes and land cover as seen on historical maps in Denmark A.D. 1800. PhD Thesis. University of Copenhagen, Denmark.

Prentice, I.C., 1985, Pollen representation, source area, and basin size: Toward a unified theory of pollen analysis. *Quaternary Research*, **23**, pp. 76–86, 10.1016/0033-5894(85)90073-0.

Prentice, I.C., 1988, Records of vegetation in time and space: the principles of pollen analysis. In *Vegetation History*, edited by Huntley, B. and Webb, T.III, (Dordrecht: Kluwer Academic Publishers), pp. 17–42.

Räsänen, S., Suutari, H. and Nielsen, A.B., 2007, A step further towards quantitative reconstruction of past vegetation in Fennoscandian boreal forests: Pollen productivity estimates for six dominant taxa. *Review of Palaeobotany and Palynology*, **146**, pp. 208–220, 10.1016/j.revpalbo.2007.04.004.

Soepboer, W., Sugita, S., Lotter, A.F., van Leeuwen, J.F.N. and van der Knaap, W.O., 2007, Pollen productivity estimates for quantitative reconstruction of vegetation cover on the Swiss plateau. *The Holocene*, **17**(1), pp. 65–77, 10.1177/0959683607073279.

Sokal, R.R. and Rohlf, F.J., 1969, *Biometry: The Principles and Practice of Statistics in Biological Research*, 4th edition, (New York: W.H Freeman and Company).

Sugita, S., 1993, A model of pollen source area for an entire lake surface. *Quaternary Research*, **39**, pp.239–244, 10.1006/qres.1993.1027.

Sugita, S., 1994, Pollen representation of vegetation in Quaternary sediments: theory and method in patchy vegetation. *Journal of Ecology*, **82**, pp. 881–897, 10.2307/2261452.

Sugita, S., 2007a, Theory of quantitative reconstruction of vegetation I: Pollen from larges sites REVEALS regional vegetation composition. *The Holocene*, **17**(2), pp. 229–241, 10.1177/0959683607075837.

Sugita, S., 2007b, Theory of quantitative reconstruction of vegetation II: All you need is LOVE. *The Holocene*, **17**(2), pp. 243–257, 10.1177/0959683607075838.

Sugita, S., Gaillard, M-J. and Broström, A., 1999, Landscape openness and pollen records: a simulation approach. *The Holocene*, **9**, pp. 409–421, 10.1191/095968399666429937.

Vincens, A., Bremond, L., Brewer, S., Buchet, G. and Dussouillez, P., 2006, Modern pollen-based biome reconstructions in East Africa expanded to southern Tanzania. *Review of Palaeobotany and Palynology*, **140**(3–4), pp.187–212, 10.1016/j.revpalbo.2006.04.003.

von Stedingk, H., Fyfe, R. and Allard, A., 2008, Pollen productivity estimates for the reconstruction of past vegetation at the forest-tundra ecotone. *The Holocene*, **18**, pp. 323–332, 10.1177/0959683607086769.

CHAPTER 17

Modern pollen-vegetation relationships in the Drakensberg Mountains, South Africa

Trevor R. Hill & Jemma M. Finch

Discipline of Geography, School of Agricultural, Earth and Environmental Sciences, University of KwaZulu-Natal, Pietermaritzburg, South Africa

ABSTRACT: Contemporary pollen rain studies are conducted to determine pollen-vegetation relationships and develop modern analogues to facilitate palaeoreconstruction from fossil pollen assemblages. In this study, a two year modern pollen sampling campaign was conducted for eleven vegetation communities across a range of altitudes in the Drakensberg Mountains, South Africa. The aim was to compare pollen assemblage composition, diversity, and influx between vegetation communities, and investigate representivity of taxa. Despite a high degree of within and between community homogeneity, characteristic pollen taxa are suggested as indicative of particular communities. We highlight over- and under-representation of certain taxa within the vegetation communities and list 'palynologically silent' taxa – those that appear in the vegetation but are absent from the modern pollen spectra. From the resultant modern pollen assemblages, it is not possible to distinguish the grassland types, although the possibility does exist for the forest, shrubland and wetland communities. Arboreal *Celtis* and *Podocarpus* pollen types are diagnostic for the Afrotemperate forest community, and *Leucosidea sericea* for the shrubland community which it dominates. We suggest that modern pollen analogues are feasible for the non-grassland Drakensberg communities and echo the sentiments of others that modern pollen rain-vegetation dynamics are a prerequisite to accurately constrain interpretation of fossil pollen spectra.

17.1 INTRODUCTION

The modern pollen-vegetation relationship underpins vegetation reconstructions from fossil pollen data, and yet foundational studies exploring this relationship are often lacking, particularly in the tropics and subtropics. Outside of temperate zones, traditional assumptions surrounding pollen representivity of anemophilous and zoophilous taxa have been challenged, and require further investigation to understand complex bias introduced by variable pollen production and dispersal (Watrin *et al.* 2007). Detailed studies of local pollen and parent vegetation can be used to determine pollen taphonomy, and explore representivity within the pollen rain. Few such small-scale studies exist for the African tropics and subtropics (e.g. Elenga *et al.* 2000; Henga-Botsikabobe *et al.* 2020; Julier *et al.* 2018; Marchant and Taylor 2000; Schüler *et al.* 2014), thereby limiting palaeoecological inferences and attempts to model vegetation from pollen data.

Southern Africa has an established tradition of palynological research, dating back to the pioneering efforts of van Zinderen Bakker (1950, 1951). Early palynologists often incorporated a few surface pollen spectra to support fossil pollen records (e.g. Cooremans 1989; Scott 1982a, b, 1989; Scott and Cooremans 1992; Scott *et al.* 1992), however, relatively few dedicated studies, particularly those with a temporal component, exist (Hill 1995). A recent modelling synthesis of modern pollen data from southern Africa pulled together a dataset of mostly surface soil samples from 211 sites across the subregion (Sobol *et al.* 2019), showing strong bias towards savanna (n = 67) and grassland (n = 53) relative to the other seven biomes. More detailed local studies, which employ passive pollen traps to calibrate modern-pollen vegetation relationships, include those from the Winterberg (Meadows and Meadows 1988), Nuweveldberg (Sugden and Meadows 1989), and Cederberg (Meadows and Sugden 1990; 1991a;b), all in the winter rainfall zone. In the absence of such detailed contemporary pollen studies, Meadows (1989, p. 160) commented that the reconstruction of palaeovegetation communities is 'largely guesswork'. Recent advances in the field include modelled estimates of relevant source area of pollen and relative pollen productivity undertaken in the Kruger National Park (Duffin and Bunting 2008) and Drakensberg Mountains (Duthie *et al.*, this volume). Despite these efforts, modern pollen–vegetation relationships in the subregion 'remain poorly resolved relative to other parts of the world' (Meadows 2015, p. 4).

In this paper, we present the results of a highly detailed modern pollen rain survey from the KwaZulu-Natal Drakensberg, comprising data from eleven vegetation communities and corresponding surface soil samples, in addition to seasonal pollen trap data spanning a two year period. The aim is to characterise the modern pollen rain produced by each vegetation community by (i) comparing pollen spectra, diversity and influx across the different vegetation communities; and (ii) determining which taxa are under- and overrepresented in the pollen assemblage. The research was carried out in the Cathedral Peak area of the northern KwaZulu-Natal Drakensberg Mountains (29°00′S, 29°15′E; Figure 1), in South Africa's summer rainfall zone. This mountainous environment presents the opportunity to study a range of vegetation communities within a confined region due to the altitudinal gradient, ranging from open grassland to ericaceous scrub, closed woodland and forest. The broader Maloti-Drakensberg region is recognised as having the preservation potential for late Quaternary palynological studies due to the availability of high altitude polliniferous deposits (e.g. Fitchett *et al.* 2016; 2017; Lodder *et al.* 2018; Neumann *et al.* 2014; Norström *et al.* 2009; van Zinderen Bakker 1955).

17.2 MATERIALS AND METHODS

Sampling design was based on eleven vegetation communities from three altitudinal zones (Table 1; Figure 2), *viz.* the montane (1280–1829 m asl), subalpine (1830–2865 m asl) and the alpine belts (2866–3353 m asl) (Killick 1963). Selected communities represented either (i) the dominant community within a vegetation belt; (ii) communities which follow a particular successional pathway; or (iii) communities that characterise a distinct topographic or geomorphological feature within the altitudinal vegetation belt. Killick (1963) and Hill (1996) describe the species composition and abundance within each of these vegetation communities in detail.

Vegetation surveys were undertaken within each of the communities and species present and percentage aerial cover were recorded (Table 1). The percentage aerial cover was deemed more appropriate than the more conventional basal cover (Bradshaw 1981) as aerial cover is an estimate of the dominance of a particular species within the community. As the pollen is produced from the upper aerial part of the plant it was presumed that this measure would provide a better representation of the amount of pollen produced. Percentage aerial cover was measured as the proportion of the ground occupied by perpendicular projections onto it of the aerial parts of the individuals of a species under consideration (Kershaw and Looney 1985).

Meadows (1989) advocates for the collection of surface samples, and corresponding trap data, from as many plant communities as possible. Here, modern pollen rain was sampled passively using a modified version of the Oldfield pollen trap (Flenley 1973), which is easy to construct, transport, install and recollect in an inaccessible mountainous environment. Pollen traps were randomly installed within each community at a height of 1 m above the ground. The number of pollen traps placed in each vegetation community varied between 5-18 based on the

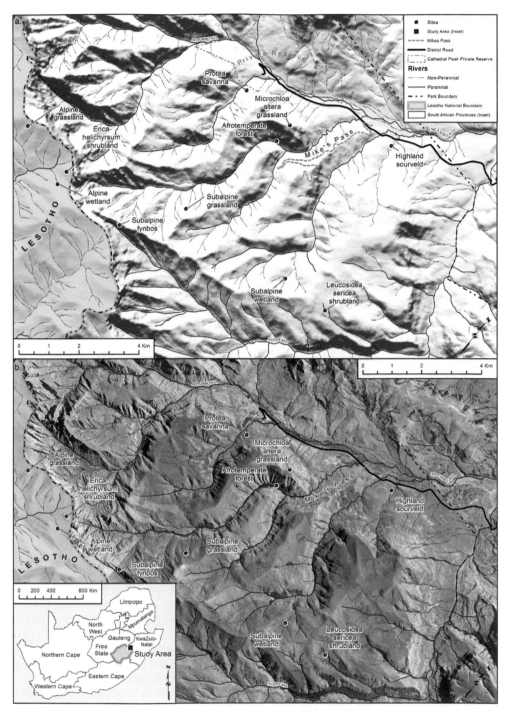

Figure 1. (a) Location map and (b) aerial photograph of the Cathedral Peak area of the KwaZulu-Natal Drakensberg, indicating the position of vegetation communities investigated relative to important topographic features.

Table 1. Distribution of pollen and vegetation samples among the eleven vegetation communities. Pollen traps were sampled on a seasonal basis across a two-year period, hence total pollen samples exceeds individual samples per community. Alpine communities were sampled using soils, while subalpine and montane communities used pollen traps.

Vegetation community (code)	Geographical position (GPS)	Elevation (m asl)	No. pollen samples	Total pollen samples (all seasons)	No. of vegetation quadrats	Quadrat size
Alpine						
Alpine grassland (AG)	−28.98; 29.13	3000	20	20	20	1 × 1 m
Alpine wetland (AW)	−29.01; 29.18	2965	20	20	20	1 × 1 m
Erica Helichrysum shrubland (EH)	−29.00; 29.18	3010	20	20	20	1 × 1 m
Subalpine						
Leucosidea sericea shrubland (LS)	−28.99; 29.26	1895	10	63	20	10 × 10 m
Microchloa altera grassland (MA)	−28.95; 29.22	1810	5	32	20	1 × 1 m
Subalpine fynbos (SF)	−29.01; 29.20	2690	10	37	20	10 × 10 m
Subalpine grassland (SG)	−28.99; 29.21	1975	25	175	60	1 × 1 m
Subalpine wetland (SW)	−28.99; 29.25	1890	6	31	25	1 × 1 m
Montane						
Afrotemperate forest (AF)	−28.96; 29.22	1550	18	125	20	10 × 10 m
Highland sourveld (HS)	−28.93; 29.25	1350	5	32	25	1 × 1 m
Protea savanna (PS)	−28.95; 29.20	1545	18	115	20	10 × 10 m
Total			**159**	**670**	**270**	

size and complexity of the community (Table 1). The traps have proved successful collectors of pollen rain and, in some instances, provide a better reflection of the vegetation community prevailing in the catchment area than surface soil samples (Crowder and Starling 1980; Meadows and Sugden 1991b; Sugden 1990). Twenty surface soil samples were taken randomly from each vegetation community. A sample of approximately 100 g constituting the top 3–5 cm of the soil was collected, with overlying organic matter having been first removed. Samples were placed into paper bags, sealed and refrigerated to reduce bacterial and fungal growth, until the necessary laboratory techniques could be carried out (Hicks and Hyvärinen 1986).

Fieldwork was undertaken from June 1988 to June 1990, thereby gaining a full two years of modern pollen rain data. The traps were replaced on a tri-monthly basis until November-December 1989. A seventh sample of six months duration, December 1989 to June 1990, was collected to compare a six month sample with the usual tri-monthly samples. Two full years of data were not achieved for the subalpine fynbos and subalpine wetland communities (Table 1).

Figure 2. (a) 1 × 1 m quadrat used for aerial vegetation survey; (b) passive pollen trap; (c) subalpine grassland; (d) alpine grassland; (e) *Erica Helichrysum* shrubland; (f) alpine wetland; (g) subalpine fynbos; (h) subalpine wetland; (i) *Leucosidea sericea* shrubland; (j) Afrotemperate forest; (k) *Protea* savanna; (l) highland sourveld.

Pollen was extracted from the traps and soil samples using acetolysis (Faegri and Iversen 1989). Identification was achieved using a predominantly local pollen reference collection of *c*. 700 species (Hill 1992). Absolute counts of *c*. 500 on trap samples and *c*. 1000 on soil samples were performed on pollen grains and spores (hereafter collectively termed 'pollen' for brevity). Pollen influx values were calculated based on sample volume as grains/cm^2/30 days (Berglund and Ralska-Jasiewiczowa 1986). R-rel values were calculated (R-rel = % mean abundance in pollen rain/% aerial cover of vegetation) (Davis 1963; Gosling *et al.* 2005) for each taxon in each community to understand the representivity of pollen taxa across the landscape. The Shannon-Wiener Index was calculated for each pollen sample and each vegetation community to determine how well the pollen assemblage captured vegetation diversity. The index ranges from 0–5, with most values falling between 1.5 and 3.5, and is a comparative index with higher values indicating higher diversity (Barnes *et al.* 1998). It should be acknowledged that many pollen taxa cannot be distinguished below the family level and thus the diversity index was calculated at the taxonomic level of the identified palynomorphs. The multivariate structure of the pollen assemblage data was represented using non-metric multidimensional scaling (nMDS) using the software PRIMER v7 (Clarke and Gorley 2015). The data were 4th root transformed to down weight the dominant species and Bray-Curtis similarity matrix calculated. Vectors of important species were based on Pearsons rank correlation. A pollen diagram showing characteristic taxa for each community was plotted using C2 (Juggins 2007).

17.3 RESULTS

From a total of *c*. 700 pollen traps deployed over the two year period, 16 were lost or damaged by animals or fire. All 36 traps deployed in the alpine belt over two attempts were stolen, thus surface soil samples were relied upon as a substitute within this vegetation belt. Pollen assemblage count size varied with an average of 1120 grains counted per soil sample in the alpine communities (ranging from 965-1272 grains per sample), and an average of 560 grains counted per trap in the subalpine and montane communities (ranging from 207-3392 grains per sample). A complete list of all pollen taxa recorded together with potential parent vegetation based on vegetation survey data is provided in supplementary information and used to infer likely parent taxa, which are indicated in parentheses (Hill 1992; Table S1).

17.3.1 Characterisation of vegetation communities by pollen assemblage

A total of 25 characteristic taxa were designated on the basis of being recorded from at least 80% of the samples from a vegetation community, and at >3% abundance in one sample or more (Table S1). Asteraceae, Cyperaceae, Poaceae and Pteridophyta are ubiquitous and characteristic across all altitudinal zones and vegetation communities, and are therefore deemed to be of limited value as diagnostic taxa.

For pollen trap data, relative abundance data are averaged across the seven time intervals due to a focus on spatial rather than temporal assemblage variability. Spatial variability in the pollen assemblage, both between and within vegetation communities, is depicted according to relative abundance of characteristic taxa, pollen influx, and assemblage diversity (Figure 3a-c). Influx values are indicated for all subalpine and montane belt communities together with relative abundance data for each trap (Figure 3b-c). It is worth noting that while some vegetation communities were represented by a high number of samples (up to 25 traps or 20 soil samples), the *M. altera* grassland and highland sourveld contained only five traps each, which may contribute towards their apparent homogeneity.

17.3.1.1 Alpine belt
The alpine vegetation communities display a relatively high degree of homogeneity within each community, and all three are distinctive for the presence of bryophyte spores which are absent from other altitudinal zones (Figure 3a). Within the alpine grassland the diversity index ranged from 1.2–1.5 with an average of 1.4 (Figure 4), and the most abundant taxa are Poaceae and Cyperaceae. Similar to the alpine grassland, the alpine wetland is dominated by Poaceae and Cyperaceae. The consistent presence of Umbelliferae (Apiaceae) in the pollen assemblage distinguishes the alpine wetland from other communities across the altitudinal range. The alpine wetland diversity ranged from 1.6–2.2 with an average of 1.9 (Figure 4). Distinctive taxa within the *Erica Helichrysum* shrubland include Ericaceae and Thymelaeaceae. Diversity ranged from 1.1 to 1.9, with an average of 1.6 (Figure 4).

17.3.1.2 Subalpine belt

Aside from the subalpine fynbos, the other subalpine communities are relatively homogenous within individual communities.

Leucosidea sericea shrubland – *L. sericea* shrubland is distinctive for dominant *L. sericea* pollen (30-60%), and the presence of Ericaceae. Diversity ranged from 1.2-2, with an average of 1.7. Influx rates varied from 65 to 275 grains/cm^2/month, with an average of 107 grains/cm^2/month (Figure 3b).

Microchloa altera grassland – The *M. altera* pollen assemblage is fairly uniform, with the main distinction being the low presence of extra-local forest taxa *Celtis* (<5%) and *Podocarpus* (<5%), due to the close proximity of this community to the edge of an Afrotemperate forest patch (<300 m). In the *M. altera* grassland, diversity varied from 2-2.1 with an average of 2.1 (Figure 4). Influx rates varied from 104 to 239 grains/cm^2/month, with an average of 145 grains/cm^2/month.

Subalpine fynbos – In the subalpine zone, the subalpine fynbos community displays a high degree of heterogeneity across the community, largely centred around variations in three taxa, viz. Ericaceae, Proteaceae and Scrophulariaceae (Figure 3b). In the subalpine fynbos, diversity ranged from 1.8–2.5 (average 2.3) (Figure 4). Influx rates varied from 101 to 269 grains/cm^2/month, with an average of 172 grains/cm^2/month.

Subalpine grassland – The subalpine grassland and *M. altera* grassland communities are similar, with the exception of forest taxa presence in the latter. In the subalpine grassland, pollen assemblage diversity ranged from 1.6 to 1.8, with an average of 1.7 (Figure 4). Influx rates varied from 118 to 409 grains/cm^2/month, with an average of 199 grains/cm^2/month (Figure 3b).

Subalpine wetland – As with subalpine grassland communities, the subalpine wetland pollen assemblage is Poaceae-Asteraceae dominated, and includes relatively high Cyperaceae (10–15%) and Pteridophyta (10–20%) frequencies. Diversity ranged from 1.8–2.1 (average 1.9) (Figure 4). Influx rates varied from 140 to 274 grains/cm^2/month, with an average of 198 grains/cm^2/month.

17.3.1.3 Montane belt

Afrotemperate forest – In the montane belt, there is a high degree of heterogeneity across the Afrotemperate forest community, particularly in the relative dominance of *Podocarpus* (5–40%) and *Celtis* (5–40%) (Figure 3c). This community displays a relatively strong arboreal pollen assemblage, muting the ubiquitous Asteraceae/Poaceae signal evident in other communities. The Afrotemperate forest diversity varied from 1.8–2.5 (average 2.1) (Figure 4). Pollen influx was highly variable, ranging from 111 to 493 grains/cm^2/month, with an average of 264 grains/cm^2/month (Figure 3c).

Protea savanna – The *Protea* savanna is distinctive for its high level of Proteaceae pollen (15–50%). Pollen assemblage diversity varied from 1.7 to 2 with an average of 1.9 (Figure 4). Influx rates varied from 68 to 275 grains/cm^2/month, with an average of 129 grains/cm^2/month.

Highland sourveld – The highland sourveld is similar to the subalpine grassland community, with no distinctive characteristic taxa evident from the modern pollen data. In the highland sourveld, diversity varied between 1.5–1.6, with an average of 1.5 (Figure 4). Pollen influx varied from 94 to 185 grains/cm^2/month (average 127 grains/cm^2/month) (Figure 3c).

17.3.2 Representivity of the pollen assemblage

Where characteristic pollen taxa are recorded together with the corresponding parent vegetation type in the surrounding community, the relative frequencies of pollen and vegetation can be directly compared to provide an indication of the degree of correspondence between pollen rain and parent vegetation composition (Figure S1). The R-rel value can be used as an indicator of representivity, with values <1 indicating underrepresentation and values >1 indicating overrepresentation (Figure 5). Furthermore, we can plot pollen abundance against vegetation

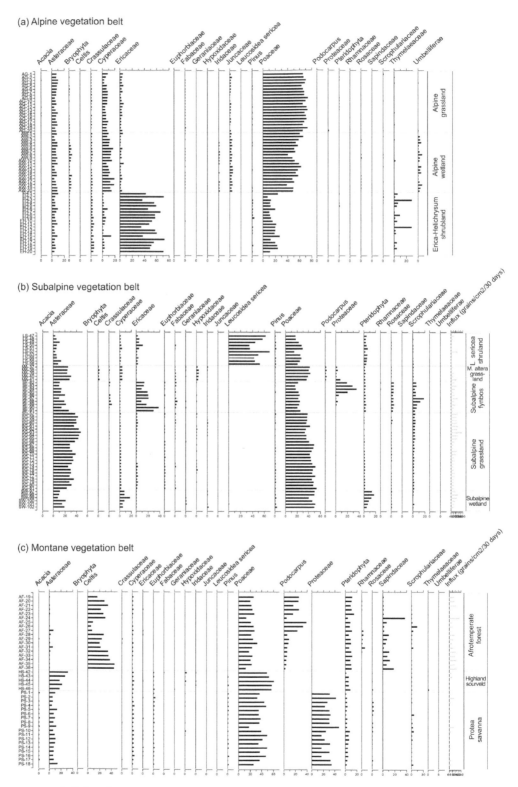

Figure 3. Pollen diagram showing pollen assemblages with samples grouped by vegetation community in the (a) alpine belt; (b) subalpine belt; and (c) montane belt, along with total influx values for each trap. For the three alpine vegetation communities, pollen data are derived from surface soil samples rather than traps.

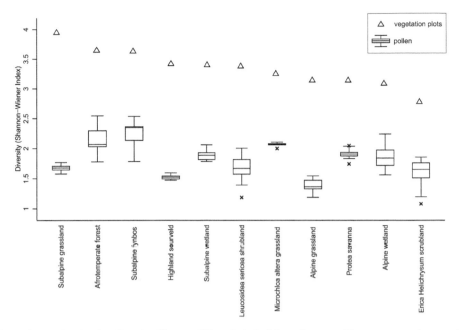

Figure 4. Boxplot showing diversity (Shannon-Wiener Index) of the pollen assemblages across each vegetation community. Communities are ranked according to species level diversity of the plots (triangle symbols), in order of decreasing diversity. For alpine grassland, alpine wetland, and *Erica Helichrysum* shrubland, pollen data are derived from surface soil samples rather than traps.

abundance using a R-rel reference line of 1 as a guide to over- and underrepresentation (Figure S2). Results show that for many taxa the representivity of the pollen type varies between vegetation communities.

In both the alpine and subalpine wetland communities, Cyperaceae is notably underrepresented, with a R-rel of 0.2–0.4 (Figure 5), whilst Poaceae is overrepresented. Some dominant taxa, such as Poaceae, are highly variable in their degree of representivity between communities (Figure S2). In general, Asteraceae appear to be overrepresented with the exception of the *Erica Helichrysum* shrubland where they are notably underrepresented (R-rel = 0.2) (Figure 5, S2). Based on high R-rel values exceeding 2 (Figure 5), Scrophulariaceae pollen appears to be highly overrepresented in all vegetation communities where it is recorded, with the exception of the *L. sericea* shrubland and subalpine wetland (Figure 5). Similarly, Crassulaceae are overrepresented in the alpine wetland and *Erica Helichrysum* communities (Figure 5).

Ericaceae pollen appears to be fairly representative of the vegetation cover in the subalpine fynbos and *L. sericea* shrubland, with a R-rel of 1.1–1.2 (Figure 5). However, in the *Erica Helichrysum* shrubland Ericaceae pollen is overrepresented with a R-rel of 1.7 (Figures 5, S2). *L. sericea* pollen abundance mirrors that of the parent vegetation in the *L. sericea* shrubland community, with a R-rel of *c.* 1.1 (Figures 5, S2).

When comparing aerial vegetation cover of *Celtis* with the pollen assemblage in the Afrotemperate forest, *Celtis* pollen is found to be overrepresented with a R-rel value of 1.7 (Figures 5, S2). *Podocarpus* pollen is underrepresented relative to the vegetation cover in the Afrotemperate forest, with a R-rel of 0.4 (Figures 5, S2). In the *Protea* savanna, Proteaceae pollen is underrepresented relative to the surrounding vegetation, with a R-rel of 0.6 (Figures 5, S2).

In the alpine grassland, Brassicaceae (*Heliophila*) pollen was absent, despite the parent taxon was recorded in the surrounding vegetation. These 'palynologically silent' taxa are

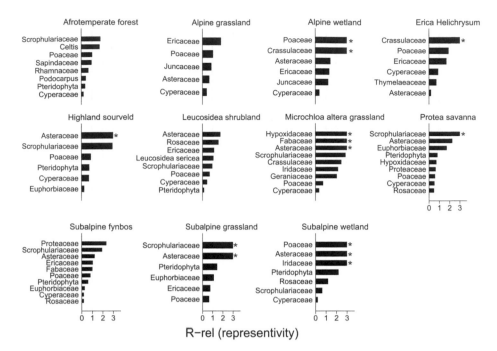

Figure 5. Bar chart showing R-rel (representivity) values of taxa represented in both pollen and vegetation. Asterisks indicate that the R-rel values were too high to show on the figure; For the alpine wetland Crassulaceae = 4.05, Poaceae = 3.35; for *Erica Helichrysum* Crassulaceae = 3.35; For Highland sourveld Asteraceae = 3.32; For *M. altera* grassland Hypoxidaceae = 5.06, Asteraceae = 4.43, Fabaceae = 3.52, for *Protea* savanna Scrophulariaceae = 3.60; for subalpine grassland Scrophulariaceae = 5.78, Asteraceae = 3.07; for subalpine wetland Asteraceae = 8.09, Poaceae = 3.99. For alpine grassland, alpine wetland, and *Erica Helichrysum* belt, pollen data are derived from surface soil samples rather than traps.

unrepresented in the pollen assemblage. For the *Erica Helichrysum* shrubland, this included Santalaceae (*Thesium*). In the alpine wetland community, all recorded vegetation types were represented in the pollen spectra, at least at family level. Afrotemperate forest appears to have the largest number of 'silent taxa', including *Ilex mitis*, Gesneriaceae (*Streptocarpus*), *Impatiens*, Apiaceae (*Conium, Sanicula*) and Vitaceae (*Rhoicissus*). In the *L. sericea* shrubland, such taxa included Sterculiaceae (*Hermannia*), Apiaceae (*Allipedia*), Malvaceae (*Hibiscus*), and Melianthaceae (*Melianthus*). Orchidaceae (*Disa*) pollen was absent from the *M. altera* grassland despite being recorded in the vegetation plots. In the subalpine fynbos, *Gunnera perpensa*, Melianthaceae (*Melianthus*), *Olea* and Primulaceae (*Lysimachia*) were absent from the pollen spectra, however, were recorded in the surrounding vegetation. Apiaceae (*Allepidea*) pollen was similarly absent from the subalpine grassland. Illecebracaceae (*Silene*) pollen was palynologically silent in both the subalpine grassland and wetland communities. Finally, Valerianaceae (*Valeriana*) pollen was recorded from the highland sourveld community but absent from the pollen spectra. In the *Protea* savanna, all plant taxa present were recorded in the pollen traps.

17.3.3 Multivariate analysis

The nMDS ordination plot depicts the distribution of the samples relative to vegetation community groupings (Figure 6). The ordination shows discrete clustering of samples within vegetation

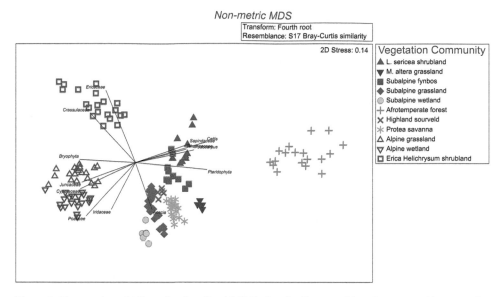

Figure 6. Non-metric multidimensional scaling (nMDS) plot of pollen assemblage data, grouped by vegetation community.

communities, with Afrotemperate forest and *Erica Helichrysum* shrubland being the most distinctive. Overall, the alpine communities are clearly distinct from the other vegetation belts, illustrating the importance of altitude. The alpine grassland and alpine wetland communities cluster fairly closely together, albeit with no direct overlap. The *L. sericea* shrubland and subalpine fynbos show some degree of overlap. Subalpine grassland, subalpine wetland, subalpine fynbos, highland sourveld and *Protea* savanna cluster fairly closely together, but with very limited direct overlap of ordination space.

17.4 DISCUSSION

17.4.1 Modern pollen-vegetation relationships

The results provide an indication of the relative uniformity of each vegetation community. Furthermore, they allow us to assess whether these vegetation communities can be qualitatively distinguished by their modern pollen assemblages. Here we evaluate whether there is a distinctive pollen signal or are distinctive 'indicator' taxa that can be used to differentiate each community.

Overall, we find that the grassland vegetation communities are relatively difficult to be separated by their pollen assemblage, which can be attributed to the limited taxonomic resolution achievable in identifying grass pollen types (i.e. family level). This is supported by the diversity data, which show that the grassland pollen assemblages consistently underrepresent the diversity of their parent vegetation, when compared with other communities. One exception to this is the alpine grassland community, which the nMDS picks out as a separate cluster from other grassland types.

The alpine vegetation communities were easily distinguished from communities in other vegetation belts using nMDS. A combination of bryophytes and Crassulaceae (likely from the common *Crassula setulosa* or possibly the less common *Crassula natans*) in the pollen assemblage appears to be relatively characteristic of the alpine communities, in particular within the alpine wetland. The alpine wetland community was rather similar to the alpine grassland, with the only differences being a higher frequencies of bryophytes and Juncaceae (*Juncus* or *Lezula*), and the presence of Umbelliferae in the alpine wetland. Umbelliferae pollen appears to be restricted in its presence to the alpine wetland community, and therefore serves as a useful indicator taxon. van Zinderen Bakker (1955) recorded Umbelliferae frequencies of 2–8% in four wetland surface samples in Lesotho.

Scott (1982b) associates the Umbelliferae with damp conditions, however its source at Cathedral Peak is unknown as no members of the Umbelliferae family were recorded in the vegetation. Similar to the situation in the alpine belt, the subalpine wetland and subalpine grassland communities showed a high degree of overlap, with differences being higher Cyperaceae and Pteridophyta in the former, and higher Scrophulariaceae (*Selago*, *Zaluziantskya*, *Diascia*, *Manulea*, *Nemesia*, *Hebenstretia*, *Sopubia*, *Sutera*) and Ericaceae (*Erica* spp.) in the latter.

The *Erica Helichrysum* shrubland is distinctive in the pollen diagram and the nMDS for consistent, and sometimes relatively high, frequencies of Thymelaeaceae (*Passerina montana* or *Gnidia polystachya*), and dominated by high representation of Ericaceae, occasionally exceeding 80% abundance. In contrast, subalpine fynbos is dominated by a mix of Proteaceae (*Protea subvestita*), Ericaceae (*Erica* spp.) and Poaceae, with Rosaceae (*Cliffortia* or *Alchemilla*) and Scrophulariaceae (*Bowkeria* or *Halleria*) consistently recorded.

The *L. sericea* shrubland has a distinctive pollen signature, with pollen of the shrub *L. sericea* dominating the assemblage (40–60%), and relatively low Poaceae and Asteraceae frequencies compared with other communities. In the nMDS, however, this community showed a slight overlap with the subalpine fynbos community. Ericaceae is prominent in *L. sericea* shrubland, as are the Pteridophyta.

The Afrotemperate forest has a highly unique pollen signature, clearly observed in the nMDS, with relatively high arboreal pollen frequencies including *Celtis* (10–45%) and *Podocarpus* (10–40%). Sapindaceae pollen, probably from the tree *Allophylus melanocarpus*, is a good indicator for this community. The *Protea* savanna community can be distinguished by high Proteaceae pollen frequencies (20–50%). Although subalpine fynbos records high Proteaceae, the *Protea* savanna assemblage can be separated by a lack of Ericaceae pollen, and the two are clearly separated in the nMDS.

The above comparison highlights the challenges of pollen analytical research in a grass dominated environment, and the need for additional supporting proxies such as grass phytoliths to detect palaeovegetation change (e.g. Alexandre *et al.* 1997). With the potential for C_3/C_4 grass species migration with climate change along a clear altitudinal gradient (e.g. Roberts *et al.* 2013) it could be of some value to attempt to quantitatively distinguish C_3 from C_4 vegetation. Despite these challenges, the nMDS was able to clearly separate almost all vegetation communities sampled, with limited overlap between samples from different communities.

Those communities with a significant tree component i.e. Afrotemperate forest, *Protea* savanna and *L. sericea* shrubland, are more easily distinguished by their pollen assemblage. The *Erica Helichrysum* shrubland and subalpine fynbos, could be distinguished by their dominant fynbos elements, although the latter overlapped with *L. sericea* in the nMDS. The two wetland communities displayed a strong degree of overlap with the surrounding grassland vegetation, but with some distinctive taxa. The presence of less abundant but nevertheless useful aquatic and 'damp habitat' indicators (*sensu* Scott 1982b) such as Dipsacaceae (*Scabiosa drakensbergensis*) and Gunneraceae (*Gunnera perpensa*) may assist in distinguishing wetland pollen communities. In the subalpine wetland community, Cyperaceae and Pteridophyta were the only pollen taxa to record higher abundances than in the surrounding subalpine grassland community. The reality is that fossil pollen preserving archives in these summer rainfall mountain environments tend to be restricted to wetland sediments (e.g. Fitchett *et al.* 2016 2017, Lodder *et al.* 2018, Neumann *et al.* 2014, Norström *et al.* 2009), thus the fossil pollen signal is usually overrepresented by local wetland taxa, usually excluded from the pollen sum. In the Cathedral Peak area, wetlands representing potential fossil pollen archives are present across a range of altitudes (e.g. Finch *et al.* 2021; Lodder *et al.* 2018), from the montane to subalpine to alpine (e.g., this study), although the depth and organic content of these sites is variable.

17.4.2 Representivity of taxa

The modern pollen-vegetation comparison allows for some general recommendations regarding interpretation of key taxa based on their relative representivity in the pollen assemblage. These are compared with existing information in the available literature.

Podocarpus – *Podocarpus* pollen is an important arboreal pollen type in southern African pollen records, and also constitutes an key component of Afromontane vegetation in across Subsaharan Africa (Gajewski *et al.* 2002; Verlhac *et al.* 2018; Vincens *et al.* 2006). Previous studies designate *Podocarpus* as a very well dispersed anemophilous pollen type (Hamilton 1972), that tends to be overrepresented within the pollen signal (Marchant and Taylor 2000; Schüler *et al.* 2014). In terms of dispersal, a West African biome-scale study (Verlhac *et al.* 2018) found *Podocarpus* pollen distribution to be similar to that of the parent plant, and identified this taxon as a good indicator of modern vegetation. Coetzee (1967) developed a guideline for interpreting *Podocarpus* pollen frequencies with 10–20% indicating close proximity of the parent vegetation, and frequencies exceeding 20% indicative of local presence of *Podocarpus* forest. Schüler *et al.* (2014) commented that such overrepresented taxa are 'probably significantly less important components of the surrounding vegetation than indicated by the modern pollen-rain' (Schüler *et al.* 2014, p. 710). In the Cathedral Peak area, *Podocarpus* pollen is likely derived predominantly from *Podocarpus latifolius*, and to a lesser degree from *Afrocarpus* (formerly *Podocarpus*) *falcatus* and *Podocarpus henkelii* (Killick 1963). Interestingly, our data suggest that *Podocarpus* pollen is underrepresented relative to the vegetation cover in the Afrotemperate forest community (R-rel = 0.4), and indeed, with the exception of the *M. altera* grassland, *Podocarpus* pollen does not appear to be represented outside of the forest community. This is surprising given the prevalence of small *Podocarpus* forest patches which form a mosaic within the grassland community (Adie *et al.* 2017; Figure 1b)). The underrepresentation of *Podocarpus* within the forest community could be explained by (i) high pollen production by other taxa e.g. *Celtis*; and/or (ii) the size and complexity of the Afromontane forest community, which appears to be characterised by a heterogeneous pollen signal, with *Celtis* dominant in some areas and *Podocarpus* in others. Averaging *Podocarpus* abundance for both the pollen traps and the parent vegetation across such a heterogeneous community may explain these anomalous results.

Celtis – as with *Podocarpus*, *Celtis* pollen is restricted to the Afrotemperate forest community, with the exception of the nearby *M. altera* community (<300 m upward of the forest boundary). Within the forest community, *Celtis* pollen is overrepresented (R-rel = 1.7), aligning with previous studies (e.g. Marchant and Taylor 2000), with *Celtis* being a very well dispersed taxon (Hamilton 1972; Verlhac *et al.* 2018). Previous studies found *Celtis* to be abundant pollen producers, present outside of the parent vegetation distribution (Watrin *et al.* 2007). Interestingly, studies from east and west Africa (e.g., Livingstone 1967; Verlhac *et al.* 2018) have shown *Celtis* to be characteristic of lower altitudes than *Podocarpus*, which is not the case in southern Africa.

Ilex mitis – it was found to be 'palynologically silent' in the Afrotemperate forest community. As a highly distinctive pollen type, this is unlikely to have been missed during analysis. Hamilton (1972) commented that *I. mitis* has low to poor pollen dispersal capacity, while Marchant and

Taylor (2000) concluded that *Ilex mitis* pollen was representative of the density of the parent vegetation.

Cyperaceae – A surprising outcome of this study is that Cyperaceae pollen was found to be underrepresented in both wetland communities, with frequencies consistently below 20%. Cyperaceae pollen often dominates fossil pollen spectra due to the local nature of the taxon. Scott (1982b) and van Zinderen Bakker (1955) found Cyperaceae to be a dominant pollen taxon in modern surface samples, with high frequencies recorded. Lodder *et al.* (2018) investigated a 5000-yr sediment core retrieved from the subalpine wetland community at Cathedral Peak (Catchment VI), and recorded Cyperaceae abundance of 40–90% in the fossil assemblages, yet in the current study the modern pollen abundance for the same site remained <20% for the two year sampling period. The apparent underrepresentation of Cyperaceae pollen in the current study might be attributed to pollen trap height (1 m from the ground), above the parent vegetation, whereas the other modern sampling campaigns used surface samples. This inference is supported by relatively higher representation of Cyperaceae pollen in surface soil samples as compared with traps in the subalpine wetland (17% *vs.* 9%) (Hill 1992). In a comparison of surface and trap data from the Winterberg, Meadows and Meadows (1988) also found that sedges were more well represented in the surface samples than in the traps, which may be attributed to trap height. Irrespective of differences in apparent representivity between studies, Cyperaceae are ubiquitous across a range of habitat types and climate zones (e.g. Gajewski *et al.* 2002), and have limited applicability as a palaeoenvironmental indicator.

Ericaceae – The representivity of Ericaceae pollen varied from representative in the subalpine fynbos (R-rel = 1.1) and *L. sericea* (R-rel = 1.2) communities to overrepresented in the *Erica Helichrysum* shrubland (R-rel = 1.7). Marchant and Taylor (2000) and Schüler *et al.* (2014) designated Ericaceae pollen as overrepresented, although the dispersal characteristics of the pollen taxon are debated, ranging from poorly dispersed (due to large grain size; Verlhac *et al.* 2018), to moderately dispersed (Hamilton 1972), and well dispersed (Schüler *et al.* 2014). Dispersal is one facet of pollen representivity, but so too is pollen production, and to a lesser degree in modern samples, preservation.

Poaceae – The results indicate a wide degree of variability in the representivity of Poaceae pollen between different communities, with few obvious patterns apparent. The two wetland communities showed a high degree of overrepresentation of Poaceae pollen relative to vegetation cover. In the closed vegetation communities, Poaceae pollen varied from representative in the Afrotemperate forest (R-rel = 1.0) to underrepresented in the *L. sericea* shrubland (R-rel = 0.7) and *Protea* savanna (R-rel = 0.6). Within open ecosystems and grassland communities, the degree of Poaceae representivity was highly variable. Previous studies have designated Poaceae pollen as a generally well dispersed taxon linked to it's anemophilous pollination syndrome (Hamilton 1972), consequently often overrepresented within the pollen signal.

17.5 CONCLUSIONS

The original palynologists' dream was to develop a 'signature' modern pollen assemblage for each distinct vegetation community within the three recorded vegetation belts, *viz.* montane, subalpine and alpine, in the KwaZulu-Natal Drakensberg. With this achieved, the aim was to correlate, or cross-reference, dated fossil pollen spectra against these modern analogues to infer past vegetation communities, which could then be extended to interpret palaeoenvironments. As with any palaeoreconstruction proxy, the outcome is less straightforward and open to interpretation.

We identified eleven vegetation communities across the three altitudinal belts, undertook a vegetation survey, deployed pollen traps and/or collected surface soil samples. Modern pollen sampling was conducted over two years and influx values were calculated. Overall, a high degree of homogeneity exists both within and between communities with regards to modern pollen, although characteristic taxa could be identified for certain communities. Dominant pollen taxa were often over- or underrepresented in comparison with the surrounding vegetation composition and abundance, taking cognisance of the taxonomic resolution. We note and justify these characteristic taxa and suggest reasons why they could prove useful as signature taxa during fossil pollen interpretation. Perhaps unsurprisingly, there appears to be little variation across the various grassland communities, greater opportunity exists to recognise the forest and shrubland types, and to distinguish the communities from different altitudinal vegetation belts, e.g. alpine communities.

The Drakensberg is a region with a high palynological potential, however it is imperative that any palaeoreconstruction be based on a sound understanding of the modern pollen processes. Therefore, contemporary pollen rain-vegetation relationships and modern analogues should be considered and refined. Thus, in conjunction with other palaeoecological proxies and dating methods, one is able to build a clearer picture of past environments in a mountainous region which has a high biodiversity and displays clear altitudinal and latitudinal gradients, imperative to improving our understanding of the past environments in southern Africa.

ACKNOWLEDGEMENTS

TRH was funded by the Foundation for Research and Development and Rhodes University. Brice Gijsbertsen professionally drafted Figure 1. The staff of Cathedral Peak Forestry and Research Station are thanked for their hospitality, and for allowing the opportunity to work in such tranquil surroundings. Thank you to Colin and Terry Everson, and Martin Hill for assisting with fieldwork. Andrew Steele assisted with data entry. Elodie Heyns-Veale kindly assisted with the multivariate analysis. We thank Henry Lamb and one anonymous reviewer for constructive feedback on the manuscript.

REFERENCES

Adie, H., Kotze, D.J. and Lawes, M.J., 2017, Small fire refugia in the grassy matrix and the persistence of Afrotemperate forest in the Drakensberg mountains. *Scientific Reports*, **7**, article: 6549, 10.1038/s41598-017-06747-2.

Alexandre, A., Meunier, J.-D, Lezine, A.M., Vincens, A. and Schwartz, D. 1997, Phytoliths: indicators of grassland dynamics during the late Holocene in intertropical Africa. *Palaeogeography, Palaeoclimatology, Palaeoecology*, **136**, pp. 213–229, 10.1016/S0031-0182(97)00089-8.

Barnes, B.V., Zak, D.R., Denton, S.R. and Spurr, S.H. 1998, *Forest Ecology*, (New York: Wiley).

Berglund, B.E. and Ralska-Jasiewiczowa, M. 1986, Pollen analysis and pollen diagrams. In: *Handbook of Holocene Palaeoecology and Palaeohydrology* edited by Berglund B.E., pp. 455–484 (Chichester: John Wiley & Sons).

Bradshaw, R.H.W., 1981, Modern pollen-representation factors for woods in South-East England. *Journal of Ecology*, **69**, pp. 45–70, 10.2307/2259815.

Clarke, K.R. and Gorley, R.N. 2015. PRIMER v7: User Manual / Tutorial (1st edn). Plymouth: PRIMER-E ltd.

Coetzee, J.A. 1967, *Pollen analytical studies in East and southern Africa*. (Cape Town: Balkema).

Cooremans, B. 1989, Pollen production in central southern Africa. *Pollen et Spores*, **36**, pp. 61–78.

Crowder, A. and Starling, R.N. 1980, Contemporary pollen in the Salmon River Basin, Ontario. *Review of Palaeobotany and Palynology*, **30**, pp. 11–26, 10.1016/0034-6667(80)90003-2.

Davis, M.B. 1963, On the theory of pollen analysis. *American Journal of Science*, **261**, pp. 897–912, 10.2475/ajs.261.10.897.

Duffin, K.I. and Bunting, J. 2008, Relative pollen productivity and fall speed estimates for southern African savanna taxa. *Vegetation History and Archaeobotany*, **17**, pp. 507–525, 10.1007/s00334-007-0101-2.

Elenga, H., de Namur, C., Vincens, A., Roux, M., and Schwartz, D., 2000, Use of plots to define pollen-vegetation relationships in densely forested ecosystems of Tropical Africa. *Review of Palaeobotany and Palynology*, **112**, 79–96, 10.1016/S0034-6667(00)00036-1.

Faegri, K. and Iversen, J. 1989, *Textbook of Pollen Analysis*, 4th Edition, (Chichester: John Wiley & Sons).

Finch, J.M., Hill, T.R., Meadows, M.E., Lodder, J. and Bodmann, L., 2021, Fire and montane vegetation dynamics through successive phases of human occupation in the northern Drakensberg, South Africa. *Quaternary International*, 10.1016/j.quaint.2021.01.026.

Fitchett, J.M., Grab, S.W., Bamford, M.K. and Mackay, A.W., 2016, A multi-proxy analysis of late Quaternary palaeoenvironments, Sekhokong Range, eastern Lesotho. *Journal of Quaternary Science*, **31**, pp. 788–798, 10.1002/jqs.2902.

Fitchett, J.M., Mackay, A.W., Grab, S.W. and Bamford, M.K. 2017, Holocene climatic variability indicated by a multi-proxy record from southern Africa's highest wetland. *The Holocene*, **27**, pp. 638–650.

Flenley, J.R. 1973, The use of modern pollen rain samples in the study of the vegetational history of tropical regions. In: *Quaternary Plant Ecology*, edited by Birks, H.J.B. and West R.G., (Oxford: Blackwell), pp. 131–141.

Gajewski, K., Lézine, A-M., Vincens, A., Delestan, A., Sawada, M. and the African Pollen Database, 2002, Modern climate-vegetation-pollen relations in Africa and adjacent areas. *Quaternary Science Reviews*, **21**, 1611-1631, 10.1016/S0277-3791(01)00152-4.

Gosling, W.D., Mayle, F.E., Tate, N.J. and Killeen, T.J. 2005, Modern pollen-rain characteristics of tall terra firme moist evergreen forest, southern Amazonia. *Quaternary Research*, **64**, pp. 284–297, 10.1016/j.yqres.2005.08.008.

Hamilton, A.C. 1972, The interpretation of pollen diagrams from highland Uganda. *Palaeoecology of Africa*, **7**, pp. 45–149.

Henga-Botsikabobe, K., Ngomanda, A., Oslisly, R., Favier, C., Muller, S.D., and Bremond, L., 2020, Modern pollen-vegetation relationships within tropical marshes of Lopé National Park (Central Gabon). *Review of Palaeobotany and Palynology*, **275**, 10.1016/j.revpalbo.2020.104168, article: 104168.

Hicks, S. and Hyvärinen, V., 1986, Sampling modern pollen deposition by means of "Tauber traps": Some considerations. *Pollen et Spores*, **28**, pp. 219–242.

Hill, T.R., 1992, *Contemporary pollen spectra from the Natal Drakensberg and their relation to associated vegetation communities*. PhD Thesis, Rhodes University, Grahamstown.

Hill, T.R., 1995, Analysis of contemporary pollen rain in South Africa – A review. *Transactions of the Royal Society of southern Africa*, **50**, pp. 27–39, 10.1080/00359199509520327.

Hill, T.R., 1996, Description, classification and ordination of the dominant vegetation communities, Cathedral Peak, KwaZulu Natal Drakensberg. *South African Journal of Botany*, **62**, pp. 263–269, 10.1016/S0254-6299(15)30655-4.

Hill, T.R., Duthie, T.J. and Bunting, M.J. this volume, Relevant source area of pollen and pollen productivity estimates from KwaZulu-Natal Drakensberg, South Africa. *Palaeoecology of Africa*, **35**, chapter: 16, 10.1201/9781003162766-16.

Juggins S., 2007, C2 Version 1.5 Software for ecological and palaeoecological data analysis and visualisation (Newcastle upon Tyne: Newcastle University).

Julier, A.C.M., Jardine, P.E., Adu-Bredud, S., Coea, A.L., Duah-Gyamfid, A., Fraser, W.T., Lomax, B.H., Malhi, Y., Moore, S., Owusu-Afriyied, K. and Gosling, W.D., 2018, The modern pollen–vegetation relationships of a tropical forest–savannah mosaic landscape, Ghana, West Africa. *Palynology*, **42**, pp. 324–338, 10.1080/01916122.2017.1356392.

Kershaw, K.A. and Looney, J.H.H., 1985, Quantitative and dynamic plant ecology (London: Edward Arnold).

Killick, D.J.B., 1963, *An account of the plant ecology of the Cathedral Peak Area of the Natal Drakensberg*, Memoirs of the Botanical Society of South Africa, Number 34, (Praetoria: The Government Printer).

Livingstone, D.A., 1967, Postglacial vegetation of the Ruwenzori Mountains in equatorial Africa. Ecological Monographs, **37**(1), pp. 25–52, 10.2307/1948481.

Lodder, J., Hill, T.R. and Finch, J.M. 2018, A 5000-yr record of Afromontane vegetation dynamics from the Drakensberg Escarpment, South Africa. *Quaternary International*, **470**, pp. 119–129, 10.1016/j.quaint.2017.08.019.

Marchant, R. and Taylor, D. 2000, Pollen representivity of montane forest taxa in south-western Uganda. *New Phytologist*, **146**, pp. 515–525, 10.1046/j.1469-8137.2000.00662.x.

Meadows, M.E., 2015, Seven decades of Quaternary palynological studies in southern Africa: a historical perspective, *Transactions of the Royal Society of South Africa*,**70**(2), pp. 103–108, 10.1080/0035919X.2015.1004139.

Meadows, M.E. and Meadows, K.F., 1988, Late Quaternary vegetation history of the Winterberg Mountains, Eastern Cape, South Africa. *South African Journal of Science*, **84**, pp. 253–259.

Meadows, M.E. and Sugden, J.M., 1990, Late Quaternary vegetation history of the Cederberg, south-western Cape. *Palaeoecology of Africa*, **21**, pp. 269–281.

Meadows, M.E. and Sugden, J.M., 1991a, A vegetation history of the last 14 000 years on the Cederberg, south-western Cape Province. *South African Journal of Science*, **87**, pp. 34–43.

Meadows, M.E. and Sugden, J.M., 1991b, The application of multiple discriminant analysis to the reconstruction of the vegetation history of Fynbos, southern Africa. *Grana*, **30**, pp. 325–336, 10.1080/00173139109431987.

Neumann, F.H., Botha, G.A. and Scott, L. 2014, 18,000 years of grassland evolution in the summer rainfall region of South Africa: evidence from Mahwaqa Mountain, KwaZulu-Natal. *Vegetation History and Archaeobotany*, **23**, pp. 665–681, 10.1007/s00334-014-0445-3.

Norström, E., Scott, L., Partridge, T.C., Risberg, J. and Holmgren, K. 2009, Reconstruction of environmental and climatic changes at Braamhoek wetland, eastern escarpment South Africa, during the last 16,000 years with emphasis on the Pleistocene-Holocene transition. *Palaeogeography Palaeoclimatology Palaeoecology*, **271**, pp. 240–258, 10.1016/j.palaeo.2008.10.018.

Roberts, P., Lee-Thorp, J.A., Mitchell, P.J. and Arthur, C., 2013, Stable carbon isotopic evidence for climate change across the Late Pleistocene to early Holocene from Lesotho, southern Africa. *Journal of Quaternary Science*, **28**, 360–369, 10.1002/jqs.2624.

Schüler, L., Hemp, A. and Behling, H., 2014, Relationship between vegetation and modern pollen-rain along an elevational gradient on Kilimanjaro, Tanzania. *The Holocene*, **24**, pp. 702–713, 10.1177/0959683614526939.

Scott, L., 1982a, A late Quaternary pollen record from the Transvaal bushveld, South Africa. *Quaternary Research*, **17**, pp. 339–170, 10.1016/0033-5894(82)90028-X.

Scott, L., 1982b, Late Quaternary fossil pollen grains from the Transvaal, South Africa. *Review of Palaeobotany and Palynology*, **36**, pp. 241–18, 10.1016/0034-6667(82)90022-7.

Scott, L. and Cooremans, B., 1992, Pollen in recent *Procavia* (hyrax), *Petromus* (dassie rat) and bird dung in South Africa. *Journal of Biogeography*, **19**, pp. 205–215, 10.2307/2845506.

Scott, L., Cooremans, B. and Maud, R.R., 1992, Preliminary palynological evaluation of the Port Durnford formation at Port Durnford, Natal coast, South Africa. *South African Journal of Science*, **88**, pp. 470–474.

Sobol, M.K., Scott, L. and Finkelstein, S.A., 2019, Reconstructing past biomes states using machine learning and modern pollen assemblages: A case study from Southern Africa. *Quaternary Science Reviews*, **212**, pp. 1–17, 10.1016/j.quascirev.2019.03.027.

Sugden, J.M., 1990, *Late Quaternary palaeoecology of the central and marginal Uplands of the Karoo, South Africa*. PhD thesis, University of Cape Town.

Sugden, J.M. and Meadows, M.E., 1989, The use of multiple discriminant analysis in reconstructing recent vegetation changes on the Nuweveldberg, South Africa. *Review of Palaeobotany and Palynology*, **60**, pp. 131–147, 10.1016/0034-6667(89)90073-0.

van Zinderen Bakker, E.M., 1950, Palynology in Africa. First report, covering the year 1950. *Palaeoecology of Africa*, **1**, pp. 1–4.

van Zinderen Bakker, E.M., 1951, Palynology in Africa. Second report, covering the year 1951. *Palaeoecology of Africa*, **1**, pp. 5–8.

van Zinderen Bakker, E.M., 1955, A preliminary survey of the peat bogs of the Alpine belt of northern Basotholand. *Acta Geographica*, **14**, pp. 413–422.

Verlhac, L., Izumi, K., Lézine, A-M., Lemonier, K., Buchet, G., Achoundong, G., and Tchiengué, B. 2018. Altitudinal distribution of pollen, plants and biomes in the Cameroon Highlands. *Review of Palaeobotany and Palynology*, **259**, pp. 21–28, 10.1016/j.revpalbo.2018.09.011.

Vincens, A., Bremond, L., Brewer, S., Buchet, G. and Dussouillez, P. 2006. Modern pollen-based biome reconstructions in East Africa expanded to southern Tanzania. *Review of Palaeobotany and Palynology*, **140**, pp. 187–212, 10.1016/j.revpalbo.2006.04.003.

Watrin, J., Lézine, A-M., Gajewski, K. and Vincens, A. 2007, Pollen–plant–climate relationships in sub-Saharan Africa. *Journal of Biogeography*, **34**, pp. 489–499, 10.1111/j.1365-2699.2006.01626.x.

CHAPTER 18

A Late Holocene pollen and microcharcoal record from Eilandvlei, southern Cape coast, South Africa

Nadia du Plessis

Department of Environmental and Geographical Science, University of Cape Town, Rondebosch, South Africa

Brian M. Chase[1]

Institut des Sciences de l'Evolution-Montpellier (ISEM), University of Montpellier, Centre National de la Recherche Scientifique (CNRS), EPHE, IRD, Montpellier, France

Lynne J. Quick

African Centre for Coastal Palaeoscience, Nelson Mandela University, Port Elizabeth, South Africa

Michael E. Meadows[2]

Department of Environmental and Geographical Science, University of Cape Town, Rondebosch, South Africa

18.1 SITE DETAILS

Eilandvlei forms part of the Wilderness Lakes system located along the southern Cape coast of South Africa (Figure 1A, B). These lakes are located behind Pleistocene dune ridges that run parallel along the coastline. Situated within South Africa's aseasonal rainfall zone, climate is influenced by both temperate and tropical circulation systems.

The regional vegetation (Figure 1B, C) is most noteworthy for the extensive development of Southern Afrotemperate Forest (generally *Afrocarpus falcatus*, *Podocarpus latifolius*, *Ocotea bullata* (Lauraceae) and *Olea capensis* spp. *marcocarpa* (Oleaceae)), which is found in valleys and on the south-facing slopes of the adjacent river catchments A variety of fynbos types occupy

[1]Other affiliation: *Department of Environmental and Geographical Science, University of Cape Town, Rondebosch, South Africa*
[2]Other affiliation: *School of Geographic Sciences, East China Normal University, Shanghai, China; and College of Geography and Environmental Sciences, Zhejiang Normal University, China*

DOI 10.1201/9781003162766-18

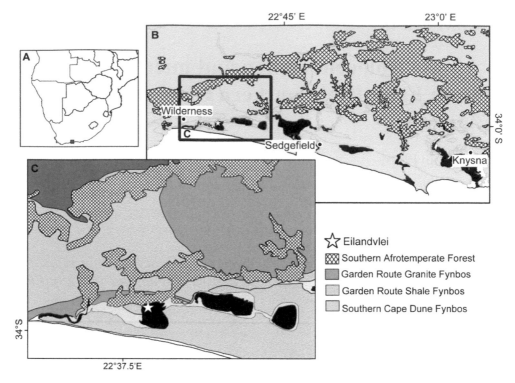

Figure 1. (A) Map of Africa showing the location of the southern Cape coast. (B) A section of the southern Cape coast between the towns of Wilderness and Knysna indicating the location of Eilandvlei and the current extent of Afrotemperate forest in the region. (C) The location of sediment core EV11 and the contemporary distribution of the dominant vegetation types (Mucina and Rutherford 2006).

this region including Garden Route Shale Fynbos (distinguished by ericaceous and tall dense proteoid fynbos), Knysna Sand Fynbos (primarily *Erica curvifolia*, *Metalasia densa* (Asteraceae) and *Passerina rigida*) and Southern Cape Dune Fynbos (predominantly *Olea exasperata*, *Phylica litoralis* and a variety of *Searsia* species).

18.2 SEDIMENT DESCRIPTION AND METHODS

The EV11 core (33°59′23.10″S, 22°38′17.60″E, 0 m asl) was retrieved using a portable vibracorer mounted on a floating platform. The final core length measured ~1.54 m. The mean grain size varies from 5.89 to 7.27 ϕ with the standard deviation ranging from 1.85 to 2.33 indicating very poorly sorted, fine to coarse silts to be the principal sedimentological components of EV11.

Forty-seven subsamples, with a minimum weight of 2 g, were analysed using standard palynological methods as per Faegri and Iversen (1989) and Moore *et al.* (1991) with adaptations for dense media separation (Nakagawa *et al.* 1998). Pollen grains were examined and counted using a Zeiss Axiostar Plus microscope. Identifications were achieved through comparison with reference material from the Environmental and Geographical Science department at the University of Cape Town and published material (Van Zinderen Bakker 1953, Van Zinderen Bakker and Coetzee 1959, Welman and Kuhn 1970, Scott 1982). Charcoal particles were counted together with the pollen grains using the particle count method (Tinner and Hu 2003). Fragments were classified according to size: 10–50 μm and 50–100 μm. The exotic marker *Lycopodium* was added during preparation to enable pollen and charcoal concentrations to be calculated (one tablet per sample; Lund University, Batch # 483216, 18583 ± 1708 spores per tablet). Two samples were omitted due to insufficient pollen concentrations; depths 80 and 138 cm. A total pollen sum of 500 grains, or three slides, was achieved for each sample. On considering the results, it was deemed appropriate to remove the local aquatic components and Amaranthaceae from the pollen sum. Pollen assemblage zones were determined by the use of cluster analysis by the application of CONISS (Constrained Incremental Sum of Squares) (Grimm 1987) – all identified taxa were included in this analysis.

18.3 DATING

Seven organic sediment samples were selected for AMS-^{14}C dating at Beta Analytic Inc (USA) (Table 2). The two top samples returned pre-bomb ages and were calibrated using CALIBomb

Table 1. Sediment description for core EV11.

Depth from surface (cm)	Sediment colour	Munsell notation	Sedimentology
0–16	very dark brown	10YR 4/2	sandy clay
16–36	dark greyish brown	10YR 3/2	fine sand and silt
36–82	very dark greyish brown	2.5Y 4/1	fine sand and silt
82–115	very dark grey	2.5Y 3/1	fine silt, sand and clay
115–154	very dark grey	2.5Y 3/1	fine silt, sand and clay

Table 2. Radiocarbon and calibration details for EV11. Samples were data at Beta Analytic Inc. as indicated by the laboratory identification.

Lab ID (Beta-)	Depth (cm)	[14]C age (yr BP)	1 σ error	Calib. data set	ΔR	2σ cal age range (cal BP)	Median probability (cal BP)
EV11-1	0	–	–	CALIBomb	–	–	−58.7
EV11-2	20	–	–	CALIBomb	–	–	−4.8
EV11-3	40	470	30	Marine13	252 ± 64	340–352 451–526	497
EV11-7	78	1290	30	Marine13	252 ± 64	1072–1192 1206–1268	1171
EV11-4	102	2140	30	Marine13	252 ± 64	2002–2154 2276–2287	2074
EV11-6	134	3300	30	Marine13	252 ± 64	3397–3568	3485
EV11-5	151	3620	30	Marine13	252 ± 64	3726–3751 3791–3794 3820–3980	3880

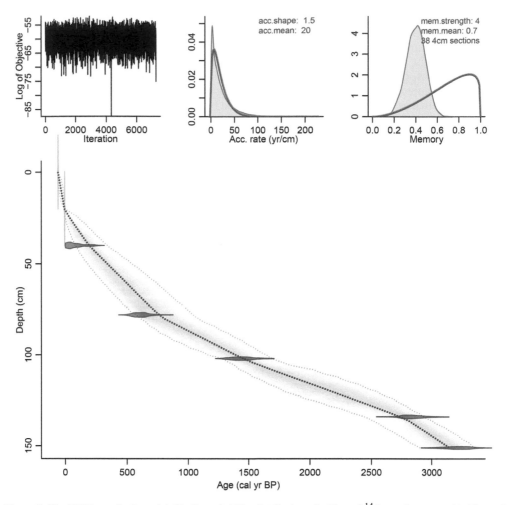

Figure 2. The EV11 age-depth model. The 2σ probability distribution of calibrated ^{14}C ages is presented in blue and the 95% confidence intervals are represented by the grey dotted line. The red line represents the best model according to the weighted mean age at each depth.

(Reimer *et al.* 2004). The remainder of the samples were calibrated with the Marine13 data set (Reimer *et al.* 2013) with a marine reservoir correction of $\Delta R = 252 \pm 64$ (Wündsch *et al.* 2016). The age-depth model (Figure 2) was developed with the R software package Bacon (v2.2) (Blaauw and Christen 2011).

18.4 INTERPRETATION

In total 65 taxa were identified. Pollen concentrations range from 6.26×10^3 grains·g^{-1} near the base, to a maximum of 7.86×10^4 grains·g^{-1} around the top of the assemblage, with an average of 2.72×10^4 grains·g^{-1}. The EV11 record extends from *c.* 3100 cal yr BP to present.

The sequence is divided into five pollen zones labelled EV11-A to E as set out below.

18.4.1 EV11-A (3100 − 2560 cal yr BP; 148−130 cm; 8 samples): Drought resistant − Afrotemperate forest taxa

Elevated levels of drought resistant taxa (i.e. Aizoaceae, *Pentzia*-type and *Euphorbia*) are seen at the onset of record, with *Euphorbia* present throughout the zone. *Stoebe*-type percentages are similarly increased, though decline towards the top of zone. Ericaceae pollen percentages are somewhat higher, at an average of 5.1%, in comparison to the zone above. Afrotemperate forest pollen (primarily *Podocarpus/Afrocarpus*) increase after 3000 cal yr BP, from 7.9 to 12.7% at the top of the zone, with coastal thicket taxa (mostly *Olea, Searsia* and *Euclea*) well represented from the start of the record with values ranging from 12 to 19.8%. The maximum number of large charcoal fragments (50–100 μm), interpreted to be indicative of a large more local fire event, is noted at 2600 cal yr BP. These trends suggest that moisture availability was fairly limited, but rainfall was likely gradually increasing over this period.

18.4.2 EV11-B (2560 − 1560 cal yr BP; 128−106 cm; 11 samples): Amaranthaceae

Zone EV11-B is strongly dominated by the halophytic taxon Amaranthaceae with a distinct increase after 2600 cal yr BP from 13.7 to 43.7%. This could be representative of expanding salt marsh vegetation in response to lower sea levels. Fluctuating forest pollen percentages are noted from around 2400 until 1600 cal yr BP with maximum charcoal concentrations concomitant with the increased presence of *Podocarpus/Afrocarpus* at 2260 cal yr BP. Restionaceae is more prominent in this zone than elsewhere in the record.

18.4.3 EV11-C (1560 − 670 cal yr BP; 104−76 cm; 9 samples): Afrotemperate Forest

Podocarpus/Afrocarpus pollen is more prominent in comparison to the zones below, likely representing a phase of increasing forest expansion. At the same time, fynbos (largely Ericaceae) show a moderate decline towards the top of the zone, from 32.9 to 24.5%, indicating a shift towards a more mesic environment with reduced rainfall seasonality.

18.4.4 EV11-D (670 − 70 cal yr BP/AD 1880; 70−30.5 cm; 13 samples): Afrotemperate forest − drought resistant taxa; *Stoebe*-type

Forest development continues until 360 cal yr BP, with a sudden break in this trend around 620 cal yr BP. After 360 cal yr BP a rather abrupt decline in Afrotemperate forest pollen is noted − from 17.5 to 10.7% in a period of *c*. 50 years. Drought resistant taxa (mostly *Euphorbia*) become more prevalent from this point with a rapid increase in *Stoebe*-type pollen (13.6%) around 130 cal yr BP. These vegetation responses are likely a result of the cooler and drier conditions experienced in the region during the latter part of the so called Little Ice Age (650 − 100 cal yr BP) (Jones *et al.* 2001; Matthews and Briffa 2005).

18.4.5 EV11-E (70 to −30 cal yr BP/AD 1880 to 1980; 26−12 cm; 4 samples): *Pinus*

The most recent part of the record is characterised by the appearance of *Pinus* pollen *c*. AD 1910 marking the onset of the anthropogenic influence in the region.

These results are broadly consistent with the pollen record from EV13. For a fuller discussion on the palaeoclimatic inferences please see (Quick *et al.* 2018).

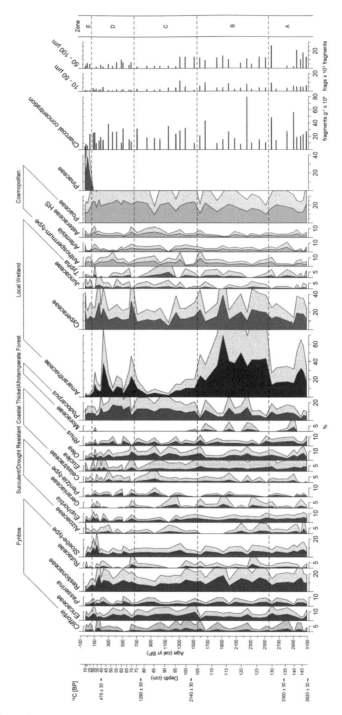

Figure 3. Relative pollen percentage diagram for EV11 organised according to ecological affinity, with charcoal concentrations and charcoal fragment counts. Pollen taxa occurring at less than 5% are not shown. Exaggeration curves are 2x for the taxa presented here. Charcoal and pollen concentrations were calculated in the same manner using *Lycopodium* counts.

DATA AVALIABILITY

The pollen and microcharcoal data presented here is available at:
http://apps.neotomadb.org/Explorer/?datasetid=48877

ACKNOWLEDGEMENTS

This study was funded in part by the German Federal Ministry of Education and Research (BMBF). The investigations were conducted as part of a pilot study for the collaborative project 'Regional Archives for Integrated Investigations' (RAiN), which is embedded in the international research programme SPACES (Science Partnership for the Assessment of Complex Earth System Processes). We also thank the anonymous reviewer and William Gosling for their constructive comments.

REFERENCES

Blaauw, M. and Christen, J.A., 2011, Flexible paleoclimate age-depth models using an autoregressive gamma process. *Bayesian Analysis*, **6**(3), pp.457–474, 10.1214/ba/1339616472.

Faegri, K. and Iversen, J., 1989. *Textbook of Pollen Analysis*, Chichester: John Wiley & Sons Ltd.

Grimm, E.C., 1987, CONISS: A Fortran 77 program for stratigraphically constrained cluster analysis by the method of incremental sum of squares. *Computers & Geosciences*, **13**, pp.13–35, 10.1016/0098-3004(87)90022-7.

Jones, P.D., Osborn, T.J. and Briffa, K.R., 2001, The evolution of climate over the last millennium. *Science*, **292**(5517), pp.662–667, 10.1126/science.1059126.

Matthews, J.A. and Briffa, K.R., 2005, The' Little Ice Age': Re-Evaluation of an Evolving Concept. *Geografiska Annaler. Series A, Physical Geography*, **87**(1), pp.17–36.

Moore, P.D., Webb, J.A. and Collinson, M.E., 1991, *Pollen Analysis* 2nd edition, (Oxford: Blackwell Scientific Publications).

Mucina, L. and Rutherford, M.C., 2006, *The Vegetation of South Africa, Lesotho and Swaziland*, (Pretoria: South African National Biodiveristy Institute, Sterlitzia).

Nakagawa, T., Brugiapaglia, E., Digerfeldt, G., Reille, M., De Beaulieu, J.L. and Yasuda, Y., 1998, Dense media separation as a more efficient pollen extraction method for use with organic sediment/deposit samples: comparison with the conventional method. *Boreas*, **27**, pp.15–24, 10.1111/j.1502-3885.1998.tb00864.x.

Quick, L.J., Chase, B.M., Wundsch, M., Kirsten, K.L., Chevalier, M., Mäusbacher, R., Meadows, M.E. and Haberzettl, T., 2018, A high-resolution record of Holocene climate and vegetation dynamics from the southern Cape coast of South Africa: pollen and microcharcoal evidence from Eilandvlei. *Journal of Quaternary Science*, **33**(5), pp.487–500, 10.1002/jqs.3028.

Reimer, P.J., Bard, E., Bayliss, A., Beck, J.W., Blackwell, P.G., Ramsey, C.B., Buck, C.E., Cheng, H., Edwards, R.L., Friedrich, M., Grootes, P.M., Guilderson, T.P., Haflidason, H., Hajdas, I., Hatté, C., Heaton, T.J., Hoffmann, D.L., Hogg, A.G., Hughen, K.A., Kaiser, K.F., Kromer, B., Manning, S.W., Niu, M., Reimer, R.W., Richards, D.A., Scott, E.M., Southon, J.R., Staff, R.A., Turney, C.S.M. and van der Plicht, J., 2013, IntCal13 and Marine13 Radiocarbon Age Calibration Curves 0–50,000 Years cal BP. *Radiocarbon*, **55**(04), pp.1869–1887, 10.2458/azu_js_rc.55.16947.

Reimer, P.J., Brown, T.A. and Reimer, R.W., 2004, Discussion: Reporting and Calibration of Post-Bomb [14]C Data. *Radiocarbon*, **46**(3), pp.1299–1304, 10.1017/S0033822200033154.

Scott, L., 1982, Late Quaternary fossil pollen grains from the Transvaal, South Africa. *Review of Palaeobotany and Palynology*, **36**(3–4), pp.241–278, 10.1016/0034-6667(82)90022-7.

Tinner, W. and Hu, F.S., 2003, Size parameters, size-class distribution and area-number relationship of microscopic charcoal: Relevance for fire reconstruction. *The Holocene*, **13**(4), pp.499–505, 10.1191/0959683603hl615rp.

Welman, W.G. and Kuhn, L., 1970, *South African Pollen Grains and Spores*, Volume VI, (Amsterdam-Cape Town: AA Balkema).

Wündsch, M., Haberzettl, T., Meadows, M.E., Kirsten, K.L., Kasper, T., Baade, J., Daut, G., Stoner, J.S. and Mäusbacher, R., 2016, The impact of changing reservoir effects on the ^{14}C chronology of a Holocene sediment record from South Africa. *Quaternary Geochronology*, **36**, pp.148–160, 10.1016/j.quageo.2016.08.011.

Van Zinderen Bakker, E.M., 1953, *South African Pollen Grains and Spores*, Volume I, (Cape Town: AA Balkema).

Van Zinderen Bakker, E.M. and Coetzee, J.A., 1959, *South African Pollen Grains and Spores*, Volume III, (Cape Town: AA Balkema).

CHAPTER 19

A *c.* 650 year pollen and microcharcoal record from Vankervelsvlei, South Africa

Nadia du Plessis

Department of Environmental and Geographical Science, University of Cape Town, Rondebosch, South Africa

Brian M. Chase[1]

Institut des Sciences de l'Evolution-Montpellier (ISEM), University of Montpellier, Centre National de la Recherche Scientifique (CNRS), EPHE, IRD, Montpellier, France

Lynne J. Quick

African Centre for Coastal Palaeoscience, Nelson Mandela University, Port Elizabeth, South Africa

Paul Strobel

Physical Geography, Institute of Geography, Friedrich Schiller University Jena, Germany

Torsten Haberzettl

Physical Geography, Institute of Geography and Geology, University of Greifswald, Germany

Michael E. Meadows[2]

Department of Environmental and Geographical Science, University of Cape Town, Rondebosch, South Africa

[1] Other affiliation: *Department of Environmental and Geographical Science, University of Cape Town, Rondebosch, South Africa*
[2] Other affiliation: *School of Geographic Sciences, East China Normal University, Shanghai, China; and College of Geography and Environmental Sciences, Zhejiang Normal University, China*

DOI 10.1201/9781003162766-19

19.1 SITE DETAILS

The Vankervelsvlei wetland is situated along the southern Cape coast of South Africa, about *c.* 5 km inland at an elevation of 152 m asl, surrounded by a lithified aeolian dune of Middle to Late Pleistocene age (Illenberger 1996) (Figure 1A, B). The site falls within the year round rainfall zone, with moisture being delivered from both temperate and tropical climate systems.

Vankervelsvlei is an enclosed and endorheic wetland that is today covered with a floating vegetation mat primarily comprising several species of Cyperaceae, as well as some Bryophytes and Pteridophytes (Irving and Meadows 1997; Quick *et al.* 2016). The surrounding dune(s) are covered by pine plantations (*Pinus*) with scrub forest elements (e.g. *Cassine, Euclea, Kiggelaria*) occupying the area between the plantations and the wetland. Along the wetland edges, vegetation predominantly consists of fynbos pioneer communities (*Erica*, Restionaceae, *Leucadendron* (Proteaceae), *Passerina*) (Quick *et al.* 2016). Northward of Vankervelsvlei, Southern Afrotemperate forest is present in patches, largely represented by *Ocotea bullata, Olea capensis, Afrocarpus falcatus* and *Podocarpus latifolius* (in the pollen record we cannot differentiate between these species, as such these are all labelled *Podocarpus* for the purpose of this paper) (Midgley *et al.* 2004) (Figure 1C).

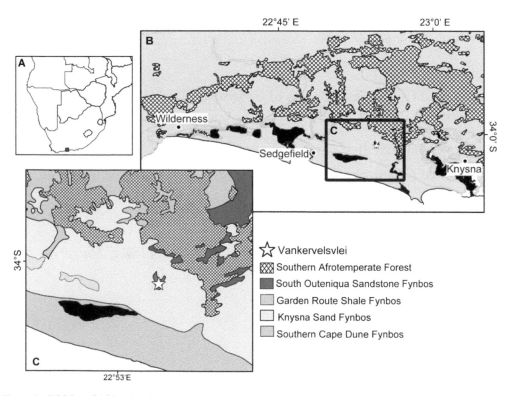

Figure 1. (A) Map of Africa showing the location of the southern Cape coast; (B) A section of the southern Cape coast between the towns of Wilderness and Knysna indicating the location of Vankervelsvlei and the current extent of Afrotemperate forest in the region; (C) The location of the sediment core VVV16 and the contemporary distribution of the dominant vegetation types (Mucina and Rutherford 2006).

Table 1. Sediment description for core sections VVV16-4 and VVV16-1-1-1/2

Core section (VVV)	Depth from surface (cm)	Sediment colour	Other observations
16-4	0–41	brownish	mainly roots, macro plant remains
	41–55	brownish	more sediment than above
16-1-1-1	80–90	dark brown to black	vertical plant roots wetter than above
	90–137	darker than above	vertical plant roots drier than above
	137–147	dark brown to redish	finer roots than above
16-1-1-2	148–183	dark brown to black	fine roots high water content
	183–205	dark brown to redish	fine roots

19.2 SEDIMENT DESCRIPTION AND METHODS

The sediment cores VVV16-4 (55 cm), VVV16-1-1-1 (67 cm) and VVV16-1-1-2 (57 cm) (34°0′46.8″S, 22°54′14.4″E) were both recovered using a UWITEC piston corer. Combined, the core sections are used here to create a 205 cm long record. Due to the nature of this water-body, VVV16-4 was a push core from the surface while VVV16-1 started at 80 cm depth below the surface, resulting in the *c.* 25 cm gap between the two sections.

A total of 24 samples, with a minimum weight of 2 g, were processed using standard paly-nological methods as per Faegri and Iversen (1989) and Moore *et al.* (1991) with adaptations for dense media separation (Nakagawa *et al.* 1998). LacCore's polystaene microsphere pollen spike (0.5 ml per sample; 5.0×10^4 sph/ml $\pm$ 7%) was added to each sample to determine pollen con-centrations. Pollen grains were examined and counted using a Zeiss Axiostar Plus microscope. Identification of pollen were based on comparison with reference material from the Environ-mental and Geographical Science department at the University of Cape Town, and published images (Van Zinderen Bakker 1953, Van Zinderen Bakker and Coetzee 1959, Welman and Kuhn 1970 and Scott 1982). Charcoal particles were counted together with pollen grains using the particle count method (Tinner and Hu 2003), and were classified according to size: 10–100 μm and >100 μm. Counts of 300 terrestrial pollen grains, or three slides, were performed for each sample. Three samples were excluded due to insufficient pollen concentrations; depths 50, 54 and 182 cm.

19.3 DATING

The age-depth model for this section of the VVV16 composite record is based on three AMS ^{14}C ages from organic macro-particles in VVV16-1-1-1 and VVV16-1-1-2 (Table 2) and compliments an age-depth model previously published by Strobel *et al.* (2019) (Figure 2). The samples were dated at the Poznan Radiocarbon Laboratory (Poland). The resultant ages were calibrated using the SHCal13 curve (Hogg *et al.* 2013). The age-depth model was developed with the R software package Bacon (v2.3) (Blaauw and Christen 2011).

Table 2. Radiocarbon and calibration details for VVV16. The designation Poz indicates that the samples were dated at the Poznan Radiocarbon Laboratory. All samples were calibrated using the SHCal12 curve (Hogg *et al.* 2013).

Lab ID (Poz-)	Depth (cm)	Core section (VVV-)	[14]C age (BP)	1σ error	2σ cal age range (cal BP)	Median cal age (cal BP)	*c.* AD
92269	83	16-1-1-1	10	30	21–240	57	1893
92270	143	16-1-1-1	545	30	504–551	527	1423
102442	182	16-1-1-2	60	30	505–555	530	1420

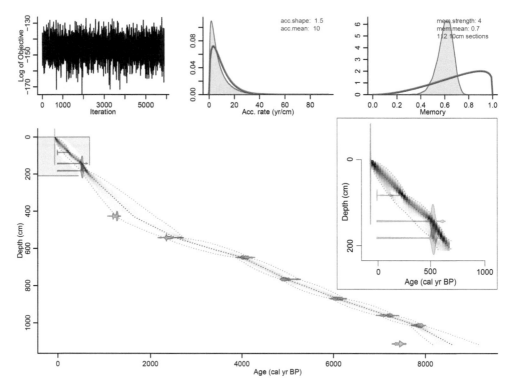

Figure 2. The age-depth model for composite core VVV16. The grey box indicates the part of the sequence presented in this paper (enlarged in the inset); the lower part of the record was previously published by Strobel *et al.* (2019) (10–4 m composite depth). The 2σ probability distribution of calibrated ^{14}C ages is presented in blue and the 95% confidence intervals are represented by the grey dotted line. The dotted red line represents the best model according to the weighted mean age at each depth.

19.4 INTERPRETATION

Fifty-nine different taxa were identified from this section of the VVV record, spanning the period *c.* AD 1300 to present. Due to the core composition, the assemblage was divided into two zones – the first comprising cores VVV16-1-1-1 and 16-1-1-2 and the second being VVV16-4. Pollen concentrations in VVV16-1-1-1/2 vary from 2.34×10^3 to 1.15×10^4 grains·g^{-1}, and in VVV16-4 they range between 2.56×10^3 and 6.20×10^4 grains·g^{-1}.

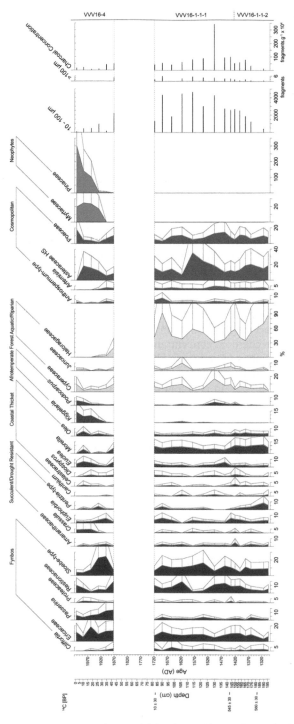

Figure 3. Relative pollen percentage diagram for VVV16-4 and VVV16-1-1-1/2 organized according to ecological affinity, with charcoal concentrations and charcoal fragment counts. Pollen taxa occurring at less than 5% are not shown. Exaggeration curves 2× for the taxa presented here. Charcoal and pollen concentrations were calculated in the same manner using microsphere counts.

19.4.1 VVV16-1-1-1/2 (AD 1300–1720; 196–82 cm; 15 samples): Coastal thicket – Afrotemperate Forest

Coastal thicket is prominent from the start of the record remaining elevated until *c*. AD 1420. This group is mainly represented by *Morella* – likely the dune, scrub and heath species *M. quercifolia* and *M. cordifolia*. Afrotemperate forest pollen percentages are low, increasing after AD 1400 to near maximum values (5.2%) at *c*. AD 1420. Haloragaceae is similarly elevated here. Fynbos pollen percentages increase towards *c*. AD 1460, Restionaceae displays a similar trend. Ericaceae generally follows this pattern with *Stoebe*-type also present. The thermophilous taxon *Pentzia*-type, is present at maximum percentages near the start of the sequence at *c*. AD 1310, followed by a decline in both this taxon and the drought/resistant group, to near minimum values at *c*. AD 1460. *Euphorbia* is present in relatively substantial proportions throughout the record (with peaks at *c*. AD 1430, 1580 and 1690). The sustained presence of *Euphorbia* could be related to enhanced dune movement in the region as opposed to a climatic response. These trends might indicate generally warmer and drier conditions around *c*. AD 1300 with moisture availability increasing towards *c*. AD 1420, and a decline in temperature moving into the Little Ice Age (LIA) – a cooling period identified in the Northern Hemisphere at this time (*c*. AD 1300 – 1850) (Jones *et al.* 2001; Matthews and Briffa 2005).

After *c*. AD 1420, coastal thicket percentages are lower until *c*. AD 1580. Afrotemperate forest taxa are more prominent in the assemblage during this period which could reflect a stage of vegetation succession and/or climatic conditions more conducive to enhanced forest spread. Both Cyperaceae and *Podocarpus* are present at maximum percentages at *c*. AD 1500 which could suggest a preceding period of enhanced moisture availability and/or reduced rainfall seasonality. Maximum charcoal concentrations at *c*. AD 1460 are followed by a notable increase in drier asteraceous fynbos (Asteraceae HS; high-spined) until *c*. AD 1580. This could indicate a progressively drier environment or alternatively a local vegetation response to a large fire event, as inferred from the peak in charcoal concentrations. *Stoebe*-type values increase after *c*. AD 1500, remaining elevated towards the top of this section of the record, *c*. AD 1690, probably reflecting a colder LIA environment.

Throughout this part of the record Cyperaceae and Restionaceae display similar trends, while Haloragaceae (likely *Myriophyllum spicatum*), the most prominent taxon in the record, displays an opposite pattern. These trends are possibly related to changing water levels in Vankervelsvlei.

19.4.2 VVV16-4 (AD 1870–2010/Present; 42–1 cm; 6 samples): *Pinus* and Myrtaceae

Pinus and Myrtaceae (most likely *Eucalyptus*) are first seen in this section, reflecting the influences of the forestry industry that became established in the late AD 1700's and still remains active in the region today.

Kiggelaria, largely absent from the pollen assemblage, becomes more prevalent in this section. As *Kiggelaria* percentages increase, other scrub forest elements, i.e. *Euclea* and *Morella*, decline. This vegetation succession was probably triggered by the establishment of the pine plantations in the area. A significant decline in Haloragaceae is further noted in this section of the record.

ACKNOWLEDGEMENTS

This study was funded by the German Federal Ministry of Education and Research (BMBF). The investigations were conducted as part of a pilot study for the collaborative project 'Regional

Archives for Integrated Investigations' (RAiN), embedded in the international research programme SPACES (Science Partnership for the Assessment of Complex Earth System Processes). We also thank the anonymous reviewer and William Gosling for their helpful comments.

DATA AVALIABILITY

The pollen and microcharcoal data presented here is available at:
http://apps.neotomadb.org/Explorer/?datasetid=48875

REFERENCES

Blaauw, M. and Christen, J.A., 2011, Flexible paleoclimate age-depth models using an autoregressive gamma process. *Bayesian Analysis*, **6**(3), pp.457–474, 10.1214/ba/1339616472.

Faegri, K. and Iversen, J., 1989. *Textbook of Pollen Analysis*, (Chichester: John Wiley & Sons Ltd).

Hogg, A.G., Hua, Q., Blackwell, P.G., Niu, M., Buck, C.E., Guilderson, T.P., Heaton, T.J., Palmer, J.G., Reimer, P.J., Reimer, R.W., Turney, C.S.M. and Zimmerman, S.R.H., 2013, SHCal13 Southern Hemisphere Calibration, 0–50 000 Years Cal BP. *Radiocarbon*, **55**(4), pp.1889–1903, 10.2458/azu_js_rc.55.16783.

Illenberger, W.K., 1996, The Geomorphological Evolution of the Wilderness Dune Cordons, South Africa. *Quaternary International*, **33**, pp.11–20, 10.1016/1040-6182(95)00099-2.

Irving, S.J.E. and Meadows, M.E., 1997, Radiocarbon Chronology and Organic Matter Accumulation at Vankervelsvlei, near Knysa, South Africa. *South African Geographical Journal*, **79**(2), pp.101–105, 10.1080/03736245.1997.9713630.

Jones, P.D., Osborn, T.J. and Briffa, K.R., 2001. The evolution of climate over the last millennium. *Science*, 292(5517), pp.662–667, 10.1126/science.1059126.

Matthews, J.A. and Briffa, K.R., 2005, The' Little Ice Age': Re-Evaluation of an Evolving Concept. *Geografiska Annaler. Series A, Physical Geography*, **87**(1), pp.17–36, 10.1111/j.0435-3676.2005.00242.x.

Midgley, J.J., Cowling, R.M., Seydack, A.H.W. and van Wyk, G.F., 2004, Forest. In *Vegetation of Southern Africa*, edited by Cowling, R.M., Richardson, D.M., and Pierce, S.M., (Cambridge: Cambridge University Press), pp. 278–296.

Moore, P.D., Webb, J.A. and Collinson, M.E., 1991, *Pollen Analysis* 2nd edition (Oxford: Blackwell Scientific Publications).

Mucina, L. and Rutherford, M.C., 2006, *The Vegetation of South Africa, Lesotho and Swaziland*, (Pretoria: South African National Biodiveristy Institute, Sterlitzia).

Nakagawa, T., Brugiapaglia, E., Digerfeldt, G., Reille, M., De Beaulieu, J.-L. and Yasuda, Y., 1998, Dense media separation as a more efficient pollen extraction method for use with organic sediment/deposit samples: Comparison with the conventional method. *Boreas*, **27**, pp.15–24, 10.1111/j.1502-3885.1998.tb00864.x.

Quick, L.J., Meadows, M.E., Bateman, M.D., Kirsten, K.L., Mäusbacher, R., Haberzettl, T. and Chase, B.M., 2016, Vegetation and climate dynamics during the last glacial period in the fynbos-afrotemperate forest ecotone, southern Cape, South Africa. *Quaternary International*, **404**, pp.136–149, 10.1016/j.quaint.2015.08.027.

Scott, L., 1982, Late Quaternary fossil pollen grains from the Transvaal, South Africa. *Review of Palaeobotany and Palynology*, **36**(3–4), pp.241–278, 10.1016/0034-6667(82)90022-7.

Strobel, P., Kasper, T., Frenzel, P., Schittek, K., Quick, L.J., Meadows, M.E., Mäusbacher, R. and Haberzettl, T., 2019, Late Quaternary palaeoenvironmental change in the year-round rainfall zone of South Africa derived from peat sediments from Vankervelsvlei. *Quaternary Science Reviews*, **218**, pp. 200–214, 10.1016/j.quascirev.2019.06.014.

Tinner, W. and Hu, F.S., 2003, Size parameters, size-class distribution and area-number relationship of microscopic charcoal: relevance for fire reconstruction. *The Holocene*, **13**(4), pp.499–505, 10.1191/0959683603hl615rp.

Welman, W.G. and Kuhn, L., 1970, *South African Pollen Grains and Spores*, Volume VI, (Amsterdam-Cape Town: AA Balkema).

Van Zinderen Bakker, E.M., 1953, *South African Pollen Grains and Spores*, Volume I, (Cape Town: AA Balkema).

Van Zinderen Bakker, E.M. and Coetzee, J.A., 1959, *South African Pollen Grains and Spores*, Volume III, (Cape Town: AA Balkema).

CHAPTER 20

Pollen records of the 14th and 20th centuries AD from Lake Tsizavatsy in southwest Madagascar

Estelle Razanatsoa & Lindsey Gillson

Plant Conservation Unit, Department of Biological Sciences, University of Cape Town, South Africa

Malika Virah-Sawmy

Humboldt-Universität, Geography, Berlin, Germany

Stephan Woodborne

iThemba LABS, Johannesburg, South Africa

20.1 SITE DETAILS

Lake Tsizavatsy is located at the north of the Mangoky River, in the southwest of Madagascar (21.780°S, 43.897°E, at 45 m asl, Figure 1). The region has an extensive formation of sandstone eroded from the Precambrian basement and also less eroded tertiary limestones of marls and chalks from marine facies formed during the Eocene (Du Puy and Moat 1996; Moat and Smith 2007) that occur near the coast between 0 and 300 m asl. The region's climate is characterised by semi-arid conditions with pronounced seasonality (Donque 1972) and annual precipitation ranging from 400–600 mm per year near the coast (Stiles 1998). The site's vegetation is classified as tropical dry forests with patches of savanna and woodland savanna and some agricultural areas (Figure 1). These savannas include a small number of tree species belonging to the Arecaceae family that are highly adapted to fire, such as cf. *Medemia nobilis* (Grubb 2003). In addition, there is a patchy sclerophyllous forest which contains taxa such as *Leptolaena* spp. and a high number of dry adapted species such as those belonging to the Euphorbiaceae and Didiereaceae families (Moat and Smith 2007). Lake Tsizavatsy is a shallow lake of approximately 500 m in diameter. The local community reports that the lake recedes to half its area during the dry season. The lake did not exceed a depth of 0.5 m during our field work in September 2015. The studied sediment core provides a record of at least 10 km^2 surrounding the lake representing local to landscape scale of vegetation change. The lake is surrounded by Cyperaceae at the margins and is encompassed by wooded savanna with the presence of degraded and intact dry forest in the wider landscape. Within and around the lake, taxa such as *Phragmites mauritianum* (Poaceae), *Cryptostegia madagascariensi* (Apocynaceae), *Acacia morondavensis* (Fabaceae), *Hyphaene shatan* (Arecaceae), and *Euphorbia* spp. (Euphorbiaceae) are found. The communities surrounding this area comprise foragers and maize horticulturalists (*Mikea*), agropastoralists (*Masikoro*), and fishers (*Vezo*).

DOI 10.1201/9781003162766-20

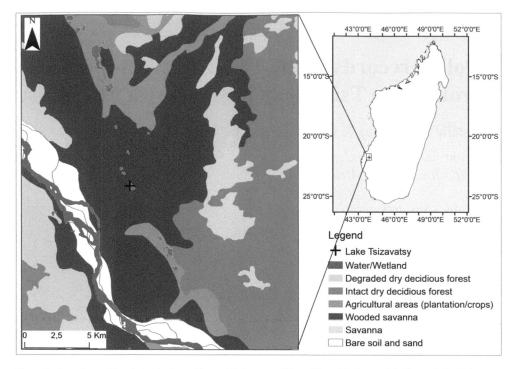

Figure 1. Location of the study site in southwest Madagascar. The red box (black cross) indicates Lake Tsizavatsy. Source: Moat and Smith (2007).

20.2 SEDIMENT DESCRIPTION AND METHODS

A 48 cm sediment core was obtained from the edge of Lake Tsizavatsy using a Russian corer in August 2015. Lithostratigraphic units were described using a modified version of the Troels-Smith classification (Kershaw 1997; Table 1). Twelve sub-samples of 1 cm^3 were taken from the core for pollen extraction following Bennett and Willis (2001). Pollen was counted using a Leica microscope DM750 and identification was based on published pollen references (Gosling *et al.* 2013; Vincens *et al.* 2007). A pollen sum of at least 250 terrestrial taxa was achieved for each sample and relative abundances were calculated and plotted using C2 Version 1.7.7. (Juggins 2003). Cluster analysis was performed using the CONISS method (Analogue package) in R (Grimm 1987; R core Team 2018) to identify zones related to temporal change of the vegetation. Diversity was measured through rarefaction analysis and beta diversity (Birks and Line 1992).

20.3 DATING

Four AMS radiocarbon dates were obtained from sub-samples (4–10 g) of bulk sediment extracted from the Lake Tsizavatsy core (Table 2). Measurements were made at the iThemba LABS facility in Johannesburg (South Africa), Beta analytic Inc, Laboratory in Florida (USA), and 14-CHRONO labs in Belfast (UK) (Table 2). Dates were calibrated using the southern hemisphere calibration curve (Hogg *et al.* 2020) and the age-depth model plotted based on the linear interpolations of age point estimates for depths on weighted means of the dated levels.

Table 1. Lithostratigraphic units of the sediment core from Lake Tsizavatsy.

Depth (cm)	Unit	Troels-Smith	Description	Munsel colour code	Colour
16-0	TSZ-strat 3	As3 Ld1	Clayey and not very humified deposits	2.5Y 3//1 or 5Y 3//1	Very dark grey
28-17	TSZ-strat 2	As2 Ga1 Ld1	Clayey and fine sand deposits, few organic materials	10YR 4//4	dark yellowish brown
48-29	TSZ-strat 1	As2 Ge1 Ld1	Clayey and coarse sand deposits, low humicity deposits	5Y 3//2	Dark olive grey

Table 2. Radiocarbon age of the Lake Tsizavatsy (TSZ) sediment core. Measurements were conducted at 14-CHRONO Centre Queens University (code AUB), iThemba Labs (IT) and Beta Analytic (Beta) and raw radiocarbon dates were calibrated using the southern hemisphere calibration curve (Hogg *et al.* 2020). Uncal. = uncalibrated. Cal. = calibrated. Prob. = probability. BP = before present (AD 1950).

Depth (cm)	Code	Lab ID	Uncal. ^{14}C dates	δ^{13}C ‰	Calibrated ages	Prob. (%)
24-25	TSZ 24	UBA-35667	106 ± 28 yr BP	−18.55	AD 1804–1935	67.4
29-30	TSZ 29	IT-C-1739	640 ± 62 yr BP	−19.8	AD 1378–1406 (572–544 cal. yr BP)	68.3
35-36	TSZ 35	IT-C-1608	680 ± 40 yr BP	−20.0	AD 1289–1396 (661–554 cal. yr BP)	94.5
43-44	TSZ 43	Beta-435564	640 ± 30 yr BP	−18.2	AD 1301–1365 (649–585 cal. yr BP)	94.4

20.4 INTERPRETATION

We identified 65 pollen taxa from 33 families, excluding broken and unidentified grains. The pollen analysis revealed two pollen zones, TSZ_1, and TSZ_2. Radiocarbon dates revealed a hiatus at about 24 cm dividing the core into two parts covering the 14th and the 20th century (Figure 2). The hiatus is probably caused by sediment scouring from the floodwaters of the Mangoky River during wet periods similar to those recorded for example at the Lake Ihotry in the south of the river (Vallet-Coulomb *et al.* 2006). The two pollen zone assemblages cover the periods from AD 1300–1420 (48–25 cm, TSZ_1) and AD 1910–2010 (24–0 cm, TSZ_2). We use these data to compare the 14th century vegetation with that of the 20th century (Figure 3).

20.4.1 Tsizavatsy Zone 1 (TSZ_1): 5 samples, 48–28 cm depth, AD 1300 to 1420 – Palm-*Pandanus* wooded savanna

The TSZ_1 zone is characterised as Palm-*Pandanus* wooded savanna, as indicated by a high abundance of Poaceae, Arcaceae and tree taxa, including *Acacia* and *Pandanus*. The abundance

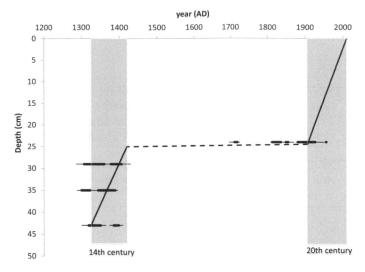

Figure 2. Age depth model of the core from Tsizavatsy adapted from Razanatsoa *et al.* (in press). Horizontal lines indicate the 1-sigma calibration intervals for the radiocarbon dates done at the depth indicated in the y-axis. Dotted lines indicate the 500-year hiatus period from AD 1420 to 1910.

of Poaceae between AD 1300 and 1420 ranged between 45% and 71% (mean *c.* 60%), with Arecaceae (*c.* 5%) and *Acacia* (*c.* 3%) the most abundant trees. Other notable tree taxa are *Pandanus*, which reached a maximum of *c.* 5% around AD 1330 (40 cm) but decreased in relative abundance until around AD 1420 (28 cm), while *Celtis*, *Gnidia*, and *Trema* have respective mean values of 3%, 2%, and 4%. All other herbaceous (Asteraceae, Urticaceae and Amaranthaceae type 1) and arboreal (Phyllanthaceae, Myrtaceae, Malvaceae and Moraceae) taxa were present at low abundances in Zone TSZ_1; while aquatic taxa, mostly Cyperaceae, are equivalent to *c.* 10% of the pollen sum.

20.4.2 Tsizavatsy Zone 2 (TSZ_2): 7 samples, 24–0 cm depth, AD 1910 to 2010 – Palm-xerophytic degraded wooded savanna

The TSZ_2 zone is characterised as Palm-xerophytic degraded wooded savanna, marked by the gradual increase in the abundance of xerophytic, herbaceous, and tree taxa, despite the high abundance of Poaceae. Poaceae pollen is highly variable with a minimum value of 45% at AD 1940 (16 cm), and a maximum value of 71% around AD 2000 (2 cm, mean 58%). The most abundant herbaceous taxon is Asteraceae (mean 4%) which reaches its highest value around AD 2010 (*c.* 10%). All other herbaceous taxa have relatively low abundances. The minimum value of Arecaceae was recorded around AD 2000 (2%) while the maximum value was recorded at AD 1940 (16 cm, *c.* 12%). Some taxa, such as *Celtis* and *Gnidia* also increased in abundance during this period with a mean value respectively of *c.* 3% and *c.* 2%. Dry adapted taxa such as *Securinega* (mean *c.* 2%), peak in this zone with a maximum value recorded at AD 1990 (4 cm, 7%). The aquatic taxa had a mean abundance of *c.* 10%; with *Myriophyllum* (mean *c.* 1%), *Colocasia* (mean *c.* 2%) and Cyperaceae (mean *c.* 6%) being important components.

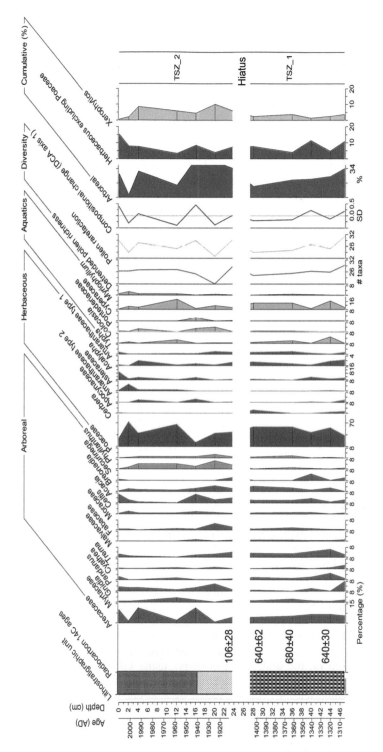

Figure 3. Lake Tsizavatsy summary pollen diagram, showing only taxa with relative abundance >2% of the terrestrial pollen sum. Colours indicate taxonomic groupings: green = arboreal taxa, yellow = xerophytic taxa, orange = herbaceous taxa, and blue = aquatic taxa.

20.5 COMPARISON OF THE COMPOSITIONAL TURNOVER OF THE VEGETATION DURING THE 14TH AND 20TH CENTURY

During the two periods, compositional change was recorded around AD 1360 (36cm, 2.65 SD), AD 1920 (22 cm, 2.6 SD). During the 14th century, taxa such as *Pandanus* and *Trema* were abundant at the beginning but decrease throughout the period which explains the massive turnover around AD 1360 (36 cm). The turnover recorded around AD 1920 (22 cm) is linked to the increasing dominance of pioneer and xerophytic taxa between AD 1910 to 1940 (24–16 cm). These include Asteraceae, *Gnidia,* and *Securinega* in the 20th-century record compared to the 14th century. Overall, the data show a high rate of grass and palm in the surrounding of the lake during both periods. The main difference in the 14th and 20th century pollen records is a slight increase and more variable abundance of arboreal taxa during the latter. An increase in the abundance of xerophytics taxa, possibly linked to decreasing rainfall, was also recorded during the 20th century. Similarly, although tree abundance was higher in the 20th century compared with the 14th century, it began to decrease from the middle of the 20th century. This reduction in trees was associated with a rise in pioneer taxa such as Asteraceae. The pollen data suggest a combined effect of ecosystem degradation and increasing climate impact on the vegetation which requires further investigation.

ACKNOWLEDGEMENT

We would like to thank the editor and the reviewers for providing helpful feedback to improve the manuscript. We would also like to thank the team in Madagascar that helped with the field work and Tsilavo Razafimanantsoa for helping with the map. This project has been funded as part of the Faculty PhD fellowship (University of Cape Town, R.E.) 2015-2018 and the Applied Centre for Climate and Earth Systems Science (ACCESS NRF UID 98018, R.E.) project, the UCT University Research Committee accredited (URC) from the University of Cape Town [URC, 2019-2020] and SASSCAL (Southern African Science Service Centre for Climate Change and Adaptive Land Management) [466418 c/c BIO1056, 2020].

REFERENCES

Bennett, K.D., and Willis, K.J., 2001, Pollen. In *Tracking Environmental Change Using Lake Sediments Volume 3: Terrestrial, Algal and Siliceous Indicators*, edited by Smol, J.P., Birks, H.J.B., and Last, W.M., (Dordrecht, Boston and London: Kluwer Academic Press), pp. 5–32.

Birks, H.J.B., and Line, J.M., 1992, The use of rarefaction analysis for estimating palynological richness from Quaternary pollen analytical data. *The Holocene* **2**, pp. 1–10, 10.1177/095968369200200101.

Donque, G., 1972, The climatology of Madagascar. In *Biogeography and Ecology of Madagascar*, edited by Battistini, R., and Richard-Vindard, G. (Junk, The Hague), pp. 87–144.

Du Puy, D.J., and Moat, J., 1996, A refined classification of the primary vegetation of Madagascar based on the underlying geology: using GIS to map its distribution and to assess its conservation status. In *Proceedings of the International Symposium on the 'Biogeography of Madagascar'* + 3 maps, edited by Lourenço W.R. (Editions de l'ORSTOM, Paris), pp. 205–218.

Gosling, W.D., Miller, C.S. and Livingstone, D.A., 2013, Atlas of the tropical West African pollen flora. *Review of Palaeobotany and Palynology.* **199**, pp. 1–135, 10.1016/j.revpalbo.2013.01.003.

Grimm, E.C., 1987, CONISS: a FORTRAN 77 program for stratigraphically constrained cluster analysis by the method of incremental sum of squares. *Computers and Geosciences*, **13**(1), pp. 13–35, 10.1016/0098-3004(87)90022-7.

Grubb, P. J., 2003, Interpreting some outstanding features of the flora and vegetation of Madagascar. *Perspectives in Plant Ecology, Evolution and Systematics*, **6**(1–2), pp. 125–146, 10.1078/1433-8319-00046.

Hogg, A., Heaton, T., Hua, Q., Palmer, J., Turney, C., Southon, J., Bayliss, A., Blackwell, G.P., Boswijk, J., Ramsey, C.B., Pearson, C., Petchey, F., Reimer, P., Reimer, R., Wacker, L., 2020, SHCal20 Southern Hemisphere Calibration, 0–55,000 Years cal BP. *Radiocarbon*, **62**(4), pp. 759–778, 10.1017/RDC.2020.59.

Juggins, S., 2003, *C2 Version 1.5 User guide: Software for ecological and palaeoecological data analysis and visualization*, (Newcastle upon Tyne: University of Newcastle).

Kershaw, A.P., 1997, A modification of the Troels-Smith system of sediment description and portrayal. *Quaternary Australasia*, **15**(2), pp.63–68.

Moat, J. and Smith, P., 2007, *Atlas de la Végétation de Madagascar*, (Richmond: Royal Botanic Garden-Kew).

Razanatsoa, E., Virah-Sawmy, M., Woodborne, S., and Gillson, L., in press, Subsistence strategies and adaptation of the Mikea foragers from southwest Madagascar in the face of climate change. *Malagasy Nature*.

R Core Team., 2018, *R: A Language and Environment for Statistical Computing* R. Foundation for Statistical Computing, Vienna, https://www.R-project.org.

Stiles, D., 1998, The Mikea hunter-gatherers of southwest Madagascar: Ecology and socioeconomics. *African Studies Monographs*, **19**, pp. 127–148.

Vallet-Coulomb, C., Gasse, F., Robison, L. and Ferry, L., 2006, Simulation of the water and isotopic balance of a closed tropical lake at a daily time step (Lake Ihotry, South-West of Madagascar). *Journal of Geochemical Exploration*, **88**(1–3), pp. 153–156, 10.1016/j.gexplo.2005.08.103.

Vincens, A., Lézine, A.-M., Buchet, G., Lewden, D. and Le Thomas, A., 2007, African pollen database inventory of tree and shrub pollen types. *Review of Palaeobotany and Palynology*. **145**, pp. 135–141, 10.1016/j.revpalbo.2006.09.004.

CHAPTER 21

Modern pollen studies from tropical Africa and their use in palaeoecology

Adele C.M. Julier, Saúl Manzano, Estelle Razanatsoa & Andriantsilavo H.I. Razafimanantsoa

Plant Conservation Unit, University of Cape Town, Cape Town, South Africa

Esther Githumbi[1]

Department of Physical Geography and Ecosystem Science, Lund University, Lund, Sweden

Donna Hawthorne

School of Geography and Sustainable Development, Irvine Building, University of St Andrews, United Kingdom

Glory Oden

Department of Plant and Ecological Studies, University of Calabar, Calabar Nigeria

Lisa Schüler

Department of Palynology and Climate Dynamics, Albrecht-von-Haller Institute for Plant Sciences, Göttingen University, Göttingen, Germany

Monique Tossou

Department of Plant Biology, University of Abomey-Calavi, Benin

Jane Bunting

Department of Geography, Geology and Environment, Faculty of Science and Engineering, University of Hull, United Kingdom

ABSTRACT: Modern pollen studies are valuable for the calibration of pollen records and contribute to the understanding of past vegetation dynamics. Here, we present a qualitative review of available published and (where possible) unpublished modern pollen studies conducted in tropical Africa since pollen analysis emerged as a discipline in the early 20th century. At present,

[1] Other affiliation: *Department of Biology and Environmental Science, Linnaeus University, Kalmar, Sweden*

DOI 10.1201/9781003162766-21

modern pollen rain studies are geographically unevenly distributed across the continent. We found that most countries across tropical Africa have some modern pollen records, with East African countries being particularly well represented in both older and more recent literature. Many countries, arid regions and transitional phytochoria, however, require further study. This review is intended to guide palaeoecologists and palynologists embarking on new studies by bringing together the history of modern pollen studies conducted to date. Targeting new studies to areas where data are currently lacking will help to build a better understanding of modern pollen deposition on the continent. Moreover, we provide recommendations for designing studies so that their results can be used in quantitative modelling techniques for climate or vegetation reconstructions.

21.1 INTRODUCTION

Palaeoecological studies on sedimentary deposits use proxies such as pollen, charcoal, fungal spores and diatoms to reconstruct changes in past environments. These proxies can reveal crucial information about ecosystem responses to climate change, fire occurrence, herbivore activity, and impacts of humans on landscapes on centennial to millennial time scales (Gillson 2015). As such, palaeoecological data are hugely important in understanding the past dynamics of ecosystems and are increasingly being used to inform management decisions and restoration targets (Manzano *et al.* 2020). In order to interpret and validate records from the past, however, these proxies need to be calibrated.

Calibration most commonly takes the form of studies of modern systems; for instance, pollen from surface lake mud samples can be compared to the vegetational composition of the area around those lakes (Sugita 2007a, 2007b). These comparisons allow palaeoecologists to determine which plants produce lots of pollen grains relative to their vegetational abundance and which produce very little. Modern pollen datasets are increasingly being used to calibrate models of pollen dispersal and deposition, allowing for more quantitatively rigorous reconstructions of past vegetation to be developed (Fang *et al.* 2019).

In temperate regions, there has been an abundance of studies conducted on modern pollen rain (Davis *et al.* 2013), and these results have been used to calibrate models of pollen dispersal and deposition (Bunting and Middleton 2005; Jackson and Lyford 1999; Theuerkauf *et al.* 2016). These models have allowed increasing accuracy and precision of vegetation reconstructions, but often rely on extensive datasets, which are relatively rare in the tropics (Bonnefille *et al.* 2004; Gajewski *et al.* 2002; Vincens *et al.* 2006). Gajewski *et al.* is a particularly important work as it used 1170 modern pollen samples available at the time of publication to reconstruct climate across all major phytochoria and climatic zones (albeit with patchy coverage) in Africa. This demonstrates the importance and potential utility of modern pollen work in Africa, and highlights the need to target under-studied geographical regions and vegetation types.

Here, we present qualitative accounts of modern pollen work (published and unpublished, where available) conducted to date in tropical Africa with the aim of summarising key details from the studies, and highlighting those geographical regions and vegetation types that require further targeted work. This review will provide a useful starting point for researchers looking to ground their palaeoecological work in modern data, and will also highlight potential ways forward to improve our interpretations of African palaeoecological data.

21.2 METHODS AND TERMINOLOGY

The geographical extent of this study is the tropics of Africa, the region between 23.5° N and 23.5° S (Figure 1). This is due to the global importance of the tropical regions in terms of

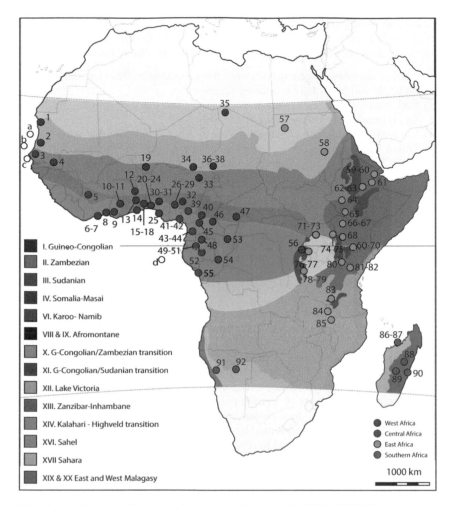

Figure 1. Map showing locations of modern pollen studies and phytochoria (White 1983). Numbers refer to studies in Table 2.

biodiversity and conservation status. Pollen studies were located by searching for country specific papers in, for example, Google Scholar, Microsoft Academic, ResearchGate and Web of Science, using terms such as '[country name] modern pollen' '[country] aeropalynology' and equivalent terms in French. References to unpublished theses were found in existing literature; Table 2 shows whether they are publicly available. Existing modern pollen rain studies from each country are summarised using vegetation type descriptions as they are reported in publications so that the studies retain their local relevance. We then use phytochoria as established by White (1983) (eg. Guineo-Congolian Regional Centre of Endemism) to group studies. We chose this approach to retain the diversity of vegetation classifications used in the studies, whilst also capturing continent-scale patterns.

For each of the reviewed papers, we identified (where possible) the methods used to collect the modern pollen data. Melissopalynological studies are excluded from this review as they are of limited relevance to palaeoecological studies. Sampling techniques vary in how closely they resemble the sediment records, the length of time over which they collect pollen, and which types of pollen they are biased towards collecting. There are a variety of methods used to collect modern pollen data in the tropics (see Jantz *et al.* 2013). For details of the specific methods, refer to the individual studies listed in Table 2. General terms used in this paper are defined as follows:

(i) Surface samples: A general term referring to samples from the uppermost sediment layer, typically soil, peat, lake or marine sediment.
(ii) Soil samples: Samples taken from an area from the top layer of soil (including leaf litter and humus), not from within a water body.
(iii) Core top samples: samples from the top of a sediment core that are assumed to be modern.
(iv) Pollen Traps: Artificial pollen traps come in a variety of formats (Bush, 1992; Gosling *et al.*, 2003; Tauber, 1967) and capture pollen over set periods of time.
(v) Moss/Lichen polsters: Samples of moss and/or lichen are harvested from the ground or trees.
(vi) Glycerol/Silicon oil slides: These are used to sample pollen over a very short period of time by catching grains directly onto a slide
(vii) Filters: Air filters can be used to trap pollen grains; mainly used in offshore sampling.

21.3 WEST AFRICA

West Africa has been the site of a large number of studies (Figure 1). Multiple studies have been carried out in Senegal, Côte d'Ivoire and Togo, (Edorh 1986; Lézine 1988; Ybert 1980), with studies covering the main vegetation types of that region. Nigeria has a large number of pollen rain studies, with many conducted for aeropalynological purposes. Several countries do not appear to have any published (or unpublished but cited elsewhere) modern pollen records, namely; Guinea, Guinea-Bissau, Sierra Leone, Liberia, Gambia, Mali, and Burkina Faso.

An extensive summary of modern pollen data for west Africa was compiled by Lézine *et al.* (2009), who compared modern pollen spectra to climate and vegetation. This work found that biomes are generally well represented by their pollen spectra, which capture the main components of the vegetation and can be used to reconstruct climatic parameters. The exceptions to this were tropical forest biomes, likely due to their high levels of entomophily, and human-impacted dry forest at the Sudanian-Sahelian ecotone.

Off-shore pollen studies have been conducted to complement work on marine cores in the Gulf of Guinea (Calleja *et al.* 1993; Caratini and Cour 1980; Dupont and Agwu 1991; Hooghiemstra *et al.* 1988; Melia 1984; Romero *et al.* 2003). Marine studies off West Africa are synthesised in Hooghiemstra *et al.* (2006) and in this review we focus on continental records.

21.3.1 Mauritania

Only five modern pollen samples have been recorded from Mauritania, (Lézine and Hooghiemstra 1990). These samples are soil samples and record high levels of Amaranthaceae and Brassicaceae pollen, reflecting the Saharan environment. Due to the dry environment, these samples likely represent a maximum of one year of pollen deposition, despite being collected from near to water sources.

21.3.2 Senegal

Modern pollen from soil and mud samples from Senegal was first studied by Lézine (1987) with the majority of samples published in later manuscripts (Lézine 1988; Lézine and Edorh 1991; Lézine and Hooghiemstra 1990). Samples were collected from Sudanian forests and demonstrated that pollen and vegetation from this zone are relatively well correlated (see also Togo). A strong

Sahelian influence was observed in the northernmost samples, typified by taxa such as *Acacia* and *Balanites*. In southern samples, Combretaceae represents the Sudanian vegetation type. Lake surface samples from Lake Guiers (Lézine 1988), demonstrated that the pollen record from the lake, which is located in the Sahelian zone but has a drainage basin extending across vegetation types, is dominated by relatively local taxa such as Poaceae, with a strong Sahelian element (7%) and less of a Sudanian element (2%).

21.3.3 Côte d'Ivoire

Work has been conducted terrestrially and in mangrove swamps in Côte d'Ivoire. Ybert (1975) covered the savannah-forest boundary using monthly sampled pollen traps across transects of forest to savannah. These revealed high levels of temporal variability in pollen rain, but showed that Poaceae dominated in the savannah sections (at percentages of >40%), along with Ulmaceae-Moraceae type and *Lippia multiflora* Moldenke. *Alchornea* was reported as being abundant across the transects. Fredoux (1978) presented modern samples from Mangrove swamps which showed low levels of *Rhizophora* and other species typical of Mangrove swamps, instead being dominated by dry land pollen types (Lézine 1997).

21.3.4 Ghana

Three modern pollen studies are currently known from Ghana. Maley (1983) included some modern surface samples in a preliminary study of Lake Bosumtwi (a site which went on to yield a *c.* 500,000 year record (Miller and Gosling 2014)). These surface samples were taken from small ponds and lakes in the semi-deciduous forest region of the country and showed high levels of *Musanga, Celtis* and *Triplochiton scleroxylon* pollen and low levels of Poaceae. These findings correlated with those of Julier *et al.* (2019) which showed consistently high levels of *Celtis* and *Triplochiton* in pollen trap samples from transects across 100 m by 100 m square vegetation monitoring plots, with traps placed every 20 m along the transect, near to Kumasi, also in the semi-deciduous vegetation type. Traps from a wet evergreen forest plot showed that variability in modern pollen rain was high. Unlike traps from the semi-deciduous plot, however, the wet evergreen traps demonstrated high spatial variability, with traps 20 m apart producing drastically different pollen spectra.

Trap data from the savannah-forest boundary in Ghana (Julier *et al.* 2017) showed that Poaceae abundances of >40% could be considered indicative of a savannah landscape. These two studies (Julier 2017; Julier 2019) were intended to inform palaeoecological interpretations, specifically of Lake Bosumtwi. The data gathered cover a range of vegetation types but lack temporal resolution, with the maximum number of sampled years being three for the forested ecosystems and one for the savannah transitional mosaic.

21.3.5 Togo

The first, and most extensive, study of modern pollen conducted in Togo was by Edorh (1986) who collected 125 soil and river surface samples from a range of vegetation types across Togo, encompassing forest and savannah. It was found that pollen broadly reflected vegetation, but at a local scale. Some of these samples were later published in Lézine and Edorh (1991), who demonstrate that some Sudanian vegetation types such as Combretaceae woodlands and *Isoberlina* woodlands are distinguishable on the basis of their pollen spectra. Anemophilous non-arboreal taxa frequently dominated the signal, and that samples may be biased towards local representation, meaning that a high spatial density of samples is necessary to reflect regional vegetation composition. The pollen spectra were also correlated successfully with rainfall, highlighting the difference between the Guinea-Congolian and Sudanian vegetation zones (Chalié *et al.* 1990).

A study of modern pollen from Mangroves in Togo (Edorh and Afidegnon 2008) demonstrated that the modern pollen spectra, on the whole, represented their parent vegetation types well and were able to distinguish between *Avicennia* and *Rhizophora* mangrove environments.

21.3.6 Bénin

The first known pollen rain study in Bénin is Tossou *et al.* (2012) which focuses on southern Bénin. Surface sediments were taken from the main plant formations: semi-deciduous dense humid forest, floodplain forest, mangrove, swampy forest, swampy meadow and plantation with *Acacia auriculiformis*. Pollen analysis of 38 soil samples identified a total of 126 pollen taxa from 65 families. The most abundant pollen taxa broadly reflect the different plant formations recorded, therefore suggesting that they could be used to reconstruct changes in landcover from palaeo-ecological data in Southern Bénin. A further study (Zanou *et al.* (2020)) focused on the modern pollen rain in the Lama classified forest (south Bénin) from surface sediments. This study allowed a comparison between the pollen content of the sediments and the botanical diversity of the formation. Pollen analysis of 50 sediment samples identified a total of 62 pollen taxa belonging to 31 families and 44 genera. The dominant taxa are Poaceae, *Cassia*, and Combretaceae. The index of similarity between the species inventoried and those contained in the sediments is greater than 50%. Aeropalynological work using traps has also been conducted in Bénin (Tchabi *et al.* 2017; Tossou *et al.* 2016), finding higher pollen content in the air with higher winds and temperature, but lower pollen content with high rainfall, and the most abundant pollen counts in the dry season (November to March).

21.3.7 Niger

The only known study from Niger is an aeropalynological study from Caratini and Frédoux (1988), which presents 27 pollen trap samples with pollen data presented as monthly values from a savannah area in Niamey, south Niger. Around 60% of the annual pollen rain was Poaceae, which peaked in August and September, the rainy season. Cyperaceae pollen was found to peak in October to December and Chenopodiaceae pollen peaked at 5% in December through January, the dry season.

21.3.8 Nigeria

In Nigeria, a large number of aeropalynological studies have been conducted (Abdulrahaman *et al.* 2015; Adekanmbi and Ogundipe 2010; Adeniyi *et al.* 2018; Adeonipekun *et al.* 2016; Agwu 2001 1986; Agwu and Osibe 1992; Alebiousu *et al.* 2018 2018; Ezikanyi *et al.* 2016; Ibigbami and Adeonipekun 2020; Njokuocha 2006). These studies all use some form of pollen trap (modified Tauber traps or 'Gbenga 2' aerofloral samplers) and are primarily for the study of allergenic particles in the atmosphere. They typically record results monthly, with time periods recorded varying from a few months (Abdulrahaman *et al.* 2015) to more than a year (Adeniyi *et al.* 2018). The studies generally found high pollen abundances in the wet season (aside from (Adeniyi *et al.* 2018) who found lower abundances of pollen in the wet season) and high abundances of Poaceae, regardless of their location and year of sampling. In some instances, forest elements such as *Celtis*, *Milicia*, and *Berlinia* are found, indicating long distance transport from forested areas (Ezikanyi *et al.* 2016; Njokuocha 2006). The studies are almost exclusively located in urban areas and are therefore of limited use for palaeoecological reconstructions. Modern pollen work on soil and trap samples from the north of Nigeria showed extremely high levels of Poaceae and Cyperaceae (>95%), but also that modern pollen assemblages primarily reflect their local vegetation, not long-distance transport (Salzmann 2000; Salzmann and Waller 1998).

21.3.9 Chad (including Lake Chad)

The first modern pollen study conducted in Chad was by Maley (1972) who took surface samples from Lake Chad and nearby areas. These results were reproduced as part of Maley's account of the palaeoecology of Lake Chad (Maley 1981). These samples demonstrated the importance of rivers in the long distance transport of pollen, but that the surface samples did capture vegetation changes from the area around the lake. Schulz (1976) also reported modern samples from Chad and compared these with 8000–9000 yr BP samples. The study showed relatively high levels of Poaceae pollen in both modern and fossil samples, but higher levels of montane pollen in the fossil samples. Ybert (1980) used pollen filters to capture pollen rain from November 1971 through December 1972. It was shown that Poaceae is the most abundant taxon, and peaks in September. Amaral *et al.* (2013) used modern sample data from Maley (1972) and other, previously unpublished work to perform biomisation, thereby assigning estimates of mean annual precipitation to time periods covered by a core from Lake Chad covering the period from 6700–500 yr BP.

21.4 CENTRAL AFRICA

In Central Africa, Cameroon, Gabon and the Republic of Congo contain the largest number of studies focusing on modern pollen-vegetation relationships (Elenga 1992; Elenga *et al.* 2000; Henga-Botsikabobe *et al.* 2020; Jolly *et al.* 1996; Lebamba *et al.* 2009; Reynaud-Farrera 1995; Tovar *et al.* 2019; Vincens *et al.* 2000). Other countries in the region lack records; the Central African Republic yields just one study (Aleman *et al.* 2012), the Democratic Republic of Congo has a single study spanning the Uganda-Congo border (Beuning and Russell 2004), and Equatorial Guinea has one study of marine surface pollen records, offshore in the Gulf of Guinea (Dupont and Agwu 1991).

21.4.1 Cameroon

Modern pollen studies in Cameroon are currently concentrated in the south and south-east of the country, and focus on pollen-vegetation relationships. Vincens *et al.* (2000) examine the pollen in soil and litter samples from 26 contiguous plots along a forest-savannah transect. They conclude that the two main ecosystems, forest and savannah, are clearly represented in the surface samples at the local scale, with few pollen grains found outside their producing communities. The density of the forest canopy largely inhibits regionally wind-dispersed pollen from entering the forest and in the savannah these values are typically <1% (Vincens *et al.* 2000).

A similar study by Lebamba *et al.* (2009) examines 24 surface samples from the southern Cameroon lowlands, in forest and savannah environments. They demonstrate that major vegetation types can be differentiated using modern pollen percentages, and a distinction made between secondary and mature forests and well drained and hygrophilous forest types, therefore the pollen analysis is a robust way of discriminating between the maturity, structure and composition of Central African forests. This is reinforced by a study of 50 surface samples from the forests and savannahs of southern Cameroon which defined distinct plant communities (Reynaud-Farrera (1995)). Verlhac *et al.* (2018) studied modern soil samples from across an elevational gradient in the Cameroon highlands and found that modern pollen assemblages reconstruct the upper and lower limits of the Afromontane forests well.

Vincens *et al.* (2000) note that there are individual taxa which can be misrepresented in the pollen spectra, including markers of the main community types (e.g. *Albizia* and *Rinorea*). The differences between the observed vegetation and the pollen analysed in surface samples is often a function of the production and dispersal of individual taxa. However, botanical survey methods, such as only recording trees of more than 10 cm in diameter at breast height (DBH) will also influence the record.

A number of modern pollen studies from Cameroon examine the transportation and dispersal of pollen throughout the Sanaga River Basin (Bengo 1996, 1992; Bengo *et al.* 2020) and off-shore in marine surface sediments on the continental shelf (Van Campo and Bengo 2004). Thirty-eight surface samples from the banks of the Sanaga River and twelve of its main tributaries were collected to characterize the origin of the pollen and its mode of transport to the continental shelf. Bengo *et al.* (2020) demonstrate that pollen analysed in the riverbank sediment clearly reflects the surrounding vegetation cover, and transport to the coast of the Gulf of Guinea is essentially through the drainage of large basins. Examining seventy-one modern samples from the continental shelf off the coast of Cameroon (Van Campo and Bengo 2004), they found a dominance of the mangrove pollen type *Rhizophora,* which is located along the shoreline, a high pollen producer and effectively transported by water. They also note the dominance of Poaceae and *Alchornea,* which likely arrive on the shelf by riverine transport from the grasslands of northern Cameroon and riverbanks and forest edges throughout the Sanaga Basin, respectively. Analysing modern surface samples along river basins therefore greatly supports the interpretation of source area of pollen captured in marine and off-shore sediments.

21.4.2 Central African Republic

There is a paucity of modern pollen studies in the Central African Republic. A single study by Aleman *et al.* (2012) investigates modern pollen (identified as arboreal and non-arboreal), phytoliths and stable carbon isotopes in 17 soil samples from a forest-savannah environment. Combining these results with field Leaf Area Index (LAI) measurements, they calibrate a multi-proxy model to reconstruct LAI in palaeo-sequences, providing a long-term estimate of tree cover. This approach provides a useful method for assessing past changes in vegetation cover, and transitions from forest to savannah and *vice versa*.

21.4.3 Equatorial Guinea

There is a lack of records from terrestrial Equatorial Guinea, however, one study exists which examines the distribution of pollen grains within the Gulf of Guinea and the influence of wind and water on pollen transport (Dupont and Agwu 1991), this study is reviewed in Hooghiemstra *et al.* (2006)

21.4.4 Gabon

Gabon has a number of studies incorporating modern pollen samples. An early study by Jolly *et al.* (1996) examined 16 surface soil samples from the rainforest of north-eastern Gabon to test pollen-vegetation relationships along a number of transects. They establish pollen markers for a number forest types, and explore the pollen production of individual taxa (e.g. *Alchornea* and *Macaranga*) which showed the largest pollen abundance. Ngomanda (2005) carried out research in the forests of Gabon, focusing on the past 5000 years, as part of a doctoral dissertation. Later publishing, with others, palynological data from two lake cores, the tops of which represent the modern pollen signal (Ngomanda *et al.* 2007).

Lebamba *et al.* (2009) analysed 57 surface soil and litter samples from the forests of north-eastern and south western Gabon. They demonstrated that major vegetation types can be differentiated by their modern pollen percentages, and distinctions made between different forest types (e.g. secondary and mature forests).

More recently in Gabon, Henga-Botsikabobe (2015) completed a doctoral dissertation examining surface samples and palaeo-cores from Lope National Park, to reconstruct the past forest composition and analyse its response to changing environmental conditions. Henga-Botsikabobe *et al.* (2020) then later published detailed results from 23 tropical marshes within the Lope National Park examining modern pollen-vegetation relationships. Surface samples were collected from 50 random points across each marsh environment and homogenised to obtain one representative sample per marsh. They found that the modern pollen signal is highly influenced by the local marsh vegetation and a very high diversity of forest taxa are recorded, making it possible to quantify changes between adjacent mature, secondary and pioneer forests. Low proportions of Poaceae were found in the surface samples, which is thought to be influenced by their mode of deposition. This challenges common interpretations, which use these proportions as a measure of savannah *vs.* forest. This is a unique study from an alternative depositional context and provides unique insights for interpreting modern samples and core sequences from marsh environments.

21.4.5 Republic of the Congo

Modern palynological studies in the Republic of Congo are concentrated in the South and South West of the country, within the Batéké Plateau and Mayombe massif regions and along the littoral coast. Elenga *et al.* (1994) examine two cores from a swamp in the Batéké Plateaux region of the Congo, within the Guineo-Congolian region (White 1983) to illuminate palynological evidence of past vegetation change. While this study does not focus on modern pollen-vegetation relationships, it utilises insights and results from a previous doctoral study by Elenga (1992) incorporating modern pollen rain from the same region. For example, Elenga *et al.* (1994) comment on the large local pollen production of Gramineae.

Elenga *et al.* (2000) later carry out a more focused study on the pollen-vegetation relationships in the same region. Palynological data was analysed from 12 sites, each comprising 20-30 surface subsamples and compared to forest inventory data. At the local scale, the modern pollen rain largely reflected the forest taxa observed, and due to the dense canopy, only a few grains originated from extra-local sources. There were variations in the pollen production of individual species and Elenga *et al.* (2000) found that certain species, which are characteristic of these forests, are found in low percentages in the surface samples, for instance, species in the families Caesalpiniaceae, Euphorbiaceae and Annonaceae. Equally, a number of taxa may be overrepresented in the pollen spectra due to high pollen productivity (e.g. *Treculia, Syzygium* and *Dacryodes*).

Tovar *et al.* (2019) examined 12 surface samples and a core sequence from a monodominant *Gilbertiodendron dewevrei* (De Wild.) J. Léonard forest in the northern Republic of Congo. The authors found that *Gilbertiodendron* represented between 3.8% and 6% of the pollen count in the surface samples from the monodominant forest patches. The most abundant taxa in these samples were *Lophira alata*-type and *Celtis* spp. No *Gilbertiodendron* pollen was found in samples outside the monodominant patches. These studies provide important insights into the pollen production and dispersal of key Congo Basin taxa which are important for many palaeoecological interpretations in the region.

While currently, modern pollen-vegetation studies are relatively scarce across the Congo Basin, the recent discovery of a large area of peatland in the Cuvette Centrale (Dargie *et al.* 2017) has sparked new research in the region. The peatland represents an important palaeoecological archive, which may holds insights into additional pollen-vegetation relationships in this alternative depositional context.

21.4.6 Democratic Republic of the Congo

There are no modern pollen studies solely within the Democratic Republic of Congo. A single palaeoecological study exists from the Lake Edward Basin, which spans the Ugandan-Congo border (Beuning and Russell 2004). This study examines a lacustrine core to reconstruct the climatic and vegetation history of the area, with reference to a number of modern pollen assemblages collected from crater lakes within Maramagambo forest reserve, on the Ugandan side of the border. While the authors found that the modern pollen assemblages contained taxa similar to those found in the basal metres of the Lake Edward Basin core, the percentages of Poaceae varied, indicating that the basal metres of the core were more similar to the pollen spectra of the semi-deciduous forest to the east of Lake Victoria, potentially representing a comparable modern environment.

21.5 EAST AFRICA

East Africa is where some of the continent's foundational modern pollen work was conducted (Coetzee 1967; Hedberg 1954; Livingstone 1967). An early review of modern pollen work conducted in the tropics included a summary of East African montane work (Flenley 1973). Several countries, however, do not appear to have any studies of modern pollen rain; Eritrea, Somalia and South Sudan. Laseski (1983) analysed 59 surface samples from Uganda, Rwanda, Tanzania, Zambia, Kenya and Tanzania. These samples were used to derive relationships between pollen and rainfall, and are included in analyses such as Gajewski *et al.* (2002) but the original thesis was not available for this review and is therefore not covered in detail here. Vincens (2006) analysed a large number (150) of modern pollen samples across seven phytogeographical regions. These are used in biomisation with a high level of success (82.6% of sites correctly assigned). These studies by Gajewski and Vincens demonstrate the usefulness and potential of accessible, broad-geographical and climatic analytical studies on modern pollen in the tropics.

21.5.1 Sudan

Eleven soil samples were analysed by Ghazali and Moore (1998) from an initial set of 30, the majority of which did not contain pollen. These were taken from desert, semi-desert, scrub and grassland ecosystems. Poaceae was, by far, the most abundant pollen type identified in the samples. Non-arboreal pollen dominated generally, which is a true reflection of the nature of the ecosystems sampled. A higher proportion of entomophilous pollen than might usually be expected was found in some samples, including *Acacia* pollen despite no *Acacia* trees being present within a 100 m radius of the sample site. Further work by Ghazali (2002) resulted in seven more samples (including one collected in a moving vehicle) from along a transect in the Nile Valley, encompassing mainly desert ecosystems. The assemblages from these samples were mostly dominated by anemophilous taxa such as Poaceae, although only 8 out of 40 pollen taxa identified were anemophilous. It was noted by the authors that, as in their previous study, a large number of samples of those originally collected did not contain pollen. This is likely due to the highly arid climate and sometimes alkaline soils of Sudan.

21.5.2 Djibouti

Three modern samples, from river sediment, lake surface mud, and soil, from desert sites in Djibouti are presented in Bonnefille *et al.* (1980). These samples are presented for comparison with the results from a core dated to 45,000 yr BP, although only three samples possessed high concentrations of pollen, likely due to arid climate conditions. The modern samples showed higher levels of Cyperaceae and forest taxa, but lower Poaceae than those from the core.

21.5.3 Ethiopia

Ethiopia has been the site of many studies of early hominin evolution, including some pollen work from site Fejej FJ-1 establishing palaeoecological histories for archaeological finds (Mohamed Umer *et al.* 2004).

Bonnefille *et al.* (1987) studied modern pollen samples from the Awash Basin from an elevational range of 500–3000 m asl and four vegetation types; subdesert steppe, riparian, evergreen bushland and montane forest. River and soil samples were taken and used to inform interpretations of the Hadar formation, from between 2.9–3.3 million years ago. These data are used to infer climate conditions during the evolution of early hominins. Another study from Bonnefille *et al.* (1993) found that modern pollen samples separated vegetation communities from different altitudes well, although due to high levels of entomophily, the taxonomic composition was not accurately reflected.

Recently, 20 modern pollen (and phytolith) samples from springs in Awash Valley have been published (Barboni *et al.* 2019) which allowed records from fossil springs to be interpreted. These spring environments are thought to be important environments for early hominins, as constant supplies of water in otherwise fluctuating or arid landscapes, therefore playing an important role in early hominin evolution and dispersal.

A large number of modern pollen samples from Ethiopia (with fewer samples from Kenya and Burundi) were included in studies that aimed to develop transfer functions to reconstruct rainfall (Bonnefille *et al.* 1992; Roeland *et al.* 1988) and temperature in fossil records. It was found that it is possible to reconstruct modern temperatures and precipitation with a high degree of accuracy based on the large and geographically widespread datasets used in the analyses.

21.5.4 Uganda

Hedberg (1954) sampled moss and lichen, predominantly from *Carex* bogs, within the Alpine and Ericaceous belts of the Ruwenzori Mountains in Uganda. It was found that, within the Alpine belt, pollen dispersal is predominantly local. Ericaceous pollen is found to be, although relatively local in its dispersal, a potentially useful indicator taxon. For other taxa, similar conclusions were drawn for these Ugandan samples as for those from Mount Kenya.

Osmaston (1958) conducted palaeoecological work in the same region as Hedberg (1954) and Livingstone (1967) took cores from Lakes Bujuku, Kitandra and Mahoma, whose top-most sediments can be considered surface samples. These core-top samples showed low levels of Poaceae, even in alpine belts, leading Livningstone dismiss Poaceae as an indicator of cooler alpine conditions in the past.

Hamilton (1972) analysed modern samples from Kigezi (largely samples from swamps) and Ruwenzori (various environments) and Mount Elgon (within the 'Montane Forest' and 'Ericaceous and Afroalpine' Belts). These samples demonstrated that, in general, transport of pollen to higher altitudes from low altitudes is more common than in the opposite direction. Flenley later hypothesised that this was due to wind directions generally being up slope in the day and downslope at night, and that pollen is generally released in the day (Flenley 1973). Hamilton used the modern samples to develop lists of pollen types by dispersal propensity, and from these lists made suggestions on the interpretation of pollen diagrams, including the fact that aquatic taxa and taxa that likely represent long distance transport are omitted from pollen sums.

Vincens *et al.* (1997) used modern data from previous studies, as well as including original data from surface samples from their own research and previously unpublished data to investigate pollen rain across a wide environmental and elevational gradient in Uganda. Their study encompassed multiple vegetation types including savannah, forest-savannah mosaic, montane forest, the Ericaceous belt and the alpine belt. They found that modern pollen rain can successfully separate different vegetation types but that characteristic plant taxa of certain ecosystems may be completely lacking from their pollen signals.

In their paper from Mubwindi Swamp, south-west Uganda, Marchant and Taylor (2000) present data of modern pollen spectra recovered from surface samples of sediment. Pollen types encountered were grouped into five categories, according to their level of representivity when compared with abundances of the same taxa in the sampled vegetation: under representative, moderately under representative, representative, moderately over representative, and over representative. The results of this research show that most of the dryland pollen incorporated into swamp sediments is produced by local plants close to the site of deposition. Generally, abundances of arboreal taxa such as *Alchornea*, *Anthocleista*, *Croton*, *Dombeya*, *Ilex*, Myrtaceae, *Newtonia*, *Polyscias*, *Prunus*, *Rapanea* and *Zanthoxylum* in the modern pollen appear to closely reflect the occurrence of their parent taxa within the surrounding vegetation. This close relationship between the representation of the parent taxa in dryland vegetation and the pollen percentage in surface samples allows the incorporation of these taxa in the nonlocal pollen sum (Marchant *et al.* 1997).

21.5.5 Kenya

Hedberg (1954) conducted the first modern pollen analyses for Kenya, using moss and lichen samples taken from the Teleki Valley, Mount Kenya. He also sampled ice from the peak of Mount Kenya for pollen, which was found to contain many pollen grains and spores, assumed to be from the preceding years to decades. *Podocarpus* was found to be abundant in the moss samples, demonstrating long-distance transport of 6-10 km. Samples from the Bamboo forest zone were found to contain little Poaceae pollen, which was attributed to Bamboo's irregular flowering and death after flowering. *Alchemilla* was found to be locally abundant, with poor dispersal. Hedberg also determined that it was not possible to determine *Senecio* forest from its pollen spectrum alone, as similar amounts of '*Compositae: Tubiflorae*' (the Asteraceae pollen type used that included *Senecio*) were found in samples from other vegetation types. Undifferentiated *Lobelia* pollen was found but was not considered to be of any use as an indicator taxon due to the occurrence in the area of various species of *Lobelia* with different environmental preferences. Coetzee (1967) also examined modern pollen samples from Mount Kenya, and drew similar conclusions to Hedberg (1954) around the abundance of *Podocarpus*. Rucina (2000) also notes over-abundance of *Podocarpus* from modern samples from the Aberdare Mountains, south-west of Mount Kenya. Hamilton (1972) noted that the analyses of Coetzee and Herberg possessed relatively low taxonomic resolution. This issue was somewhat addressed by Livingstone (1967) who worked from a better reference collection and published results including core-top samples from Uganda (see Uganda).

Modern pollen on Mount Elgon (which experiences different weather patterns on its southern and eastern flanks, with the southern receiving higher rainfall) was studied by Hamilton and Perrott (1980) who used both soil samples and pollen traps along two elevational transects. Of these transects, one was on the wetter southern side of the mountain, and the other was on the drier, eastern side. Due to their transects being elevational, a number of different vegetation types were sampled; Montane forest, Ericaceae thicket, High altitude grassland, *Alchemilla* scrub, Rocky ground communities, Mires, and agricultural land. Their findings indicated that Montane forest produced the most pollen, and that it was represented in other areas both up and downslope of its extent. Poaceae pollen abundances were found to reflect the local abundance of grasses in the vegetation. Many pollen taxa, including *Schefflera* and Amaranthaceae, were found to disperse very poorly and do not travel far from their parent plants, whereas some, such as *Acalypha*, Urticaceae, *Juniperus*-type and *Podocarpus*-type, demonstrated long-distance transport. Sediment samples taken nearly a decade apart (1967 and 1976) showed very similar pollen assemblages, leading the authors to conclude that surface samples represent longer-term modern pollen deposition well. Vincens (1984) investigated modern pollen from Lake Turkana and concluded that only 7% of pollen in the lake was aerially deposited.

A biome reconstruction using modern pollen data from East Africa used 40 sites from Northern Kenya (Vincens *et al.* 2006), the modern pollen rain was assigned into plant functional types then biomes. The study indicated that modern pollen rain correctly predicted modern vegetation types except in transition zones where the drier/more open vegetation type was predicted. This Vincens (2006) study includes the results and data from previous unpublished studies by the same author. A study running between 2014 and 2015 set up pollen traps in the Mau forest (around Nyabuiyabuyi swamp) and the Amboseli Park at Kimana, Ormakau and Isinet swamps. 30 pollen traps at each site (total 120) were left at each site for a period of 1 year (Githumbi 2017). During the recovery of the traps a year later only at Nyabuiyabuyi were 21 of the 30 traps recovered, at the other sites the recovery rate was less than 20%. The major cause of trap disappearance/selection was wildlife. Due to this, new initiatives in Amboseli are in progress where new traps are to be located in areas that can be continually monitored to ensure survival of the traps.

21.5.6 Burundi

Bonnefille and Riollet (1988) conducted a study of modern pollen as part of work on a core from Kashiru Swamp, in the Burundi highlands. They collected soil samples from a range of different altitudes in the Teza forest, between 2100 and 2500 m asl. These samples were taken across the montane forest-Ericaceous moorland transition, and reflected this vegetational change. Ericaceous pollen did not disperse far from its vegetation zone, and as altitude increased, a sharp decrease in arboreal pollen was observed. *Macaranga* pollen accounts for around 40% of the pollen from the montane forest samples, and the authors interpret this as a sign that these forests may have been previously disturbed. They also note that the trees that are characteristic of the forest (*Entandrophragma, Prunus, Chrysophyllum and Neoboutonia*) are extremely rare in the modern pollen spectra.

21.5.7 Tanzania

Modern pollen work in Tanzania was conducted by Vincens (1987) on Lake Natron. It was established that modern pollen assemblages recovered from the lake were relatively uniform across the lake except from where rivers emptied into the lake. Pollen assemblages from the lake surface samples captured the important vegetation types in the catchment basin, but not in a representative manner.

From a modern pollen-rain study along the elevational gradient on Kilimanjaro Schüler *et al.* (2014) show that it is crucial to establish a modern pollen-rain-vegetation relationship for the calibration and interpretation of a fossil pollen record from mountain sites across a gradient of alpine vegetation types including savannah, lower montane, middle montane and upper montane forest types. On Kilimanjaro, the authors analysed modern pollen-rain on plant family level to derive the elevational forest zone (lower, mid, and upper forest zone as per Hemp (2006)) of the surrounding vegetation. Climatic conditions of the study sites were further assessed and put into relation with pollen-rain and forest vegetation (Schüler 2013) which is important for the interpretation of palaeo-records. However, it was shown that the occurrence of plant families along the altitudinal gradient are differently represented in the modern pollen depending on their various reproduction factors. This can further be quantified as a transfer factor as introduced for montane forest taxa in south west Uganda by Marchant and Taylor (2000). The diversity trend captured in the modern pollen-rain on the elevational gradient on Mt Kilimanjaro reflects the plant diversity in the vegetation. However, Schüler *et al.* (2014) observed differences in the taxa richness as the pollen and fern spore dispersal seems to be strongly influenced by the regional wind patterns.

Ivory and Russell (2016) collected modern pollen samples from Lake Tanganyika, which showed consistently high levels of Poaceae pollen (>60% in all samples) and low levels (<6%) of forest taxa even when intact forest was present on the shoreline very close to the sampling site. They conclude that modern pollen assemblages in Lake Tanganyika do reflect, to some extent, the shoreline vegetation, and that depth is not a significant factor in the assemblage composition.

21.5.8 Malawi

Modern pollen samples in Malawi were collected by Meadows (1984), who used a combination of pollen traps and soil samples on the Nyika Plateau. Pollen traps placed at ground level and one metre height were compared and found to produce extremely similar assemblages. The author found that pollen trap and surface samples produced very similar pollen spectra, even with the amount of re-worked pollen being similar. Re-worked, in this instance, implied re-floated into the atmosphere. The three vegetation types investigated in the study, Dambo (shallow wetlands), Grassland and Forest were able to be distinguished by their pollen spectra. Some pollen taxa were found to be 'well-dispersed', including Ericaceae (in contrast to other studies from East Africa eg. (Hamilton 1972)), Poaceae and Cyperaceae. Further samples were collected and analysed from surface samples of Lake Malawi itself by DeBusk (1995, 1997), who found that high levels of mixing, and different inputs (e.g. wind versus riverine) led to a very coarse signal within the lake, from which it was not possible to distinguish distinct pollen assemblages corresponding to vegetation types on the shores.

21.6 SOUTHERN AFRICA

Within the tropics of Southern Africa, a relatively large number of studies have been carried out in some countries, including in Madagascar and South Africa. Angola, Botswana, Mozambique and Zimbabwe lack any modern pollen studies in their tropical regions.

21.6.1 Madagascar

Modern pollen samples from Madagascar mostly covered the sub-humid forest grassland mosaic of Central Highlands with little to no representation of the dry deciduous forest, savannah wood-land, and the spiny thickets in the south western region. Early surface sample analysis for palaeoecological purposes was conducted by Burney (1988) based on 13 perennial lakes and bogs in the sub-humid forest grassland mosaic of Central Highlands. These wetlands were sampled directly, by pushing a vial into the surface of the bog using various devices, depending on the water depth and type of sediment (Burney 1988). Based on the rate between arboreal pollen (AP) and non-arboreal pollen (NAP) distinction of taxa associated with grasslands (with grass pollen ≥55%), forest/grassland mosaic (grass pollen between 22–38%), and montane shrub grassland (Ericaceae pollen ≥40%) were identified. The presence of Ericaceae-dominated montane shrub grasslands and the forest formations of the northern and eastern parts of the island were recorded. Along these, ecosystems such as riparian woodlands on the lakeshores, patches of montane were also represented as the case of the sample collected at Ankazominady in the central highlands. These findings enabled interpretations of palaeoecological work and established at least a qualitative distinction between Malagasy paleoenvironments in the region. However, Burney suggested the need for larger modern pollen datasets and the inclusion of more phytogeographic information combined with multivariate statistical methods for better interpretations.

Straka (1991) published 89 samples from mosses and surface sediments within which 60 samples were from Madagascar. Straka's samples represented seven vegetation types. The humid forest included seven samples with three samples in the littoral forests, three in the humid forest at low altitude, and one in the secondary forest called *Savoka*. However, most of the samples were collected in the subhumid forest in the Central Highlands with a total of 48 samples. The ericoid thickets were represented with four samples which were characterised by a high percentage of Ericaceae/Vaccinaceae (45–98%) with low diversity or with a moderate percentage of Ericaceae/Vaccinaceae combined with Asteraceae type Aster (30%). For all the samples analysed by Straka, only one from the Spiny thicket in the south of the island was collected. Yet, the pollen results presented taxa typical of this vegetation such as type Euphorbia (24,5%), type Acacia and they even found one pollen grain of an *Alluaudia ascendens* (Drake) Drake (Didiereaceae) and also a high percentage of Asteraceae (52%) (Straka 1991). Straka's analysis was however only qualitative and more advanced quantitative analysis are needed.

After almost three decades, seven samples from the top of sediment cores and surface samples were collected and analysed by Razafimanantsoa (2015) from the Northwest dry deciduous forest in Madagascar. These samples represented taxa from different soil types within the dry deciduous forest, but also the degraded dry forest and savannah ecosystems. Results showed the presence of high pollen rain concentration in the dry forest and savannah ecosystem compared to the degraded forest (Razafimanantsoa 2015). In the pollen results, the dry deciduous forest is characterised by *Commiphora* and *Dalbergia* as well as Rubiaceae type, while Poaceae and *Ziziphus* were abundant in the savannah. The degraded forest was on the other hand characterised by the abundance of *Eucalyptus* sp. These initial surface sample results by Burney (1988), Straka (1991) and Razafimanantsoa (2015) in addition to recently collected samples (fieldwork session in 2015, 2016 and 2019) from the succulent woodland, dry deciduous and sub-humid forest could form the basis to establish a reference based on the distribution of pollen and their representation in the various vegetation types in Madagascar.

Similar to these surface samples destined for palaeoecological interpretations, several investigations of pollen rain based on samples from spore Brutayert traps are important tools to help for the interpretation of past pollen record. Analysis of pollen samples from the Capital of Madagascar in the Central Highlands was first published as a thesis in the early 80' covering the period of 1979, 1980, 1981 (Rajeriarison 1983). This study found that the pollen in the region contains a high percentage of Poaceae which were interpreted by Straka (1991) to have reflected rice fields and savannah that are abundant in the region (Straka 1991).

21.6.2 Namibia

Namibian modern pollen has been recently studied by Tabares *et al.* (2018) who used soil surface samples to test the correspondence between vegetation and pollen in savannah ecosystems. They found that pollen is able to separate out distinct vegetation types such as *Acacia* woodland and open mixed woodland well, although overall species richness is not reflected well by pollen assemblages. Grazing pressure was found to be indicated by *Alternanthera*, *Tribulus*, and *Limeum* and the co-occurrence of *Dichrostachys*, *Phyllanthus* and *Crotalaria* is an indicator of environmental encroachment.

Hyrax dung middens, one of the few pollen records available from extreme arid environments such as the Namib desert, were investigated by Gil-Romera *et al.* (2006) and included four modern dung samples. These modern samples were found to be more widely variable than their fossilised equivalents, suggesting that they represented shorter time sampling periods than older dung extracted from middens.

21.7 SPATIAL SPREAD OF MODERN POLLEN STUDIES

The spatial distribution of modern pollen studies in tropical Africa is not even, with work being concentrated in certain areas and countries (Figure 1). Table 1 shows a breakdown of number

Table 1. Phytochoria and number of modern pollen studies that include samples from each Phytochoria (White 1983).

Phytochoria (White 1983)	Number of studies including samples from this region
I. Guineo-Congolian regional centre of endemism	39
II. Zambezian regional centre of endemism	5
III. Sudanian regional centre of endemism	16
IV. Somalia-Masai regional centre of endemism	14
VI. Karoo-Namib regional centre of endemism	2
VII. Mediterranean regional centre of endemism	1
VIII and IX. Afromontane archipelago-like regional centre of endemism and extreme floristic impoverishment	7
X. Guineo-Congolia/Zambezia regional transition zone.	0
XI. Guinea-Congolia/Sudania regional transition zone.	1
XII. Lake Victoria regional mosaic.	8
XIII. Zanzibar-Inhambane regional mosaic.	0
XIV. Kalahari-Highveld regional transition zone.	0
XVI. Sahel regional transition zone.	11
XVII. Sahara regional transition zone.	5
XIX and XX East and West Malagasy regional centres of endemism	4

of studies by Phytochoria, which demonstrates that within the tropical region of Africa, the Guineo-Congolian region is most represented, followed by the Sudanian region and the Sahel. These regions cover large areas of land, and are also important biologically, harbouring a very large number of species within multiple vegetation types. There are other comparably extensive phytochoria which are much less well represented. The Zambezian region, for instance, has very few studies but covers a large part of southern Africa. The Guineo-Congolian/Zambezian regional transition zone and the Kalahari-Highveld regional transition zone do not have any studies at all, and the Guinea-Congolia/Sudania regional transition zone has just one study.

It is also apparent, from Figure 1 (and Table 2) that large swathes of Africa have few to no studies, including much of central and central Southern Africa. Coastal regions are better represented than inland areas, and studies tend to be clustered in countries where foundational palaeoecological work had taken place, such as Uganda and Cameroon.

21.8 THE FUTURE OF MODERN POLLEN STUDIES IN AFRICA

21.8.1 Geographical and vegetation priority areas

Although the number of modern pollen rain studies in Africa has increased over the past few decades, many countries and vegetation types are still under-investigated (Figure 1). Some of these instances of under-representation, for instance in the Saharan region, are likely due to environmental factors like aridity, meaning that gathering modern pollen samples from soil is impractical and often unsuccessful. Any modern samples that are collected in the course of palaeoecological fieldwork will not necessarily coincide with flowering of ephemeral species that thrive in arid environments, potentially producing pollen that is dispersed rapidly and potentially

Table 2. Table showing studies consulted in this study including their number for Figure 1.

Number	Area	Reference	Data collection	Focus of the Study
	WEST AFRICA			
1	Mauritania	Romero *et al.* (2003)	Pollen traps	Palaeoecology
2	Senegal, Mauritania	Lézine and Hooghiemstra (1990)	Lake surface	Palaeoecology
3	Senegal, Mauritania, Ocean	Lézine and Hooghiemstra (1990)	Soil, river, marine mud surface samples	Palaeoecology
4	Senegal, Togo	Lézine and Edorh (1991)	Soil, river surface samples	Palaeoecology
-	Senegal, Mauritania	Lézine (1987)	Soil and river samples	Palaeoecology
5	Côte d'Ivoire	Ybert (1975)	Pollen traps	Palaeoecology
6	Côte d'Ivoire	Fredoux (1978)		
7	Côte d'Ivoire	Caratini *et al.* (1987)		Palaeoecology
8	Côte d'Ivoire, Chad	Ybert (1980)	Glycerine slides	Aeropalynology
9	Ghana	Julier *et al.* (2019)	Pollen traps	Palaeoecology
10	Ghana	Maley (1983)	Lake surface samples	
11	Ghana	Julier *et al.* (2017)	Pollen traps	Palaeoecology
12	Togo	Chalié *et al.* (1990)	Soil surface	
13	Togo	Edorh (1986)	Soil surface	Palaeoecology
14	Togo	Edorh and Afidegnon (2008)	Soil surface	Palaeoecology
15	Bénin	Zanou *et al.* (2020)	Soil surface	
16	Bénin	Tossou *et al.* (2012)	Soil surface	Palaeoecology
17	Bénin	Tchabi *et al.* (2017)	Pollen traps	Aeropalynology
18	Bénin	Tossou *et al.* (2016)	Pollen traps	Aeropalynology
19	Niger	Caratini (1988)	Pollen traps	Aeropalynology
20	Nigeria	Adeniyi *et al.* (2018)	Pollen traps	Aeropalynology

(continued)

Table 2. Continued.

Number	Area	Reference	Data collection	Focus of the Study
21	Nigeria	Adekanmbi and Ogundipe (2010)	Pollen traps	Aeropalynology
22	Nigeria	Ajikah et al. (2015)	Pollen traps	Aeropalynology
23	Nigeria	Ibigbami and Adeonipekun (2020)	Pollen traps	Aeropalynology
24	Nigeria	Adeonipekun et al. (2016)	Pollen traps	Aeropalynology
25	Nigeria	Agwu (2001)		Aeropalynology
26	Nigeria	Abdulrahaman et al. (2015)	Pollen traps	Aeropalynology
27	Nigeria	Ezikanyi et al. (2016)	Pollen traps	Aeropalynology
28	Nigeria	Agwu (1986)	Soil surface/ Water samples	Palaeoecology
29	Nigeria	Agwu and Osibe (1992)		Aeropalynology
30	Nigeria	Njokuocha (2006)	Pollen traps	Aeropalynology
31	Nigeria	Adekanmbi et al. (2018)	Pollen traps	Aeropalynology
32	Nigeria	Alebiosu et al. (2018)	Pollen traps	Aeropalynology
33	Nigeria	Salzmann (2000)	Pollen traps/ Soil surface	Palaeoecology
34	Nigeria	Salzmann and Waller (1998)	Pollen traps/ Soil surface	Palaeoecology
35	Chad	Schulz (1976)	Pollen traps	Palaeoecology
36	Chad	Maley (1972)	Soil surface	Palaeoecology
37	Chad	Amaral et al. (2013)	Soil surface	Palaeoecology
38	Chad	Maley (1981)	Soil surface	Palaeoecology
–	Various	Melia (1984)	Filters and Marine	Aerobiology, Palaeoecology
–	Various	Hooghiemstra (1988)	Littoral and marine surface	Palaeoecology
–	Various	Lézine et al. (2009)	Various	Palaeoecology
–	Multiple	Dupont (1991)	Marine surface	Palaeoecology
–	Ocean	Calleja et al. (1993)	Filters	Palaeoecology

(continued)

Table 2. Continued.

Number	Area	Reference	Data collection	Focus of the Study
a	Various	Lézine (1997)	Littoral and Marine	Palaeoecology
b	Various	Caratini and Cour (1980)	Filters with silicon oil	Aeropalynology, Palaeoecology

CENTRAL AFRICA

Number	Area	Reference	Data collection	Focus of the Study
39	Cameroon	Verlhac *et al.* (2018)	Soil surface	Palaeoecology
40	Cameroon	Bengo (1992)		Palaeoecology
41	Cameroon	Van Campo (2004)	Soil surface	
42	Cameroon	Reynaud-Farrera (1995)	Marine surface	Palaeoecology Palynology
43	Cameroon	Lebamba *et al.* (2009)	Core top	Palaeoecology
44	Cameroon	Bengo (1996)	Soil surface	Palaeoecology, Palynology
45	Cameroon	Bengo *et al.* (2020)	Soil surface	Palynology, Hydrology
46	Cameroon	Vincens *et al.* (2000)		Modern pollen-vegetation relationships, Palynology
47	Central African Republic	Aleman *et al.* (2012)	Soil surface and superficial litter	Modern pollen-vegetation relationships, Palaeoecology
48	Gabon, Cameroon	Lebamba *et al.* (2009)	Soil surface	Palaeoecology, Palynology
49	Gabon	Ngomanda *et al.* (2007)	Soil surface and superficial litter	Palaeoecology, Palynology
50	Gabon	Henga-Botsikabobe (2015)	Soil surface	Modern pollen-vegetation relationships, Palaeoecology
51	Gabon	Henga-Botsikabobe *et al.* (2020)	Soil surface	Modern pollen-vegetation relationships, Palaeoecology, Palynology
52	Gabon	Jolly *et al.* (1996)	Soil surface	Modern pollen-vegetation relationships

(continued)

Table 2. Continued.

Number	Area	Reference	Data collection	Focus of the Study
-	Gabon	Ngomanda (2005)		
53	Republic of Congo	Tovar *et al.* (2019)	Soil surface	
54	Republic of Congo	Elenga (1992)	Surface samples	
55	Republic of Congo	Elenga *et al.* (2000)	Soil surface	Palaeoecology
56	Democratic Republic of Congo	Beuning *et al.* (2004)	Soil surface and core top	Palaeoecology
d	Marine, Gulf of Guinea	Dupont *et al.* (1998)	Soil surface and core top	Palaeoecology

EAST AFRICA

Number	Area	Reference	Data collection	Focus of the Study
57	Sudan	El Ghazali (2002)	Soil surface	Palacoccology
58	Sudan	El Ghazali and Moore (1998)	Soil surface	Palaeoecology
59	Djibouti	Bonnefille *et al.* (1980)	Soil surface	Palaeoecology
60	Ethiopia	Bonnefille (2004)	Soil surface	Archaeology
61	Ethiopia, Tanzania	Barboni *et al.* (2019)	Soil surface	Palaeoecology
62	Ethiopia	Bonnefille *et al.* (1987)	Soil surface	Palaeoecology
63	Ethiopia, Kenya, Burundi	Roeland *et al.* (1988)	Soil surface	Palaeoecology
64	Ethiopia	Bonnefille *et al.* (1993)	Soil surface	Palaeoecology
65	Ethiopia	Mohamed Umer *et al.* (2004)		Archaeology
66	Kenya	Vincens (1984)		Palaeoecology
67	Kenya	Vincens *et al.* (2006)		Palaeoecology
68	Kenya	Tiercelin *et al.* (1987)		
69	Kenya	Rucina (2000)	Soil surface/ Moss polster	Palaeoecology
70	Kenya, Tanzania	Coetzee (1967)		Palaeoecology
71	Uganda	Vincens *et al.* (1997)	Soil surface	Palaeoecology

(continued)

Table 2. Continued.

Number	Area	Reference	Data collection	Focus of the Study
72	Kenya, Uganda	Hedberg (1954)	Moss/Lichen polsters	Palaeoecology
73	Uganda	Hamilton (1972)	Soil surface	Palaeoecology
74	Uganda, Kenya	Hamilton and Perrott (1980)		Palaeoecology
75	Tanzania, Kenya, Uganda, Ethiopia, Somalia	Bonnefille *et al.* (1992)	Soil surface	Palaeoecology
76	Uganda	Marchant and Taylor (2000)	Soil surface	Palaeoecology
77	Uganda	Marchant *et al.* (1997)	Soil surface	
78	Tanzania	Schüler *et al.* (2014)	Pollen traps	Palaeoecology
79	Tanzania	Ivory and Russell (2016)	Soil surface	Palaeoecology
80	Tanzania	Vincens (1987)	Core top	Palaeoecology
81	Burundi	Bonnefille and Riollet (1988)	Soil surface	Palaeoecology
82	Tanzania	Schüler (2013)		
83	Tanzania	Vincens *et al.* (2006)	Surface samples	Palaeoecology
84	Malawi	Meadows (1984)	Soil surface, pollen trap, core top	Palaeoecology
85	Malawi	DeBusk (1997)	Lake surface	Palaeoecology

SOUTHERN AFRICA

Number	Area	Reference	Data collection	Focus of the Study
86	Madagascar	Razafimanantsoa (2015)	Soil surface	Modern pollen-vegetation relationships
87	Madagascar	Ramavovololona (1986)	Pollen traps	Aeropalynology
88	Madagascar	Rajeriarison (1983)	Pollen traps	Aerobiology
89	Madagascar and Mascareings	Straka (1991)	Soil surface/ moss polster	Modern pollen-vegetation relationships
90	Madagascar	Burney (1988)	Soil surface	Palaeoecology
91	Namibia	Tabares *et al.* (2018)	Soil surface	Palaeoecology
92	Namibia	Gil-Romera *et al.* (2006)	Dung	Palaeoecology

* Note that publications that are inaccessible but whose existence is confirmed by citations, or their inclusion in revision works are in italics.

destroyed due to dryness and heat. This problem of aridity could be addressed by setting artificial pollen traps. Transitional zones between different phytochoria are also under-represented, possibly due to their relatively small geographical extent. These areas should, however, receive attention in future studies due to their potential value in developing understanding of vegetation responses to climate change over time. The transitions between phytochoria such as the Guineo-Congolian region, the Sudanian region, and the Sahel, for instance, are likely to have shifted in the past in response to climate drivers (Salzmann and Hoelzmann 2005; Watrin *et al.* 2007). The pollen assemblages at the boundaries of these phytochoria are, therefore, important for the interpretation of shifts observed in fossil pollen records such as that from Lake Bosumtwi (Miller *et al.* 2016) and marine cores (Dupont and Agwu 1992).

21.8.2 Methodological recommendations for pollen dispersal and deposition modelling

The goal of pollen analysis has since the earliest days of the discipline been the reconstruction of past landcover, but the relationship between pollen assemblage at a point and surrounding vegetation is not simple. There are three broad methods for using modern pollen and vegetation data to improve interpretation of sedimentary pollen records. The first method, biomisation, assigns pollen types to plant functional types, then using the plant functional type spectrum to assign a biome, on the assumption that biomes are persistent features of the vegetation (Jolly *et al.* 1998; Vincens *et al.* 2006). Biomisation results are used in the validation of vegetation models such as Dynamic Global Vegetation Models (Hély *et al.* 2006). The second method, the modern analogue technique, statistically matches fossil pollen assemblages to the nearest modern pollen assemblage, and the past environment then assumed to be the same as the present one (Peyron *et al.* 2006). The third method, pollen-vegetation relationship modelling, uses a model of pollen dispersal and deposition (pdd) to establish a relationship between landcover and pollen assemblage, then applies an inverse form of this to reconstruct past land cover (Duffin and Bunting 2008; Gillson and Duffin 2007; Sugita 2007a, 2007b).

Here, we focus on the pdd model and how studies of modern pollen rain in tropical Africa could improve its applicability and relevance. The pdd modelling approach to landcover reconstruction reconstructs the wider past landscape (Bunting and Middleton 2009; Sugita 2007a, 2007b), by assuming that the relationship between plant and pollen is constant over time. The technique has been developed extensively over the last decade and is becoming a valuable tool for investigations of patterns of and controls on land cover change, especially in the northern temperate zone (e.g. Marquer 2017; Li *et al.* 2020).

Challenges in extending the pdd approach (and the modern analogue approach) to the tropics are two-fold-collecting calibration datasets of modern pollen and associated vegetation data, and the development of appropriate pollen dispersal models. Identifying appropriate trapping techniques in tropical ecosystems can be difficult; many ecosystems lack natural pollen traps such as moss polsters, and some ecosystems are too arid or fire-impacted to allow multi-year trap placement and recovery, which has been shown to be advantageous in tropical systems (Haselhorst *et al.* 2020). Collecting vegetation data for comparison with the modern pollen data is also often a challenge (Bunting *et al.* 2013). Existing vegetation maps may not be at appropriate scale, woody plants may only be recorded above 10 cm DBH, which excludes some that may flower thereby contributing to the pollen signal, and vegetation surveys hampered by difficult terrain and incomplete or challenging taxonomy. Dispersal models often fail to incorporate zoophilous taxa, although reducing the number of taxa used in studies can go some way to solving this issue (Whitney *et al.* 2019).

Despite these challenges, research is underway in tropical and sub-tropical areas to develop means of applying the pdd modelling approach to reconstruction of land cover from pollen records. It has considerable potential in improving the integration of palaeoecology with ecology, archaeology and climate modelling (e.g. Harrison *et al.* (2020)), adding a long term perspective critical to meet socioecological challenges such as planning for climate change adaptation and mitigation.

This review can support the use of pdd modelling in African palynology by showing the range of data available across the continent. Existing pollen datasets can be identified in locations where it may be possible to extract suitable vegetation data from remote sensing and data collected for other purposes, and achieve initial model calibration for major habitats. We also hope to alert colleagues to the potential of modern pollen samples, and therefore encourage study design which can be used for model testing and calibration as well as for the analyst's primary purpose.

21.9 CONCLUSIONS

In this review, we have collated and described modern pollen work conducted in tropical Africa to date. We have provided country-scale summaries of the literature and identified potential areas for future work, including arid regions and transitional regions. If these areas are focussed on in forthcoming work, a more complete understanding of pollen-vegetation relationships in tropical Africa will be developed, which will in turn contribute to improved interpretations of fossil pollen records. Further, targeted modern pollen studies, designed with quantitative modelling approaches in mind will also allow more rigorous interpretations of pollen fossil records to be developed for Africa.

21.10 ACKNOWLEDGEMENTS

We would like to thank Dr. Anne-Marie Lézine and Prof. Henry Hooghiemstra for kindly sharing papers and information that greatly helped this review. Prof. Christelle Hély and an anonymous reviewer also provided helpful and detailed comments that significantly improved the work, for which we are extremely grateful. ACMJ acknowledges funding from the NRF African Origins Platform (117666). SM from the NRF Competitive Programme for Rated Researchers (118538). ER acknowledges The Southern African Science Service Centre for Climate Change and Adaptive Land Management (SASSCAL, 466418 c/c BIO1056) and the University Research Committee accredited (URC) from the University of Cape Town (2020). DH acknowledges funding by CongoPeat, a NERC Large Grant (NE/R016860/1) to Simon L. Lewis.

REFERENCES

Abdulrahaman, A.-A., Aruofor, O.S., Taofik, G., Kolawole, O.S., Olahan, G.S. and Oladele, F.A., 2015, Aeropalynological Investigation of the University of Ilorin, Ilorin, Nigeria. *Journal of Applied Sciences and Environmental Management* **19**, pp. 53–63, 10.4314/jasem. v19i1.7.

Adekanmbi, O., Ogundipe, O., 2010, Aeropalynological Studies of the Univesity of Lagos Campus, Nigeria. *Notulae Scientia Biologicae,* **2**, pp. 34–39, 10.15835/nsb 245393.

Adeniyi, T., Olowokudejo, J.D. and Adeonipekun, P.A., 2018, Annual records of airbone pollen of Poaceae in five areas in Lagos, Nigeria. *Grana*, **57**, pp. 284–291, 10.1080/00173134.2017. 1356865.

Adeonipekun, A.P., Agbalaya, A.E. and Adeniyi, T., 2016, Aeropalynology of Ayetoro-Itele, Ota Southwestern Nigeria: A preliminary study. In: *Human Palaeoecology in Africa: Essays in Honour of M. Adebisi Sowunmi*, edited by Oyelaran, P.A., Alabi, R.A. and Adeonipekun, P.A., pp. 130–155 (Ibadan: PAN and University of Ibadan Press).

Agwu, C.O.C., 2001, A study of Niger Delta environment through air-borne palynomorphs, Port Harcourt, Nigeria. *Palaeoecology of Africa*, **27**, pp. 191–205.

Agwu, C.O.C., 1986. History of climate and vegetation of east central Nigeria as deduced from pollen analysis: preliminary investigations. *Bulletin de liaison de l'Association Sénégalaise pour l'Etude du Quaternaire Africain* **76** pp. 5–8.

Agwu, C.O.C. and Osibe, E.E., 1992, Airborne palynomorphs of Nsukka during the months of Febuary-April, 1990. *Nigerian Journal of Botany,* **5**, pp. 177–185.

Alebiousu, O.S., Adekanmbi, O.H., Nodza, G.I. and Ogundipe, O.T., 2018, Aeropalynological study of two selected locations in North-Central Nigeria. *Aerobiologica*, **34**, pp. 187–202, 10.1007/s10453-017-9506-2.

Aleman, J., Leys, B., Apema, R., Bentaleb, I., Dubois, M.A., Lamba, B., Lebamba, J., Martin, C., Ngomanda, A., Truc, L., Yangakola, J.-M., Favier, C., Bremond, L. and Woods, K., 2012, Reconstructing savanna tree cover from pollen, phytoliths and stable carbon isotopes. *Journal of Vegetation Science* **23**, pp. 187–197, 10.1111/j.1654-1103.2011.01335.x.

Amaral, P.G.C., Vincens, A., Guiot, J., Buchet, G., Deschamps, P., Doumnang, J.-C. and Sylvestre, F., 2013, Palynological evidence for gradual vegetation and climate changes during the African Humid Period termination at 13° N from a Mega-Lake Chad sedimentary sequence. *Climate of the Past* **9**, pp. 223–241, 10.5194/cp-9-223-2013.

Barboni, D., Ashley, G.M., Bourel, B., Arráiz, H. and Mazur, J.-C., 2019, Springs, palm groves, and the record of early hominins in Africa. *Review of Palaeobotany and Palynology* **266**, pp. 23–41, 10.1016/j.revpalbo.2019.03.004.

Bengo, M.D., 1996, *La sédimentation pollinique dans le sud-Caméroun et sur la plateforme marine à l'époque actuelle et au Quaternaire récent?: études des paléoenvironnements*, PhD Thesis, Université de Montpellier 2.

Bengo, M.D., 1992, *Etude palynologique et statistique de quelques echantillons fluviatiles actuels du Bassin de la Sanaga*. D.E.A. Université de Montpellier 2.

Bengo, M.D., Elenga, H., Maley, J. and Giresse, P., 2020, Evidence of pollen transport by the Sanaga River on the Cameroon shelf. *Comptes Rendus. Géoscience,* **352**, pp. 59–72, 10.5802/crgeos.1.

Beuning, K.R.M. and Russell, J.M., 2004, Vegetation and sedimentation in the Lake Edward Basin, Uganda-Congo during the Late Pleistocene and early Holocene. *Journal of Paleolimnology* **32**, pp. 118, 10.1023/B:JOPL.0000025253.51135.c4.

Bonnefille, R., Buchet, G., Friis, I., Kelbessa, E. and Mohammed, M.U., 1993, Modern pollen rain on an altitudinal range of forests and woodlands in South West Africa. *Opera Botanica,* **121**, pp. 71–84.

Bonnefille, R., Chalié, F., Guiot, J. and Vincens, A., 1992, Quantitative estimates of full glacial temperatures in equatorial Africa from palynological data. *Climate Dynamics*, **6**, pp. 251–257, 10.1007/BF00193538.

Bonnefille, R., Gasse, F., Azema, C. and Denefle, M., 1980, Palynologie et interprétation palaéoclimatique de trois niveaux pléitocene supérieur d'un sondage du Lac Abhé (Afar, Territoire de Djibouti). *Mémoires du Muséum national d'Histoire naturelle,* B, **27** pp. 149–164.

Bonnefille, R., Potts, R., Chalié, F., Jolly, D. and Peyron, O., 2004, High-resolution vegetation and climate change associated with Pliocene Australopithecus afarensis. *Proceedings of the National Academy of Sciences of the United States of America* **101**, pp. 12125–12129, 10.1073/pnas.0401709101.

Bonnefille, R. and Riollet, G., 1988, The Kashiru Pollen Sequence (Burundi) Palaeoclimatic Implications for the Last 40,000 yr B.P. in Tropical Africa. *Quaternary Research*, **30**, pp. 19–35, 10.1016/0033-5894(88)90085-3.

Bonnefille, R., Vincens, A. and Buchet, G., 1987, Palynology, Stratigraphy and Palaeoenvironment of a Pliocene Hominid Site (2.9-3.3 M.Y.) at Hadar, Ethiopia. *Palaeogeography, Palaeoclimatology, Palaeoecology* **60**, pp. 249–281, 10.1016/0031-0182(87)90035-6.

Bunting, M.J., Farrell, M., Broström, A., Hjelle, K.L., Mazier, F., Middleton, R., Nielsen, A.B., Rushton, E., Shaw, H. and Twiddle, C.L., 2013, Palynological perspectives on vegetation survey: A critical step for model-based reconstruction of Quaternary land cover. *Quaternary Science Reviews* **82**, pp. 41–55, 10.1016/j.quascirev.2013.10.006.

Bunting, M.J. and Middleton, D., 2005, Modelling pollen dispersal and deposition using HUMPOL software, including simulating windroses and irregular lakes. *Review of Palaeobotany and Palynology* **134**, pp. 185–196, 10.1016/j.revpalbo.2004.12.009.

Bunting, M.J. and Middleton, R., 2009, Equifinality and uncertainty in the interpretation of pollen data: the Multiple Scenario Approach to reconstruction of past vegetation mosaics. *The Holocene* **19**(5), pp. 799–803, 10.1177/0959683609105304.

Burney, D.A., 1988, Modern pollen spectra from Madagascar. *Palaeogeography, Palaeoclimatology, Palaeoecology* **66**, pp. 63–75, 10.1016/0031-0182(88)90081-8.

Calleja, M., Rossignol-Strick, M. and Duzer, D., 1993, Atmospheric pollen content off West Africa. *Review of Palaeobotany and Palynology*, **79**, pp 335–368, 10.1016/0034-6667(93)90029-T.

Caratini, C. and Frédoux, A., 1988, Caractérisation des aérosols désertiques à Niamey (Niger) par leur contenu pollinique. *Institut Français de Pondichéry, Travaux de la section scintifique et technique*, **25**, pp. 251–268.

Caratini, C. and Cour, P., 1980, Aéropalynologie en Atlantique Orientale au large e la Mauritanie, du Sénégal et de la Gambie. *Pollen et Spores* **22**, pp. 245–255.

Caratini, C., Tastet, J.-P., Tissot, C. and Fredoux, A., 1987, Sédimentation palynologique actuelle sur le plateau continental de Côte d'Ivoire. In *Palynologie et Milieux Tropicaux*, edited by Cambon, G., Richard, P. and Suc, J.P., Mémoires et Travaux de l'Institut de Montpellier, 17, pp. 69–100.

Chalié, F., Edorh, T.M., Bonnefille, R., Guiot, J. and Roeland, J.-C., 1990, Signal de la pluviosité dans les données polliniques actuelles du Togo. *Comptes rendus de l'Académie des Sciences, Paris*, 2, **311**, pp. 893–899.

Coetzee, J.A., 1967. Pollen analytical studies in East and Southern Africa. In Palaeoecology of Africa. *Palaeoecology of Africa*, **3**, pp. 11–46.

Dargie, G.C., Lewis, S.L., Lawson, I.T., Mitchard, E.T.A., Page, S.E., Bocko, Y.E. and Ifo, S.A., 2017, Age, extent and carbon storage of the central Congo Basin peatland complex. *Nature* **542** pp. 86–90, 10.1038/nature21048.

Davis, B.A.S., Zanon, M., Collins, P., Mauri, A., Bakker, J., Barboni, D., Barthelmes, A., Beaudouin, C., Bjune, A.E., Bozilova, E., Bradshaw, R.H.W., Brayshay, B.A., Brewer, S., Brugiapaglia, E., Bunting, J., Connor, S.E. de Beaulieu, J.-L., Edwards, K., Ejarque, A., Fall, P., Florenzano, A., Fyfe, R., Galop, D., Giardini, M., Giesecke, T., Grant, M.J., Guiot, Jöel, Jahns, S., Jankovská, V., Juggins, S., Kahrmann, M., Karpińska-Kołaczek, M., Kołaczek, P., Kühl, N., Kuneš, P., Lapteva, E.G., Leroy, S.A.G., Leydet, M., Guiot, José, Jahns, S., Jankovská, V., Juggins, S., Kahrmann, M., Karpiñska-Kołaczek, M., Kołaczek, P., Kühl, N., Kuneš, P., Lapteva, E.G., Leroy, S.A.G., Leydet, M., López Sáez, J.A., Masi, A., Matthias, I., Mazier, F., Meltsov, V., Mercuri, A.M., Miras, Y., Mitchell, F.J.G., Morris, J.L., Naughton, F., Nielsen, A.B., Novenko, E., Odgaard, B., Ortu, E., Overballe-Petersen, M.V., Pardoe, H.S., Peglar, S.M., Pidek, I.A., Sadori, L., Seppä, H., Severova, E., Shaw, H., Święta-Musznicka, J., Theuerkauf, M., Tonkov, S., Veski, S., van der Knaap, W.O., van Leeuwen, J.F.N., Woodbridge,

J., Zimny, M. and Kaplan, J.O., 2013, The European Modern Pollen Database (EMPD) project. *Vegetation History and Archaeobotany,* **22**, pp. 521–530.

DeBusk, G.H., 1997, The distribution of pollen in the surface sediments of Lake Malawi, Africa, and the transport of pollen in large lakes. *Review of Palaeobotany and Palynology,* **97**, pp. 123–153, 10.1016/S0034-6667(96)00066-8.

Debusk, G.H.J., 1995, *Transport and stratigraphy of pollen in Lake Malawi, Africa.* PhD Thesis, Duke University.

Duffin, K.I. and Bunting, M.J., 2008, Relative pollen productivity and fall speed estimates for southern African savanna taxa. *Vegetation History and Archaeobotany* **17**, pp. 507–525, 10.1007/s00334-007-0101-2.

Dupont, L.M. and Agwu, C.O.C., 1992, Latitudinal shifts of forest and savanna in N. W. Africa during the Brunhes chron: further marine palynological results from site M 16415 (9° N 19° W). *Vegetation History and Archaeobotany* **1**, pp. 163–175, 10.1007/BF00191556.

Dupont, L.M. and Agwu, C.O.C., 1991, Environmental control of pollen grain distribution patterns in the Gulf of Guinea and offshore NW-Africa. *Geologische Rundschau* **80**, pp. 567–589, 10.1007/BF01803687.

Edorh, T.M., 1986, *Végétation et pluie pollinique actuelles au Togo.* PhD Thesis; Université d'Aix-Marseille 3.

Edorh, T.M. and Afidegnon, D., 2008, Représentation pollinique de la végétation actuelle des zones humides du Sud-Est du Togo. *Annales des Sciences Agronomiques* **10**, pp. 51–70.

El Ghazali, G.E.B., 2002, Modern Pollen spectra and contemporary Vegetation in South-eastern Sahara, Sudan. *Sudan Journal of Scientific Research,* 81.

El Ghazali, G.E.B. and Moore, P.D., 1998, Modern lowland pollen spectra and contemporary vegetation in the eastern Sahel Vegetation Zone, Sudan. *Review of Palaeobotany and Palynology* **99**, pp. 235–246, 10.1016/S0034-6667(97)00042-0.

Elenga, H., 1992, *Végétation et climat du Congo depuis 24 000 ans B. P: analyse palynologique de séquences sédimentaires du Pays Bateke et du littoral.* PhD Thesis, Université d'Aix-Marseille 3.

Elenga, H., de Namur, C., Vincens, A., Roux, M. and Schwartz, D., 2000, Use of plots to define pollen-vegetation relationships in densely forested ecosystems of Tropical Africa. *Review of Palaeobotany and Palynology* **112**, pp. 79–96, 10.1016/S0034-6667(00)00036-1.

Elenga, H., Schwartz, D. and Vincens, A., 1994, Pollen evidence of late Quaternary vegetation and inferred climate changes in Congo. *Palaeogeography, Palaeoclimatology, Palaeoecology,* **109**, pp. 345–356, 10.1016/0031-0182(94)90184-8.

Ezikanyi, D.N., Nnamani, C.V. and Osayi, E.E., 2016, Relationship between Vegetation and Pollen Spectrum in South East Nigeria. *International Research Journal of Biological Sciences* **5**, pp. 57–66.

Fang, Y., Ma, C. and Bunting, M.J., 2019, Novel methods of estimating relative pollen productivity: A key parameter for reconstruction of past land cover from pollen records. *Progress in Physical Geography: Earth and Environment,* **43**, pp. 731–753, 10.1177/03091333198 61808.

Flenley, J.R., 1973, Use of modern pollen rain samples in the study of the vegetational history of tropical regions. *Symposium of the British Ecological Society* (Oxford: Blackwell).

Fredoux, A., 1978, *Pollens et spores d'espèces actuelles et quaternaires de région périlagunaires de Côte d'Ivoire.* PhD Thesis, Université de Montpellier 2.

Gajewski, K., Lézine, A.-M., Vincens, A., Delestan, A. and Sawada, M., 2002, Modern climate-vegetation-pollen relations in Africa and adjacent areas. *Quaternary Science Reviews* **21**, pp. 1611–1631, 10.1016/S0277-3791(01)00152-4.

Gillson, L., 2015, *Biodiversity, Conservation and Environmental Change: Using palaeoecology to manage dynamic landscapes in the Anthropocene* (Oxford University Press).

Gillson, L. and Duffin, K.I., 2007, Thresholds of potential concern as benchmarks in the management of African savannahs. *Philosophical Transactions of the Royal Society B: Biological Sciences* **362**, pp. 309–319, 10.1098/rstb.2006.1988.

Gil-Romera, G., Scott, L., Marais, E. and Brook, G.A., 2006, Middle-to late-Holocene moisture changes in the desert of northwest Namibia derived from fossil hyrax dung pollen. *The Holocene* **16**, pp. 1073–1084, 10.1177/0959683606069397.

Githumbi, E.N., 2017, *Holocene Environmental and Human Interactions in East Africa*. PhD Thesis, University of York.

Hamilton, A.C., 1972, The interpretation of pollen diagrams from highland Uganda. *Palaeoecology of Africa*, **7**, pp. 45–149.

Hamilton, A.C. and Perrott, R.A., 1980, Modem pollen deposition on a tropical African mountain. *Pollen et Spores* **22**, pp. 437–468.

Harrison, S.P., Gaillard, M.-J., Stocker, B.D., Vander Linden, M., Klein Goldewijk, K., Boles, O., Braconnot, P., Dawson, A., Fluet-Chouinard, E., Kaplan, J.O., Kastner, T., Pausata, F.S.R., Robinson, E., Whitehouse, N.J., Madella, M. and Morrison, K.D., 2020, Development and testing scenarios for implementing land use and land cover changes during the Holocene in Earth system model experiments. *Geoscientific Model Development*, **13**, pp. 805–824, 10.5194/gmd-13-805-2020.

Haselhorst, D.S., Moreno, J.E. and Punyasena, S.W., 2020, Assessing the influence of vegetation structure and phenological variability on pollen-vegetation relationships using a 15-year Neotropical pollen rain record. *Journal of Vegetation Science* **31**, pp. 606–615, 10.1111/jvs.12897.

Hedberg, O., 1954, A Pollen-Analytical Reconnaissance in Tropical East Africa. *Oikos* **5**, pp. 137–166, 10.2307/3565157.

Hély, C., Bremond, L., Alleaume, S., Smith, B., Sykes, M.T. and Guiot, J., 2006, Sensitivity of African biomes to changes in the precipitation regime. *Global Ecology and Biogeography* **15**, pp. 258–270, 10.1111/j.1466-8238.2006.00235.x.

Hemp, A., 2006, Continuum or zonation? Altitudinal gradients in the forest vegetation of Mt. Kilimanjaro. *Plant Ecology*, **184**, pp. 27–42, 10.1007/s11258-005-9049-4.

Henga-Botsikabobe, K..B., 2015, *Fonctionnement écologique et évolution des écosystèmes dans un contexte de mosaïque foret-savane et de présence humaine depuis 4000 BP au Parc National de la Lopé (Gabon)*. PhD Thesis, Paris.

Henga-Botsikabobe, K., Ngomanda, A., Oslisly, R., Favier, C., Muller, S.D. and Bremond, L., 2020. Modern pollen-vegetation relationships within tropical marshes of Lopé National Park (Central Gabon) *Review of Palaeobotany and Palynology* **275**, pp. 104–168, 10.1016/j.revpalbo.2020.104168.

Hooghiemstra, H., Lézine, A.-M., Leroy, S A.G., Dupont, L. and Marret, F., 2006, Late Quaternary palynology in marine sediments: A synthesis of the understanding of pollen distribution patterns in the NW African setting. *Quaternary International* **148**, pp. 29–44, 10.1016/j.quaint.2005.11.005.

Hooghiemstra, H., Maizels, J., Ward, R.G.W., Shackleton, N.J., West, R.G. and Bowen, D.Q., 1988, Palynological records from northwest African marine sediments: A general outline of the interpretation of the pollen signal. *Philosophical Transactions of the Royal Society of London. B, Biological Sciences* **318**, pp. 431–449.

Ibigbami, T. and Adeonipekun, O., 2020, Comparative aeropalynology of two communities in Lagos State, southwestern Nigeria. *Notulae Scientia Biologicae* **12**, pp. 729740, 10.15835/nsb12310768.

Ivory, S.J. and Russell, J., 2016, Climate, herbivory, and fire controls on tropical African forest for the last 60ka. *Quaternary Science Reviews* **148**, pp. 101–114, 10.1016/j.quascirev.2016.07.015.

Jackson, S.T. and Lyford, M.E., 1999, Pollen Dispersal Models in Quaternary Plant Ecology: Assumptions, Parameters, and Prescriptions. *Botanical Review* **65**, pp. 39–75, 10.1007/BF02856557.

Jantz, N., Homeier, J., León-Yánez, S., Moscoso, A. and Behling, H., 2013, Trapping pollen in the tropics – Comparing modern pollen rain spectra of different pollen traps and surface samples across Andean vegetation zones. *Review of Palaeobotany and Palynology* **193**, pp. 57–69, 10.1016/j.revpalbo.2013.01.011.

Jolly, D., Bonnefille, R., Burcq, S. and Roux, M., 1996, Representation pollinique de la forêt dense humide du Gabon, tests statistiques. *Comptes rendus de l'Académie des Sciences, Paris, 2, A* **322**, pp. 63–70.

Jolly, D., Prentice, I.C., Bonnefille, R., Ballouche, A., Bengo, M., Brenac, P., Buchet, G., Burney, D., Cazet, J.-P., Cheddadi, R., Edorh, T., Elenga, H., Elmoutaki, S., Guiot, J., Laarif, F., Lamb, H., Lézine, A.-M., Maley, J., Mbenza M., Peyron, O., Reille, M., Reynaud-Farrera, I., Riollet, G., Ritchie, J.C., Roche, E., Scott, L., Ssemmanda, I., Straka, H., Umer, M., Van Campo, E., Vilimumbalo, S., Vincens, A.and Waller, M., 1998, Biome reconstruction from pollen and plant macrofossil data for Africa and the Arabian peninsula at 0 and 6000 years. *Journal of Biogeography* **25**, pp. 1007–1027, 10.1046/j.1365-2699.1998.00238.x.

Julier, A.C.M., Jardine, P.E., Adu-Bredu, S., Coe, A.L., Duah-Gyamfi, A., Fraser, W.T., Lomax, B.H., Malhi, Y., Moore, S., Owusu-Afriyie, K. and Gosling, W.D., 2017, The modern pollen-vegetation relationships of a tropical forest-savannah mosaic landscape, Ghana, West Africa. *Palynology,* **42**(3), pp. 1–15.

Julier, A.C.M., Jardine, P.E., Adu-Bredu, S., Coe, A.L., Fraser, W.T., Lomax, B.H., Malhi, Y., Moore, S. and Gosling, W.D., 2019, Variability in modern pollen rain from moist and wet tropical forest plots in Ghana, West Africa. *Grana* **58**, pp. 45–62, 10.1080/00173134.2018.1510027.

Laseski, R., 1983, *Modern pollen data and Holocene climate change in eastern Africa.* PhD Thesis, Brown University.

Lebamba, J., Ngomanda, A., Vincens, A., Jolly, D., Favier, C., Elenga, H. and Bentaleb, I., 2009, Central African biomes and forest succession stages derived from modern pollen data and plant functional types. *Climate of the Past* **5**, pp. 403–429, 10.5194/cp-5-403-2009.

Lézine, A.-M., 1997, Evolution of the West African mangrove during the Late Quaternary: A review. *Géographie physique et Quaternaire* **51**, pp. 405–414, 10.7202/033139ar.

Lézine, A.-M., 1988, New pollen data from the Sahel, Senegal. *Review of Palaeobotany and Palynology*, **55**, pp. 141–154, 10.1016/0034-6667(88)90082-6.

Lézine, A.-M., 1987, *Paléoenvironnements végétaux d'Afrique Nord-Tropicale depuis 12 000 B.P.: analyse pollinique de séries sédimentaires continentales (Sénégal-Mauritanie).* PhD Thesis, Université Aix-Marseille 2.

Lézine, A.-M. and Edorh, T.M., 1991, Modern pollen deposition in West African Sudanian environments. *Review of Palaeobotany and Palynology* **67**, pp. 41–58, 10.1016/0034-6667(91)90015-U.

Lézine, A.-M. and Hooghiemstra, H., 1990, Land-sea comparisons during the last glacial-interglacial transition: pollen records from West Tropical Africa. *Palaeogeography, Palaeoclimatology, Palaeoecology* **79**, pp 313–331, 10.1016/0031-0182(90)90025-3.

Lézine, A.-M., Watrin, J., Vincens, A. and Hély, C., 2009, Are modern pollen data representative of West African vegetation? *Review of Palaeobotany and Palynology* **156**, pp. 265–276, 10.1016/j.revpalbo.2009.02.001.

Li, F., Gaillard, M.-J., Cao, X., Herzschuh, U., Sugita, S., Tarasov, P.E., Wagner, M., Xu, Q., Ni, J., Wang, W., Zhao, Y., An, C., Beusen, A.H.W., Chen, F., Feng, Z., Goldewijk, C.G.M.K., Huang, X., Li, Yuecong, Li, Yu, Liu, H., Sun, A., Yao, Y., Zheng, Z. and Jia, X., 2020, Towards quantification of Holocene anthropogenic land-cover change in temperate China: A review in the light of pollen-based REVEALS reconstructions of regional plant cover. *Earth-Science Reviews,* **203**, pp. 103–119, 10.1016/j.earscirev.2020.103119.

Livingstone, D.A., 1967, Postglacial Vegetation of the Ruwenzori Mountains in Equatorial Africa. *Ecological Monographs* **37**, pp. 25–52, 10.2307/1948481.

Maley, J., 1981, *Etudes Palynologiques dans le Bassin du Tchad et Paléoclimatologie de l'Afrique Nord-Tropicale de 30,000 ans a l'Epoque Actuelle* (Paris: ORSTOM).

Maley, J., 1972, La sedimentation pollinique actuelle dans la zone du Lac Tchad. *Pollen et Spores* **14**, pp. 263–397.

Manzano, S., Julier, A.C.M., Dirk-Forbes, C.J., Razafimanantsoa, A.H.I., Samuels, I., Petersen, H., Gell, P., Hoffman, M.T. and Gillson, L., 2020, Using the past to manage the future: the role of palaeoecological and long-term data in ecological restoration. *Restoration Ecology*, **28** 6, pp. 1335–1342, 10.1111/rec.13285.

Marchant, R. and Taylor, D., 2000, Pollen representivity of montane forest taxa in south-west Uganda. *New Phytologist* **146**, pp. 515525, 10.1046/j.1469-8137.2000.00662.x.

Marchant, R., Taylor, D. and Hamilton, A., 1997, Late Pleistocene and Holocene History at Mubwindi Swamp, Southwest Uganda. *Quaternary Research*, **47**, pp. 316–328, 10.1006/qres.1997.1887.

Marquer, L., Gaillard, M.-J., Sugita, S., Poska, A., Trondman, A.-K., Mazier, F., Nielsen, A.B., Fyfe, R.M., Jönsson, A.M., Smith, B., Kaplan, J.O., Alenius, T., Birks, H.J.B., Bjune, A.E., Christiansen, J., Dodson, J., Edwards, K.J., Giesecke, T., Herzschuh, U., Kangur, M., Koff, T., Latałowa, M., Lechterbeck, J., Olofsson, J. and Seppä, H., 2017, Quantifying the effects of land use and climate on Holocene vegetation in Europe. *Quaternary Science Reviews* **171**, pp. 20–37, 10.1016/j.quascirev.2017.07.001.

Meadows, M.E., 1984, Contemporary Pollen Spectra and Vegetation of the Nyika Plateau, Malawi. *Journal of Biogeography* **11**, pp. 223–233, 10.2307/2844641.

Melia, M.B., 1984, The distribution and relationship between palynomorphs in aerosols and deep-sea sediments off the coast of Northwest Africa. *Marine Geology* **58**, pp. 345–371, 10.1016/0025-3227(84)90208-1.

Miller, C.S. and Gosling, W.D., 2014, Quaternary forest associations in lowland tropical West Africa. *Quaternary Science Reviews* **84**, pp. 7–25, 10.1016/j.quascirev.2013.10.027.

Miller, C.S., Gosling, W.D., Kemp, D.B., Coe, A.L. and Gilmour, I., 2016, Drivers of ecosystem and climate change in tropical West Africa over the past ~540 000 years. *Journal of Quaternary Science* **31**, pp. 671–677, 10.1002/jqs.2893.

Mohamed Umer, M., Kaniewski, D. and Renault-Miskovsky, J., 2004, Etude palynologique du site de Fejej FJ-1. In: *Les Sites Préhistoriques de La Région de Fejej, Sud-Omo, Ethiopie, Dans Leur Contexte Stratigraphique et Paléontologique*, editted by de Lumley, H. and Beyene, Y., pp. 173202, (Paris: Association Pour La Diffusion de La Pensée Française, Editions Recherche Sur Les Civilisations).

Ngomanda, A., 2005, *Dynamique des écosystèmes forestiers du Gabon au cours des cinq derniers millénaires*. PhD Thesis, Université de Montpellier 2.

Ngomanda, A., Jolly, D., Bentaleb, I., Chepstow-Lusty, A., Makaya, M., Maley, J., Fontugne, M., Oslisly, R. and Rabenkogo, N., 2007, Lowland rainforest response to hydrological changes during the last 1500 years in Gabon, Western Equatorial Africa. *Quaternary Research*, **67**, pp. 411–425, 10.1016/j.yqres.2007.01.006.

Njokuocha, R.C., 2006, Airborne pollen grains in Nsukka, Nigeria. *Grana* **45**, pp. 73–80, 10.1080/00173130600555797.

Osmaston, H.A., 1958, *Pollen analysis in the study of the past vegetation and climate of Ruwenzori and its neighbourhood*. BSc Thesis, University of Oxford.

Peyron, O., Jolly, D., Braconnot, P., Bonnefille, R., Guiot, J., Wirrmann, D. and Chalié, F., 2006. Quantitative reconstructions of annual rainfall in Africa 6000 years ago: Model-data comparison. *Journal of Geophysical Research: Atmospheres* **111**, 10.1029/2006JD007396.

Rajeriarison, C., 1983, Etude du contenu pollinique de l'atmosphere de la région d'Antananarivo (Madagascar): Bilan d'une année d'observations. *Pollen et Spores* **25**(1), pp. 75–90.

Razafimanantsoa, A.H.I., 2015, *Etude des pluies polliniques de la Région de Boeny et des autres formations végétales malgaches*. MS Thesis, Université d'Antananarivo.

Reynaud-Farrera, I., 1995, *Histoire des paléoenvironnements forestiers du sud-Cameroun à partir d'analyses palynologiques et statistiques de dépôts holocènes et actuels*. PhD Thesis, Universté de Montpellier 2.

Roeland, J.-C., Guiot, J. and Bonnefille, R., 1988, Pollen et reconstruction quantitative du climat. Validation des données d'Afrique orientale. *Comptes rendus de l'Académie des Sciences, Paris* **307**, pp. 1735–1740.

Romero, O.E., Dupont, L., Wyputta, U., Jahns, S. and Wefer, G., 2003, Temporal variability of fluxes of eolian-transported freshwater diatoms, phytoliths, and pollen grains off Cape Blanc as reflection of land-atmosphere-ocean interactions in northwest Africa. *Journal of Geophysical Research: Oceans* **108**(C5), 10.1029/2000JC000375.

Rucina, S.M., 2000, *Pollen vegetation relationships in the Aberdare Mountains, Kenya*. MS Thesis, University of Wales, Aberystwyth.

Salzmann, U., 2000, Are modern savannas degraded forests? – A Holocene pollen record from the Sudanian vegetation zone of NE Nigeria. *Vegetation History and Archaeobotany* **9**, pp. 1–15, 10.1007/BF01295010.

Salzmann, U. and Hoelzmann, P., 2005, The Dahomey Gap: an abrupt climatically induced rain forest fragmentation in West Africa during the late Holocene. *The Holocene* **15**, pp. 190–199, 10.1191/0959683605hl799rp.

Salzmann, U. and Waller, M., 1998, The Holocene vegetational history of the Nigerian Sahel based on multiple pollen profiles. *Review of Palaeobotany and Palynology* **100**, pp. 39–72, 10.1016/S0034-6667(97)00053-5.

Schüler, L., 2013, *Studies on late Quaternary environmental dynamics (vegetation, biodiversity, climate, fire and human impact) on Mt. Kilimanjaro, comparing the dry northern with the wet southern slopes*. PhD Thesis, Georg-August-Universität Göttingen.

Schüler, L., Hemp, A. and Behling, H., 2014, Relationship between vegetation and modern pollen-rain along an elevational gradient on Kilimanjaro, Tanzania. *The Holocene,* **24**, pp. 702–713, 10.1177/0959683614526939.

Schulz, E., 1976, Aktueller Pollenniederschlag in der zentralen Sahara und Interpretationsmoglichkeiten quartarer Pollenspektren. *Palaeoecology of Africa*, **9**, pp. 8–14.

Straka, H., 1991, Palynologia Madagassica et Mascarenica. 2nd partie. Echantillons de surface. *Akademie der Wissenschaften und der Litteratur* (Stuttgart: Franz Steiner Verlag).

Sugita, S., 2007a, Theory of quantitative reconstruction of vegetation I: pollen from large sites REVEALS regional vegetation composition. *The Holocene* **17**, pp. 229–241, 10.1177/0959683607075837.

Sugita, S., 2007b, Theory of quantitative reconstruction of vegetation II: all you need is LOVE. *The Holocene* **17**, pp. 243–257, 10.1177/0959683607075838.

Tabares, X., Mapani, B., Blaum, N. and Herzschuh, U., 2018, Composition and diversity of vegetation and pollen spectra along gradients of grazing intensity and precipitation in southern Africa. *Review of Palaeobotany and Palynology* **253**, pp. 88–100, 10.1016/j.revpalbo.2018.04.004.

Tchabi, F.L., Tossou, G.M., Akoegninou, A. and Trigo, M.M., 2017, Étude aéropalynologique de la commune d'Abomey-Calavi (Benin) au cours de la grande saison des pluies. *Revue Française d'Allergologie* **57**, pp. 308–316, 10.1016/j.reval.2017.03.004.

Theuerkauf, M., Couwenberg, J., Kuparinen, A. and Liebscher, V., 2016. A matter of dispersal: REVEALSinR introduces state-of-the-art dispersal models to quantitative vegetation reconstruction. *Vegetation History and Archaeobotany*, **25**, pp. 541–553, 10.1007/s00334-016-0572-0.

Tossou, G.M., Chabi, L.F., Akoègninou, A., Ballouche, A. and Akpagana, K., 2016, Analyse pollinique de l'atmosphère du campus d'Abomey-Calavi (Bénin). *Revue Française d'Allergologie* **56**, pp. 65–75, 10.1016/j.reval.2015.11.002.

Tossou, G.M., Yedomonhan, H.Y., Adomou, C.A., Akoegninou, A., Akpagana, K., 2012, Dépôt pollinique actuel et végétation du sud du Bénin. *International Journal of Biological and Chemical Sciences* **6**, pp. 1647–1668, 10.4314/ijbcs.v6i4.23.

Tovar, C., Harris, D.J., Breman, E., Brncic, T. and Willis, K.J., 2019, Tropical monodominant forest resilience to climate change in Central Africa: A Gilbertiodendron dewevrei forest pollen record over the past 2,700 years. *Journal of Vegetation Science* **30**, pp. 575–586, 10.1111/jvs.12746.

Van Campo, E. and Bengo, M.D., 2004, Mangrove palynology in recent marine sediments off Cameroon. *Marine Geology*, **208**, pp. 315–330, 10.1016/j.margeo.2004.04.014.

Verlhac, L., Izumi, K., Lézine, A.-M., Lemonnier, K., Buchet, G. and Achoundong, G., Tchiengué, B., 2018, Altitudinal distribution of pollen, plants and biomes in the Cameroon highlands. *Review of Palaeobotany and Palynology,* **259**, pp. 21–28, 10.1016/j.revpalbo.2018.09.011.

Vincens, A., 1987, Recent pollen sedimentation in Lake Natron, Tanzania: a model for the interpretation of fossil spectra in arid region. Sédimentation pollinique actuelle dans le lac Natron, Tanzanie?: un modèle pour l'interprétation de spectres fossiles en région aride. *Sciences Géologiques, bulletins et mémoires,* **40**, pp. 155–165, 10.3406/sgeol.1987.1757.

Vincens, A., 1984, Environnement végétal et sédimentation pollinique lacustre actuelle dans le bassin du lac Turkana (Kenya). *Revue de Paléobiologie*, **vs**, pp. 235–242.

Vincens, A., Bremond, L., Brewer, S., Buchet, G. and Dussouillez, P., 2006, Modern pollen-based biome reconstructions in East Africa expanded to southern Tanzania. *Review of Palaeobotany and Palynology* **140**, pp. 187–212, 10.1016/j.revpalbo.2006.04.003.

Vincens, A., Dubois, M.A., Guillet, B., Achoundong, G., Buchet, G., Kamgang Kabeyene Beyala, V., de Namur, C. and Riera, B., 2000, Pollen-rain-vegetation relationships along a forest-savanna transect in southeastern Cameroon. *Review of Palaeobotany and Palynology* **110**, pp. 191–208, 10.1016/S0034-6667(00)00009-9.

Vincens, A., Ssemmanda, I., Roux, M. and Jolly, D., 1997, Study of the modern pollen rain in Western Uganda with a numerical approach. *Review of Palaeobotany and Palynology* **96**, pp. 145–168, 10.1016/S0034-6667(96)00022-X.

Watrin, J., Lézine, A.-M., Gajewski, K., Vincens, A., 2007, Pollen-plant-climate relationships in sub-Saharan Africa. *Journal of Biogeography* **34**, pp. 489–499, 10.1111/j.1365-2699.2006.01626.x.

White, F., 1983, The vegetation of Africa. (Paris: UNESCO).

Whitney, B.S., Smallman, T.L., Mitchard, E.T., Carson, J.F., Mayle, F.E. and Bunting, M.J., 2019, Constraining pollen-based estimates of forest cover in the Amazon: A simulation approach. *The Holocene* **29**, pp. 262–270, 10.1177/0959683618810394.

Ybert, J.-P., 1980, Le contenu pollinique de l'atmosphère en Côte d'Ivoire et au Tchad. *Grana* **19**, pp. 31–46, 10.1080/00173138009424985.

Ybert, J.-P., 1975, Emissions polliniques actuelles dans la zone du contact forêt-savane en Côte d'Ivoire *Bulletin de la Société Botanique de France* **122**, pp. 251–268, 10.1080/00378941.1975.10835659.

Zanou, A.R.S., Tossa-Dognon, A.D., Batawila, K., Akoegninou, A. and Akpagana, K., 2020, Étude de la pluie pollinique de la forêt classée de la Lama. *Journal de la Recherche Scientifique de l'Université de Lomé (Togo)* **22**, pp. 63–76.

CHAPTER 22

Vegetation response to millennial- and orbital-scale climate changes in Africa: A view from the Ocean

Ilham Bouimetarhan[1]

Faculté des Sciences appliquées, CUAM, Ibn Zohr University, Agadir, Morocco

Lydie Dupont

MARUM – Center for Marine Environmental Sciences, University of Bremen, 28359 Bremen, Germany

Hanane Reddad

L'Ecole Supérieure de Technologie & Faculté des Lettres et des Sciences Humaines, Sultan Moulay Slimane University, Beni Mellal, Morocco

Asmae Baqloul

Faculté des Sciences, Ibn Zohr University, Agadir, Morocco

Anne-Marie Lézine

Laboratoire d'Océanographie et du Climat, Expérimentation et Approche numérique/IPSL, Sorbonne University, CNRS-IRD-MNHN, Paris, France

ABSTRACT: Pollen from deep-sea sedimentary archives have proved to be a particularly useful tool to provide an integrated regional reconstruction of vegetation and climate (temperature, precipitation, and seasonality) on the adjacent continent throughout the Earth's climate history. In this paper, we have compiled marine pollen records from the African margin in order to assess long-term patterns of vegetation changes during climate cycles. We investigate the changing and complex interplay between African climate and high- and low-latitude forcing at orbital and millennial timescales. More importantly, the study of those records has shown the extent of different biomes during the last ten million years covering the Plio-Miocene, glacial-interglacial cycles, as well as eight Heinrich Stadials, the last deglaciation and the Holocene. In the West African records, arboreal pollen expanded during most interglacials and during the early Holocene. The

---

[1] Other affiliation: *MARUM-Center for Marine Environmental Sciences, University of Bremen, 28359 Bremen, Germany*

DOI 10.1201/9781003162766-22

savannah and semi-desert/desert vegetation expanded abruptly during glacials, Heinrich Stadials and the Last Glacial Maximum. Most eastern African pollen records have shown a clear dominance of arboreal taxa throughout all climate cycles except during MIS32 suggesting that a decoupling between eastern and western Africa took place. However, the scarcity of eastern African marine pollen records hampers a reliable comparison between these two sectors.

22.1 INTRODUCTION

The distribution and composition of the African vegetation is largely controlled by climate primarily through temperature ranges, radiation and precipitation (e.g. Box 1981; Holdridge 1947; Woodward and Williams 1987; Woodward and McKee 1991). In turn, vegetation can exert an important influence on surface energy fluxes and the hydrological cycle through alteration of the surface albedo and biogeochemical processes, thus affecting climate locally (Denman *et al.* 2007; Pielke *et al.* 1998). In Africa, a continent that represents around 20% of the Earth's surface, and where climate and vegetation data are sparse, our understanding of the vegetation-climate interaction and its consequences on the composition, structure and maintenance of terrestrial ecosystems remains poor. Spanning the equator and encompassing tropical, sub-tropical and temperate climate systems of both hemispheres, Africa provides a unique opportunity to study not only local and regional climate dynamics, but also to investigate and understand vegetation-climate interactions. Thus, allowing feedback effects to be identified on different spatial and temporal scales, and to assess the inter-hemispheric teleconnections in the global climate system. In this paper, we review the current state of knowledge related to African vegetation change over the entire Quaternary (2.6 million years) and older at orbital, millennial and centennial timescales by reviewing a large set of fossil pollen records obtained from marine sediments.

It is well established that the study of fossil pollen contained in sediments facilitates the reconstruction of terrestrial palaeoenvironmental changes through geological times, and constitutes the most widely used quantitative proxy of past vegetation changes (Gajewski 1993; Webb 1986). Pollen grains can be transported over long distances by winds and/or by rivers, and consequently reach the sea floor. Therefore, pollen are present in marine sediments from estuarine to abyssal environments and from the tropics to the Polar Regions. Consequently, their broad occurrence makes them valuable palaeoclimatic and palaeoenvironmental markers especially along the coast of arid environments where terrestrial records are scarce and organic microfossils are badly preserved. Pollen data derived from marine sediments integrate palynological information on large shifts in vegetation over long and continuous periods, with individual sequences of sediments often spanning more than one glacial/interglacial climatic cycle. Pollen records obtained from marine sediments have been shown to be suitable for tracing large-scale climatically related vegetation changes (Hooghiemstra 1988), hydrological variability (Dupont 1999), and for correlations between change on land and in the oceans (Heusser and Shackleton 1979; Hooghiemstra *et al.* 2006). However, the interpretation of pollen records from marine sediments should be done carefully accounting for several aspects. These include: (i) source and production of pollen grains, because it varies remarkably from one species to another, (ii) transport to the ocean floor and through the water column, (iii) displacement by ocean currents, (iv) sedimentation process, (v) taphonomic processes and early diagenesis, and (vi) fossilization in the sediment (Dupont 1999).

Pollen grains are dispersed over long distances (up to 200 km) reflecting the ease with which they can be transported from their source area to the site of sedimentation (Wright 1952). Therefore, the evaluation of transport agents is particularly important in the interpretation of marine pollen records. Aeolian transport of pollen grains predominates in deep-sea sediments located far from the coast and along arid areas with no or small river discharge (Heusser and

Morley 1985; Hooghiemstra *et al.* 1986), whereas fluvial transport is especially dominant in humid areas and at sites close to river mouths (Hooghiemstra *et al.* 1992).

By reviewing a large set of African vegetation reconstructions based on deep-sea pollen records along the African ocean margins both in low- and mid-latitudinal climatic regions of both hemispheres, we intend to: (i) assess the impact, timing and amplitude of climate changes on the regional vegetation, (ii) study the large-scale vegetation changes, (iii) compare the hydrological variability between the northern and southern hemisphere, and (iv) map the tree pollen distribution through geological times. Specifically, this synthesis will allow us to trace past changes in the African vegetation and to investigate the difference between the variability of tropical/equatorial vegetation and subtropical vegetation (north and south of the equator).

22.2 CLIMATIC AND OCEANOGRAPHIC SETTINGS AND THEIR IMPACTS ON POLLEN TRANSPORT

The climate of Africa includes a range of several different types that vary more in rainfall amount than in temperatures (Nicholson and Grist 2003). Owing to its position across equatorial and subtropical latitudes in both the northern and southern hemisphere we can distinguish the equatorial climate, the tropical wet and dry climates, the tropical monsoon climate, the semi-desert climate (semi-arid), the desert climate (hyper-arid and arid), the subtropical climate and highland climate.

Most of the continent lies within the intertropical zone between the Tropic of Cancer and the Tropic of Capricorn. The climate of Africa depends on the seasonal migration of the tropical rainbelt associated to the Intertropical Convergence Zone (ITCZ). The ITCZ is a seasonally migrating low-pressure belt that forms where the Northeast (NE) Trade Winds converge with the Southeast (SE) Trade Winds (Nicholson and Grist 2003). During boreal summer, between July and September, the ITCZ reaches its northernmost position bringing moist air upward. This causes water vapour to condense resulting in a band of heavy precipitation over the northern tropical-subtropical areas as a humid monsoon flow bringing most of the annual rainfall (Nicholson and Grist 2003). A contrasting dry season develops during the boreal winter, from December to February, when the ITCZ reaches its southernmost position causing dry conditions in the northern subtropics associated with strong NE Trade Winds and humid conditions in the southern subtropical regions.

The northernmost and the southernmost edges of the continent have a Mediterranean climate controlled respectively by variations in the North Atlantic Oscillation (NAO) (Le Houérou *et al.* 1986) and by the position and strength of the South Atlantic anticyclone and the Indian Ocean anticyclone (Shannon 1985; Shannon and Nelson 1996). The climate of Africa is predominantly tropical due to its latitudinal position in relation to the equator and the tropics (Nicholson 2000). Temperate climates are rare across the continent except at high elevations and along the edges.

African deserts are the sunniest and the driest parts of the continent, owing to the prevailing presence of the subtropical ridge with subsiding, hot, dry air masses such as the Saharan Air Layer (SAL), a mid-tropospheric zonal wind system occurring at higher altitudes (1500 – 5500 m), incorporating the African Easterly Jet (AEJ). SAL is responsible for transporting dust and terrestrial remains such as pollen grains from the Sahara and Sahel belt to the Atlantic Ocean (Colarco *et al.* 2003; Hooghiemstra *et al.* 1986; Prospero and Nees 1986; Prospero *et al.* 2002; Stuut *et al.* 2005).

The coast of the Gulf of Guinea, in tropical and equatorial Africa, is characterized by high rainfall and consequently several river systems that complement the aeolian transport of pollen grains. Therefore, fluvial-transported pollen is expected to dominate the pollen assemblages found in marine sediments offshore tropical rivers such as Senegal, Niger, Congo and Cunene Rivers. Further south, the climate system of southern Africa is controlled by the position and

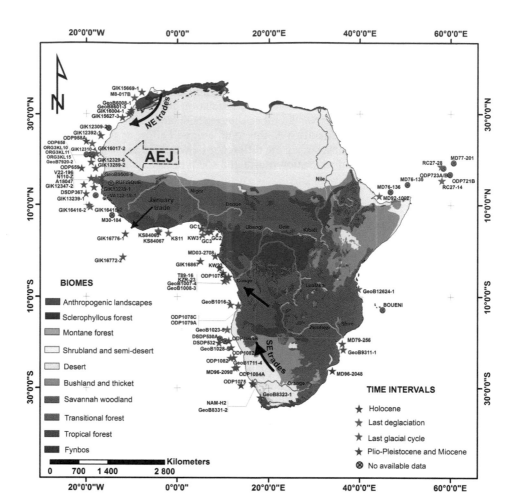

Figure 1. Map of the modern vegetation simplified after White (1983) with the locations of all African marine sites.

strength of the South Atlantic anticyclone and the Indian Ocean anticyclone (Shannon 1985; Shannon and Nelson 1996). The pressure difference between the South Atlantic anticyclone and the continental pressure field causes alongshore SE trade winds (Figure 1). SE trade winds are the main transport agent of pollen from Namibia and western South Africa. Orange River has shown little to no contribution to the pollen content of the adjacent marine sites (Meadows *et al.* 2002). Along the SE African coast, conditions are dominantly warm and moist, due to the influence of the warm Indian Ocean and Agulhas Western Boundary Current. Moisture is advected from the tropical Indian Ocean mainly during summer months when precipitation often exceeds 100 mm per month (Nash and Meadows 2012) in north-eastern South Africa (Tyson 1986; Tyson and Preston-Whyte 2000). Fluvial transport of pollen grains by Limpopo and Zambezi Rivers is also considerable in this area. Regarding the Gulf of Aden site, the area is dominated by the Indian monsoon (Almogi-Labin *et al.* 2000).

The wind systems influence the most important large-scale oceanographic features along the African continent. The Canary Current, a wind-driven surface current that is part of the North Atlantic Gyre, dominates the ocean circulation in northwest Africa. This eastern boundary current branches south from the North Atlantic Current due to intensified NE trade winds and flows southwest about as far as Senegal where it turns west and later joins the Atlantic North Equatorial Current. Further south, the west coast of southern Africa is dominated by the northward flowing Benguela Coastal Current (BCC) stretching from 34°S to 15°S. The SE African continental shelf is dominated by the Agulhas Current. Hooghiemstra *et al.* (2006) suggests that the distribution of pollen over the ocean surface is reflected in the marine sediments without substantial displacement by marine currents. This might not be fully true (Fischer *et al.* 2009), but isopollen maps show that on continental scales the displacement by ocean currents is not large enough to bias the trends seen in the pollen distribution (Dupont and Agwu 1991; Hooghiemstra *et al.* 1986). We therefore assume that no large-scale transport has displaced the palynomorphs and other terrigenous material while sinking through the water column.

Modern vegetation distribution in Africa reflects the north–south precipitation gradient (Figure 1). The present-day African vegetation around the equator consists of tropical forests including evergreen rainforest in the Congo basin and in the moist part of West Africa and semi-evergreen and deciduous forest in northern Cameroon and southern Congo. North and south of the forest fringes there is a narrow belt of natural grass savannah. North of the xerophytic shrubland of the Sahel zone and along the coast of Namibia and southwest Angola, desert vegetation occurs (White 1983). Mediterranean vegetation containing trees and shrubs dominate the northern and the southernmost regions of Africa. Coastal vegetation is represented by mangrove, occurring in tropical/subtropical estuaries and near the river mouths (White 1983). The distribution of mangrove depends on water salinity, river runoff, and humidity (Blasco 1984).

22.3 COMPILATION OF THE RECORDS

In this study, we use all the available marine pollen records from Africa with an appropriate age model stored in the African Pollen Database (APD) and Pangaea as well as those delivered by authors upon request. Out of 95 marine sites illustrated in Figure 1, only 60 pollen records have the potential to document the vegetation response to climate variability and, therefore, have been used in this paper (Supporting Online Material [SOM] Appendix Table 1).

All of the radiocarbon-dated records considered here have age models in calendar years (cal kyr BP) after conversion of ^{14}C dates to calendar age according to Stuiver *et al.* (2020). The age models were directly retrieved from the published articles or from the APD (i.e. based on linear interpolation between the dated samples) taking into account the marine reservoir age of 400 years, which reflects the present-day global average (Hughen *et al.* 2004). Age models beyond the limit of radiocarbon dating were based on the marine oxygen isotope stratigraphy. Selected sites with sufficient temporal resolution were used to illustrate the different patterns of the vegetation response to major climatic events in a centennial-, millennial- and orbital-time scale (Figures 2, 3 and 4). The whole dataset was then used to map tree pollen distribution in marine sediments off African shores at a continental scale throughout different climate cycles (Figure 5). Tree pollen types include trees, shrubs, palms and lianas. Percentages are calculated against a sum of total pollen excluding those of aquatics (including Cyperaceae) and *Pinus* (SOM Appendix Table 2).

22.4 PATTERN OF VEGETATION CHANGES IN AFRICA

Here, we focus on changes in pollen records during four main time intervals; the last deglaciation, the last glacial cycle, the mid-Pleistocene and the Plio-Miocene. To be selected for further analysis, the sites had to cover at least parts of these intervals, with a sufficient temporal resolution for a substantial part of the record, and sufficient age control to pinpoint centennial- or millennial-scale changes. Only eight records of the last deglaciation, eight records of the last glacial cycle,

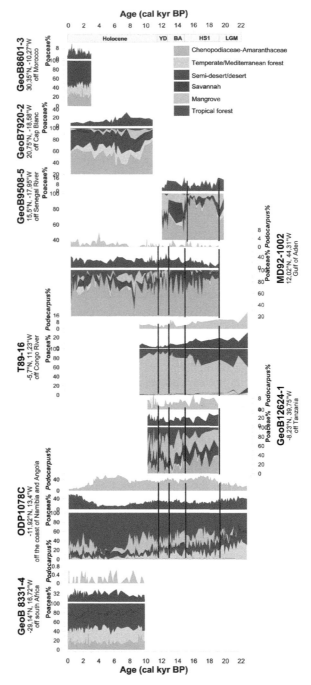

Figure 2. Summary diagrams showing a south-north transect of vegetation changes during the last deglaciation, using the groups given in Table 1, for African marine sites (right hand axes: eastern African margin, left hand axes: western African margin). Additional indicators such as Poaceae (corn yellow) and *Podocarpus* pollen (lila) are shown for each site in percentages of total pollen (Y-axes). Note the absence of *Podocarpus* pollen in the northern records. YD: Younger Dryas, BA: Bølling-Allerød warm event, HS1: Heinrich Stadial 1, LGM: Last Glacial Maximum. Black vertical lines are denoting the LGM, HS1, BA, YD and the Holocene for each record.

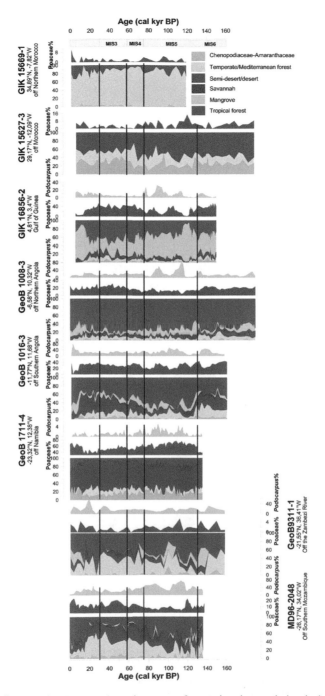

Figure 3. Summary diagrams showing a south-north transect of vegetation changes during the last glacial cycle, using the groups given in Table 1, for African marine sites (right hand axes: eastern African margin, left hand axes: western African margin). Additional indicators such as Poaceae (corn yellow) and *Podocarpus* pollen (lila) are shown for each site in percentages (Y-axes). Note the absence of *Podocarpus* pollen in the northern records. MIS: Marine isotope stage. Black vertical lines are delimiting MIS5, MIS4 and MIS3 for each record.

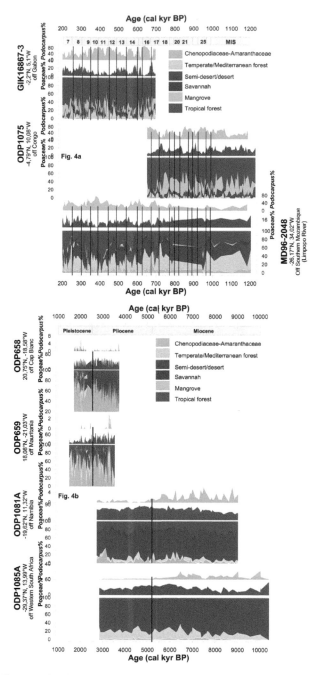

Figure 4. Summary diagrams showing a south-north transect of vegetation changes during (A) the Mid-Pleistocene, and (B) the Plio-Miocene using the groups given in Table 1, for African marine sites (right hand axes: eastern African margin, left hand axis: western African margin. Additional indicators such as Poaceae (corn yellow) and *Podocarpus* pollen (lila) are shown for each site in percentages (Y-axes). MIS: Marine isotope stage. Black vertical lines are delimiting the different MISs in Figure 4A and are separating the Miocene, the Pliocene and the early Pleistocene in Figure 4B.

three mid-Pleistocene and four Plio-Miocene records met these criteria (Figure 2, 3 and 4). Differences in regional floras make it difficult to compare pollen records from different regions directly. To facilitate comparisons between sites, the vegetation records are summarized in terms of major vegetation types (biomes) (Table 1). This naturally involves a certain loss of information, but allows us to describe the main vegetation changes for each record (Figure 2, 3 and 4) in a consistent way. It also should be kept in mind that the selection of pollen types reduces the sums on which percentage calculation is based.

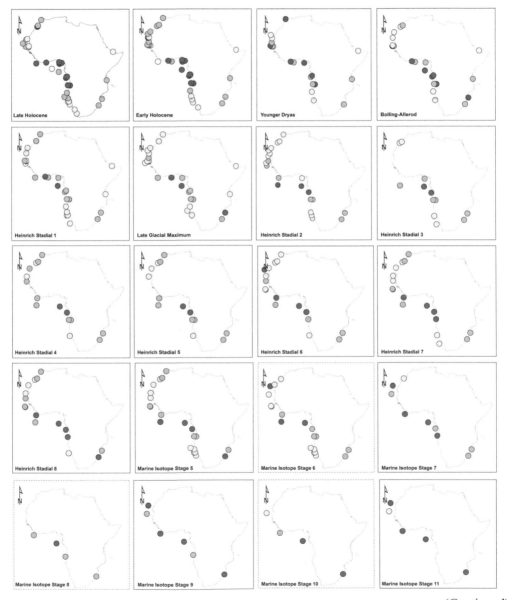

(Continued)

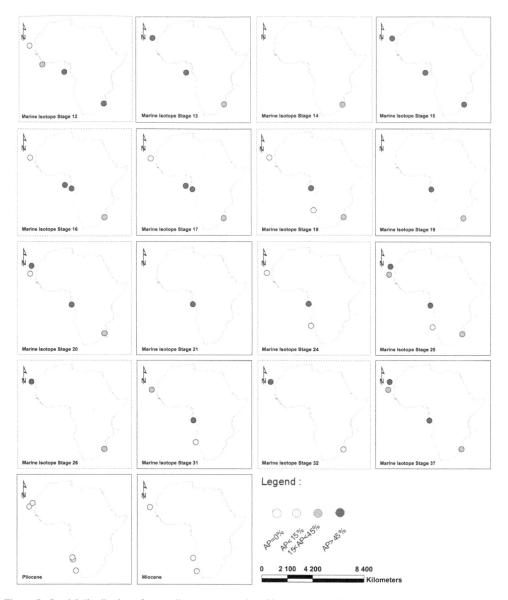

Figure 5. Spatial distribution of tree pollen percentages in African marine sediments. Four categories are considered. Black dot: AP>45% showing high percentages of arboreal taxa, grey dot: 15<AP<45% intermediate percentages of arboreal taxa, light grey dot: AP<15% weak percentages of arboreal taxa and white dot: AP=0% absence of arboreal taxa. Dashed boxes, indicate cold stages and solid boxes denote warm stages.

22.4.1 Last deglaciation and Holocene

The ODP1078C from southwestern (SW) Africa (Dupont *et al.* 2008) and MD92-1002 from eastern Africa in the Gulf of Aden (Fersi *et al.* 2016) provide continuous pollen records of the last 20,000 years with high temporal resolution (SOM Appendix Table 1, Figure 2). The other pollen records cover either the early part of the last deglaciation or the Holocene (Figure 2).

Table 1. List of the African biomes and characteristic pollen taxa grouped in major vegetation types according to the phytogeographical assignment.

Mega-Biome	Equivalent local vegetation names	Characteristic pollen taxa
Temperate/	Cool temperate rain forest	Azioaceae, *Cedrus*, Ericaceae, *Juniperus, Olea, Pinus, Pistacia, Quercus,* Restionaceae
Mediterranean forest	Warm-temperate forest Fynbos	
Semi-desert/Desert	Desert shrubs	*Artemisia*, Asteroideae, Cichoroideae, *Ephedra, Plantago, Ziziphus, Zygophyllum*
	Grass Herbs	
CCA	Desert and halophytic vegetation	Amaranthaceae, Caryophyllaceae
Savannah	Dry forest	*Acacia, Spermacoce, Boscia, Celtis, Cleome,* Mimosaceae, *Mitracarpus*
	Steppe Savannah grassland	
Mangrove	Mangrove	*Avicennia, Rhizophora*
Tropical forest	Tropical lowland forest	*Alchornea, Allophylus,* Arecaceae,
	Tropical evergreen forest	*Berlina/Isoberlina,* Bignoniaceae, *Blighia,* Bombacaceae, *Borassus, Brachystegia, Butyrospermum, Canthium, Capparis, Cassia,* Combretaceae, *Croton, Cuviera, Daniellia, Diospyros,*
	Tropical semi-deciduous forest	*Dodonaea, Elaeis, Ficus, Gaertnera,* Hippocrateaceae, *Hygrophila, Hymenocardia, Ilex, Indigofera, Ixora, Lannea, Leea, Mallotus, Manilkara, Melochia, Ochna, Parinari, Phoenix, Phyllanthus, Piliostigma, Pterocarpus,* Rhamnaceae, *Rhus, Schreberea, Sterculia, Stereospermum, Syzygium, Tamarindus, Tarenna, Trichilia, Uapaca,* Urticaceae, Vitaceae, *Ximenia, Zanthoxylum*

In Equatorial Africa, the pollen record of T89-16 (Marret *et al.* 2001) (Figure 2) provides information into the equatorial vegetation development reflecting environmental changes in the Congo basin prior to the Holocene. We excluded core KW 31, located south of the Niger Delta (Lézine and Cazet 2005), despite a sufficient time resolution due to substantial gaps in the pollen record (SOM Appendix Table 1). The GeoB9508-5 off the Senegal River and GeoB12624-1 off the Rufiji Delta, (Bouimetarhan *et al.* 2012; 2015) show high-resolution vegetation change during the northern hemisphere cold Heinrich Stadial 1 (HS1; *c.* 18-15 kyr BP) (McManus *et al.* 2004) and the Younger Dryas (YD; *c.* 12.9-11.6 kyr BP) (Rasmussen and Thomsen 2015) (Figure 2). The Holocene records are represented by the high-resolution GeoB8331-4 and GeoB7920-2 pollen records (Dupont and Schefuß 2018; Zhao *et al.* 2016) covering the last 10,000 years and the GeoB8601-3 pollen record covering the last 3000 years (Zhao *et al.* 2019) (Figure 2).

In southern Africa, the ODP1078C pollen record off the coast of Namibia and Angola shows a clear dominance of semi-desert vegetation and Poaceae during the Last Glacial Maximum (LGM), which decreases slightly during the Heinrich Stadial (HS) 1 while tropical forest and mangrove as well as *Podocarpus* increased, indicating wetter and cooler climate conditions. This trend is almost stable until the Holocene. However, from 10 to 8 kyr BP, a slight increase in mangrove is observed reflecting a continuous input freshwater supply, which decreases in favour of the tropical forest and *Podocarpus* (reaching up to 54%) from 8 until 3 kyr BP. After 3 kyr BP Poaceae increase rapidly up to 40% possibly reflecting a shift towards drier conditions, although human impact cannot be excluded.

In eastern Africa, the MD92-1002 pollen record from the Gulf of Aden shows rather a contrasting trend. The last deglacial vegetation is dominated mainly by CCA (Caryophyllaceae and Amaranthaceae including Chenopodiaceae) and Poaceae. CCA reach up to ~58% and 45% during HS1 and YD respectively along with semi-desert/desert vegetation indicating the extremely dry climatic conditions during the North Atlantic cold stadials. A slight increase in the savannah is observed during the time period equivalent to the northern hemisphere warm event known as the Bølling-Allerød (BA; *c.* 14.7 to 12.8 kyr BP) (Rasmussen *et al.* 2006). The onset of the Holocene (*c.* 11.7 kyr BP) exhibits a substantial decrease in CCA and semi-desert representation along with an increase in savannah elements as well as the tropical forest and mangrove indicating a shift towards relatively wetter climatic conditions until 5 kyr BP. *Podocarpus* pollen is present in very low relative abundances during the BA and late Holocene testifying rather the long-distance wind transport.

In equatorial Africa, the pollen record off the Congo River T89-16 shows a constant dominance of the mangrove forest throughout the whole sequence reaching up to 78% during the YD. High abundances of mangrove pollen have been associated with transgressive phases and large flood of the Congo River (Scourse *et al.* 2005). Poaceae and *Podocarpus* exhibit simultaneously higher percentages during the LGM indicating cooler climate and decrease dramatically during the YD.

In western and eastern Africa, pollen records off Senegal (GeoB9508-5) and off Tanzania (GeoB12624-1) cover partly the period between *c.* 19 and 12 kyr BP, at centennial- to subcentennial-resolution, and provide new information on the deglacial vegetation of these regions, which shows contrasting vegetation changes. Using last deglaciation pollen records from the continental slope off Senegal, the data show that the Sahel drought remained severe during HS1 with maximum CCA representation along with elements from the savannah and semi-desert/desert vegetation, while Poaceae are represented before and after HS1. The marine pollen record off the Rufiji River shows low percentages of CCA during HS1 along with high percentages of tropical forest, savannah elements and mangrove as well as *Podocarpus* indicating less severe HS1 droughts in contrast to West Africa.

The Holocene records off northwest (NW) Africa (GeoB7920-2) and South Africa (GeoB8331-4) show similar trends with a clear dominance of three biomes, the semi-desert/desert vegetation and CCA, the Mediterranean forest, and fynbos. While the latter is rather constant throughout the Holocene, the former shows a clear decrease in CCA and semi-desert/desert vegetation during the early Holocene followed by a remarkable increase after *c.* 5 kyr BP. Poaceae reach their maximum values between *c.* 10 and 6 kyr BP. The short record of GeoB8601-3 offshore Morocco shows a constant dominance of semi-desert/desert vegetation with a clear increase in Poaceae pollen during the last 1000 years.

22.4.2 Last glacial cycle

Eight sites, of which two sites from the eastern African margin have been selected for this time interval (Figure 3). In northern Africa, a record of maximum extension of the Mediterranean forest is found at the northernmost site off northern Morocco, GIK15669-1 at 34°N (Hooghiemstra *et al.* 1992) during the last glacial cycle, which is coincident with a similar forest expansion in Europe (e.g. Sánchez Goñi *et al.* 1999). The GIK15627-3 pollen record off southern Morocco at 29°N (Hooghiemstra *et al.* 1992) depicts a different pattern with a clear dominance of CCA and semi-desert/desert vegetation. Although lower than the previous record, a slight extent of the Mediterranean forest is observed for Marine Isotope Stage (MIS) 5 that continues until the present along with Poaceae that decrease however, rapidly after MIS5.

In equatorial Africa, the GIK16856-2 pollen record offshore Guinea at 5°N (Dupont and Weinelt 1996) shows an extension of mangrove forest during MIS5 and the last 20,000 years along with a minimum representation of Poaceae pollen reflecting moist continental conditions that coincide with interglacial and last deglacial conditions, respectively. *Podocarpus* percentages show maxima during the latter part of MIS5 indicating a well-developed Afromontane forest as a response to a cooler climate during the stadials MIS5b and MIS5d.

Further south, maximum expansion of the tropical forest is observed at the complete equatorial site GeoB1008-3 at 6°S offshore northern Angola (Jahns 1996). MIS 5 exhibit low Poaceae percentages while *Podocarpus* reach their highest values indicating a cooler climate during the stadials of MIS5. In contrast to the previous record, mangrove forest is less represented along with other biomes. The pollen record GeoB1016-3 off Southern Angola at 11°S (Shi and Dupont 1997) shows a contrasting pattern compared to the previous records. Here, we see a clear dominance of the semi-desert/desert vegetation along with Poaceae pollen except for MIS5 where tropical forest and to a lesser extent mangrove are well developed. *Podocarpus* is well developed during MIS5c-d and the last 10,000 years of the record. The record of GeoB1711-4 (Shi *et al.* 2001) from the Namib Desert shows a constant dominance of semi-desert/desert vegetation and Poaceae with the presence of CCA. However, a small but significant increase in tropical forest and *Podocarpus* pollen is observed during the stadials of MIS5 reflecting slightly cooler conditions.

In eastern Africa, the MD96-2048 pollen record from southern Mozambique at 26°S shows a clear shift from tropical forest dominance during MIS5 to semi-desert/desert vegetation and Mediterranean forest dominance afterwards (Dupont *et al.* 2011). In contrast, the GeoB9311-1 pollen record off the Zambezi catchment at 21°S (Dupont and Kuhlmann 2017) shows a co-dominance of tropical and temperate forests for the entire record. On the other hand, *Podocarpus* pollen is very well represented during MIS5 in the southern record while they do not exceed 20% in the northern record. This suggest that on a regional scale climatic perturbations of the vegetation cover in the Zambezi catchment are less severe than in southern Mozambique.

22.4.3 Mid-Pleistocene

Three pollen records were selected to assess glacial-interglacial fluctuations in the vegetation records around Africa during the Mid-Pleistocene (between *c.* 1.2 and 0.4 Million years (Myr) BP; Berger and Jansen 1994) (Figure 4A). In eastern Africa, the recently published pollen record off southern Mozambique MD96-2048 at 26°S (Dupont *et al.* 2019) represents a unique and continuous high-resolution African record from the Indian Ocean, to study the long-term evolution of orbital-scale vegetation changes. The terrestrial pollen assemblages indicate that during glacials, specifically during MIS 7, 9, 11, and 17, the vegetation of eastern South Africa and southern Mozambique largely consisted of tropical forest vegetation indicating relatively wet conditions. During glacials, Mediterranean elements (here represented by the ericaceous vegetation in the mountains) along with semi-desert/desert elements dominated the pollen assemblage. *Podocarpus* pollen does not display variability related to glacials or interglacials.

In equatorial Africa, represented by the ODP1075 pollen record off Congo at 4°S (Dupont *et al.* 2001) and GIK16867-3 record off Gabon at 2°S (Dupont *et al.* 1998), vegetation is mainly dominated by tropical forest and mangrove. The tropical forest expands widely during interglacials (MIS 7, 9, 11, 13, 15, and 17) along with mangrove in sites offshore Gabon, while the semi-desert/desert vegetation slightly increases during the glacials. *Podocarpus* shows higher percentages during glacials indicating the expansion of mountainous forest during cold periods. The ODP1075 pollen record shows a much higher representation of mangrove forest along with tropical forest increasing strongly during periods of sea-level rise (Scourse *et al.* 2005). Poaceae show rather lower values during interglacials MIS17, 19, 21, and 25 while *Podocarpus* does not show a clear trend.

22.4.4 Plio-Miocene

Four records are presented here to reconstruct Miocene (*c.* 23 to 5.3 Myr BP) to Pliocene (*c.* 5.3 to 2.5 Myr BP) changes of African vegetation and rainfall regime (Figure 4B). In SW Africa, the South Atlantic ODP1085A pollen record at 29°S (Dupont *et al.* 2019) consists mainly of a strong representation of desert and semi-desert vegetation and Poaceae pollen implying that conditions were already arid 10 Myr ago. *Podocarpus* exhibit elevated relative abundances during the Miocene while it disappears quasi-completely during the Pliocene. Authors have inferred changes in the relative amount of precipitation and indicated a shift of the main moisture source from the Atlantic to the Indian Ocean during the onset of a major aridification 8 Myr ago (Dupont *et al.* 2013). The ODP1081A pollen record (Hötzel *et al.* 2015) offshore Namibia covering the time interval between 9 and 2.7 Myr BP consists mainly of pollen from semi-desert/desert vegetation. While, *Podocarpus* occurred during the Miocene, grasses expanded during the Pliocene.

In NW Africa, the Plio-Pleistocene pollen record ODP659A at 18°N (Vallé *et al.* 2017) shows more variability in the pollen content. After 3.5 Myr BP, the abrupt increase of CCA and Poaceae suggests larger climate variability during that interval implying continent-wide aridity during eccentricity maxima between 3.2 and 2.5 Myr BP. From 2.5 Myr BP on, the trend in ODP Site 659 is characterized by a slight increase of Mediterranean forest and semi-desert/desert vegetation and a clear decrease in Poaceae pollen. *Podocarpus* pollen are present during the Pliocene with very low relative abundances but quasi-absent during the early Pleistocene. The ODP658 pollen record off Cap Blanc at 20°N (Leroy and Dupont 1994) shows a similar trend as ODP659 during the late Pliocene. However, Poaceae pollen that reach up to 59% during the Late Pliocene continue to be well represented through the early Pleistocene. *Podocarpus*, similarly to the ODP site 659, is present with very low percentages during the late Pliocene and between 1.7 and 2 Myr BP.

22.5 POLLEN-CLIMATE RELATIONSHIP IN AFRICA

Today, Africa is mainly dominated by open landscapes between 30°S and 30°N with the exception of the Mediterranean, equatorial and mangrove forests and the afromontane forests. These open landscapes range from desert, dry forests, to woodlands, wooded grasslands, grasslands, and fynbos (White 1983). Transitions between forest, savannah and desert vegetation types as well as Mediterranean forest in African fossil pollen records are often poorly understood due to the scarcity of modern pollen-vegetation studies (Julier *et al.* 2018; Lézine and Hooghiemtra, 1990; Vincens *et al.* 2000) and the over-representation/under-representation of certain taxa. Shifts in biome distributions imply significant changes in climate, especially the precipitation gradient throughout several time scales, reflecting a change in monsoon extent combined with a southward expansion of the ITCZ (Shanahan *et al.* 2015). Outside the influence of the monsoon, westerly storm tracks bring rainfall to the northernmost and southernmost parts of Africa (Jolly *et al.* 1998). Additionally, biomes are differently represented, because the biome characteristic taxa have very different pollen production rates with different pollination strategies (entomophily (insects); lepidopterophily (butterflies); anemophily (wind); chiropterophily (bats); and ornithophily (birds)), and different dispersal mechanisms. Forest taxa which are frequently entomophilous are under-represented (Vincens *et al.* 2000), while anemophilous taxa such as Poaceae are generally over-represented relative to their abundance in the vegetation (Bush and Rivera 1998). Pollen of Combretaceae and *Alchornea* are better represented in the marine sediments than other trees due to the better pollen dispersal compared to many other tropical forest trees (Dupont 2011; Watrin *et al.* 2007). Although, the relative abundances of pollen grains cannot be simply translated into vegetation composition, it could be interpreted as relative changes of biomes through time. In marine sediments, this issue becomes even more serious as fossil pollen records effectively show only a part of the vegetation on the adjacent continent. The reduced representation of the assemblage is less in pollen data from terrestrial deposits, which give more local information. However, due to the relatively good preservation of pollen grains in certain marine records, good age control and coverage of long periods, marine archives allow us to study vegetation changes over several continuous climate cycles. Here, we present distribution maps of arboreal pollen percentages (AP%) from available marine pollen records off African shores for different time-slices (Figure 5) in order to discuss changes in regional vegetation and environmental conditions over orbital and glacial-interglacial timescales and to assess the low- to mid-latitude interhemispheric climatic implications and mechanisms.

22.6 PATTERN OF LANDSCAPE CHANGES

We show four arboreal taxa categories (AP%), as also used by Lézine *et al.* (this issue), to characterize the African landscapes through several climate cycles (Figure 5): (i) AP higher than 45% corresponding to more or less closed canopy, (ii) 15<AP<45% intermediate between closed canopy and open landscapes, (iii) AP lower than 15% indicating grass-dominated landscapes, and (iv) AP%=0.

During the early Holocene, marine pollen records show high percentages (AP>45%) of arboreal taxa in most of Africa. Exceptions are found offshore of SW Africa (South Africa and Namibia), Gulf of Aden and Mauritania. This indicates the prevalence of forested vegetation types over tropical and Mediterranean Africa linked to humid conditions. Late Holocene records show a more open landscape as many sites contain lower relative abundances of arboreal taxa (AP<15%) reflecting grass dominated landscape, thus drier conditions than in the early Holocene.

Between 21 and 11.8 kyr BP, last deglaciation marine pollen records show a more open vegetation with widespread grassland, as well as savannah and xerophytic scrubland vegetation (Figure 2) indicating dry climate especially during the last glacial maximum (LGM). AP% are higher (>45%) in Equatorial Africa and in NW and SE Africa even during HS1 and LGM whereas the rest of the marine African pollen records are characterized by a strong representation of grasses.

Between 50 and 21 kyr BP, open grasslands and savannahs dominate NW and SW Africa, the tropical regions and SE Africa show a mixture between closed canopy and open landscapes, while equatorial Africa is characterized by a more or less dense vegetation during HS2 and HS3. In general, HS4 and HS5 display intermediate AP% (15<AP<45%) indicating relatively wetter conditions.

Between 80 and 50 kyr BP, pollen records show that HS6 is characterized by a return to a more open vegetation and thus, drier conditions in NW Africa and Gulf of Guinea compared to the previous HS. HS7 and HS8 exhibit similar trends as HS4 and HS5 with particularly higher AP% (>45%) during HS8 in SE Africa. Taken together the AP% distributions over the last 80,000 years show, to a certain extent, a north-south rainfall gradient consisting of lower AP% (<15%) in NW and SW Africa, intermediate AP% in tropical Africa and SE Africa and higher AP% in equatorial Africa. Although marine pollen records from eastern Africa are extremely understudied compared to western Africa, we could observe a contrast between the relatively denser SE and the open SW African landscape; forests only occurred in western equatorial Africa between 2°N and 12°S. The AP% distribution maps show that the intervals spanning the LGM, HS1 HS2 and HS6 are marked by an increase in open vegetation, mainly savannah and xerophytic vegetation as shown in Figure 5. These changes occur mainly in NW and SW Africa indicating that major changes in biome distributions occurred north of 10°N and to a certain extent south of 20°S.

22.6.1 Interglacials *vs.* glacials

For MIS5, one of the best studied warm stages in Africa represented by 21 records, open landscapes dominate the Sahel/Sahara region from 20° to 30°N and SW Africa and decline in 12 sites located in western equatorial Africa, north and SE Africa where intermediate and higher AP% are more prominent. This indicates a relatively wetter climate during MIS5 compared to the previous period. During MIS6 (represented by 19 records), open landscapes dominate the coast between 30°N and 9°N indicating a southward shift of the savannah biome during this glacial period. MIS7 (represented by nine records) shows a clear dominance of arboreal taxa except one site offshore Morocco indicating a return to wetter conditions. The next time slices, which are represented with only few sites each (average of 3 to 5 records) seem to follow the pattern of odd- and even-numbered MIS and show an alternation between dense arboreal dominance during MIS9, 11, 13, 15, 17, 19, 21, 25, 31, and 37 and more open vegetation during MIS 10, 12, 14, 16, 18, 20, 24, and 32, especially in NW and SW Africa.

MIS20 and MIS26 form an exception showing higher arboreal pollen percentages offshore of the Sahara. Moreover, the eastern African site that exhibited intermediate to higher arboreal taxa values throughout all timescales indicated an open-dominated landscape during MIS32 reflecting dry conditions in this area compared to previous climate cycles. The only few available records representing the Pliocene and the Miocene show mostly grass-dominated landscapes and dry forests.

Taken together, the AP% distribution maps show that dense forests and intermediate tree cover were able to persist in equatorial and eastern Africa through all marine isotope stages shown in Figure 5 (except MIS32). The scarcity of eastern African marine sites, however, hampers a reliable comparison in order to understand the response of the eastern African biomes to climate variability in this climatically highly complex region.

22.6.2 Low *vs.* mid-latitudinal variability

The influence of obliquity, the tilt of the Earth's rotational axis, on incoming solar radiation at low latitudes is small, yet many tropical and subtropical palaeoclimate records reveal a clear obliquity signal (Dupont 2011; Dupont *et al.* 1989). This is particularly important since in terms of mean energy balance the integrated summer insolation is more important than maximum insolation, which is influenced by obliquity and not by precession (Huybers 2011). Additionally, the insolation gradient between high and low-latitudes depending on obliquity controls the atmospheric meridional flux of heat, moisture, and latent energy (Raymo and Nisancioglu 2003). Obliquity's control on meridional insolation gradients weaken, or strengthen, the mid-latitude westerlies and the subtropical trade winds (Lee and Poulsen 2005). Yet, model simulations suggest that the *c*. 100 kyr cycle signal was not a robust feature of tropical vegetation, which is subject to stronger direct forcing by the precessional (*c*. 21 kyr) orbital cycle (Gosling and Holden 2011). However, Dupont (2011) has shown significant power around 100 kyr at all records she has analysed in her review paper, which is interpreted as an effect of glacial boundary conditions. Thus, the correlation of palaeoenvironmental changes from African pollen records with global temperature change remains ambiguous and must be done with great caution. The AP% distribution maps (Figure 5) and pollen diagrams (Figure 2) indicate that during the Holocene, YD and HS1, the response of arboreal vegetation reversed at subtropical latitudes between hemispheres, whereby NW African sites exhibit higher AP% than those of SW Africa. Arboreal taxa values show maxima at equatorial and tropical sites from the northern and southern hemispheres during warm stages (Figure 5) and a clear reduction during cold stages (Figures 3 and 4A) probably driven by summer insolation, which is expected to be the main forcing mechanism of rain forest variability. Summarizing the marine palynological results covering several glacial-interglacial climate cycles, it is suggested that the equatorial rain forest extent in tropical Africa varies with summer insolation; with boreal summer insolation in the northern part and with austral summer insolation in the southern part.

22.7 PERSPECTIVES

This paper attempts to review and synthetize all available marine fossil pollen records from offshore marine cores around the African continent to explore the vegetation response to climate changes in millennial- and orbital-timescales. There is a clear need for more and higher resolution records especially from the eastern African shores. There is also a need to improve dating of the key records and create age models. Marine pollen records may be difficult to interpret as they record an integrated mixture of pollen coming from large source areas. Nevertheless, if the dispersal and transport mechanisms as well as the diagenesis and preservation are taken into account, they can be very useful to document latitudinal shifts in biomes and to complement the existing information from terrestrial records. This review was possible thanks to the available data sets of pollen records submitted to different databases and to personal communications with some authors. However, there is still a big amount of missing data that have not been submitted in any database although the corresponding articles have been recently published. Therefore, there is an urgent need to encourage colleagues all over the world to make their data public for the general scientific interest.

22.8 CONCLUSIONS

Palynological approaches are moderately useful in distal offshore ocean settings characterized by oligotrophic conditions because of low concentration of palynomorphs, making it difficult to do statistically representative counts. On the other hand, combined with the fundamental problems of tropical palynology (the under-representation of many species in the pollen record), the pollen grains found in marine sediments are typically transported over long distances, through winds and water, making it difficult to identify the precise source area. However, one of the main advantages of pollen records from marine sediment sequences is that they provide access to exceptionally long palaeoecological records that are crucial to understand the vegetation response to climate forcing over time and allow direct correlations of climate change over land and in the ocean, with minimal chronological uncertainty. Marine pollen records offer a unique opportunity to associate continental environmental changes with marine environmental changes inferred from the study of other palynomorphs (dinoflagellate cysts) or other sedimentary, geochemical or micropalaeontological proxies.

The overview presented here has shown that African vegetation and climate responded to long-term and shorter-term climate changes. There was a dynamic equilibrium between vegetation and climate for short periods of forcing such as the Holocene and HEs. Within the Interglacials (especially MIS5e), the African margin shows a clear dominance of arboreal taxa in most sites (with some exceptions) indicating moist conditions. A different trend has been observed between low- and mid-latitudes. The forest expansion shows maxima during interglacials and a clear reduction during glacials at equatorial and tropical sites from the African northern and southern hemispheres.

This overview has also shown that the western coast of Africa is well documented with continuous pollen records dating back to the Miocene thanks to the commitment of some researchers over the last few decades. The atmospheric circulation conditions that favour pollen transport towards the ocean, the latitudinal distribution of the major biomes as well as the reconstruction of regional vegetation and its response to climate changes are one of the most important outcomes of this research. However, the eastern coast of Africa, a climatically highly complex region, is poorly documented. With the exception of few sites at the mouths of large rivers such as Limpopo, Zambezi and Rufiji, marine pollen records are rare in this part of Africa but extremely important to understand the response of the ecosystems to climate variability. More marine pollen records are therefore, necessary to provide independent evidence into the timing of vegetation changes in a regional context and their connection to global climate. Moreover, they would offer an important complement to previously published palaeorecords from continental archives such as the Great Rift Lakes (Tanganyika, Malawi, Victoria).

ACKNOWLEDGEMENTS

The authors wish to express their gratitude to all African Pollen Database (APD) contributors and colleagues who made their data available for this study and for the excellent discussions during the APD meeting in Bondy. We thank Henry Hooghiemstra and William D. Gosling for their constructive comments on the manuscript. IB, HR, AB are funded by the German Federal Ministry of Education and Research (BMBF), through the PMARS III grant (PMARS2015-100)". LD by the University of Bremen, and AML by CNRS.

REFERENCES

Adojoh, O., Marret-Davies, F., Duller, R., Osterloff, P., Oboh-Ikuenobe, F., Hart, M. and Smart, C., 2020, The biostratigraphy of the offshore Niger delta during the Late Quaternary: Complexities and progress of dating techniques. *Quaternary Science Advances*, **1**, pp. 1–7, 10.1016/j.qsa.2020.100003.

Agwu, C.O.C. and Beug, H.-J., 1982, Palynological studies of marine sediments off the West African coast. *'Meteor' Forschungsergebnisse, Deutsche Forschungsgemeinschaft, Geologie und Geophysik, Gebrüder Bornträger*, **36**, pp. 1–30.

Almogi-Labin, A., Schmiedl, G., Hemleben, C., Siman-Tov, R., Segl, M. and Meischner, D., 2000, The influence of the NE winter monsoon on productivity changes in the Gulf of Aden, NW Indian Ocean during the last 530 kyr as recorded by foraminifera. *Marine Micropaleontology*, **40**, pp. 295–319, 10.1016/S0377-8398(00)00043-8.

Bayon, G., Schefuß, E., Dupont, L., Borges, A. V., Dennielou, B., Lambert, T. and André, L., 2019, The roles of climate and human land-use in the late Holocene rainforest crisis of Central Africa. *Earth and Planetary Science Letters*, **505**, pp. 30–41, 10.1016/j.epsl.2018.10.016.

Berger, W.H. and Jansen, E., 1994, Mid-Pleistocene Climate Shift, The Nansen Connection. The Polar Oceans and Their Role in Shaping the Global Environment, *Geophysical Monograph Series*, **85**, pp. 295–311.

Blasco, F., 1984, Climatic factors and the biology of mangrove plants. *Monographs on Oceanographic Methodology*, **8**, pp. 18–35.

Bouimetarhan, I., Dupont, L., Kuhlmann, H., Pätzold, J., Prange, M., Schefuß, E. and Zonneveld, K., 2015, Northern Hemisphere control of deglacial vegetation changes in the Rufiji uplands (Tanzania). *Climate of the Past*, **11**, pp. 751–764, 10.5194/cp-11-751-2015.

Bouimetarhan, I., Dupont, L., Schefuß, E., Mollenhauer, G., Mulitza, S. and Zonneveld, K., 2009, Palynological evidence for climatic and oceanic variability off NW Africa during the late Holocene. *Quaternary Research*, **72**, pp. 188–197, 10.1016/j.yqres.2009.05.003.

Bouimetarhan, I., Groeneveld, J., Dupont, L. and Zonneveld, K., 2013, Low-to high-productivity pattern within Heinrich Stadial 1: Inferences from dinoflagellate cyst records off Senegal. *Global and Planetary Change*, **106**, pp. 64–76, 10.1016/j.gloplacha.2013.03.007.

Bouimetarhan, I., Prange, M., Schefuß, E., Dupont, L., Lippold, J., Mulitza, S. and Zonneveld, K., 2012, Sahel megadrought during Heinrich Stadial 1: Evidence for a three-phase evolution of the low-and mid-level West African wind system. *Quaternary Science Reviews*, **58**, pp. 66–76, 10.1016/j.quascirev.2012.10.015.

Box, E. O., 1981, Predicting physiognomic vegetation types with climate variables. *Vegetatio*, **45**, pp. 127–139, 10.1007/BF00119222.

Bush, M.B., and Rivera R., 1998, Pollen dispersal and representation in a neotropical rain forest. *Global Ecology and Biogeography Letters*, **7**, pp. 379–392, 10.2307/2997685.

Colarco, P. R., Toon, O. B., Reid, J. S., Livingston, J. M., Russell, P. B., Redemann, J. and Campbell, J. R., 2003, Saharan dust transport to the Caribbean during PRIDE: 2. Transport, vertical profiles, and deposition in simulations of in situ and remote sensing observations. *Journal of Geophysical Research: Atmospheres*, **108**, article: 8590.

Denman, K. L., Brasseur, G., Chidthaisong, A., Ciais, P., Cox, P. M., Dickinson, R. E., Hauglus-taine, D., Heinze, C., Holland, E., Jacob, D., Lohmann, U., Ramachandran, S., Leite da Silva Dias, P., Wofsy, S.C. and Zhang, X., 2007, Couplings between changes in the climate system and biogeochemistry. In *Climate Change 2007: The Physical Science Basis*, (Cambridge University Press, Cambridge, United Kingdom and New York, NY, USA), pp. 499–587.

Dupont, L. M., 2006, Late Pliocene vegetation and climate in Namibia (southern Africa) derived from palynology of ODP Site 1082. *Geochemistry, Geophysics, Geosystems*, **7**, 10.1029/2005GC001208, article: Q05007.

Dupont, L.M., 2011, Orbital scale vegetation change in Africa. *Quaternary Science Reviews*, **30**, pp. 3589–3602, 10.1016/j.quascirev.2011.09.019.

Dupont, L.M., 1999, Pollen and spores in marine sediments from the east Atlantic-A view from the ocean into the African Continent. In Use of Proxies in Paleoceanography, edited by Fischer, G. and Wefer, G. (Springer, Berlin, Heidelberg), pp. 523–546.

Dupont, L.M., 2009, The Congo deep-sea fan as an archive of Quaternary change in Africa and the eastern tropical South Atlantic (A review). In *External Controls on Deep-Water Depositional Systems, SEPM Society for Sedimentary*, Volume 92, edited by Kneller, B., Martinsen, O.J. and McCaffrey, B. (McLean, Va: SEPM Society for Sedimentary Geology), pp. 79–87.

Dupont, L. M. and Agwu, C. O., 1991, Environmental control of pollen grain distribution patterns in the Gulf of Guinea and offshore NW-Africa. *Geologische Rundschau*, **80**, pp. 567–589, 10.1007/BF01803687.

Dupont, L.M. and Kuhlmann, H., 2017, Glacial-interglacial vegetation change in the Zambezi catchment. *Quaternary Science Reviews*, **155**, pp. 127–135, 10.1016/j.quascirev.2016.11.019.

Dupont, L.M. and Schefuß, E., 2018, The roles of fire in Holocene ecosystem changes of West Africa. *Earth and Planetary Science Letters*, **481**, pp. 255–263, 10.1016/j.epsl.2017.10.049.

Dupont, L.M. and Weinelt, M., 1996, Vegetation history of the savanna corridor between the Guinean and the Congolian rain forest during the last 150,000 years. *Vegetation History and Archaeobotany*, **5**, pp. 273–292, 10.1007/BF00195296.

Dupont, L. M. and Wyputta, U., 2003, Reconstructing pathways of aeolian pollen transport to the marine sediments along the coastline of SW Africa. *Quaternary Science Reviews*, **22**, pp. 157–174, 10.1016/S0277-3791(02)00032-X.

Dupont, L. M., Behling, H. and Kim, J. H., 2008, Thirty thousand years of vegetation development and climate change in Angola (Ocean Drilling Program Site 1078). *Climate of the Past*, **4**, pp. 107–124, 10.5194/cp-4-107-2008.

Dupont, L. M., Behling, H., Jahns, S., Marret, F. and Kim, J. H., 2007, Variability in glacial and Holocene marine pollen records offshore from west southern Africa. *Vegetation History and Archaeobotany*, **16**, pp. 87–100, 10.1007/s00334-006-0080-8.

Dupont, L. M., Beug, H. J., Stalling, H. and Tiedemann, R., 1989, First palynological results from Site 658 at 21°N off Northwest Africa: Pollen as climate indicators. *Proceedings of Ocean Drilling Program Scientific Results*, **108**, pp. 93–112.

Dupont, L. M., Caley, T., Kim, J. H., Castañeda, I., Malaizé, B. and Giraudeau, J., 2011, Glacial-interglacial vegetation dynamics in South Eastern Africa coupled to sea surface temperature variations in the Western Indian Ocean. *Climate of the Past*, **7**, pp. 1209–1224, 10.5194/cp-7-1209-2011.

Dupont, L. M., Donner, B., Schneider, R. and Wefer, G., 2001, Mid-Pleistocene environmental change in tropical Africa began as early as 1.05 Ma. *Geology*, **29**, pp. 195–198, 10.1130/0091-7613(2001)029#amp;lt;0195:MPECIT#amp;gt;2.0.CO;2.

Dupont, L. M., Marret, F. and Winn, K., 1998, Land-sea correlation by means of terrestrial and marine palynomorphs from the equatorial East Atlantic: Phasing of SE trade winds and the oceanic productivity. *Palaeogeography, Palaeoclimatology, Palaeoecology*, **142**, pp. 51–84, 10.1016/S0031-0182(97)00146-6.

Dupont, L. M., Rommerskirchen, F., Mollenhauer, G. and Schefuß, E., 2013, Miocene to Pliocene changes in South African hydrology and vegetation in relation to the expansion of C4 plants. *Earth and Planetary Science Letters*, **375**, pp. 408–417, 10.1016/j.epsl.2013.06.005.

Dupont, L. M., Caley, T. and Castañeda, I. S., 2019, Effects of atmospheric CO_2 variability of the past 800 kyr on the biomes of southeast Africa. *Climate of the Past*, **15**, pp. 1083–1097, 10.5194/cp-15-1083-2019.

Fersi, W., Lézine, A.-M. and Bassinot, F., 2016, Hydro-climate changes over southwestern Arabia and the Horn of Africa during the last glacial–interglacial transition: A pollen record from the Gulf of Aden. *Review of Palaeobotany and Palynology*, **233**, pp. 176–185, 10.1016/j.revpalbo.2016.04.002.

Fischer, G., Karakas, G., Blaas, M., Ratmeyer, V., Nowald, N., Schlitzer, R. and Wefer, G., 2009, Mineral ballast and particle settling rates in the coastal upwelling system off NW Africa and the South Atlantic. *International Journal of Earth Sciences*, **98**, pp. 281–298, 10.1007/s00531-007-0234-7.

Gajewski, K., 1993, The role of paleoecology in the study of global climatic change. *Review of Palaeobotany and Palynology*, **79**, pp. 141–151, 10.1016/0034-6667(93)90044-U.

Gosling, W.D., and Holden, P.B., 2011, Precessional forcing of tropical vegetation carbon storage. *Journal of Quaternary Science*, **26**, pp. 463–467, 10.1002/jqs.1514.

Hessler, I., Dupont, L., Handiani, D., Paul, A., Merkel, U. and Wefer, G., 2012, Masked millennial-scale climate variations in South West Africa during the last glaciation. *Climate of the Past*, **8**, pp. 841–853, 10.5194/cp-8-841-2012.

Heusser, L. E. and Morley, J. J., 1985, Pollen and radiolarian records from deep-sea core RC14-103: climatic reconstructions of Northeast Japan and Northwest Pacific for the last 90,000 years. *Quaternary Research*, **24**, pp. 60–72, 10.1016/0033-5894(85)90083-3.

Heusser, L.E. and Shackleton, N.J., 1979, Direct marine-continental correlation: 150,000-year oxygen isotope-pollen record from the North Pacific. *Science*, **204**, pp. 837–839.

Hötzel, S., Dupont, L. M. and Wefer, G., 2015, Miocene–Pliocene vegetation change in southwestern Africa (ODP Site 1081, offshore Namibia). *Palaeogeography, Palaeoclimatology, Palaeoecology*, **423**, pp. 102–108, 10.1016/j.palaeo.2015.02.002.

Holdridge, L. R., 1947, Determination of world plant formations from simple climatic data. *Science*, **105**, pp. 367–368, 10.1126/science.105.2727.367.

Hooghiemstra, H., 1988, Palynological records from northwest African marine sediments: a general outline of the interpretation of the pollen signal. *Philosophical Transactions of the Royal Society of London. B,Biological Sciences*, **318**, pp. 431–449.

Hooghiemstra, H. and Agwu, C. O., 1988, Changes in the vegetation and trade winds in equatorial northwest Africa 140,000–70,000 yr BP as deduced from two marine pollen records. *Palaeogeography, Palaeoclimatology, Palaeoecology*, **66**, pp. 173–213, 10.1016/0031-0182(88)90199-X.

Hooghiemstra, H., Agwu, C. O. C. and Beug, H. J., 1986, Pollen and spore distribution in recent marine sediments: a record of NW-African seasonal wind patterns and vegetation belts. *Meteor Forschungsergeb*nisse, *Deutsche Forschungsgemeinschaft, Reihe C Geologie und Geophysik, Gebrüder Bornträger*, **40**, pp. 87–135.

Hooghiemstra, H., Bechler, A. and Beug, H. J., 1987, Isopollen maps for 18,000 years BP of the Atlantic offshore of northwest Africa: Evidence for paleowind circulation. *Paleoceanography*, **2**, pp. 561–582, 10.1029/PA002i006p00561.

Hooghiemstra, H., Lézine, A.-M., Leroy, S. A., Dupont, L. and Marret, F., 2006, Late Quaternary palynology in marine sediments: a synthesis of the understanding of pollen distribution patterns in the NW African setting. *Quaternary International*, **148**, pp. 29–44, 10.1016/j.quaint.2005.11.005.

Hooghiemstra, H., Stalling, H., Agwu, C. O. and Dupont, L. M., 1992, Vegetational and climatic changes at the northern fringe of the Sahara 250,000–5000 years BP: evidence from 4 marine pollen records located between Portugal and the Canary Islands. *Review of Palaeobotany and Palynology*, **74**, pp. 1–53, 10.1016/0034-6667(92)90137-6.

Hughen, K. A., Baillie, M. G., Bard, E., Beck, J. W., Bertrand, C. J., Blackwell, P. G. and Edwards, R. L., 2004, Marine04 marine radiocarbon age calibration, 0–26 cal kyr BP. *Radiocarbon*, **46**, pp. 1059–1086, 10.1017/S0033822200033002.

Huybers, P., 2011, Combined obliquity and precession pacing of Late Pleistocene deglaciations. *Nature*, **480**, pp. 229–232, 10.1038/nature10626.

Jahns, S., 1996, Vegetation history and climate changes in West Equatorial Africa during the Late Pleistocene and Holocene, based on a marine pollen diagram from the Congo fan. *Vegetation History and Archaeobotany*, **5**, pp. 207–213, 10.1007/BF00217498.

Jahns, S., Hüls, M. and Sarnthein, M., 1998, Vegetation and climate history of west equatorial Africa based on a marine pollen record off Liberia (site GIK 16776) covering the last 400,000 years. *Review of Palaeobotany and Palynology*, **102**, pp. 277–288, 10.1016/S0034-6667(98)80010-9.

Jolly, D., Prentice, I. C., Bonnefille, R., Ballouche, A., Bengo, M., Brenac, P. and Edorh, T., 1998, Biome reconstruction from pollen and plant macrofossil data for Africa and the Arabian peninsula at 0 and 6000 years. *Journal of Biogeography*, **25**, pp. 1007–1027, 10.1046/j.1365-2699.1998.00238.x.

Julier, A.C.M., Jardine, P.E., Adu-Bredu, S., Coe, A.L., Duah-Gyamfi, A., Fraser, W.T., Lomax, B.H., Malhi, Y., Moore, S., Owusu-Afriyie, K. and Gosling, W.D., 2018, The modern pollen-vegetation relationships of a tropical forest-savannah mosaic landscape, Ghana, West Africa. *Palynology*, **42**, pp. 324–338, 10.1080/01916122.2017.1356392.

Le Houérou, H.N., Evenari, M. and Goodall, D.W., 1986, The desert and arid zones of Northern Africa. *Ecosystems of the World*, **12**, pp. 101–147.

Lee, S. Y. and Poulsen, C. J., 2005, Tropical Pacific climate response to obliquity forcing in the Pleistocene. *Paleoceanography*, **20**, article: PA4010, 10.1029/2005PA001161.

Leroy, SAG., Dupont, LM., 1994, Development of vegetation and continental aridity in northwestern Africa during the Late Pliocene: the pollen record of ODP 658. *Palaeogeography, Palaeoclimatology, Palaeoecology*, **109**, pp. 295–316, 10.1016/0031-0182(94)90181-3.

Lézine, A.-M., 1991, West African paleoclimates during the last climatic cycle inferred from an Atlantic deep-sea pollen record. *Quaternary Research*, **35**, pp. 456–463, 10.1016/0033-5894(91)90058-D.

Lézine, A.-M. and Cazet, J.P., 2005, High-resolution pollen record from core KW31, Gulf of Guinea, documents the history of the lowland forests of West Equatorial Africa since 40,000 yr ago. *Quaternary Research*, **64**, pp. 432–443, 10.1016/j.yqres.2005.08.007.

Lézine, A.-M. and Hooghiemstra, H., 1990, Land-sea comparison during the last glacial-interglacial transition: Pollen records from west tropical Africa. *Palaeogeography, Palaeoecology, Palaeoclimatology*, **79**, pp. 313–331, 10.1016/0031-0182(90)90025-3.

Lézine, A.-M. and Vergnaud-Grazzini, C., 1993, Evidence of forest extension in West Africa since 22,000 BP: A pollen record from the eastern tropical Atlantic. *Quaternary Science Reviews*, **12**, pp. 203–210, 10.1016/0277-3791(93)90054-P.

Lézine, A.-M., Lemonnier, K., Waller, M.P., Bouimetarhan, I., Dupont, L. and APD contributors, this volume, Changes in the West African landscape at the end of the African Humid Period. *Palaeoecology of Africa*, **35**, chapter: 6, 10.1201/9781003162766-6.

Lézine, A.-M., Leroux, M., Turon, J.L., Buchet, G. and Tastet, J.P., 1995, Transport pollinique et circulation atmospherique au large de l'Afrique tropicale occidentale au cours de la derniere deglaciation. *Bulletin de la Société Géologique de France*, **166**, pp. 247–257.

Lézine, A.-M., Tastet, J.P. and Leroux, M., 1994, Evidence of atmospheric paleocirculation over the Gulf of Guinea since the Last Glacial Maximum. *Quaternary Research*, **41**, pp. 390–395, 10.1006/qres.1994.1043.

Marret, F., 1994, Evolution paléoclimatique et paléohydrologique de l'Atlantique est-équatorial et du proche continent au Quaternaire terminal: contribution palynologique, kystes de dinoflagelles, pollen et spores. PhD thesis, University of Bordeaux I, 271 pp.

Marret, F., Scourse, J.D., Versteegh, G., Fred Jansen, J.H. and Schneider, R., 2001, Integrated marine and terrestrial evidence for abrupt Congo River palaeodischarge fluctuations during the last deglaciation. *Journal of Quaternary Science*, **16**, pp. 761–766, 10.1002/jqs.646.

Marret, F., Scourse, J., Fred Jansen, J.H. and Schneider, R., 1999, Climate and palaeooceanographic changes in west central Africa during the last deglaciation: palynological investigation. In *Comptes Rendus de l'Académie des Sciences. Série 2, Sciences de la Terre et des Planètes*, **1**, pp. 721–726.

McGregor, H.V., Dupont, L., Stuut, J. B.W. and Kuhlmann, H., 2009, Vegetation change, goats, and religion: A 2000-year history of land use in southern Morocco. *Quaternary Science Reviews*, **28**, pp. 1434–1448, 10.1016/j.quascirev.2009.02.012.

McManus, J.F., François, R., Gherardi, J.-M., Keigwin, L.D. and Brown-Leger, S., 2004, Collapse and rapid resumption of Atlantic meridional circulation linked to deglacial climate changes. *Nature*, **428**, pp. 834–837, 10.1038/nature02494.

Meadows, M.E., Rogers, J., Lee-Thorp, J.A., Bateman, M.D. and Dingle, R.V., 2002, Holocene geochronology of a continental shelf mud belt off southwestern Africa. *The Holocene*, **12**, pp. 59–67, 10.1191/0959683602hl521rp.

Nash, D.J. and Meadows, M.E., 2012, *Africa*. In *Quaternary Environmental Change in the Tropics*, edited by Metcalfe, S.E., Nash, D.J., (Wiley-Blackwell, Oxford, UK), pp. 79–150.

Nicholson, S.E., 2000, The nature of rainfall variability over Africa on time scales of decades to millennia. *Global and Planetary Change*, **26**, pp. 137–158, 10.1016/S0921-8181(00)00040-0.

Nicholson, S.E. and Grist, J.P., 2003, The seasonal evolution of the atmospheric circulation over West Africa and equatorial Africa. *Journal of Climate*, **16**, pp. 1013–1030, 10.1175/1520-0442(2003)016#amp;lt;1013:TSEOTA#amp;gt;2.0.CO;2.

Pielke, R.A., Avissar, R., Raupach, M., Dolman, A.J., Zeng, X. and Denning, A. S., 1998, Interactions between the atmosphere and terrestrial ecosystems: influence on weather and climate. *Global Change Biology*, **4**, pp. 461–475, 10.1046/j.1365-2486.1998.t01-1-00176.x.

Prospero, J.M. and Nees, R.T., 1986, Impact of the North African drought and El Nino on mineral dust in the Barbados trade winds. *Nature*, **320**, pp. 735–738, 10.1038/320735a0.

Prospero, J.M., Ginoux, P., Torres, O., Nicholson, S.E. and Gill, T.E., 2002, Environmental characterization of global sources of atmospheric soil dust identified with the Nimbus 7 Total Ozone Mapping Spectrometer (TOMS) absorbing aerosol product. *Reviews of Geophysics*, **40**, pp. 1–31, 10.1029/2000RG000095.

Rasmussen, T.L. and Thomsen, E., 2015, Palaeoceanographic development in Storfjorden, Svalbard, during the deglaciation and Holocene: Evidence from benthic foraminiferal records. *Boreas*, **44**, pp. 24–44, 10.1111/bor.12098.

Rasmussen, S.O., Andersen, K.K., Svensson, A.M., Steffensen, J.P., Vinther, B.M., Clausen, H.B., Siggaard-Andersen, M.-L., Johnsen, S.J., Larsen, L.B., Dahl-Jensen, D., Bigler, M., 2006, A new Greenland ice core chronology for the last glacial termination. *Journal of Geophysical Research*, **111**, D06102, 10.1029/2005JD006079.

Raymo, M.E. and Nisancioglu, K.H., 2003, The 41 kyr world: Milankovitch's other unsolved mystery. *Paleoceanography*, **18**, PA1011, 10.1029/2002PA000791.

Rossignol-Strick, M. and Duzer, D., 1979, West African vegetation and climate since 22,500 BP from deep-sea cores palynology. *Pollen et Spores*, **21**, pp. 105–134.

Sánchez Goñi, M.F., Eynaud, F., Turon, J.L. and Shackleton, N.J., 1999, High resolution palynological record off the Iberian margin: Direct land-sea correlation for the Last Interglacial complex. *Earth and Planetary Science Letters*, **171**, pp. 123–137, 10.1016/S0012-821X(99)00141-7.

Scourse, J., Marret, F., Versteegh, G.J., Jansen, J.F., Schefuss, E. and van der Plicht, J., 2005, High-resolution last deglaciation record from the Congo fan reveals significance of mangrove pollen and biomarkers as indicators of shelf transgression. *Quaternary Research*, **64**, pp. 57–69, 10.1016/j.yqres.2005.03.002.

Shannon, L.V., 1985, The Benguela ecosystem. I: Evolution of the Benguela physical features and processes. *Oceanography and Marine Biology*, **23**, pp. 105–182.

Shannon, L.V. and Nelson, G., 1996, The Benguela: Large-scale features and processes and system variability. In *The South Atlantic*, edited by Wefer, G.W.H., Berger and Webb, D.J. (Springer, Berlin, Heidelberg), pp. 163–210.

Shi, N. and Dupont, L.M., 1997, Vegetation and climatic history of southwest Africa: a marine palynological record of the last 300,000 years. *Vegetation History and Archaeobotany*, **6**, pp. 117–131, 10.1007/BF01261959.

Shi, N., Dupont, L.M., Beug, H. J. and Schneider, R., 1998, Vegetation and climate changes during the last 21 000 years in SW Africa based on a marine pollen record. *Vegetation History and Archaeobotany*, **7**, pp. 127–140, 10.1007/BF01374001.

Shi, N., Schneider, R., Beug, H. J. and Dupont, L.M., 2001, Southeast trade wind variations during the last 135 kyr: evidence from pollen spectra in eastern South Atlantic sediments. *Earth and Planetary Science Letters*, **187**, pp. 311–321, 10.1016/S0012-821X(01)00267-9.

Stuut, J.B., Zabel, M., Ratmeyer, V., Helmke, P., Schefuß, E., Lavik, G. and Schneider, R., 2005, Provenance of present-day eolian dust collected off NW Africa. *Journal of Geophysical Research: Atmosphere*s, **110**, D04202, 10.1029/2004JD005161.

Tyson, P.D., 1986, *Climatic Change and Variability in Southern Africa*, (Cape Town: Oxford University Press).

Tyson, P.D. and Preston-Whyte, R.A., 2000, *The Weather and Climate of South Africa*, (Cape Town: Oxford University Press).

Vallé, F., Westerhold, T. and Dupont, L.M., 2017, Orbital-driven environmental changes recorded at ODP Site 959 (eastern equatorial Atlantic) from the Late Miocene to the Early Pleistocene. *International Journal of Earth Sciences*, **106**, pp. 1161–1174, 10.1007/s00531-016-1350-z.

Vincens, A., Dubois, M.A., Guillet, B., Achoundong, G., Buchet, G., Kamgang Kabeyene Beyala, V., de Namur, C. and Riera, B., 2000, Pollen-rain-vegetation relationships along a forest-savanna transect in southeastern Cameroon. *Review of Palaeobotany and Palynology*, **110**, pp. 191–208, 10.1016/S0034-6667(00)00009-9.

Watrin, J., Lézine, A.-M., Gajewski, K. and Vincens, A., 2007, Pollen–plant–climate relationships in sub-Saharan Africa. *Journal of Biogeography*, **34**, pp. 489–499, 10.1111/j.1365-2699.2006.01626.x.

Webb, T.I., 1986, Is vegetation in equilibrium with climate? How to interpret Late-Quaternary pollen data. *Vegetatio*, **67**, pp. 75–91, 10.1007/BF00037359.

White, F., 1983, The vegetation of Africa, In *Natural Resources Research*, Vol.20, (UNESCO, Paris), 384 pp.

Woodward, F.I. and McKee, I.F., 1991, Vegetation and climate. *Environment International*, **17**, pp. 535–546, 10.1016/0160-4120(91)90166-N.

Woodward, F.I. and Williams, B.G., 1987, Climate and plant distribution at global and local scales, *Vegetatio*, **69**, pp. 189–197, 10.1007/BF00038700.

Wright, J.W., 1952, Pollen dispersion of some forest trees, In *Research paper 46* (Upper Darby, US Department of Agriculture Service-Forest Service Northeastern Forest, Experiment Station), pp. 1–45.

Zhao, X., Dupont, L., Cheddadi, R., Kölling, M., Reddad, H., Groeneveld, J. and Bouimetarhan, I., 2019, Recent climatic and anthropogenic impacts on endemic species in southwestern Morocco. *Quaternary Science Reviews*, **221**, pp. 1–15, 10.1016/j.quascirev.2019.105889.

Zhao X., Dupont L., Meadows, M.E. and Wefer, G., 2016, Pollen distribution in the marine surface sediments of the mudbelt along the west coast of South Africa. *Quaternary International*, **404**, pp. 44–56, 10.1016/j.quaint.2015.09.032.

CHAPTER 23

Inside-of-Africa: How landscape openness shaped *Homo sapiens* evolution by facilitating dispersal and gene-flow in Middle and Late Pleistocene Africa

Mick N.T. Bönnen, William D. Gosling & Henry Hooghiemstra

Institute for Biodiversity and Ecosystems Dynamics, University of Amsterdam, Amsterdam, The Netherlands

ABSTRACT: *Homo sapiens* as a clade originated *c*. 500 thousand years before present (500 ka) as it diverged from *Homo neanderthalensis*. The topic of early *H. sapiens* evolution and dispersal since this cladogenesis has long been of interest in scientific literature and public debate. The development of this field has been significantly accelerated in recent years by the advances made in the scientific fields of archaeological, anthropological and genetic research; exemplified by the publication of the earliest observed fossil belonging to the archaic *H. sapiens* clade at Jebel Irhoud, Morocco, dated at *c*. 315 ka in 2017. Recent evidence from these fields opposes the long-held view that anatomically modern humans (AMH) evolved linearly from a single population. Instead, a pan-African model of evolution is proposed, whereby geographically isolated *H. sapiens* populations, possibly shaped and maintained by ecological boundaries, evolved independently with fluctuating degrees of gene-flow over time. A thorough understanding of the ecological context these hominins experienced has long been hampered by spatial and temporal gaps in the African palaeovegetation record. Records of past vegetation that cover timescales relevant to the emergence of AMHs now exist that are relevant to environmental change in northen, southern, eastern, western and central Africa. This means it is becoming possible to explore how hominin evolutionary development coincided with the changing vegetational (habitat) context. We present the idea of a three-stage 'Inside-of-Africa' environmental framework for hominin evolution: (i) a predominance of hospitable vegetation 500–400 ka facilitating initial dispersal of archaic *H. sapiens*, (ii) a predominance of ecological barriers (e.g. deserts and rainforests) 400–250 ka limiting dispersal and gene-flow, causing independent evolution, and (iii) a predominance of hospitable vegetation 250–100 ka (re-)connecting populations and resulting in the combination of the full suite of contemporary AMH characteristics. To test this framework we review and synthesize all available long (>*c*. 100 ka) lacustrine palynological records relevant to past vegetation change across Africa. We find the past vegetation data supports the waxing and waning of hospitable vegetation and ecological barriers during the last 500 ka in line with the timings proposed in our environmental framework.

DOI 10.1201/9781003162766-23

23.1 INTRODUCTION

23.1.1 Archaic *Homo sapiens*

Africa is regarded as the cradle of our own species (Mounier and Mirazón Lahr 2016) because it is where the archaic *Homo sapiens* clade diverged from the *H. heidelbergensis c.* 700-400 thousands of years before present (ka) (Stringer 2016), and where we shared our last common ancestor with *H. neanderthalensis c.* 440-410 ka (Endicott *et al.* 2010) (Figure 1). Although the exact location of cladogenesis remains unknown, these archaic forms of *H. sapiens* radiated all over Africa. Currently, the earliest 'archaic' manifestations of the *H. sapiens* clade are found in Jebel Irhoud, Morocco, comprising of cranial fossils dated to be about 315 ka (Hublin *et al.* 2017; Richter *et al.* 2017). Later dated archaic *H. sapiens* remains include fossils from Florisbad in South Africa (*c.* 260 ka) (Grun *et al.* 1996) and Omo Kibish and Herto, both Ethiopia, which have estimated ages of *c.* 195 and *c.* 160 ka respectively (McDougall *et al.* 2005; White *et al.* 2003). Even though all these archaic humans belong to the *H. sapiens* clade, and share a fairly recent last common ancestor, their fossilized remains show distinct spatial and temporal morphological variation when compared with each other, and with anatomically modern humans (AMH; Gunz *et al.* 2009).

The pan-African spread of hominins, in conjunction with the array of morphological variation (Hammond *et al.* 2017; Lieberman *et al.* 2002; Neubauer *et al.* 2018; Pearson 2008), has lead the modern scientific discourse to consider a mosaic-like pattern for the evolution of modern human traits (Gunz *et al.* 2009; Scerri *et al.*, 2017a; Stringer 2016). A mosaic-like pan-African pattern could have been created if populations became (semi-)isolated for millennia by distance and/or ecological barriers (such as deserts or dense rainforests). The consequence could be that specific modern human traits might have evolved independently in different regions and became combined later when populations reconnected. The amalgamation of our modern characteristics could have therefore potentially been strongly mediated by past environmental changes. For example, periods of increased landscape connectivity between hominin populations, brought about by an increasing predominance of open vegetation, could have facilitated gene flow between small semi-isolated populations across Africa (Bertola *et al.* 2016). Recent genetic studies have suggested that between *c.* 250-100 ka most extant lineages emerged (Lipson *et al.* 2020). The eventual admixture of individual traits associated with these populations led to the homogenization of the human phenotype, thereby refuting the notion that our lineage can be traced back to one specific region, but favouring a pan-African view of *H. sapiens* evolution instead. Analyses of whole-genome sequences of contemporary African populations provide a time-dependent estimate of gene flow that suggests a pattern of population separation, followed by isolation, and then subsequent recombination (Wang *et al.* 2020).

Similar to morphological characteristics, spatiotemporal diversifications since our last common ancestor with *H. neanderthalensis* are also reflected in behavioural practices such as material cultures. An abandonment of large and crude cutting tools (e.g. hand axes) in favour of more sophisticated techniques for tool-carving ('prepared-core' technologies) occurred almost simultaneously on a pan-African scale. These innovations in stone tool technology followed a long period of stasis (the Early Stone Age [ESA] *c.* 3 Myr to 200 kyr BP) and mark the start of the Middle Stone Age (MSA) period (*c.* 280-20 kyr BP, Galway-Witham *et al.* 2019). The MSA is also when regionally distinctive material cultures involving complex stone tools first become apparent in the archaeological records. For example during the MSA features such as tangled implements (securing stone tools into wooden hafts) emerged in northern African populations (Scerri 2017b), central African materialistic culture started to be characterized by distinctive

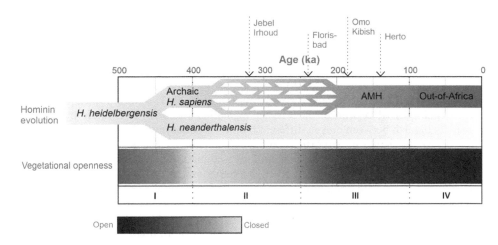

Figure 1. 'Inside-of-Africa' conceptual environmental framework related to the evolution and development of hominins. Vegetation openness relates to the proportion of open (savannah) or closed (forest) vegetation on the continent. General characteristics of vegetation for each period are indicated by (I) open and hospitable to hominins, (II) closed with more environmental barriers to hominins, and (III) open and hospitable for hominins. AMH = Anatomically Modern Humans. Out-of-Africa = the period of human dispersal from the African continent (IV). Key archaeological sites yielding *H. sapiens* fossils indicated are Jebel Irhoud, Morocco (Hublin *et al.* 2017), Florisbad, South Africa (Grun *et al.* 1996), Omo Kibish, Ethiopia (McDougall *et al.* 2005), and Herto, Ethiopia (White *et al.* 2003).

stone tools (Scerri 2017a), while southern African cultural artifacts started to include complex material aspects; such as the use of ochre, bone tools and shell beads (D'Errico *et al.* 2017).

23.1.2 Hypothesis development

In this review, we focus on how changing environments could have shaped hominin evolution inside Africa. Building on the literature, we propose here a new framework for understanding this interplay between archaic *H. sapiens* and their contemporary landscapes. Our 'Inside-of-Africa' framework includes alternating periods of environmental hospitability and ecological barriers that either limit or encourage dispersal (Figure 1):

  I. Initial dispersal of archaic *H. sapiens* across Africa (*c.* 500-400 ka) was facilitated by increased landscape openness, which made areas generally more hospitable (hereafter defined as landscapes not dominated by dense (rain)forests or deserts, but rather by open ecosystem mosacis, including: (wooded) grasslands, savannahs, and gallery forests).
  II. Intraspecific variation was promoted and cultural development fragmented between (*c.* 400-250 ka) by a rise in the number of environmental barriers (e.g. arid deserts and impenetrable forests), which (semi-)isolated *H. sapiens* populations to environmentally constrained hospitable regions (refugia).
  III. The (re-)combination of traits into the full suite of AMH characteristics (*c.* 250-100 ka) was facilitated by increased landscape openness.

This review collates lacustrine palynological records with a pan-African spatial coverage from the period relevant to hominin evolution as outlined above (Figure 1). These data assess the current state of available information on past vegetation from terrestrial sites. We use the synthesized past vegetation information to test ideas on the evolution, development and dispersal of hominins. Specifically, we classify the long palynological records as indicative of relatively more open, or closed, vegetation structure (grasslands to dense forests) and identify any directional vegetation trends coincident with hominin evolution.

23.2 AFRICAN ENVIRONMENTS DURING THE MIDDLE AND UPPER PLEISTOCENE

23.2.1 Current ecosystems

The wide range in topography and climatic conditions across Africa caused varied landscapes that are home to a plethora of ecosystems (Dupont 2011; White 1983). Following Dupont's (2011) simplified version of the African vegetation classifications made by White (1983), we can distinguish between eleven different biomes in contemporary Africa (Figure 2A).

The continent's most northern and south-western extent is typified by arid biomes. Generally speaking, we find Mediterranean forests along the north-African coast from Morocco to Egypt adjoining the Mediterranean Sea, while Africa's most southern tip, the southern-African coasts and Cape Fold Mountains, are dominated by fynbos vegetation. Mediterranean forests have a significant presence of arboreal species; typically oak (*Quercus*) and maple (*Acer*) in broadleaved forests, while pine (*Pinus*) and yew (*Taxus*) dominate the coniferous forests. The fynbos vegetation on the other hand has a low density of arboreal taxa, as it is mostly dominated by shrubby vegetation like Ericaceae. Further inland deserts occur. In the north, the Sahara (best-represented families being Asteraceae, Leguminosae and Poaceae (Ozenda 1991)) stretches from the Red Sea to the Atlantic. In the south the Namib desert covers the entire Namibian coast and parts of the Angolan and southern African coast. The transition between the Mediterranean forests and the Sahara, the Sahel region, the African Horn, and the area of the fynbos-to-desert transition in the south (the Kalahari desert) are classified as semi-deserts, where we find low densities of shrubby vegetation like Asteraceae. Poaceae-dominated systems are found throughout Africa. Grass savannahs are ubiquitous at the 15°N and the 20°S latitudes and in east African Rift Valley system. Edaphic grasslands (those controlled by soil conditions instead of climate, fire regimes,

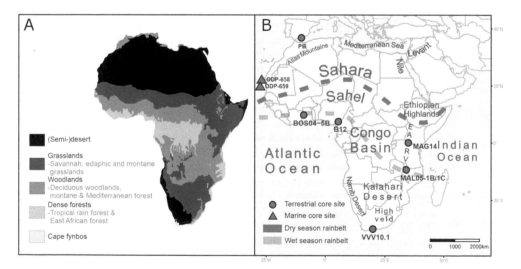

Figure 2. (A) Modern African vegetation simplified from Dupont (2011). (B) Location of the seven pollen records obtained from lake settings relevant to past vegetation change in Africa relevant to hominin evolution over the last 500 thousand years. PB = Padul Basin, Spain (Torres *et al.* 2020), VVV10.1 = Vankervelsvlei, South Africa (Quick *et al.* 2016), MAG14 = Lake Magadi, Kenya (Owen *et al.* 2018), MAL05-1B/1C (MAL05-1B and MAL05-1C) = Lake Malawi, Malawi (Beuning *et al.* 2011; Ivory *et al.* 2018), B12 = Lake Bambili, Cameroon (Lézine *et al.* 2019), and BOS04-5B = Lake Bosumtwi, Ghana (Miller and Gosling 2014). Marine cores presented for additional context are ODP-658 (Dupont 1989; Dupont and Hooghiemstra 1989) and ODP-659 (Tiedeman *et al.* 1994).

or herbivory) can be found in Angola, Zambia and the Democratic Republic of Congo. Montane grasslands are prevalent in the arid high-altitude regions of central Africa, the East-African Highlands and the South-African Highveld. Forested areas are restricted to the moist equatorial regions and the high elevation Afromontane forests found in East-Africa and South Africa. Tropical rainforests extend across the basin of the Congo River and its tributaries but are also found along the coast of western Africa, spreading from southern Guinea all the way east to southern Cameroon. At higher latitudes tropical moist forests transitions into deciduous forests/woodlands as the vegetation becomes increasingly influenced by seasonality in precipitation.

To understand how vegetation association, and thus potential habitats or barriers for early *H. sapiens*, varied over time it is important to understand the climatic factors that relate to vegetation formation and how this has changed. Orbital precession, with a *c.* 21,000 year cycle, has a strong influence on long-term climate change in the low elevation tropics (Clement *et al.* 2004). Precession directly affects when in the year the earth is closest to the sun, and thus governs the amount of insolation received during each season. The strength and the annual latitudinal reach of the tropical rain belts are governed by changes in insolation that consequently are directly influenced by changes in the precessional cycle. Specifically, at the peak in the precessional cycle, the monsoons can reach further northwards into the otherwise arid part of Africa (Dupont and Hooghiemstra 1989; Frumkin *et al.* 2011). This additional moisture delivery has a significant impact on vegetation cover, and climate-vegetation simulations suggest that this processional climate signal is a dominant control on vegetation change at low latitudes (Gosling and Holden 2011).

23.2.2 Ecological boundaries and refugia for archaic *Homo sapiens*

Environmental changes, prompted by changes in precipitation related to multi-millennial-scale climate oscillations have been shown to shape and maintain ecological boundaries that eventually limited gene-flow between populations in lions (*Panthera leo*), thereby creating distinct genetic clades (Bertola *et al.* 2016). Similar biogeographic patterns of speciation seem to exist in a wide array of African mammals (list of references in Bertola *et al.* 2016), and the same mechanism has been postulated for archaic *H. sapiens* populations (Scerri *et al.* 2018). Habitats largely inhospitable to archaic *H. sapiens*, such as arid deserts and dense rainforests (Bailey *et al.* 1989; Blome *et al.* 2012; Coulthard *et al.* 2013), expanded and retracted over time-scales of millennia in response to orbital scale vegetation change (Dupont 2011). The expansion of deserts and rainforests often occurred at the expense of open ecosystems (e.g. savannahs) which are thought to have been the preferred habitat of early hominins based on the close affinity between archaeological evidence of hominins and grasslands ecosystems (Cerling *et al.* 2011; Larrasoaña *et al.* 2013). It can therefore be hypothesized that climatic conditions that caused a decline in savannah ecosystems in favour of deserts and rainforests would have caused a loss of habitat for archaic *H. sapiens*. This loss of hospitable environment could have lead to the migration of some *H. sapiens* populations and a loss in connectivity between others and thus generated the population structure needed for independent evolution. As the remaining hominin populations became more isolated they likley retreated to 'refugia' (isolated regions where environmental conditions allowed humans to persevere amidst inhospitable conditions (Mirazón Lahr and Foley 2016)) which have also been highlighted as catalysts for evolutionary change in *H. sapiens* (Stewart and Stringer 2012).

23.2.3 Palaeoenvironmental reconstructions

Marine palynological records have been used to provide a pan-African view of vegetation change (Dupont 2011). The advantage of marine records is that they are well-dated and yield the ability to record large scale vegetation change from a wide source area (Castañeda *et al.* 2016; Dalibard *et al.* 2014; Hooghiemstra *et al.* 2006; Jahns *et al.* 1998; Zhao *et al.* 2003). However, interpretation of past vegetation cover from marine pollen records is challenging (Frédoux 1994; Hooghiemstra *et al.* 1986) because of the inherent wide pollen source area, which means that the signal is influenced by: (i) the intricate complexity of wind current and fluvial dynamics that muddle the opportunity to precisely identify the source (Dupont *et al.* 2000), and (ii) a low spatial resolution that makes it impossible to see vegetation dynamics on a local scale. For example in regions dominated by arid vegetation but where lush vegetation boarders rivers, the marine pollen signal could be biased towards the lush vegetation as it produces relatively more pollen and they are easily transported into the ocean sediments. In contrast to the marine records, the pollen found within a lake (diameter 5–20 km) likely reflects local/extra-local (10–100 s m) and regional (1–10 s km) vegetation (Jacobson and Bradshaw 1981). Therefore, lacustrine pollen records are more suitable to reconstruct vegetation changes at the local and regional scales, which allows the habitats of archaic *H. sapiens* to be investigated.

23.3 SITE SELECTION

To provide the vegetational context in which early *H. sapiens* evolution took place we selected palaeovegetation records that contain information relevant from across Africa. The most appropriate records were chosen to represent each of five African regions (northen, southern, eastern, western and central) based on two criteria:

- Sedimentary record covering at least the last *c.* 100 ka.
- At least one proxy of past vegetation change has been published from the record.

Six lacustrine records were identified meeting these criteria from in, and around, the African continent (Table 1; Figure 2B).

23.4 LONG-TERM VEGETATION CHANGES IN AFRICA

One record of past vegetation was obtained from each region, except for eastern Africa, from which two were obtained (Figure 3). The records ranged from *c.* 90 ka to 800 ka in duration and past vegetation information was obtained from fossil pollen at all sites. To place these records in the context of global climatic change we refer to the global glacial (cool) and interglacial (warm) Marine Isotope Stages (MIS) as defined by Imbrie *et al.* (1984). The six records are now reviewed in turn, in the order of the earliest archaeological evidence of *H. sapiens* in the vicinity of the core site: (i) Padul Basin, northern Africa region (Torres *et al.* 2020), (ii) Vankervelsvlei, southern Africa region (Quick *et al.* 2016), (iii) Lake Magadi, eastern Africa region (Owen *et al.* 2018), (iv and v) Lake Malawi, eastern Africa (Beuning *et al.* 2011; Ivory *et al.* 2018), (vi) Lake Bambili, central Africa (Lézine *et al.* 2019), and (vii) Lake Bosumtwi, western Africa (Miller and Gosling 2014).

23.4.1 Padul Basin, Spain

No lacustrine sedimentary record spanning *c.* 100 kyr has, to date, been recovered or analysed for pollen from continental north-west Africa. Therefore, the sediment record from the Padul Basin (PB) from the Granada province in south-east Spain (Torres *et al.* 2020) was chosen to provide the closest available vegetational context for the earliest known *H. sapiens* found to date (Figure 3A). The archaic *H. sapiens* fossils found at Jebel Irhoud (Morocco) have been dated to 315 ± 34 ka and include facial bones (Hublin *et al.* 2017). The PB sediment core was retrieved from a peat bog inside a closed flat depression, with a catchment area of 44 km^2, located 720 m

Table 1. List of selected lacustrine pollen records and metadata.

Region	Site name	Core name	Country	Latitude Longitude	Core length (m)	Dating method	# age samples	Timescale (ka)	Referenc	Earliest human evidence (ka)
Northern Africa	Padul Basin	PB	Spain	37°01′01″N 3°36′07″W	107	^{14}C/U/Th/palaeomag	27	800	Torres et al. (2020)	315
Southern Africa	Vankervelsvlei	VVV10.1	South Africa	34°044.58″S 22°54′15.06″E	4	^{14}C/Luminescence	6	110	Quick et al. (2016)	260
Eastern Africa	Lake Magadi	MAG14	Kenya	1°54′3.56″S 36°14′48.52″E	194	^{14}C/Ar-Ar/U-series	18	1000	Owen et al. (2018)	195
Eastern Africa	Lake Malawi	MAL05-1C	Malawi	11°17′37″S, 34°26′04″E	90	^{14}C/Luminescence	25	135	Beuning et al. (2011)	—
Eastern Africa	Lake Malawi	MAL05-1B/1C	Malawi	11°17′37″S, 34°26′04″E	170	^{14}C/palaeomag/Ar-Ar	41	600	Ivory et al. (2018)	—
Central Africa	Lake Bambili	B12	Cameroon	6°0′19.56″N 10°15′46.14″E	26.6	^{14}C	29	90	Lézine et al. (2019)	30
Western Africa	Lake Bosumtwi	BOS04-5B	Ghana	6°30′6.04″N 1°24′52.14″W	295	^{14}C/Luminescence/U-series/Ar-Ar	135	1000	Miller and Gosling (2014)	12

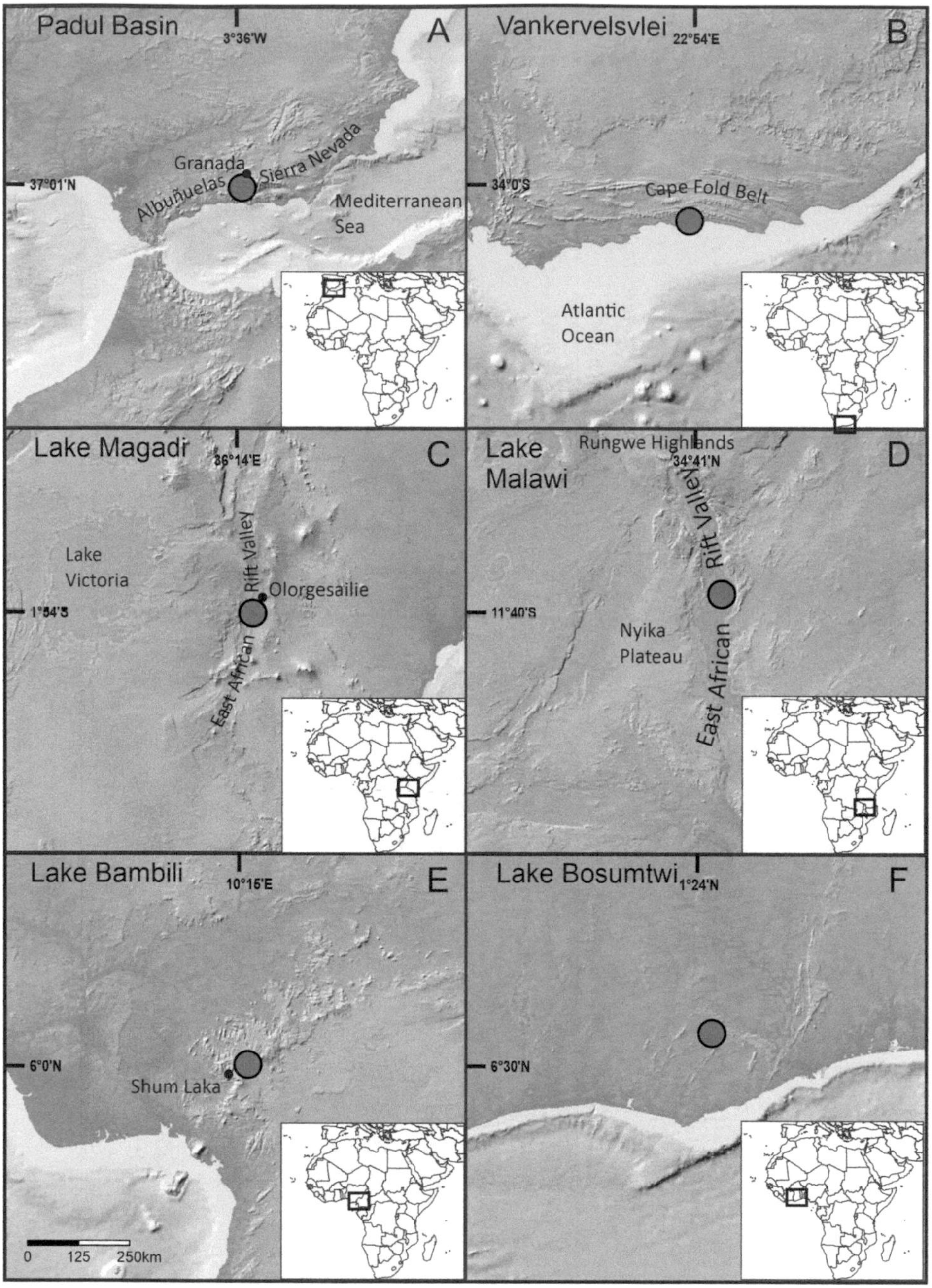

Figure 3. Regional topographic context (main panels) and continental geographic contexts (inset panels) of lacustrine records reviewed in this study. (A) Padul Basin (Spain). (B) Vankervelsvlei (South Africa). (C) Lake Magadi (Kenya). (D) Lake Malawi (Malawi). (E) Lake Bambili (Cameroon). (F) Lake Bosumtwi (Ghana).

above sea level (m asl), and enclosed by mountains to the west (Albuñuelas range) and east (the Sierra Nevada). The altitudinal range of the mountains around the Padul Basin provides habitat for a wide variety of vegetation; from evergreen oak groves below 700 m to pine forests at the higher elevations (Rivas-Martínez 1987). Importantly, even though PB and Jebel Irhoud are located *c*. 750 km apart, they still encompass the same temperate Mediterranean Köppen climate area (Chen and Chen 2013; Rubel and Kottek 2010).

Pinus dominates the pollen record from PB during MIS 13.13 (*c*. 500 ka), making up *c*. 60% of the pollen spectra while the presence of Poaceae is low and accounts for only 10%. *Pinus* and Poaceae start to diminish between MIS 13.11 and 12.1 (480-425 ka) as mesophilic taxa (*Acer, Alnus, Quercus, Corylus, Carpinus, Fraxinus, Ulmus*) and riparian taxa (*Cornus, Populus, Salix*) become more abundant. Subsequently, a period of low pollen preservation is found until the start of MIS 11.2 (385 ka), which ends with a pollen assemblage dominated by *Pinus* that, in turn, transitions in to a steppic vegetation (*Artemisia, Ephedra*) over a *c*. 45 kyr period. The transition to steppic vegetation is not smooth, with a peak in *Pinus* (70%) around the end of MIS 11 (360 ka), followed by a rapid decline to *c*. 10% at the start of MIS 10.2 (341 ka). Poaceae has a low presence throughout this transitional period, averaging *c*. 10%. Between MIS 10.2 and MIS 9.1 (341-310 kyr BP), the pollen assemblage is relatively diverse. Throughout this period Poaceae declines from *c*. 40% *c*. 310 ka to *c*. 10%, while *Pinus*, steppic taxa and Amaranthaceae all see a marked rise. From the start of MIS 8.5 (287 ka) Poaceae rises to its maximum abundance (*c*. 80%) in the entire record and *Pinus*, the most prominent taxon throughout the palynological record, is reduced to *c*. 5%. The dominance of the pollen assemblage by Poaceae continues for *c*. 5 kyr and then recedes back to *c*. 10% by 280 ka. From 280 ka until 170 ka Poaceae is *c*. 10%, and *Pinus* is *c*. 40%, while *Quercus* is most abundant between MIS 7.5-6.6 (238-183 ka). These episodes between MIS 7.5 and 6.6 (238-183 ka) are also characterized by a growing presence of steppic taxa, especially within the timeframe where *Pinus* percentages show lower values. Between *c*. 183 ka and *c*. 170 ka there is a hiatus in the pollen record. Pollen is recorded again halfway through MIS 6 (*c*. 170 ka), starting with a high abundance of *Pinus* (*c*. 80%) that declines to 30% at 130 ka, with steppic taxa and Poaceae present throughout. Steppic taxa and Poaceae then practically disappear from the record during MIS 5 (122-71 ka), but *Pinus* rises again reaching *c*. 80% at 120 ka. *Pinus* percentages keep on fluctuating around 50% for the remainder of the record up until the present, and the distinct changes in the vegetational compositions are due to fluctuations in pollen percentages of other taxa. The shrubby xerophilous taxa show a stable representation of *c*. 15% during the MIS 4-2 interval (65-12 ka), after which their presence disappears. Similar trends can be found in the steppic taxa, which also steadily increase up to 40%. When steppic taxa disappeard Poaceae shows another peak, contributing close to 50% to the pollen sum; however, Poaceae, xerophilous (Amaranthaceae) and steppic (*Artemisia, Ephedra*) taxa are no longer represented towards the end of PB record.

23.4.2 Vankervelsvlei, South Africa

The sediment core recovered from Vankervelsvlei wetlands (34°0′44.58″S, 22°54′15.06″E) (VVV10.1) is located on the southern coastal plains of South Africa (Quick *et al.* 2016), a stretch of land bordering the Atlantic Ocean in the south, isolated from inland South Africa by the mountainous Cape Fold Belt (Figure 3B). These wetlands encompass a catchment of *c*. 0.5 km^2 and are situated on the landward edge of the Wilderness embayment at an elevation of 153 m asl. Vankervelsvlei is hydrologically isolated from the regional aquifer and fed exclusively by rainfall (Roets *et al.* 2008). Today the local vegetation is dominated by Cyperaceae mats floating on the wetland, fynbos-dominated vegetation (e.g. *Passerina*, Ericaceae, *Leucadendron* and Restionaceae) on the wetland's margins, and Afrotemperate forest types to the north (including *Ocotea, Olea* and *Podocarpus*). Compton (2011) proposed that these isolated coastal plains could be the geographical point of origin for AMH, suggesting the periodic sea-level changes induced by interglacial-glacial cycles provoked competition and innovation. However, subsequent research, based on mitochondrial DNA lineages, indicated that *H. sapiens* migrated to the southern African coast from wetlands in modern-day Botswana *c*. 100 ka (Chan *et al.* 2019); although these interpretations have been subsequently challenged (Schlebusch *et al.* 2021) so uncertainty prevails. However, the region remains important from an evolutionary perspective because it contains the Blombosgrot cave which is one of the most prominent archaeological sites in southern Africa regarding the cultural development in *H. sapiens* (Douze *et al.* 2015). The Blombosgrot cave, located 160 km west of Vankervelsvlei, has been inhabited since at least 140 ka (Jacobs *et al.* 2006) and contains toolkits providing evidence for long-term planning and an elementary knowledge of chemistry *c*. 100 ka (Henshilwood *et al.* 2011).

The VVV10.1 record has a basal age in MIS 6.3 (146 ka), but no pollen was preserved until MIS 5.5 (122 ka), with another hiatus in the record during MIS 3.0-2.0 (24-12 ka). Afromontane forest taxa, predominantly *Podocarpus,* characterize the early sections but are barely found throughout the rest of the pollen record. Percentages for Afromontane taxa vary from *c*. 5 to *c*. 15% in MIS 5.4-5.3 (107-99 ka), but gradually decline until they are lost from the record in late MIS 5 and only sporadically occur in trace amounts thereafter. The most dominant vegetational elements between MIS 5.4 and 5.3 (107-99 ka) are the fynbos taxa. Among the fynbos taxa, Ericaceae is particularly abundant (reaching *c*. 40%), with *Stoebe* (20%) also significant. Subsequently, Ericaceae abundance remains high (average *c*. 60%) until MIS 3.0 (24 ka). The Ericaceae record starts declining around the transition to MIS 2.0 to about 45%. In the following *c*. 15 kyr pollen is poorly preserved, but Ericaceae ends up with percentages less than 5% in the most recent samples, which are dominated by unprecedented high values of Cyperaceae of over 40%.

23.4.3 Lake Magadi, Kenya

Lake Magadi is a seasonally flooded pan with a surface catchment of *c*. 100 km^2, situated 606 m asl in the south Kenyan part of the east African Rift Valley (Figure 3C). The Lake Magadi record (MAG14-2A) (Owen *et al.* 2018), was chosen based on its proximity (*c*. 25 km) to the Olorgesailie archaeological site (Behrensmeyer *et al.* 2002) and its length; covering the past one million years. The core was drilled as part of The Hominin Sites and Palaeolakes Drilling Project (HSPDP) initiative (Campisano *et al.* 2017). The Olorgesailie site is of particular interest because its archaeological surveys document a transition from the Acheulean to the Middle Stone Age mode of stone toolmaking between *c*. 320 and 305 ka (Brooks *et al.* 2018; Deino *et al.* 2018). This period of technological and cultural development is itself associated with the evolution of *H. sapiens*. Furthermore, this core is also the closest to the archaeological sites which have yielded the earliest recognized anatomically modern *H. sapiens* fossils in eastern-Africa, Omo Kibish (*c*. 195 kyr BP) and Herto (*c*. 160 kyr BP) at 700 and 1400 km respectively, both in Ethiopia.

The pollen record of MAG14-2A shows vegetation change over the last 500 ka, although the MIS 11.2 to 9.0 interval (375-303 ka) is devoid of pollen. Pre-500 ka vegetation was dominated locally by Poaceae and Cyperaceae, both making up 80% in at least one sample dated before the onset of MIS 13.13 (502 ka). Afromontane forest taxa make up the majority of the regional pollen signal, and *Podocarpus* dominates this regional signal with percentages of *c*. 80%, indicating a significant presence of arboreal taxa. By MIS 13 the pollen spectra seem to represent dry open grassland, almost completely comprised of Poaceae and deprived of arboreal taxa, except for low (*c*. 10%) percentages of *Podocarpus*. Poaceae shows a stable representation during the subsequent 125 ka, while both Cyperaceae and *Podocarpus* show an almost linear increase in representation. Although the following 75 kyr were not sampled, percentages of both local and regional pollen taxa seem to remain stable, potentially suggesting stable environmental conditions during this interval. Vegetation diversifies with the onset of MIS 6.6 (183 ka), during which *Olea, Juniperus, Commiphora, Acacia* and Amaranthaceae fluctuate between 0 and 40%.

23.4.4 Lake Malawi, Malawi

The sediment records MAL05-1B and -1C (Beuning *et al*. 2011; Ivory *et al*. 2018) were retrieved from Lake Malawi, the southernmost lake in the east African Rift Vally system (Figure 3D). Lake Malawi at 472 m asl has a surface area of *c*. 29,500 km² and a catchment area of *c*. 65,000 km² bordering the Rungwe volcanic highlands in the north and the Nyika plateau in the west. The mountainous topography means that the modern vegetation surrounding Lake Malawi is characterized by Afromontane forests (*Podocarpus*) and Afromontane grasslands in the west, while scrubland and woodland vegetation types can be found east of Lake Malawi. Despite the absence of AMH fossils in the vicinity of Lake Malawi, the vegetation records from the MAL05-1B/1C records are still relevant to AMH dispersal. The location of origin of AMH and the direction of their dispersal events is still debated; both southern Africa (e.g. Jacobs *et al*. 2008; Schlebusch *et al*. 2017) and north-eastern Africa (e.g. Lamb *et al*. 2018; Walter *et al*. 2000) have been proposed as regions harbouring the population of early humans that would ultimately disperse throughout the rest of the world. This makes Lake Malawi a likely passing point in the corridor used in either a northward or a southward dispersal of these AMH.

MAL05-1B and MAL05-1C together comprise 170 metres of sediment, of which 1C represents the youngest 62 metres and 1B the older 108. The MAL05-1B/1C pollen stratigraphy shows distinct antiphasing between periods of high arboreal (i.e. *Podocarpus*) pollen and periods with a distinct presence of non-arboreal (notably Poaceae) pollen. These arboreal phases are marked by analogous trends in other Afromontane taxa (e.g. *Apodytes*, Ericaceae, *Ilex, Juniperus, Kiggelaria*, Myrtaceae, *Olea* and *Syzygium*), and palynological studies on contemporary Afromontane forest suggests that *Podocarpus* percentages of >40% indicate dense forests, reaching all the way down to the shores of Lake Malawi (Ivory *et al*. 2018). The non-arboreal phases (indicated by grasses, herbs like Amarathaceae or woodland trees such as *Acacia*) on the other hand are indicative of dry woodland/wooded grassland type of ecosystems. In total the MAL05-1B/1C cores both record seven forest phases and seven grassland phases over the past 500 kyr. These start with a grassland phase (*c*. 75% Poaceae) lasting until MIS 12.4 (470 ka), when an extensive forested period is initiated for the subsequent 200 kyr. During this phase *Podocarpus* percentages fluctuate between *c*. 45% (MIS 12.32 and MIS 11.2, 450 and 375 kyr respectively) and *c*. 10% (MIS 9.2, 325 kyr), averaging on *c*. 30%. While *Podocarpus* percentages of 10% are low, arboreal taxa still maintain a relative dominance over non-arboreal pollen due to high percentages of *Celtis* (*c*. 15%), especially during MIS 11.2. Subsequent transitions between forest and open ecosystems are more frequent. Throughout the remainder of MAL05-1B we find four more grassland phases; MIS 8.4-7.5 (274-241 ka), MIS 7.2-7.0 (209-193 ka), MIS 6.6-6.5 (189-165 ka) and MIS 6.2 (142-137 ka), interspersed with forest phases at MIS 7.5-7.2 (241-209 ka), MIS 6.6 (193-189 ka) and MIS 6.5-6.2 (165-142 ka).

The base of record MAL05-1C (Beuning *et al*. 2011) is dated to the later part of MIS 6.2 (135 ka), and immediately records an increase in *Podocarpus* from near absence to *c*. 35% during the period MIS 6.2-6.0 (135-128 ka). A similar increase in *Podocarpus* is noted at the start of MIS 5 (between MIS 5.5 and 5.4 (122-107 ka)). Both these increases in *Podocarpus* are concomitant with an opposite trend in Poaceae percentages, which shows values of 80% between MIS 6 and 5.5 (122 ka), but 40-30% before and after this interval. The halfway point of MIS 5 is marked by a decline of *Podocarpus* percentages again coeval with another increase in Poaceae. Poaceae reaches its maximum at MIS 5.2 (87 ka) when its pollen make up nearly 100% of the pollen spectra. Samples from MIS 4 onwards show Poaceae stabilizing at 40%, a value that continues throughout the rest of the record, except for a short increase to *c*. 60% and a short depression to *c*. 20% at MIS 4.2 (65 ka) and 3.3 (53 ka), respectively. Meanwhile, *Podocarpus* enters the regional vegetation again in MIS 4, showing percentages of 10% over a period of *c*. 10 kyr and reaching *c*. 30% between MIS 4.2 and 3.3 before it decreases again to 10%, a value which is maintained throughout the remainder of the record. Prior to the peak in *Podocarpus,* peaks in other montane and woodland taxa (e.g. *Acalypha*, Acanthaceae, *Blighia*, Combretaceae, *Craterispermum*, *Commiphora*, Lannea, *Maclura* and Mimosaceaeas), as well as maximum values in evergreen trees (e.g. *Alchornea*, *Celtis*, *Faurea*, *Ixora*, *Macaranga*, Moraceae, *Myrica* and Trema), are shown in the pollen record (*c*. 10%, 20% and 20%, respectively), after which all three groups continue at *c*. 10% throughout the rest of the record.

23.4.5 Lake Bambili, Cameroon

The inclusion of Lake Bambili's pollen record (B12), given it has the shortest timeframe (90 ka) in this review, is justified by its close geographical proximity to the Cameroonian Shum Laka archaeological site (5°51′31″ N, 10°4′40″ E), situated *c*. 20 km southwest of Lake Bambili (Figure 3E). The oldest layer containing evidence of the presence of *H. sapiens* is dated *c*. 30 ka (Lipson *et al*. 2020), while the rest of the archaeological record also displays evidence of significant societal developments such as a transition period in tool use (sometimes referred to as the 'Stone to Metal age') between the end of the Later Stone Age at 8 ka and the beginning of the Iron Age at 2.5 ka, as well as evidence for foundation and early stages of agriculture (Lavachery 2001). Lake Bambili is a high-altitude (2273 m asl) crater lake, currently enclosed locally by Afromontane forests characterized by the conifer *Podocarpus milanijanus* (Lézine *et al*. 2019), while the vegetation transitions to Afroalpine grasslands at higher altitudes and a mosaic of submontane forests and savannahs at lower altitudes (Lézine *et al*. 2013).

The initial aim of the palynological record presented in Lézine *et al*. (2019) was to reconstruct the history of Afromontane forests and their responses to glacial-interglacial cycles. Nevertheless, the pollen assemblages at Bambili can also be used to infer the landscape at a regional scale. The base of record B12 is dated to MIS 5.3 (99 ka), and the record starts off with percentages of *Podocarpus* that decrease from *c*. 25% to <5% just before the onset of MIS 5.1 (80 ka), while Poaceae show a coeval increase from 50% to 90%. Trace amounts (<5%) of other taxa are also present during this interval, *e.g.*Ericaceae, *Olea*, Asteraceae and *Alchornea*. This is immediately succeeded by a reversal in the trends of the two dominating taxa, and the representation of *Podocarpus* reaches *c*. 60% during MIS 5.1 at 75 ka, concurrent with a drop of Poaceae to <5%. During these extreme values, *Olea, Syzygium, Schefflera* and Asteraceae are also present in low quantities. Following the start of MIS 4 (71 ka), *Podocarpus* and Poaceae trends reverse again. *Podocarpus* remain practically absent (<5%) from the record in the MIS 5.0-3.3 (71-53 ka) interval, all the while Poaceae holds a stable presence, fluctuating between 70-80% and Ericaceae and Asteraceae fluctuate around 5%. MIS 3.3 sees *Podocarpus* reappearing in the record with *c*. 10%, and Poaceae starts a slow decline to 40% at 45 ka, but both taxa return to percentages similar to those in the MIS 5.0-3.3 interval at the 37 ka mark until Poaceae starts decreasing MIS 2.2 (19 ka). The gradual decrease of Poaceae lasts until it reaches *c*. 0% at 10 ka, and stays like this until *c*. 3 ka. During this 7 kyr interval *Podocarpus, Olea,* Alchornea, *Syzygium,* and *Schefflera* reappear in the pollen record but get replaced by Poaceae in the modern vegetation (3-0 ka).

23.4.6 Lake Bosumtwi, Ghana

Core BOS04-5B was retrieved from Lake Bosumtwi, Ghana (Koeberl *et al*. 2007), with the intent of exploring the vegetational compositions of tropical western Africa in response to changes in regional climate during the Quaternary. Lake Bosumtwi (6°30′6.04″N, 1°24′52.14″W) is situated in a meteorite impact crater formed 1.08 ± 0.04 million years ago (Jourdan *et al*. 2009) with a diameter of *c*. 11 km (Figure 3F). The bedrock and crater walls formed by this impact cause the lake's isolation from any regional bodies of water, precipitation being the only influx of water to the sedimentary archive. This closed hydrology makes the lake sensitive to regional climate change and precipitation in particular (Shanahan *et al*. 2012). The current lake surface of

Bosumtwi is at 97 m asl, has a diameter of 8.5 km and a surface area of 52 km^2. Situated in the 'Tropical and Subtropical Moist Broadleaf Forest' biome, Lake Bosumtwi occupies the transition between shrubby/woody savannah ecosystem to the north and moist forest ecosystem to the south (Olson *et al.* 2001). This setting makes the pollen record ideal for western-African vegetation reconstructions. Western Africa is notably underrepresented in the archaeological record, the oldest *H. sapiens* fossil being dated to only 12 ka in southwestern Nigeria (Harvati *et al.* 2011).

An age-depth model is lacking in the initial publication presenting the pollen record of BOS04-5B (Miller and Gosling 2014). We follow the age model of Shanahan *et al.* (2013) and used by Miller *et al.* (2016) of the same core. The period MIS 14.2-12.33 (538-461 ka) is characterized by fluctuating Poaceae percentages (reaching values >60% in the beginning and <10% later in this interval) and the presence of moist broadleaf forest taxa like *Celtis* (*c.* 8%) and Moraceae (*c.* 15%). At MIS 12.3-11.3 (461-405 ka) Poaceae increase up to *c.* 80%, interrupted by a fall during MIS 12.31 (443 ka) to 40%. In this period the moist broadleaf taxa show very low percentages. Cyperaceae reach *c.* 10% and montane tree *Olea* shows a single peak of 40% during MIS 12.31. Interval MIS 11.3-11.2 (405-375 ka) shows low (*c.* 20%) Poaceae percentages, and an increase in the arboreal broadleaf taxa *Celtis* and Moraceae, as is the case for MIS 13.13-12.33 (502-461 ka), but this time there is also a notable presence of *Macaranga* (*c.* 10%). Pollen spectra during MIS 11.2-10.2 (375-341 ka) resemble those found during the period of MIS 12.33-11.3, followed by a period of large fluctuations (from <10% to >80%) in Poaceae percentages that lasts until MIS 7.3 (216 ka). During this period Moraceae continues to have a marked presence in the pollen record, whereas it seems that *Celtis* is replaced by montane *Olea* at MIS 8.5 (287 ka). Poaceae dominates the pollen signal from MIS 7.3 to MIS 6, with consistent percentages of 90% and virtually no other pollen types found in the sediment except for low presences of Cyperaceae and *Olea* (*c.* 10%). In the upper part of the sediment column, during MIS 6-5.5 (183-122 ka) and MIS 2.2-0 (19-0 ka), Poaceae pollen is almost absent whereas taxa of moist broadleaf and the montane forest dominate.

23.5 DISCUSSION

We characterized vegetation as represented by the palynological records on a scale from open to closed (Figure 4). Our classification was done visually, using the pollen diagram provided in the original publication and based on the original interpretations by the authors. These characterisations are based on several factors: (i) the ratio Poaceae/arboreal pollen, (ii) the abundance of pollen reflecting shrubby taxa, (iii) (when provided) charcoal data, and (iv) relevant notes in the original publication. We relate the vegetation changes to the timing of the archaic *H. sapiens* transition to AMH and the $\delta^{18}O$ record (Lisiecki and Raymo 2005); including the MISs notation following Imbrie *et al.* (1984) indicating either glacial or interglacial periods.

Here we interpret the vegetation structure based on the palynological record and try to environmentally contextualize the three steps in archaic *H. sapiens* migration as related to our proposed Inside-of-Africa environmental framework: (i) An initial dispersal event of archaic *H. sapiens* after our lineage split off *H. neanderthalensis* (*c.* 700-400 ka, MIS 18.2-11.3), (ii) a period of isolation imposed by regional-scale environmental conditions, limiting the exchange of both genes and cultural knowledge, thereby promoting intraspecific variation within the *H. sapiens* clade (*c.* 400-250 ka, MIS 11.2-8.2), and (iii) a period of increased interconnectivity and growing population sizes that lead to homogenization and, ultimately, the full suite of AMH characteristics (*c.* 250-100 ka, MIS 7-5.3). To complete the picture of vegetation change in Africa we briefly review the last 100 ka and discuss the relevance of vegetation openness in this period for understanding the dispersal of hominins out of Africa (Blome *et al.* 2012; Carto *et al.* 2009; Tierney *et al.* 2017).

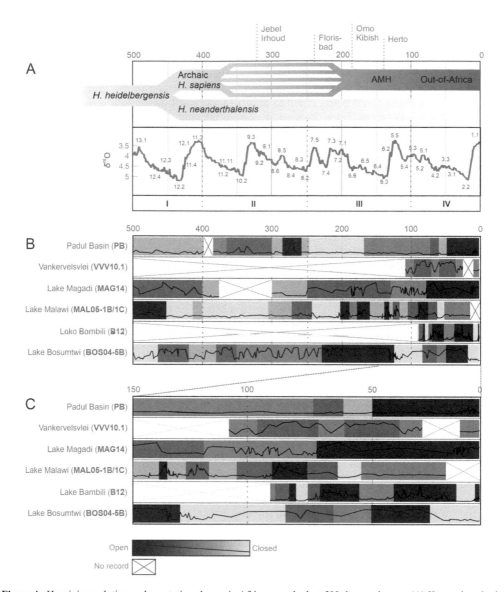

Figure 4. Hominin evolution and vegetation change in Africa over the last 500 thousand years. (A) Key archaeological sites (Grun *et al.* 1996; Hublin *et al.* 2017; McDougall *et al.* 2005; White *et al.* 2003), 'Inside-of-Africa' framework (Figure 1), and global climate change as indicated by global δ^{18}O record (Lisiecki and Raymo 2005). Numbers on the δ^{18}O panel indicate Marine Isotope Stage numbers as defined by Imbrie *et al.* (1984). (B) Degree of vegetation openness inferred from the ancient pollen records including respective open/close vegetation curve*. (C) Blown up view on the last 150 thousand years of vegetation change to show additional details. Indicated changes in the openness of the landscape are relative only to the pollen record itself. * The following data are plotted for each site: PB: *Pinus*%/Poaceae% (Torres *et al.* 2020), VVV10.1: Ericaceae% (Quick *et al.* 2016), MAG14, MAL05-1B/1C, B12: *Podocarpus*%/Poaceae% (Beuning *et al.* 2011; Ivory *et al.* 2018; Lézine *et al.* 2019; Owen *et al.* 2018), BOS04-5B: Poaceae% (Miller and Gosling 2014).

23.5.1 Vegetational trends throughout Africa 500 to 400 ka (MIS 13.1-11.3)

Currently, only four palynological records meeting our criteria are long enough to examine the vegetation change coeval with the initial radiation of archaic *H. sapiens* after their lineage split with *H. neanderthalensis*. These four records are the Padul Basin (PB; Torres *et al.* 2020) representing a change in the northern African region, Lake Magadi (MAG14; Owen *et al.* 2018), Lake Malawi (MAL05-1B; Ivory *et al.* 2018) in eastern Africa, and Lake Bosumtwi (BOS04-5B; Miller and Gosling 2014) in western Africa (Figure 4). We expected Africa in this period to host a relatively larger proportion of open vegetation that provided hospitable environments so the earliest members of the *H. sapiens* clade could disperse throughout Africa.

In northern Africa the relatively low resolution and multiple gaps in the PB pollen record make it difficult to discern variation in the vegetation cover during this period (Torres *et al.* 2020). However, the overall trend is apparent with the main arboreal taxon (*Pinus*) decreasing towards 400 ka. The loss of arboreal taxa could suggest that the vegetation structure became gradually more open; however, Poaceae does not replace the trees (also diminishing and absent 425-400 ka (MIS 12.1-11.3)). The loss of *Pinus* and Poaceae are also concomitant with a decline in total pollen influx rates, suggesting either reduced pollen productivity, a faster sediment accumulation, or poor preservation, all potentially indicative of more arid conditions. The high-resolution dust flux record from Atlantic core ODP 659 located at 18°N (Tiedeman *et al.* 1994) shows relevant information about climatic conditions in Saharan and sub-Saharan Africa. The 400-250 ka interval shows a maximum in dust outbreaks, indicative of dry conditions in Saharan Africa. It implicates that in this period the rain bringing ITCZ (where the northeast Trades meet the moisture loaden southwest Monsoon) hardly shifted into the northern hemisphere. Therefore, rains concentrated in equatorial and adjacent low southern hemisphere latitudes, allowing over large areas dense arboreal vegetation to expand. For the 400-250 ka interval the pollen record of site ODP 658 at 21°N (Dupont 1989; Dupont and Hooghiemstra 1989) is in support of substantial vegetation cover at equatorial latitudes. On the contrary, during the intervals 500-400 ka and 250-100 ka the dustflux shows significantly lower values (Tiedeman *et al.* 1994) and the monsoon front had more southern positions, implicating the monsoon front reached higher latitudes: Saharan Africa had wetter conditions and Africa at equatorial latitudes received less precipitation, allowing for a more open mosaic like vegetation pattern.

In the east African Rift Valley, between 500-400 ka, Lake Magadi and Lake Malawi both record open landscapes, savannahs in the north and wooded grasslands in the south, that transition into dense Afromontane forests over the course of the 100 kyr. Owen *et al* (2018) interpret this section of the Lake Magadi record as indicative of progressively more open vegetation, however, an examination of the pollen data indicates a wide variation in the abundance of arboreal taxa, e.g. *Podocarpus* fluctuates from near absence up to *c*. 80%. We therefore also note that there were important woody components, and high dynamism, in the east African Rift Valley vegetation for much of the pre-400 ka part of the record. The exception is between 525-425 ka (MIS 13 and 12) where an abundance of Poaceae (80%) suggests that savannah-like conditions dominated. This trend is almost exactly mirrored at Lake Malawi, where high Poaceae percentages (also 80%) suggest grasslands to be present around the lake before 470 kyr, followed by an abrupt expansion of dense Afromontane forests in the latter part of this time interval (Ivory *et al.* 2018). This grassland period in the east African Rift Valley would likely have offered the most hospitable conditions to early hominins and, with the Congo Basin rainforest to the west of the east African Rift Valley, a passageway between southern Africa and the rest of the continent.

In western Africa no archaeological evidence for archaic *H. sapiens* during the pre-400 ka period has been found in the vicinity of Lake Bosumtwi, but environmental conditions do not necessarily exclude them from inhabiting the area. Miller *et al.* (2016) find significant grassland expansions in favour of forested areas as indicated by increasing percentages of Poaceae pollen. Grassland expansion in this region seems to be driven by increased dryness prompted by orbital cycles, as Poaceae peaks (sometimes even reaching 99%) in record BOS04-5B are coeval with maxima in precessional amplitudes.

We find that during the period 500-400 ka (MIS 13.1-11.3) all of the available records show a degree of openness in the vegetation structure. The potential barriers to hominins are in western Africa in the early part of this period when the high representation of arboreal taxa shows a high forest cover (although this is not likely a dense rainforest). In the later part of this period, when poor pollen preservation prevails increased aridity is suggested in northern Africa.

23.5.2 Vegetational trends throughout Africa between 400 and 250 ka (MIS 11.2-8.2)

To date, only four pollen records have been published that are long enough to examine the period of proposed less hospitable condition that limited dispersal, and thereby gene flow, in archaic *H. sapiens*; Padul Basin (PB; Torres *et al.* 2020) representing northern Africa, Lake Magadi (MAG14; Owen *et al.* 2018) and Lake Malawi (MAL05-1B/1C; Ivory *et al.* 2018) eastern Africa, and Lake Bosumtwi (BOS04-5B; Miller and Gosling 2014) western Africa (Figure 4).

At PB, following a gap in the record, the main arboreal taxa (*Pinus*) fluctuates widely from *c.* 80% to *c.* 5% between *c.* 380 and 250 ka suggesting, apart from changes in aeolian pollen dispersal (Hooghiemstra *et al.* 1992) also major changes in the woody component. Within this period a change from Mediterranean (365-340 ka, MIS 11), to steppe vegetation (340-310 ka, MIS 10), and back to Mediterranean (310-295 ka, MIS 9.1-8.6) has been identified (Torres *et al.* 2001). Suggesting very dynamic vegetation with no discernable directional change in openness through this period.

The sedimentary record from Lake Magadi between 375 and 300 kyr BP was analysed for pollen (17 samples) but no pollen was recovered (Owen *et al.* 2018). Just before and after the hiatus in the record the percentage of aquatic taxa (e.g. Cyperaceae) and Afromontane taxa (e.g. *Podocarpus*) increase. The higher representation of aquatics could be interpreted as a lowering of the lake level providing more shallow water habitat for taxa such as Cyperaceae. The hiatus in the pollen record most likely represents a period of aridity when the basin (at least periodically) dried out and consequently little or no pollen is preserved; suggesting that through MIS 9 warmer conditions resulted in evaporation exceeding precipitation and so reduced the lake level. The more arid conditions potentially formed a barrier that disconnected northern Africa from southern Africa, thereby limiting any hominin dispersal between these two regions. In strong contrast to the arid conditions at Lake Magadi, we see that Lake Malawi hosts a vegetational community that is highly dominated by arboreal pollen from *c.* 470 to 270 ka. It is therefore suggested in Ivory *et al.* (2018) that the rigid ecosystem changeovers elucidated by MAL05-1B are a product of the response of local vegetation to the regional precipitation regime, comprising of a feedback between moisture recycling in Lake Malawi and eccentricity mediated summer insolation (Ivory *et al.* 2016). Still, since we characterise Lake Malawi as being surrounded by dense (Afromontane) forests, it is in line with our hypothesis of inhospitable vegetational cover during the period 400-250 ka, albeit for a different reason than at Lake Magadi.

The Lake Bosumtwi record shows wide fluctuations between *c.* 90% and *c.* 10% in the abundance of Poaceae between 400 and 250 ka which was initially interpreted as abrupt high-magnitude transitions between grassland and forest (Miller *et al.* 2016). Later investigations of how modern forest and grasslands are represented in the pollen rain (Julier *et al.* 2019) suggest that this magnitude of change could, for a large part, be explained by variation in the woody component of grassland ecosystems, rather than switching to a closed evergreen forest. Therefore, during the period from which the earliest archaic *H. sapiens* fossils have been found at Jebel Irhoud the landscape around Lake Bosumtwi was most likely covered with a wooded savannah.

We find that during the period 400-250 ka (MIS 11.2-8.2) the sites representative of northern and western Africa show a dynamic vegetation cover, but likely were not inhospitable to hominins.

The eastern African record was likely arid for much of this period which could have created a geographic barrier to hominins.

23.5.3 Vegetational trends throughout Africa between 250-100 ka (MIS 8.2-5.3)

Between 250 and 100 ka past vegetation data is available from the Padul Basin (PB; Torres *et al.* 2020) in northern Africa, Lake Magadi (MAG14; Owen *et al.* 2018) and Lake Malawi (MAL05-1B/1C; Ivory *et al.* 2018) in eastern Africa, and Lake Bosumtwi (BOS04-5B; Miller and Gosling 2014) in western Africa records plus, for the younger final few thousand years, parts of the Vankervelsvlei (VVV10.1; Quick *et al.* 2016) southern Africa, and Lake Malawi (1C; Beuning *et al.* 2011) eastern Africa records. In this time window, we expect a dominance of an open vegetation structure that promoted the dispersal, and consequently gene flow, between morphologically dissimilar archaic *H. sapiens* populations. This increased connectiveness would have ultimately lead to the full range of AMH associated phenotypical characteristics throughout the entire *H. sapiens* clade.

There is evidence for forest development at PB during the 250-100 ka interval (MIS 8.0-6.5) as indicated by the high abundance of evergreen taxa (*Pinus* >50% through most of this period) and deciduous taxa in parts (e.g. *Quercus*, *c*. 10%, 270-250 ka), indicating relatively stable vegetation structure with a strong woody component. The forest vegetation could have been less favourable to *H. sapiens* encouraging populations to move to find more open environment to which they were more familiar.

The high diversity of pollen taxa present in record MAG14 between 250-100 ka suggests that many different ecosystems were present along the banks of Lake Magadi. Within this time window Afromontane taxa diminishing in favour of the relatively open Zambezian Miombo woodland and humid evergreen woodlands, offering a diverse landscape vegetation mosaic with a wide variety in resources and thereby generating hospitable environments for archaic *H. sapiens*. The other lake core from eaststern Africa considered here, Lake Malawi, shows high variation in landscape openness. Five wooded grassland periods with a combined duration of *c*. 90 ka have been identified at Lake Malawi, giving ample oppurtunities for archaic *H. sapiens* to settle and disperse through the vicinity of Lake Malawi. Thus, both eastern African records indicate vegetation that would have been hospitable to archaic *H. sapiens*, and could have provided a north-south corridor allowing connectivity across the continent.

The pollen record from Lake Bosumtwi suggests an expanded savannah in western Africa, unprecedented in both stability (lasting almost 100 kyr) and cover (reaching >95% Poaceae). The consequences of this heavily grass-dominated landscape for *H. sapiens* could be interpreted in two ways: (i) as an ideal open landscape which they could occupy and traverse comfortably, or (ii) as a barrier due to high aridity and a lack of resources (absence of trees and water). The absence of any archaeological evidence of *H. sapiens* in western Africa until *c*. 12 ka (Harvati *et al.* 2011) tentatively provides support for the latter, but further evidence is clearly required.

The Vankervelsvlei sedimentary record begins *c*. 110 ka and contains pollen indicative of predominantly open woodland vegetation. This relatively open woodland could have provided a favourable to *H. sapiens* in a similar way to the vegetation of eastern Africa.

During the period 250-100 ka we find that there are potential barriers to *H. sapiens* in the north (covered by forest) and western (arid ecosystems) regions of Africa. While in the eastern and southern regions there is evidence for more hospitable vegetation configurations.

23.5.4 Vegetational trends throughout Africa after 100 ka (MIS 5.3)

The last 100 ka is an important period in hominin evolution and dispersal, particularly related to the movement of hominins out-of-Africa between 70 and 40 ka (MIS 5.0-3.2; (Blome *et al.* 2012; Carto *et al.* 2009; Tierney *et al.* 2017)). During this period an increasing amount of information on past vegetation change in Africa is available, however, a detailed exploration of all this literature is beyond the scope of this review. Instead, we limit this section to briefly contextualizing change around the period of movement of hominins out-of-Africa from just the six records used to provide information on longer-term vegetation change (Figure 4C).

The northern African vegetational makeup, as inferred from the PB pollen record, implies steppic environmental conditions with high *Artemisia* and Poaceae presence in the Mediterranean regions of northwest Africa for MIS 4 and MIS 3, indicating cold and arid climate for both stages (Torres *et al.* 2020). While in southern Africa the VVV10.1 pollen record suggests vegetation remain fairly stable in their characteristics over the last 100 ka, notwithstanding an expansion of fynbos vegetation between 70-40 ka (Quick *et al.* 2016). Similar vegetation stability is also seen in eastern Africa from the MAG14 pollen record (Owen *et al.* 2018), although the MAL 1C pollen record shows a trajectory of increasing woodland vegetation (Beuning *et al.* 2011). This increase in woodland vegetation has been interpreted as increased landscape diversity that could have aided hominin development and dispersal by providing access to new and more varied resources (Beuning *et al.* 2011). In central (record B12) and western Africa (record BOS04-5B) the degree of vegetation openness is shown to fluctuate; with the greatest depression of the treeline (hence most open vegetation being available at high elevation) occurring at Lake Bambili around 35 ka (Lézine *et al.* 2019).

Through the period of hominin dispersal 'Out-of-Africa', some portion of each of the six records indicates open vegetation that would have helped to facilitate dispersal of hominins. It is also the period in which the greatest variation in the vegetation record is seen, which could also have promoted dispersal as hominins sought to track resources. However, we should be cautious in our interpretations because the evidence is likely biased by the increased number of locations from which evidence is available, and the temporal detail of the information.

23.6 CONCLUSIONS

Growing evidence from genetic and archaeological sources paints a picture of complex pan-African hominin evolution, development and dispersal (Hublin *et al.* 2017; Lipson *et al.* 2020; Scerri *et al.* 2018; Stringer 2016; Wang *et al.* 2020). Intuitively it seems reasonable that the habitat that hominins existed within played an important role in shaping these processes and, based on archaeological evidence, it has been suggested that more open vegetation formations were the favoured habitat of early hominins (Bertola *et al.* 2016; Blome *et al.* 2012; Cerling *et al.* 2011; Larrasoaña *et al.* 2013; Scerri *et al.* 2018). However, unfortunately, evidence for vegetation, and vegetation change, over the more than half-million years over which evolution, development and dispersal have occurred remains scant at the continental scale. The terrestrial records of past vegetation change, based on the analysis of ancient pollen, obtained from six sedimentary basins that are relevant to Africa on these timescales indicate that the degree of openness varied regionally. The balance between a predominance of open *vs.* closed vegetation across Africa tends to follow a pattern of indicating more hospitable (open vegetation) during the period 500-400 ka when archaic *H. sapiens* dispersed around the continent, more inhospitable (closed vegetation/desert) between 400-250 ka when many new phenotypically distinct clades of hominins arose, and a return to more hospitable (open vegetation) conditions 250-100 ka coincident with the rise of anatomically modern humans. This pattern of vegetation change fits with our proposed 'Inside-of-Africa' environmental framework as a backdrop to hominin evolution. While these data provide some evidence for vegetation change in Africa over the last 500 ka there are clearly huge spatial and temporal gaps in the evidence available. Consequently, any inferences drawn from this must be done so with caution, however, we hope that our proposed framework will provide food for thought among researchers and be subsequently tested against new and more detailed records of past vegetation change.

ACKNOWLEDGEMENTS

We would like to thank all the authors of the original publications which we review here for publishing their studies and making their data available either through open access databases or directly following corresponance. We thank Alayne Street-Perrott (Swansea University, UK) and Anne-Marie Lézine (CNRS, France) for their thorough evaluation and feedback on our manuscript, and the PoA editorial team for the opportunity to contribute to this volume.

REFERENCES

Bailey, R.C., Head, G., Jenike, M., Owen, B., Rechtman, R. and Zechenter, E., 1989, Hunting and Gathering in Tropical Rain Forest: Is It Possible? *American Anthropologist*, **91**(1), pp. 59–82, 10.1525/aa.1989.91.1.02a00040.

Behrensmeyer, A.K., Potts, R., Deino, A.L. and Ditchfield, P., 2002, Olorgesailie, Kenya: a Million Years in the Life of a Rift Basin. *Sedimentation in Continental Rifts*, pp. 97–106.

Bertola, L.D., Jongbloed, H., Van Der Gaag, K.J., De Knijff, P., Yamaguchi, N., Hooghiemstra, H., Bauer, H., Henschel, P., White, P.A., Driscoll, C.A., Tende, T., Ottosson, U., Saidu, Y., Vrieling, K. and De Iongh, H.H., 2016, Phylogeographic Patterns in Africa and High Resolution Delineation of Genetic Clades in the Lion (Panthera leo). *Scientific Reports*, **6**, article: 30807, 10.1038/ srep30807.

Beuning, K.R.M., Zimmerman, K.A., Ivory, S.J. and Cohen, A.S., 2011, Vegetation response to glacial-interglacial climate variability near Lake Malawi in the southern African tropics. *Palaeogeography, Palaeoclimatology, Palaeoecology*, **303**(1–4), pp. 81–92, 10.1016/j.palaeo.2010.01.025.

Blome, M.W., Cohen, A.S., Tryon, C.A., Brooks, A.S. and Russell, J., 2012, The environmental context for the origins of modern human diversity: A synthesis of regional variability in African climate 150,000–30,000 years ago. *Journal of Human Evolution*, **62**(5), pp. 563–592, 10.1016/j.jhevol.2012.01.011.

Brooks, A.S., Yellen, J.E., Potts, R., Behrensmeyer, A.K., Deino, A.L., Leslie, D.E., Ambrose, S.H., Ferguson, J.R., D'Errico, F., Zipkin, A.M., Whittaker, S., Post, J., Veatch, E.G., Foecke, K. and Clark, J.B., 2018, Long-distance stone transport and pigment use in the earliest Middle Stone Age. *Science*, **360**(6384), pp. 90–94, 10.1126/science.aao2646.

Campisano, C.J., Cohen, A.S., Arrowsmith, J.R., Asrat, A., Behrensmeyer, A.K., Brown, E.T., Deino, A.L., Deocampo, D.M., Feibel, C.S., Kingston, J.D., Lamb, H.F., Lowenstein, T.K., Noren, A., Olago, D.O., Owen, R.B., Pelletier, J.D. and Potts, R., 2017, The Hominin Sites and Paleolakes Drilling Project: High resolution paleoclimate records from the East African Rift System and their implications for understanding the environmental context of hominin evolution. *PaleoAnthropology*, pp. 1–43.

Carto, S.L., Weaver, A.J., Hetherington, R., Lam, Y. and Wiebe, E.C., 2009, Out of Africa and into an ice age: on the role of global climate change in the Late Pleistocene migration of early modern humans out of Africa. *Journal of Human Evolution*, **56**(2), pp. 139–151, 10.1016/j.jhevol.2008.09.004.

Castañeda, I.S., Caley, T., Dupont, L., Kim, J.H., Malaizé, B. and Schouten, S., 2016, Middle to Late Pleistocene vegetation and climate change in subtropical southern East Africa. *Earth and Planetary Science Letters*, **450**, pp. 306–316, 10.1016/j.epsl.2016.06.049.

Cerling, T.E., Wynn, J.G., Andanje, S.A., Bird, M.I., Korir, D.K., Levin, N.E., MacE, W., MacHaria, A.N., Quade, J. and Remien, C.H., 2011, Woody cover and hominin environments in the past 6-million years. *Nature*, **476**(7358), pp. 51–56, 10.1038/nature10306.

Chan, E.K.F., Timmermann, A., Baldi, B.F., Moore, A.E., Lyons, R.J., Lee, S.S., Kalsbeek, A.M.F., Petersen, D.C., Rautenbach, H., Förtsch, H.E.A., Bornman, M.S.R. and Hayes, V.M.,

2019, Human origins in a southern African palaeo-wetland and first migrations. *Nature*, **575**(7781), pp. 185–189, 10.1038/s41586-019-1714-1.

Chen, D. and Chen, H.W., 2013, Using the Köppen classification to quantify climate variation and change: An example for 1901–2010. *Environmental Development*, **6**(1), pp. 69–79, 10.1016/j.envdev.2013.03.007.

Clement, A.C., Hall, A. and Broccoli, A.J., 2004, The importance of precessional signals in the tropical climate. *Climate Dynamics*, **22**(4), pp. 327–341, 10.1007/s00382-003-0375-8.

Compton, J.S., 2011, Pleistocene sea-level fluctuations and human evolution on the southern coastal plain of South Africa. *Quaternary Science Reviews*, **30**(5–6), pp. 506–527, 10.1016/j.quascirev.2010.12.012.

Coulthard, T.J., Ramirez, J.A., Barton, N., Rogerson, M. and Brücher, T., 2013, Were rivers flowing across the Sahara during the last interglacial? Implications for human migration through Africa. *PloS One*, **8**(9), article: e74834, 10.1371/journal.pone.0074834.

D'Errico, F., Banks, W.E., Warren, D.L., Sgubin, G., Van Niekerk, K., Henshilwood, C., Daniau, A.L. and Sánchez Goñi, M.F., 2017, Identifying early modern human ecological niche expansions and associated cultural dynamics in the South African Middle Stone Age. *Proceedings of the National Academy of Sciences*, **114**(30), pp. 7869–7876, 10.1073/pnas.1620752114.

Dalibard, M., Popescu, S.M., Maley, J., Baudin, F., Melinte-Dobrinescu, M.C., Pittet, B., Marsset, T., Dennielou, B., Droz, L. and Suc, J.P., 2014, High-resolution vegetation history of West Africa during the last 145 ka. *Geobios*, **47**(4), pp. 183–198, 10.1016/j.geobios.2014.06.002.

Deino, A.L., Behrensmeyer, A.K., Brooks, A.S., Yellen, J.E., Sharp, W.D. and Potts, R., 2018, Chronology of the Acheulean to Middle Stone Age transition in eastern Africa. *Science*, **360**, pp. 95–98, 10.1126/science.aao2216.

Douze, K., Wurz, S. and Henshilwood, C.S., 2015, Techno-cultural characterization of the MIS 5 (*c.* 105-90 Ka) lithic industries at Blombos Cave, Southern Cape, South Africa. *PLoS ONE*, **10**(11), pp. 1–29, 10.1371/journal.pone.0142151.

Dupont, L.M., 1989, Palynology of the last 680 000 years of ODP Site 658 (off NW Africa): fluctuations in paleowind systems. In: *Paleoclimatology and Paleometeorology: Modern and past patterns of global atmospheric transport* edited by Leinen, M. and Sarnthein, M., pp. 719–794, (Dordrecht: Kluwer).

Dupont, L., 2011, Orbital scale vegetation change in Africa. *Quaternary Science Reviews*, **30**(25–26), pp. 3589–3602, 10.1016/j.quascirev.2011.09.019.

Dupont, L. M. and Hooghiemstra, H., 1989, The Saharan-Sahelian boundary during the Brunhes chron. *Acta Botanica Neerlandica*, **38**(4), pp. 405–415, 10.1111/j.1438-8677.1989.tb01372.x.

Dupont, Lydie M., Jahns, S., Marret, F. and Ning, S., 2000, Vegetation change in equatorial West Africa: Time-slices for the last 150 ka. *Palaeogeography, Palaeoclimatology, Palaeoecology*, **155**(1–2), pp. 95–122, 10.1016/S0031-0182(99)00095-4.

Endicott, P., Ho, S.Y.W. and Stringer, C., 2010, Using genetic evidence to evaluate four palaeoanthropological hypotheses for the timing of Neanderthal and modern human origins. *Journal of Human Evolution*, **59**(1), pp. 87–95, 10.1016/j.jhevol.2010.04.005.

Frédoux, A., 1994, Pollen analysis of a deep-sea core in the Gulf of Guinea: vegetation and climatic changes during the last 225,000 years B.P. *Palaeogeography, Palaeoclimatology, Palaeoecology*, **109**(2–4), pp. 317–330, 10.1016/0031-0182(94)90182-1.

Frumkin, A., Bar-Yosef, O. and Schwarcz, H.P., 2011, Possible paleohydrologic and paleoclimatic effects on hominin migration and occupation of the Levantine Middle Paleolithic. *Journal of Human Evolution*, **60**(4), pp. 437–451, 10.1016/j.jhevol.2010.03.010.

Galway-Witham, J., Cole, J. and Stringer, C., 2019, Aspects of human physical and behavioural evolution during the last 1 million years. *Journal of Quaternary Science*, **34**(6), pp. 355–378, 10.1002/jqs.3137.

Gosling, W.D. and Holden, P.B., 2011, Precessional forcing of tropical vegetation carbon storage. *Journal of Quaternary Science*, **26**(5), pp. 463–467, 10.1002/jqs.1514.

Grun, R., Brink, J.S., Spooner, N.A., Taylor, L., Stringer, C.B., Franciscus, R.G. and Murray, A.S., 1996, Direct dating of Florisbad hominid. *Nature*, **382**(6591), pp. 500–501, 10.1038/382500a0

Gunz, P., Bookstein, F.L., Mittero, P., Stadlmayr, A., Seidler, H., Gerhard, W., Weber, G.W., Kelly, D.J., Poolman, B., Gavin, H., Zeniou-meyer, M., Liu, Y., Olanich, M.E., Becherer, U., Bader, M. and Vitale, N., 2009, Early modern human diversity suggests subdivided population structure and a complex out-of-Africa scenario. *Proceedings of the National Academy of Sciences*, **106**(20), pp. 8398–8398, 10.1073/pnas.0903734106.

Hammond, A.S., Royer, D.F. and Fleagle, J.G., 2017, The Omo-Kibish I pelvis. *Journal of Human Evolution*, **108**, pp. 199–219, 10.1016/j.jhevol.2017.04.004.

Harvati, K., Stringer, C., Grün, R., Aubert, M., Allsworth-Jones, P. and Folorunso, C.A., 2011, The later stone age calvaria from Iwo Eleru, Nigeria: Morphology and chronology. *PLoS ONE*, **6**(9), pp. 1–8, 10.1371/journal.pone.0024024.

Henshilwood, C.S., D'Errico, F., Van Niekerk, K.L., Coquinot, Y., Jacobs, Z., Lauritzen, S.E., Menu, M. and García-Moreno, R., 2011, A 100,000-year-old ochre-processing workshop at Blombos Cave, South Africa. *Science*, **334**(6053), pp. 219–222, 10.1126/science.1211535.

Hooghiemstra, H., Stalling, H., Agwu, C. and Dupont, L.M., 1992, Vegetational and climatic changes at the northern fringe of the Sahara 250,000-5,000 years BP. *Review of Palaeobotany and Palynology*, **74**, pp. 1–53, 10.1016/0034-6667(92)90137-6.

Hooghiemstra, H., Agwu, C.O.C. and Beug, H.-J., 1986, Pollen and spore distribution in recent marine sediments: a record of NW-African seasonal wind patterns and vegetation belts. *Meteor Forschungsergebnisse, Reihe C Geologie Und Geophysik*, **40**(40), pp. 87–135.

Hooghiemstra, H., Lézine, A.M., Leroy, S.A.G., Dupont, L. and Marret, F., 2006, Late Quaternary palynology in marine sediments: A synthesis of the understanding of pollen distribution patterns in the NW African setting. *Quaternary International*, **148**(1), pp. 29–44, 10.1016/j.quaint.2005.11.005.

Hublin, J.J., Ben-Ncer, A., Bailey, S.E., Freidline, S.E., Neubauer, S., Skinner, M.M., Bergmann, I., Le Cabec, A., Benazzi, S., Harvati, K. and Gunz, P., 2017, New fossils from Jebel Irhoud, Morocco and the pan-African origin of Homo sapiens. *Nature*, **546**(7657), pp. 289–292, 10.1038/nature22336.

Imbrie, J., Hays, J.D., Martinson, D.G., McIntyre, A., Mix, A.C., Morley, J.J., Pisias, N.G., Prell, W.L. and Shackleton, N.J., 1984, The orbital theory of Pleistocene climate: support from a revised chronology of the marine δ^{18}O record. In: *Milankovitch and Climate* edited by Berger, A., Imbrie, J., Hays, G., Kukla, G. and Saltzman, B., pp. 269–306, (Dordrecht: Reidel).

Ivory, S.J., Blome, M.W., King, J.W., McGlue, M.M., Cole, J.E. and Cohen, A.S., 2016, Environmental change explains cichlid adaptive radiation at Lake Malawi over the past 1.2 million years. *Proceedings of the National Academy of Sciences*, **113**(42), pp. 11895–11900, 10.1073/pnas.1611028113.

Ivory, S.J., Lézine, A.-M., Vincens, A. and Cohen, A.S., 2018, Waxing and waning of forests: Late Quaternary biogeography of southeast Africa. *Global Change Biology*, **24**(7), pp. 2939–2951, 10.1111/gcb.14150.

Jacobs, Z., Duller, G.A.T., Wintle, A.G. and Henshilwood, C.S., 2006, Extending the chronology of deposits at Blombos Cave, South Africa, back to 140 ka using optical dating of single and multiple grains of quartz. *Journal of Human Evolution*, **51**(3), pp. 255–273, 10.1016/j.jhevol.2006.03.007.

Jacobs, Z., Roberts, R.G., Galbraith, R.F., Deacon, H.J., Grün, R., Mackay, A., Mitchell, P., Vogelsang, R. and Wadley, L., 2008, Ages for the Middle Stone Age of southern Africa: Implications for human behavior and dispersal. *Science*, **322**(5902), pp. 733–735, 10.1126/science.1162219.

Jacobson, G.L. and Bradshaw, R.H.W., 1981, The selection of sites for paleovegetational studies. *Quaternary Research*, **96**, pp. 80–96, 10.1016/0033-5894(81)90129-0.

Jahns, S., Hüls, M. and Sarnthein, M., 1998, Vegetation and climate history of west equatorial Africa based on a marine pollen record off Liberia (site GIK 16776) covering the last 400,000 years. *Review of Palaeobotany and Palynology*, **102**(3–4), pp. 277–288, 10.1016/S0034-6667(98)80010-9.

Jourdan, F., Renne, P.R. and Reimold, W.U., 2009, An appraisal of the ages of terrestrial impact structures. *Earth and Planetary Science Letters*, **286**(1–2), pp. 1–13, 10.1016/j.epsl.2009.07.009.

Julier, A.C.M., Jardine, P.E., Adu-Bredu, S., Coe, A.L., Fraser, W.T., Lomax, B.H., Malhi, Y., Moore, S. and Gosling, W.D., 2019, Variability in modern pollen rain from moist and wet tropical forest plots in Ghana, West Africa. *Grana*, **58**(1), pp. 45–62, 10.1080/00173134.2018.1510027.

Koeberl, C., Milkereit, B., Overpeck, J.T., Scholz, C.A., Amoako, P.Y.O., Boamah, D., Danuor, S.K., Karp, T., Kueck, J., Hecky, R.E., King, J.W. and Peack, J.A., 2007, An international and multidisciplinary drilling project into a young complex impact structure: The 2004 ICDP Bosumtwi Crater Drilling Project – An overview. *Meteoritics and Planetary Science*, **42**(4–5), pp. 483–511, 10.1111/j.1945-5100.2007.tb01057.x.

Lamb, H.F., Bates, C.R., Bryant, C.L., Davies, S.J., Huws, D.G., Marshall, M.H. and Roberts, H.M., 2018, 150,000-year palaeoclimate record from northern Ethiopia supports early, multiple dispersals of modern humans from Africa. *Scientific Reports*, **8**(1), article: 1077, 10.1038/s41598-018-19601-w.

Larrasoaña, J.C., Roberts, A.P. and Rohling, E.J., 2013, Dynamics of green Sahara periods and their role in hominin evolution. *PLoS ONE*, **8**(10), e76514, 10.1371/journal.pone.0076514.

Lavachery, P., 2001, The holocene archaeological sequence of Shum Laka rock shelter (Grassfields, western Cameroon). *African Archaeological Review*, **18**(4), pp. 213–247, 10.1023/A:1013114008855.

Lézine, A.-M., Holl, A.F.C., Lebamba, J., Vincens, A., Assi-Khaudjis, C., Février, L. and Sultan, É., 2013, Temporal relationship between Holocene human occupation and vegetation change along the northwestern margin of the Central African rainforest. *Comptes Rendus – Geoscience*, **345**(7–8), pp. 327–335, 10.1016/j.crte.2013.03.001.

Lézine, A.M., Izumi, K., Kageyama, M. and Achoundong, G., 2019, A 90,000-year record of Afromontane forest responses to climate change. *Science*, **363**(6423), pp. 177–181, 10.1126/science.aav6821.

Lieberman, D.E., McBratney, B.M. and Krovitz, G., 2002, The evolution and development of cranial form in Homo sapiens. *Proceedings of the National Academy of Sciences*, **99**(3), pp. 1134–1139, 10.1073/pnas.022440799.

Lipson, M., Ribot, I., Mallick, S., Rohland, N., Olalde, I., Adamski, N., Broomandkhoshbacht, N., Lawson, A.M., López, S., Oppenheimer, J., Stewardson, K., Asombang, R.N., Bocherens, H., Bradman, N., Culleton, B.J., Cornelissen, E., Crevecoeur, I., de Maret, P., Fomine, F.L.M., and Reich, D., 2020, Ancient West African foragers in the context of African population history. *Nature*, **577**(7792), pp. 665–670, 10.1038/s41586-020-1929-1.

Lisiecki, L.E. and Raymo, M.E., 2005, A Pliocene-Pleistocene stack of 57 globally distributed benthic $\delta18O$ records. *Paleoceanography*, **20**(1), pp. 1–17, 10.1029/2004PA001071.

McDougall, I., Brown, F.H. and Fleagle, J.G., 2005, Stratigraphic placement and age of modern humans from Kibish, Ethiopia. *Nature*, **433**(7027), pp. 733–736, 10.1038/nature03258.

Miller, C.S. and Gosling, W.D., 2014, Quaternary forest associations in lowland tropical West Africa. *Quaternary Science Reviews*, **84**, pp. 7–25, 10.1016/j.quascirev.2013.10.027.

Miller, C.S., Gosling, W.D., Kemp, D.B., Coe, A.L. and Gilmour, I., 2016, Drivers of ecosystem and climate change in tropical West Africa over the past ∼540 000 years. *Journal of Quaternary Science*, **31**(7), pp. 671–677, 10.1002/jqs.2893.

Mirazón Lahr, M. and Foley, R.A., 2016, Human Evolution in Late Quaternary Eastern Africa. In: *Africa from MIS 6-2: Population Dynamics and Paleoenvironments* edited by Jones, S. and Stewart, B., pp. 215–231, (Dordrecht: Springer).

Mounier, A. and Mirazón Lahr, M., 2016, Virtual ancestor reconstruction: Revealing the ancestor of modern humans and Neandertals. *Journal of Human Evolution*, **91**, pp. 57–72, 10.1016/j.jhevol.2015.11.002.

Neubauer, S., Hublin, J.J. and Gunz, P., 2018, The evolution of modern human brain shape. *Science Advances*, **4**(1), article: eaao5961, 10.1126/sciadv.aao5961.

Olson, D.M., Dinerstein, E., Wikramanayake, E.D., Burgess, N.D., Powell, G.V.N., Underwood, E.C., D'amico, J.A., Itoua, I., Strand, H.E., Morrison, J.C., Loucks, C.J., Allnutt, T.F., Ricketts, T.H., Kura, Y., Lamoreux, J.F., Wettengel, W.W., Hedao, P. and Kassem, K.R., 2001, Terrestrial Ecoregions of the World: A New Map of Life on Earth: A new global map of terrestrial ecoregions provides an innovative tool for conserving biodiversity. *Bioscience*, **51**(11), pp. 933–938, 10.1641/0006-3568(2001)051[0933:TEOTWA]2.0.CO;2.

Osborne, A.H., Vance, D., Rohling, E.J., Barton, N., Rogerson, M. and Fello, N., 2008, A humid corridor across the Sahara for the migration of early modern humans out of Africa 120,000 years ago. *Proceedings of the National Academy of Sciences*, **105**(43), pp. 16444–16447, 10.1073/pnas.0804472105.

Owen, R.B., Muiruri, V.M., Lowenstein, T.K., Renaut, R.W., Rabideaux, N., Luo, S., Deino, A.L., Sier, M.J., Dupont-Nivet, G., Mcnulty, E.P., Leet, K., Cohen, A., Campisano, C., Deocampo, D., Shen, C.C., Billingsley, A. and Mbuthia, A., 2018, Progressive aridification in East Africa over the last half million years and implications for human evolution. *Proceedings of the National Academy of Sciences*, **115**(44), pp. 11174–11179, 10.1073/pnas.1801357115.

Ozenda, P., 1991, *Flore et Végétation du Sahara*. (Paris: Centre Nationale Recherche Scientifique).

Pearson, O.M., 2008, Statistical and biological definitions of anatomically modern humans: Suggestions for a unified approach to modern morphology. *Evolutionary Anthropology*, **17**(1), pp. 38–48, 10.1002/evan.20155.

Quade, J., Dente, E., Armon, M., Ben Dor, Y., Morin, E., Adam, O. and Enzel, Y., 2018, Megalakes in the Sahara? A Review. *Quaternary Research*, **90**(2), pp. 253–275, 10.1017/qua.2018.46.

Quick, L.J., Meadows, M.E., Bateman, M.D., Kirsten, K.L., Mäusbacher, R., Haberzettl, T. and Chase, B.M., 2016, Vegetation and climate dynamics during the last glacial period in the fynbos-afrotemperate forest ecotone, southern Cape, South Africa. *Quaternary International*, **404**, pp. 136–149, 10.1016/j.quaint.2015.08.027.

Richter, D., Grün, R., Joannes-Boyau, R., Steele, T.E., Amani, F., Rué, M., Fernandes, P., Raynal, J.P., Geraads, D., Ben-Ncer, A., Hublin, J.J. and McPherron, S.P., 2017, The age of the hominin fossils from Jebel Irhoud, Morocco, and the origins of the Middle Stone Age. *Nature*, **546**(7657), pp. 293–296, 10.1038/nature22335.

Rivas-Martínez, S., 1987, *Memoria del mapa de Series de Vegetación de España, 1:400.000*, (Madrid: Ministerio de Agricultura, Pesca y Alimentación, ICONA).

Roets, W., Xu, Y., Raitt, L., El-Kahloun, M., Meire, P., Calitz, F., Batelaan, O., Anibas, C., Paridaens, K., Vandenbroucke, T., Verhoest, N.E.C. and Brendonck, L., 2008, Determining discharges from the Table Mountain Group (TMG) aquifer to wetlands in the Southern Cape, South Africa. *Hydrobiologia*, **607**(1), pp. 175–186, 10.1007/s10750-008-9389-x.

Rubel, F. and Kottek, M., 2010, Observed and projected climate shifts 1901–2100 depicted by world maps of the Köppen-Geiger climate classification. *Meteorologische Zeitschrift*, **19**(2), pp. 135–141, 10.1127/0941-2948/2010/0430.

Scerri, E.M.L., 2017a, The West African Stone Age. *Oxford Research Encyclopedia of African History*, 10.1093/acrefore/9780190277734.013.137.

Scerri, E.M.L., 2017b, The North African Middle Stone Age and its place in recent human evolution. *Evolutionary Anthropology*, **26**(3), pp. 119–135, 10.1002/evan.21527.

Scerri, E.M.L., Thomas, M.G., Manica, A., Gunz, P., Stock, J.T., Stringer, C., Grove, M., Groucutt, H.S., Timmermann, A., Rightmire, G.P., d'Errico, F., Tryon, C.A., Drake, N.A., Brooks, A.S., Dennell, R.W., Durbin, R., Henn, B.M., Lee-Thorp, J., deMenocal, P., Petraglia, M.D., Thompson, J.C., Scally, A. and Chikhi, L., 2018, Did Our Species Evolve in Subdivided Populations across Africa, and Why Does It Matter? *Trends in Ecology and Evolution*, **33**(8), pp. 582–594, 10.1016/j.tree.2018.05.005.

Schlebusch, C., Loog, L., Groucutt, H., King, T., Rutherford, A., Barbieri, C., Barbujani, G., Chikhi, L., Stringer, C., Jakobsson, M., Eriksson, A., Manica, A., Tishkoff, S.A., Scerri, E.M.L., Scally, A., Brierley, C. and Thomas, M.G., 2021, Human origins in Southern African palaeo-wetlands? Strong claims from weak evidence. *Journal of Archaeological Science* **130**, article: 105374, 10.1016/j.jas.2021.105374.

Schlebusch, C.M., Malmström, H., Günther, T., Sjödin, P., Coutinho, A., Edlund, H., Munters, A.R., Vicente, M., Steyn, M., Soodyall, H., Lombard, M. and Jakobsson, M., 2017, Southern African ancient genomes estimate modern human divergence to 350,000 to 260,000 years ago. *Science*, **358**(6363), pp. 652–655, 10.1126/science.aao6266.

Shanahan, T.M., Beck, J.W., Overpeck, J.T., McKay, N.P., Pigati, J.S., Peck, J.A., Scholz, C.A., Heil, C.W. and King, J., 2012, Late Quaternary sedimentological and climate changes at Lake Bosumtwi Ghana: New constraints from laminae analysis and radiocarbon age modeling. *Palaeogeography, Palaeoclimatology, Palaeoecology*, **361–362**, pp. 49–60, 10.1016/j.palaeo.2012.08.001.

Shanahan, T.M., Peck, J.A., McKay, N., Heil, C.W., King, J., Forman, S.L., Hoffmann, D.L., Richards, D.A., Overpeck, J.T. and Scholz, C., 2013, Age models for long lacustrine sediment records using multiple dating approaches – An example from Lake Bosumtwi, Ghana. *Quaternary Geochronology*, **15**, pp. 47–60, 10.1016/j.quageo.2012.12.001.

Stewart, J.R. and Stringer, C.B., 2012, Human evolution out of Africa: The role of refugia and climate change. *Science*, **335**(6074), pp. 1317–1321, 10.1126/science.1215627.

Stringer, C., 2016, The origin and evolution of homo sapiens. *Philosophical Transactions of the Royal Society B: Biological Sciences*, **371**(1698), article: 20150237, 10.1098/rstb.2015.0237.

Tiedeman, R., Sarnthein, M., Shackleton, N.J., 1994, Astronomic timescale for the Pliocene Atlantic ∂18O and dust flux records of Ocean Drilling Program site 659. Paleoceanography, **9**, pp. 619–638, 10.1029/94PA00208.

Tierney, J.E., deMenocal, P. and Zander, P.D., 2017, A climatic context for the out-of-Africa migration. *Geology*, **45**(11), pp. 1023–1026, 10.1130/G39457.1.

Torres, T., Valle, M., Ortiz, J.E., Soler, V., Araujo, R., Rivas, M.R., Delgado, A., Julià, R. and Sánchez-Palencia, Y., 2020, 800 ka of Palaeoenvironmental changes in the Southwestern Mediterranean realm. *Journal of Iberian Geology*, **46**(2), pp. 117–144, 10.1007/s41513-020-00123-2.

Walter, R.C., Buffler, R.T., Bruggemann, J.H., Guillaume, M.M.M., Berhe, S.M., Negassi, B., Libsekal, Y., Cheng, H., Edwards, R.L., Von Cosel, R., Néraudeau, D. and Gagnon, M., 2000, Early human occupation of the Red Sea coast of Eritrea during the last interglacial. *Nature*, **405**(6782), pp. 65–69, 10.1038/35011048.

Wang, K., Mathieson, I., O'Connell, J. and Schiffels, S., 2020, Tracking human population structure through time from whole genome sequences. *PLoS Genetics*, **16**(3), article: e1008552, 10.1371/journal.pgen.1008552.

White, F., 1983, *The Vegetation of Africa*. (Paris: UNESCO).

White, T.D., Asfaw, B., Degusta, D., Gilbert, H., Richards, G.D., Suwa, G. and Howell, F.C., 2003, Pleistocene Homo sapiens from Middle Awash, Ethiopia. *Nature*, **423**(6941), pp. 742–747, 10.1038/nature01669.

Zhao, M., Dupont, L., Eglinton, G. and Teece, M., 2003, n-Alkane and pollen reconstruction of terrestrial climate and vegetation for N.W. Africa over the last 160 kyr. *Organic Geochemistry*, **34**(1), pp. 131–143, 10.1016/S0146-6380(02)00142-0.

CHAPTER 24

The role of palaeoecology in conserving African ecosystems

Lindsey Gillson

Plant Conservation Unit, Department of Biological Sciences, University of Cape Town, South Africa

ABSTRACT: Africa holds critical and unique biodiversity, but conservation is challenging because of the complexity of landscapes and their history, and the multitude of drivers and stakeholder perspectives. Palaeoecology can help in developing a nuanced understanding of landscape history that can inform pressing conservation issues, including climate change integrated conservation strategies, fire management, management of herbivores and rewilding, ecosystem restoration, sustainable use of natural resources, and management of cultural landscapes. To be useful, however, the data from palaeoecological proxies needs to be converted into metrics that are useful and accessible to conservationists and managers, requiring rigorous calibration against modern and instrumental data sets, alongside sound chronologies. Palaeoecologists can help in enhancing the applied value of their work by interpreting it in ways that are underpinned by sound ecological theory in ways that are relevant to pressing conservation questions. The design of projects and the interpretation and application of data can be used as starting points for interdisciplinary, collaborative approaches with stakeholders. Finally, the modelling of palaeoecological data can help stakeholders to explore landscape changes and the effect of management interventions under different scenarios of climate and land-use change in the future.

24.1 INTRODUCTION

Africa's landscapes are home to important biodiversity hotspots as well as the world's only intact megafaunal assemblage (Archer *et al.* 2018). They are also characterised by a large rural population, whose livelihoods directly depend on ecosystem services, alongside rapidly changing socio-economic contexts that include urbanisation and development. Conservation in Africa takes place both inside and outside of protected areas and both approaches present challenges and opportunities. Developing conservation strategies that benefit both biodiversity and people depends on developing and understanding of landscape history both in terms of ecological variability and cultural context.

In the early days of conservation, the idea of maintaining, preserving or restoring 'balance' was prevalent, and ideas of equilibrium were also reflected in early ecological theory. Maintaining populations at fixed carrying capacity was an approach common in southern Africa, while in East Africa strict preservation of 'wilderness' – areas perceived, often wrongly, to have been little modified by people – was a dominant conservation ideal (Gillson 2015). Added to this, many colonial governments perceived fire to be destructive and implemented policies of fire suppression, disrupting traditional burning practices, such as seasonal mosaic burning (Gillson *et al.* 2019). However, more recent approaches to conservation accept that variability, flux and disturbance are normal to most populations and landscapes, and this raises a need for a long-term

DOI 10.1201/9781003162766-24

perspective that allows managers to explore resilience and variability over centennial-millennial timescales.

Many African National Parks were founded in the early decades of the twentieth century, when ecosystems were likely to have been in an atypical state. Though most African landscapes have been managed by people for millennia, the combined impacts of fire suppression, removal or disruption of traditional management, and the decimation of megafaunal populations from the 19th century are likely to have led to atypically higher tree density in savannas. This high tree density is likely to be further exacerbated by the effects of CO_2 fertilisation, which benefits C_3 plants such as savanna trees and shrubs, more than C_4 plants, including savanna grasses, and is likely to play a role in bush encroachment (Bond and Midgley 2012; Midgley and Bond 2015; Wigley *et al.* 2009). On the other hand, some savannas were cleared of trees because of charcoal production or to make way for agropastoralism, and woody vegetation subsequently recovered following land abandonment (Blair *et al.* 2018; O'Connor *et al.* 2014). Therefore, interplay between environmental change and the role of anthropogenic influence, both positive and neg-ative, are important aspects of landscape history that needs to be considered in conservation planning and ecosystem management.

Amid all of this complexity, science is only part of the basis for conservation decisions, while stakeholder values, cultural context and local perspectives must also be considered (Rogers *et al.* 2013). A central challenge that palaeoecologists can contribute to is developing nuanced, interdisciplinary and inclusive understanding of landscape history (Marchant and Lane 2014; Marchant *et al.* 2018; Mustaphi *et al.* 2019), that can then be applied in defining and refining conservation and management goals (Gillson 2015; Githumbi *et al.* 2020; Wolfe *et al.* 2007). Here I discuss the role of palaeoecology in understanding and managing the drivers of landscape changes, then identify some ways forward in applying this knowledge to conservation practice.

24.2 UNDERSTANDING, PREDICTING AND MANAGING THE DRIVERS OF LANDSCAPE CHANGE

24.2.1 Climate

Past warm periods such as the African Humid Period and the Mediaeval Warm Period can provide valuable insights into the response of ecosystems to climate change, as well as the feedbacks between fire, climate and vegetation structure (Lüning *et al.* 2017). They have a particularly important role in informing and refining the development of models that simulate changes in vegetation cover based on climatic and other variables. This has been demonstrated in work modelling the greening of the Sahara during the mid-Holocene. Palaeoecological data indicated that models based on orbital forcing alone were under-estimating the magnitude and extent of the greening event (Lézine *et al.* 2011; Pausata *et al.* 2020; Tierney *et al.* 2017). This helped in shaping more complex models that include for example feedbacks between vegetation and climate (Chen *et al.* 2020; Pausata *et al.* 2020; Rachmayani *et al.* 2015).

Most predictions of future climate change agree that Africa's climate will get warmer, but there is a high degree of uncertainty regarding future changes in rainfall. While it is commonly expected that warmer climates bring higher rainfall due to increased ocean evaporation, this is not always the case, and some areas are expected to experience aridification and/or increased extreme events (Engelbrecht *et al.* 2013; Pinto *et al.* 2018). This is of critical importance to areas where agriculture is rain fed and already marginal (Pohl *et al.* 2017). Reconstructing past climates and particularly the interaction between temperature and rainfall can help in understanding the interactions between regional climate systems and assist in more accurate downscaling of global models (Lüning *et al.* 2018). Furthermore, palaeoecological data can help in understanding vegetation-climate-fire feedbacks that in turn shape the distribution of fire-adapted ecosystems

(Bond and Midgley 2012). It can also inform the manipulation of drivers such as fire that interact with climate. Information such as this is critical in designing conservation strategies that consider climate change, for example, potential shifts in the distribution of ecosystems and species as well as identifying populations that are likely to be most vulnerable or most resilience to climate change (Hannah *et al.* 2002). In a world of finite space and limited conservation resources, a sound basis for conservation prioritisation in a changing climate is critical (Foden *et al.* 2019).

24.2.2 Fire

Fire is a critical component of many African ecosystems, and is essential in maintaining grass-lands, savannas and heathlands in areas of sub-Saharan Africa that have rainfall that is high enough to support forest (Bond *et al.* 2005). Nevertheless, fire management remains a contested issue in many areas, due to risk to property and life, as well as due to lack of understanding of its important ecological role (Bowman *et al.* 2013; Gillson *et al.* 2019). Many colonial governments believed fire to be detrimental and implemented policies of fire suppression in ecosystems that were naturally fire adapted (Moura *et al.* 2019). Later, when attempts to suppress fire proved futile, these policies were replaced by prescribed burns at regular intervals (van Wilgen *et al.* 2014). Policies of fire suppression and prescribed burning disrupted traditional fire management and homogenised the fuel load, leading to artificially intense and widespread fires in subsequent decades (Humphrey *et al.* 2020).

People have managed fire for millennia, for example by patch mosaic burning, which involves burning small patches of vegetation starting in the early dry season (Laris 2011). As different patches are burned at different times, this creates patches of vegetation with different post-fire ages (time since burned). Species assemblages often change throughout post-fire succession, and therefore a heterogeneous mosaic of different post-fire ages provides a wider range of habitat, biodiversity and ecosystem services than would be present had only a single, large fire occurred (Parr and Brockett 1999). In addition, such skilful management of fire helps to prevent late season wildfires that homogenise the landscapes (Humphrey *et al.* 2020), and can be used to manipulate the balance of forest and savanna elements at the savanna-forest ecotone (Fairhead and Leach 1996).

Palaeoecology is essential in understanding the impacts of anthropogenic fire management and the effects of fire suppression and management. Such records can in addition contribute to debates of anthropogenic 'savannisation' and the extent of human influence on forest ecosystems (Veldman 2016; Willis *et al.* 2004). Studying fire history through the charcoal record can help in determining whether current fire policies are within the historical range of variability. They may also assist in reinstating the skilful management of fire by communities who have incorporated fire management into their livelihoods for millennia (Humphrey *et al.* 2020).

24.2.3 Herbivory

Herbivores have both direct and indirect effects on vegetation structure and composition through physical impact on plant biomass and their influence on ecosystem processes, such as nutrient cycling and seed dispersal (Cromsigt *et al.* 2018). Changes in herbivore density thus have major implications for ecological function and biodiversity conservation (Hempson *et al.* 2015). Hunting, disease and manipulation of carnivore populations have all altered wild and domestic herbivore numbers, leading to declines in population size and distribution of many wild species in the late 19th and early 20th centuries (Venter *et al.* 2017). In southern Africa, reports of vast herds of wildebeest and springbok were reported in the early 20th century, while such herds have vanished from today's landscapes (Home 1874; Skead *et al.* 2007; Venter *et al.* 2017). In the 20th century, there have also been wide fluctuations in populations of both wild and domestic herbivores in response to, for example, changes in domestic and local markets for meat and wool, intensification of agriculture, a transition to game farming, as well as illegal hunting of rhino and elephant (Beinart 2003). These changes would have directly affected vegetation structure and composition and influenced nutrient cycling, physical disturbance and fire regimes through changes in biomass accumulation.

The use of coprophilous fungal spores is commonly used to pinpoint the timing and effects of megafaunal extinctions (Burney *et al.* 2003; Feranec *et al.* 2011), but there are also possibilities for addressing ecological questions about changes in herbivore abundance and how they affected vegetation on more recent timescales of decades to centuries (Ekblom and Gillson 2010; Goethals and Verschuren 2019). This information is invaluable in informing the management and restoration of functional herbivore-driven landscapes (Fuhlendorf *et al.* 2009). It could also play a role, alongside pollen data, in addressing questions about the impacts of elephants on their habitat, specifically whether elephant impact on trees is unprecedented due to their confinement in small areas, or whether the loss of tree cover is actually a return to the less wooded landscapes of the 17th and 18th century (Gillson 2015; Veldman 2016).

With urbanisation continuing apace in Africa, and the effects of government grants on rural livelihoods, game farming and re-wilding are likely to become increasingly attractive options for conservation (Hoogendoorn *et al.* 2019) (though the effects of the current COVID-19 pandemic on these options remain to be seen (Lindsey *et al.* 2020)). Palaeoecologists can help managers decide what should the landscapes look like, and how many herbivores there should be (Hempson *et al.* 2015; Venter *et al.* 2017).

24.2.4 Anthropogenic Influence

Many African National Parks were founded in the early decades of the twentieth century, a time when many landscapes had been affected by intensive human impact including changes in fire and fauna as well as changing climate and rising CO_2, and as well as changes in land use including sedentarisation (i.e. loss of transhumant lifestyles) and land abandonment. There is therefore a role for palaeoecology, specifically fossil pollen and phytolith analysis, in exploring variability in ecosystem structure and composition, particularly in terms of woody vegetation density. These data allow ecosystem managers to distinguish, for example, whether increasing tree cover is a recovery from past tree clearance or an unprecedented loss of open, grassy savanna (Gillson 2015; Gillson and Marchant 2014; Parr *et al.* 2014). As mentioned above, palaeoecological data can also help in understanding the role of past management of fire and herbivores, and the effects of disruption of traditional fire management (Humphrey *et al.* 2020).

24.3 CHALLENGES IN APPLICATION OF PALAEOECOLOGY TO BIODIVERSITY CONSERVATION IN AFRICA

Despite the great potential for synergy between palaeoecology and conservation, practical collaboration between the two fields remains rare. Here I will identify some challenges for palaeoecology and suggest some ways forward for increasing the application of our work.

24.3.1 Calibration

Palaeoecological data as it is presented in Quaternary journals is not easy for conservation managers to use (Davies *et al.* 2014). They need to be translated into metrics that are more closely aligned with ecological parameters and ecosystem services that are usually the target of management goals (Davies and Bunting 2010; Gillson and Marchant 2014). For example, a measure of changes in tree density is much more useful in conservation than a measure of arboreal pollen abundance (Gillson and Duffin 2007). Therefore, good calibration is needed between palaeo-proxies and ecological parameters, but calibration remains a major and barely addressed issue in palaeoecological studies of most African ecosystems (Julier *et al.* 2018). Accurate calibration that includes a measure of confidence limits will allow translation of pollen data to estimates of changing vegetation composition that are much more useful to managers. Much more work is needed in this regard, especially in Africa, where quantitative, multi-proxy data could be invaluable in understanding the feedbacks occur between climate, vegetation, fire and herbivory (Julier *et al.*, 2021, this volume).

The same can be said of charcoal and spore data. Charcoal can be invaluable for reconstructing fire history, but it is also notoriously difficult to interpret, requiring comprehensive calibration efforts (Adolf *et al.* 2018). This is partly because different fire intensities affect the completeness of combustion and different fuel sources (e.g. grass versus woody vegetation) and thus their representation in the charcoal record (Leys *et al.* 2017). Amalgamation and meta-analyses of multiple charcoal data sets via the Global Charcoal data base can help in quantifying the relationship between biomass burning and charcoal representation (Hawthorne *et al.* 2018; Marlon *et al.* 2016). Emerging techniques such as Fourier Transform Infrared Spectroscopy can help in unravelling the complexity between fire intensity and charcoal representivity (Gosling *et al.* 2019). Similarly, studies suggest that *Sporomiella* alone do not seem to accurately represent herbivore density in African ecosystems (Baker *et al.* 2013). A suite of spore proxies is needed, and more quantitative work is required to relate these to herbivore biomass and if possible to different herbivore guilds (Feranec *et al.* 2011; Goethals and Verschuren 2019).

As well as the use of modern data sets, there is also potential for calibrating palaeoecological data against instrumental records of climate change (Jacques *et al.* 2015; Vallè *et al.* 2018), satellite imagery of fire, e.g. the MODIS data set (Adolf *et al.* 2018) and historical records of herbivory numbers fire (Baker *et al.* 2013, van Asperen *et al.* 2020).

24.3.2 Chronologies and influx rates

Calibration against instrumental records requires detailed and accurate chronologies, in order to compare measured changes in historical and palaeoecological records. Where such chronologies are available, pollen influx rates can be used. Influx rates have advantages over both percentage data, which are influenced by the Fagerlind effect (Fagerlind 1952), and concentration data, which are influenced by changes in sediment accumulation rate. In the case of pollen percentages, the change in the abundance of any particular plant will affect the abundance of all other taxa present, because the total pollen sum always has to be 100%, regardless of how much biomass is present in the landscape. Similarly, a faster (slower) rate of sediment accumulation may lead to reductions (increases) in pollen concentration, even if the abundance of a plant in the landscape has not changed. However, such detailed chronologies are rarely found in palaeoecological data from semi-arid areas, where there are often hiatuses in sediment accumulation due to extended dry periods, and bioturbation by large mammals. In such cases, influx rates are not reliable, but careful comparison of percentage and concentration data can assist interpretation. There can be more confidence that trends observed in both percentage and concentration data reflect real changes in plant abundance (Baker *et al.* 2013; Forbes *et al.* 2018).

24.3.3 Taxonomic resolution

A further challenge in the application of fossil pollen data in biodiversity conservation is taxonomic resolution. Important ecological changes with implications for biodiversity conservation often occur at the species and genus level. For example, loss of species is a major concern for conservationists and changes in habitat quality, such as a shift from palatable to unpalatable grasses can have conservation management implications. Complementary proxies e.g. phytoliths can help to reserve poorly resolved but ecologically important families such as the Poaceae (Breman *et al.* 2019). Further, detailed work is needed on the pollen morphology and associated ecological affinity of taxa within diverse families, such as the Asteraceae, containing both generalist and specialist species (Blackmore *et al.* 2010). Environmental DNA from sediments (sedaDNA) is increasingly used to provide accurate taxonomic resolution (Bálint *et al.* 2018; Clarke *et al.* 2019), though further work is needed to improve coverage of African species in genomic databases (Dommain *et al.* 2020).

24.3.4 Engagement with ecological theory

Increasing the use of palaeoecology in conservation requires that palaeoecologists engage with relevant theoretical and management frameworks. The shift from 'balance of nature' to 'flux of nature' as the underlying paradigm in ecology, has had profound implications for conservation (Gillson 2015; Pickett *et al.* 1997). Specifically, managing ecosystems and populations that are not at equilibrium requires a different approach to old approaches to conservation that aimed to prevent change. Ecosystem management emerged in response to develop conservation approaches that accepted the heterogeneous, dynamic nature of ecosystems (Grumbine 1997). Working with variable ecosystems in which disturbance and change are normal, raises the need for information on variability over ecologically relevant timescales, in order to define conservation goals that are based on processes rather than states (Gillson 2015). Many ecosystems are dynamic and fluctuate in response to disturbance and environmental change, however long-term data are needed to help managers to identify when ecosystems are approaching 'tipping points' or novel states. There is a natural synergy between palaeoecology and resilience theory (Davies *et al.* 2018), in that hypotheses generated regarding resilience and thresholds in ecosystems can potentially be tested using palaeoecological data (Gillson *et al.* 2020; Seddon *et al.* 2011).

24.3.5 Theory to application: relevance to current conservation questions

Once palaeoecological data is calibrated into forms that are recognisable to ecologists and managers, there still remains the challenge of how to apply these data to current conservation questions. As well as climate-changed-integrated conservation strategies, mentioned above, an important interface between conservation practice and palaeoecological data is establishing acceptable limits of variation in key parameters such as woody vegetation cover, herbivore density and fire regimes. Recent trends in such parameters can be compared against data spanning decades and centuries, helping to quantifying the historical range of variability and the extent of intensifying human management, including the impact of changes in fire and fauna as well as land use implemented by colonial governments (Gell *et al.* 2012; Hoffman *et al.* 2019). Measures of the long-term variability can help to identify unprecedented change from normal background variability (Keane *et al.* 2009). Thus, long-term data from palaeoecology can help to understand the ecological character of an ecosystem. For example, palaeoecological data could potentially be incorporated into definitions of ecological characters, limits of acceptable change and Thresholds of Potential Concern (TPCs) (Gell 2010; Gillson and Duffin 2007).

Knowledge of long-term change is central to restoration ecology and rewilding. In restoration ecology, conservationists need to have a reference state (Falk 2017; Higgs *et al.* 2014; Nogué *et al.* 2017). While this is often a nearby undisturbed or little disturbed site, such locations are not always available. Indeed, even those that are perceived to be undisturbed may have a more complex history than can be gleaned from present day conditions (Forbes *et al.* 2018; Gell 2010). Thus, restoration targets including the rewilding of ecosystems requires data from before the intensive human impact of the 18th and 19th century. On the theme of restoration, there is a tendency to view grassy and other open ecosystems as less valuable than forests or woody systems, therefore their afforestation is prioritised (Bond *et al.* 2019; Parr *et al.* 2014). This tendency has become even more pronounced given recent drives to encourage tree planting as a means of enhancing carbon storage (Bastin *et al.* 2019). In some cases, where forest cover has become denuded or degraded, reforestation with native species can enhance biodiversity conservation and carbon storage at the same time (Lewis *et al.* 2019). In others, however, ancient grasslands and heathlands have been mistakenly identified as degraded or anthropogenically derived (Bond *et al.* 2019; Parr *et al.* 2014). This has led to inappropriate afforestation schemes that threaten rare open landscapes that are home to unique biodiversity (Bond *et al.* 2019; Veldman 2016; Veldman *et al.* 2019). Palaeoecology should therefore support the protection of open ecosystems and in the design of appropriate forest restoration programmes while at the same time raising the profile of ancient grasslands and heathlands as valid conservation targets.

A further step in the translation of palaeoecological data for conservation purposes is to convert them to indices for ecosystem services, such as biodiversity, sediment regulation, soil stability, sediment quality, and water quality (Dearing *et al.* 2012). This approach can help engage stakeholders by alerting them to changes in key services, information which could potentially inform sustainable management of natural resources, but to date has rarely been applied in Africa.

24.3.6 Integrating with other disciplines and knowledge streams

A multiproxy approach that helps to elucidate the complexities of landscape history and people's role in shaping the environment can help to build a nuanced and inclusive understanding of landscape history that can provide cultural and local context for interpreting change and informing future sustainable management (Gillson 2015; Githumbi *et al.* 2020; Marchant and Lane 2014; Mustaphi *et al.* 2019; Richer and Gearey 2018). Though much palaeoecological research occurs in the lab and behind the microscope, a key component of our work can be in collaborating with stakeholders and listening to stakeholder perspective on ecosystem change (Wolfe *et al.* 2007). Projects can be co-designed with stakeholders with their management needs and goals in mind (Rogers *et al.* 2013; Wolfe *et al.* 2007). Palaeoecologists can attend conservation conferences and workshops where managers and ecologists articulate questions that can be usefully addressed using palaeoecological techniques, or arranging consultations with community groups (Davies *et al.* 2014; Wolfe *et al.* 2007). Where palaeo-projects are already underway, it is useful to meet with communities, government, parastatals and NGOs while in the field, explain the potential utility of the palaeoecological data and to explore questions of mutual interest.

After a project's completion, it is useful to provide results in more than one format. Scientific publications are not always the most accessible or useful way in communicating results to stakeholders, formats such as reports, poster, or stakeholder feedback workshop might be more convienient. Such workshops are especially valuable when stakeholders can help in shaping elements of the project e.g. in adding insights to possible feedbacks between drivers and/or in utilising project outputs such as model-based tools that allow them to explore the effects of different scenarios or fire, herbivory and climate (Capitani *et al.* 2019; Hossain *et al.* 2020). Stakeholders often have different perspectives on how ecosystem change affects their livelihoods and there are often important cultural components of landscape change that can only be captured through consultation with stakeholders.

24.3.7 The importance of modelling and future scenarios

Reconstructing the past does not always provide analogues for the future (Williams and Jackson 2007). We face unprecedented rates of climate change and human impact, as well as social and economic structures that differ vastly from the past, as well as potentially novel combinations of climate parameters and anthropogenic stressors that may produce ecological surprises (Lindsey *et al.* 2020; Williams and Jackson 2007). Nevertheless, understanding past ecological dynamics can help to elucidate the interactions and processes that drive ecosystem change. Building models based on palaeoecological data that simulate past interactions between key drivers and responders e.g. vegetation, climate, fire and herbivory provides a basis for exploring future scenarios, especially where palaeoecological proxies have been turned into Ecosystem Service Indicators (see above) (Dearing *et al.* 2012; Perry *et al.* 2016).

Models based on palaeoecological data can simulate past interactions between human and environmental factors, for example, climate, fire, herbivory and vegetation (Perry *et al.* 2016). Such models can then be parameterised using current and future scenarios. Furthermore, some modelling approaches, such as system dynamics models offer opportunities for combining insights from long-term data with local knowledge of ecosystem change, complexity and likely future scenarios (Hossain *et al.* 2020). This has the twofold advantage of improving the relevance and applicability of palaeoecology whilst also improving access to a diverse range of voices in scenario planning and conservation management. Though highly uncertain, such models can help stakeholders to explore different possible future scenarios of land cover and ecosystem services (Caves *et al.* 2013).

24.4 CONCLUSIONS

Africa's landscapes are dynamic and complex, and managing them requires flexible and adaptive conservation approaches that safeguard biodiversity as well as providing ecosystem services and respecting indigenous knowledge. This means embracing a past-present-future perspective that includes a knowledge of ecosystem variability, ecological function as well as cultural heritage and the role of traditional management. There is thus enormous potential for better integration of palaeoecological and other long-term data into biodiversity conservation and natural resource management.

Carrying out palaeoecological research in Africa has some challenges, however, including the lack of sites and poor pollen preservation in semi-arid areas and the sheer diversity of the flora in forested areas. Despite this, the discipline has enormous potential value in conserving African biodiversity, especially as landscape history is often not well understood. Fine resolution studies covering the late Holocene are especially valuable to conservationists in understanding the impacts of climate change, the role of fire and herbivory, the effects of human management and the resilience of ecosystems. To realise this value, however, we in the palaeo-community must do our best to make our data accessible, and to present it in forms that are relevant to conservation. Aligning multiple proxies can help in distinguishing climatically driven from anthropogenically driven changes and provide the basis for defining restoration targets and thresholds of potential concern. Ensure that our research process is inclusive can contribute to building a nuanced and interdisciplinary understanding of landscape history that informs biodiversity conservation and the sustainable and equitable management of natural resources.

ACKNOWLEDGEMENTS

Many thanks to Timm Hoffman and two anonymous reviewers for helpful comments on the manuscript. The NRF Competitive Programme for Rated Researcher (Grant Number 118538), African Origins Platform (Grant number 117666), and Global Change Grand Challenge/SASSCAL (Grant number 118589) provided funding for some of the research underpinning this perspective.

REFERENCES

Adolf, C., Doyon, F., Klimmek, F. and Tinner, W., 2018, Validating a continental European charcoal calibration dataset. *The Holocene* **28**, pp. 1642–1652, 10.1177/0959683618782607.

Archer, E.R., Dziba, L., Mulongoy, K., Maoela, M.A. and Walters, M., 2018, The IPBES regional assessment report on biodiversity and ecosystem services for Africa. 3947851057, Intergovernmental Science-Policy Platform on Biodiversity and Ecosystem.

Baker, A.G., Bhagwat, S.A. and Willis, K.J., 2013, Do dung fungal spores make a good proxy for past distribution of large herbivores? *Quaternary Science Reviews* **62**, pp. 21–31, 10.1016/j.quascirev.2012.11.018.

Bálint, M., Pfenninger, M., Grossart, H.-P., Taberlet, P., Vellend, M., Leibold, M.A., Englund, G. and Bowler, D., 2018, Environmental DNA time series in ecology. *Trends in Ecology & Evolution* **33**, pp. 945–957, 10.1016/j.tree.2018.09.003.

Bastin, J.-F., Finegold, Y., Garcia, C., Mollicone, D., Rezende, M., Routh, D., Zohner, C.M. and Crowther, T.W., 2019, The global tree restoration potential. *Science* **365**, pp. 76–79, 10.1126/science.aax0848.

Beinart, W. 2003, The Rise of Conservation in South Africa: Settlers, Livestock, and the Environment 1770–1950. Oxford University Press, Oxford.

Blackmore, S., Wortley, A.H., Skvarla, J.J., Gabarayeva, N.I. and Rowley, J.R., 2010, Developmental origins of structural diversity in pollen walls of Compositae. *Plant Systematics and Evolution* **284**, pp. 17–32, 10.1007/s00606-009-0232-2.

Blair, D., Shackleton, C.M. and Mograbi, P.J., 2018, Cropland abandonment in South African smallholder communal lands: Land cover change (1950–2010) and farmer perceptions of contributing factors. *Land* **7**, pp. 121, 10.3390/land7040121.

Bond, W.J., and Midgley, G.F., 2012, Carbon dioxide and the uneasy interactions of trees and savannah grasses. *Philosophical Transactions of the Royal Society B* **367**, pp. 601–612, 10.1098/rstb.2011.0182.

Bond, W.J., Stevens, N., Midgley, G.F. and Lehmann, C.E., 2019, The trouble with trees: afforestation plans for Africa. *Trends in Ecology & Evolution* **34**, pp. 963–965, 10.1016/j.tree.2019.08.003.

Bond, W.J., Woodward, F.I. and Midgley, G.F., 2005, The global distribution of ecosystems in a world without fire. *New Phytologist* **165**, pp. 525–538, 10.1111/j.1469-8137.2004.01252.x.

Bowman, D.M., O'Brien, J.A. and Goldammer, J.G., 2013, Pyrogeography and the global quest for sustainable fire management. *Annual Review of Environment and Resources* **38**, p. 57, 10.1146/annurev-environ-082212-134049.

Breman, E., Ekblom, A., Gillson, L. and Norström, E., 2019, Phytolith-based environmental reconstruction from an altitudinal gradient in Mpumalanga, South Africa, 10,600 BP–present. *Review of Paleobotany and Palynology* **263**, pp. 104–116, 10.1016/j.revpalbo.2019.01.001.

Burney, D.A., Robinson, G.S. and Burney, L.P., 2003, Sporormiella and the late Holocene extinctions in Madagascar. *Proceedings of the National Academy of Sciences* **100**, pp. 10800–10805, 10.1073/pnas.1534700100.

Capitani, C., Garedew, W., Mitiku, A., Berecha, G., Hailu, B.T., Heiskanen, J., Hurskainen, P., Platts, P.J., Siljander, M. and Pinard, F., 2019, Views from two mountains: Exploring climate change impacts on traditional farming communities of Eastern Africa highlands through participatory scenarios. *Sustainability Science* **14**, pp. 191–203, 10.1007/s11625-018-0622-x.

Caves, J.K., Bodner, G.S., Simms, K., Fisher, L.A. and Robertson, T., 2013, Integrating collaboration, adaptive management, and scenario-planning: experiences at Las Cienegas National Conservation Area. *Ecology and Society* **18**, article: 43, 10.5751/ES-05749-180343.

Chen, W., Ciais, P., Zhu, D., Ducharne, A., Viovy, N., Qiu, C. and Huang, C., 2020, Feedbacks of soil properties on vegetation during the Green Sahara period. *Quaternary Science Reviews* **240**, article: 106389, 10.1016/j.quascirev.2020.106389.

Clarke, C.L., Edwards, M.E., Brown, A.G., Gielly, L., Lammers, Y., Heintzman, P.D., Ancin-Murguzur, F.J., Bråthen, K.A., Goslar, T. and Alsos, I.G., 2019, Holocene floristic diversity and richness in northeast Norway revealed by sedimentary ancient DNA (seda DNA) and pollen. *Boreas* **48**, pp. 299–316, 10.1111/bor.12357.

Cromsigt, J.P., te Beest, M., Kerley, G.I., Landman, M., le Roux, E. and Smith, F.A., 2018, Trophic rewilding as a climate change mitigation strategy? *Philosophical Transactions of the Royal Society B: Biological Sciences* **373**, article: 20170440, 10.1098/rstb.2017.0440.

Davies, A., and Bunting, M., 2010, Applications of palaeoecology in conservation. *Open Ecology Journal* **3**, pp. 54–67, 10.2174/1874213001003020054.

Davies, A.L., Colombo, S. and Hanley, N., 2014, Improving the application of long-term ecology in conservation and land management. *Journal of Applied Ecology* **51**, pp. 63–70, 10.1111/1365-2664.12163.

Davies, A.L., Streeter, R., Lawson, I.T., Roucoux, K.H. and Hiles, W., 2018, The application of resilience concepts in palaeoecology. *The Holocene* **28**, pp. 1523–1534, 10.1177/0959683618777077.

Dearing, J.A., Yang, X., Dong, X., Zhang, E., Chen, X., Langdon, P.G., Zhang, K., Zhang, W. and Dawson, T.P., 2012, Extending the timescale and range of ecosystem services through paleoenvironmental analyses, exemplified in the lower Yangtze basin. *Proceedings of the National Academy of Sciences* **109**, pp. E1111–E1120, 10.1073/pnas.1118263109.

Dommain, R., Andama, M., McDonough, M.M., Prado, N.A., Goldhammer, T., Potts, R., Maldonado, J.E., Nkurunungi, J.B. and Campana, M.G., 2020, The challenges of reconstructing tropical biodiversity with sedimentary ancient DNA: A 2200-year-long metagenomic record from Bwindi impenetrable forest, Uganda. *Frontiers in Ecology and Evolution* **8**, article: 218, 10.3389/fevo.2020.00218.

Ekblom, A. and Gillson, L., 2010, Dung fungi as indicators of past herbivore abundance, Kruger and Limpopo National Park. *Palaeogeography, Palaeoclimatology, Palaeoecology* **296**, pp. 14–27, 10.1016/j.palaeo.2010.06.009.

Engelbrecht, C.J., Engelbrecht, F.A. and Dyson, L.L., 2013, High-resolution model-projected changes in mid-tropospheric closed-lows and extreme rainfall events over southern Africa. *International Journal of Climatology* **33**, pp. 173–187, 10.1002/joc.3420.

Fagerlind, F. 1952, The real significance of pollen diagrams. *Botanisker Notiser* **105**, pp. 185–224.

Fairhead, J. and Leach, M., 1996, Rethinking the Forest-Savanna Mosaic: Colonial Science and its Relics in West Africa. Pages 105–121 in M. Leach and R. Mearns, editors. The Lie of the Land. (James Currey: Oxford).

Falk, D.A. 2017, Restoration ecology, resilience, and the axes of change. *Annals of the Missouri Botanical Garden* **102**, pp. 201–216, 10.3417/2017006.

Feranec, R.S., Miller, N.G., Lothrop, J.C. and Graham, R.W., 2011, The Sporormiella proxy and end-Pleistocene megafaunal extinction: a perspective. *Quaternary International* **245**, pp. 333–338, 10.1016/j.quaint.2011.06.004.

Foden, W.B., Young, B.E., Akçakaya, H.R., Garcia, R.A., Hoffmann, A.A., Stein, B.A., Thomas, C.D., Wheatley, C.J., Bickford, D. and Carr, J.A., 2019, Climate change vulnerability assessment of species. Wiley Interdisciplinary Reviews: *Climate Change* **10**, e551, 10.1002/wcc.551.

Forbes, C.J., Gillson, L. and Hoffman, M.T., 2018, Shifting baselines in a changing world: Identifying management targets in endangered heathlands of the Cape Floristic Region, South Africa. *Anthropocene* **22**, pp. 81–93, 10.1016/j.ancene.2018.05.001.

Fuhlendorf, S.D., Engle, D.M., Kerby, J. and Hamilton, R., 2009, Pyric herbivory: rewilding landscapes through the recoupling of fire and grazing. *Conservation Biology* **23**, pp. 588–598, 10.1111/j.1523-1739.2008.01139.x.

Gell, P. 2010, With the benefit of hindsight: the utility of palaeoecology in wetland condition assessment and identification of restoration targets. Ecology of Industrial Pollution. Ecological Review Series, (CUP & the British Ecological Society: Cambridge), pp. 162–188.

Gell, P., Mills, K. and Grundell, R., 2012, A legacy of climate and catchment change: the real challenge for wetland management. *Hydrobiologia* **708**, pp.1–12.

Gillson, L. 2015, Biodiversity Conservation and Environmental Change: Using palaeoecology to manage dynamic landscapes in the Anthropocene. Oxford University Press, Oxford, U.K.

Gillson, L. and Duffin, K.I., 2007, Thresholds of Potential Concern as benchmarks in the management of African savannahs. *Philosophical Transactions of the Royal Society of London Series B-Biological Sciences* **362**, pp. 309–319, 10.1098/rstb.2006.1988.

Gillson, L., MacPherson, A.J. and Hoffman, M.T. 2020, Contrasting mechanisms of resilience at mesic and semi-arid boundaries of fynbos, a mega-diverse heathland of South Africa. *Ecological Complexity* **42**, pp. 100827, 10.1016/j.ecocom.2020.100827.

Gillson, L. and Marchant, R., 2014, From myopia to clarity: sharpening the focus of ecosystem management through the lens of palaeoecology. *Trends in Ecology & Evolution* **29**, pp. 317–325, 10.1016/j.tree.2014.03.010.

Gillson, L., Whitlock, C. and Humphrey, G., 2019, Resilience and fire management in the Anthropocene. *Ecology and Society* **24**(3), p.14, 10.5751/ES-11022-240314.

Githumbi, E., Marchant, R. and Olago, D., 2020, Using the Past to Inform a Sustainable Future: Palaeoecological Insights from East Africa. pp. 187–195 Africa and the Sustainable Development Goals. Springer.

Goethals, L. and Verschuren, D., 2020, Tracing ancient animal husbandry in tropical Africa using the fossil spore assemblages of coprophilous fungi: a validation study in western Uganda. *Vegetation History and Archaeobotany* **29**, pp. 509–526, 10.1007/s00334-019-00760-3.

Gosling, W.D., Cornelissen, H. and McMichael, C. 2019, Reconstructing past fire temperatures from ancient charcoal material. *Palaeogeography, Palaeoclimatology, Palaeoecology* **520**, pp. 128–137, 10.1016/j.palaeo.2019.01.029.

Grumbine, R.E. 1997, Reflections on Ecosystem Management. *Conservation Biology* **11**, pp. 41–47, 10.1046/j.1523-1739.1997.95479.x.

Hannah, L., Midgley, G.F. and Millar, D., 2002, Climate change-integrated conservation strategies. *Global Ecology & Biogeography* **11**, pp. 485–495, 10.1046/j.1466-822X.2002.00306.x.

Hawthorne, D., Mustaphi, C.J.C., Aleman, J.C., Blarquez, O., Colombaroli, D., Daniau, A.-L., Marlon, J.R., Power, M., Vanniere, B. and Han, Y., 2018, Global Modern Charcoal Dataset (GMCD): A tool for exploring proxy-fire linkages and spatial patterns of biomass burning. *Quaternary International* **488**, pp. 3–17, 10.1016/j.quaint.2017.03.046.

Hempson, G.P., Archibald, S. and Bond, W.J., 2015, A continent-wide assessment of the form and intensity of large mammal herbivory in Africa. *Science* **350**, pp. 1056–1061, 10.1126/science.aac7978.

Higgs, E., Falk, D.A., Guerrini, A., Hall, M., Harris, J., Hobbs, R.J., Jackson, S.T., Rhemtulla, J.M. and Throop, W., 2014, The changing role of history in restoration ecology. *Frontiers in Ecology and the Environment* **12**, pp. 499–506, 10.1890/110267.

Hoffman, M.T., Rohde, R.F. and Gillson, L., 2019, Rethinking catastrophe? Historical trajectories and modelled future vegetation change in southern Africa. *Anthropocene* **25**, pp. 100189, 10.1016/j.ancene.2018.12.003.

Home, D.D. 1874, Incidents in my life. (AK Butts: New York).

Hoogendoorn, G., Meintjes, D., Kelso, C. and Fitchett, J., 2019, Tourism as an incentive for rewilding: the conversion from cattle to game farms in Limpopo province, South Africa. *Journal of Ecotourism* **18**, pp. 309–315, 10.1080/14724049.2018.1502297.

Hossain, M.S., Ramirez, J., Szabo, S., Eigenbrod, F., Johnson, F.A., Speranza, C.I. and Dearing, J.A., 2020, Participatory modelling for conceptualizing social-ecological system dynamics in the Bangladesh delta. *Regional Environmental Change* **20**, pp. 1–14, 10.1007/s10113-020-01599-5.

Humphrey, G.J., Gillson, L. and Ziervogel, G., 2020, How changing fire management policies affect fire seasonality and livelihoods. *Ambio*, pp. 1–17, 10.1007/s13280-020-01351-7.

Jacques, J.-M.S., Cumming, B.F., Sauchyn, D.J. and Smol, J.P., 2015, The bias and signal attenuation present in conventional pollen-based climate reconstructions as assessed by early climate data from Minnesota, USA. *PLoS ONE* **10**, e0113806, 10.1371/journal.pone.0113806.

Julier, A.C., Jardine, P.E., Adu-Bredu, S., Coe, A.L., Duah-Gyamfi, A., Fraser, W.T., Lomax, B.H., Malhi, Y., Moore, S., Owusu-Afriyie, K. and Gosling, W.D., 2018, The modern pollen–vegetation relationships of a tropical forest–savannah mosaic landscape, Ghana, West Africa. *Palynology* **42**, pp. 324–338, 10.1080/01916122.2017.1356392.

Julier, A.C.M., Manzano, S., Razanatsoa, E., Razafimanantsoa, A.H.I., Githumbi, E., Hawthorne, D., Oden, G., Schüler, L., Tossou, M., Bunting, M.J., this volume, Modern pollen studies from tropical Africa and their use in palaeoecology. *Palaeoecology of Africa* **35**, 10.1201/9781003162766-21.

Keane, R.E., Hessburg, P.F., Landres, P.B. and Swanson, F.J., 2009, The use of historical range and variability (HRV) in landscape management. *Forest Ecology and Management* **258**, pp. 1025–1037, 10.1016/j.foreco.2009.05.035.

Laris, P. 2011, Humanizing savanna biogeography: linking human practices with ecological patterns in a frequently burned savanna of southern Mali. *Annals of the Association of American Geographers* **101**, pp. 1067–1088, 10.1080/00045608.2011.560063.

Lewis, S.L., Wheeler, C.E., Mitchard, E.T. and Koch, A. 2019, Regenerate natural forests to store carbon. *Nature* **568**, pp. 25–28, 10.1038/d41586-019-01026-8.

Leys, B.A., Commerford, J.L. and McLauchlan, K.K., 2017, Reconstructing grassland fire history using sedimentary charcoal: Considering count, size and shape. *PLoS ONE* **12**, e0176445, 10.1371/journal.pone.0176445.

Lézine, A.-M., Hély, C., Grenier, C., Braconnot, P. and Krinner, G., 2011, Sahara and Sahel vulnerability to climate changes, lessons from Holocene hydrological data. *Quaternary Science Reviews* **30**, pp. 3001–3012, 10.1016/j.quascirev.2011.07.006.

Lindsey, P., Allan, J., Brehony, P., Dickman, A., Robson, A., Begg, C., Bhammar, H., Blanken, L., Breuer, T. and Fitzgerald, K., 2020, Conserving Africa's wildlife and wildlands through the COVID-19 crisis and beyond. *Nature Ecology & Evolution*, pp. 1–11.

Lüning, S., Gałka, M., Danladi, I.B., Adagunodo, T.A. and Vahrenholt, F., 2018, Hydroclimate in Africa during the medieval climate anomaly. *Palaeogeography, Palaeoclimatology, Palaeoecology* **495**, pp. 309–322, 10.1016/j.palaeo.2018.01.025.

Lüning, S., Gałka, M. and Vahrenholt, F., 2017, Warming and cooling: the medieval climate anomaly in Africa and Arabia. *Paleoceanography* **32**, pp. 1219–1235, 10.1002/2017PA003237.

Marchant, R. and Lane, P., 2014, Past perspectives for the future: foundations for sustainable development in East Africa. *Journal of Archaeological Science* **51**, pp. 12–21, 10.1016/j.jas.2013.07.005.

Marchant, R., Richer, S., Boles, O., Capitani, C., Courtney-Mustaphi, C.J., Lane, P., Prendergast, M.E., Stump, D., De Cort, G. and Kaplan, J.O., 2018, Drivers and trajectories of land cover change in East Africa: Human and environmental interactions from 6000 years ago to present. *Earth-Science Reviews* **178**, pp. 322–378, 10.1016/j.earscirev.2017.12.010.

Marlon, J.R., Kelly, R., Daniau, A.-L., Vannière, B., Power, M.J., Bartlein, P., Higuera, P., Blarquez, O., Brewer, S. and Brücher, T., 2016, Reconstructions of biomass burning from sediment charcoal records to improve data-model comparisons. *Biogeosciences* (BG) **13**, pp. 3225–3244, 10.5194/bg-13-3225-2016.

Midgley, G.F. and Bond, W.J., 2015, Future of African terrestrial biodiversity and ecosystems under anthropogenic climate change. *Nature Climate Change* **5**, pp. 823–829, 10.1038/nclimate2753.

Moura, L.C., Scariot, A.O., Schmidt, I.B., Beatty, R. and Russell-Smith, J., 2019, The legacy of colonial fire management policies on traditional livelihoods and ecological sustainability in savannas: impacts, consequences, new directions. *Journal of Environmental Management* **232**, pp. 600–606, 10.1016/j.jenvman.2018.11.057.

Mustaphi, C.J.C., Capitani, C., Boles, O., Kariuki, R., Newman, R., Munishi, L., Marchant, R. and Lane, P., 2019, Integrating evidence of land use and land cover change for land management policy formulation along the Kenya-Tanzania borderlands. *Anthropocene* **28**, pp. 100228, 10.1016/j.ancene.2019.100228.

Nogué, S., de Nascimento, L., Froyd, C.A., Wilmshurst, J.M., de Boer, E.J., Coffey, E.E., Whittaker, R.J., Fernández-Palacios, J.M. and Willis, K.J., 2017, Island biodiversity conservation needs palaeoecology. *Nature Ecology & Evolution* **1**, pp. 0181, 10.1038/s41559-017-0181.

O'Connor, T.G., Puttick, J.R. and Hoffman, M.T., 2014, Bush encroachment in southern Africa: changes and causes. *African Journal of Range & Forage Science* **31**, pp. 67–88, 10.2989/10220119.2014.939996.

Parr, C. and Brockett, B.H., 1999, Patch-mosaic burning: a new paradigm for savanna fire management in protected areas. *Koedoe* **42**, pp. 117–130, 10.4102/koedoe.v42i2.237.

Parr, C.L., Lehmann, C.E., Bond, W.J., Hoffmann, W.A. and Andersen, A.N., 2014, Tropical grassy biomes: misunderstood, neglected, and under threat. *Trends in Ecology & Evolution* **29**, pp. 205–213, 10.1016/j.tree.2014.02.004.

Pausata, F.S., Gaetani, M., Messori, G., Berg, A., de Souza, D.M., Sage, R.F. and deMenocal, P.B., 2020, The Greening of the Sahara: Past Changes and Future Implications. *One Earth* **2**, pp. 235–250, 10.1016/j.oneear.2020.03.002.

Perry, G.L., Wainwright, J., Etherington, T.R. and Wilmshurst, J.M., 2016, Experimental simulation: using generative modeling and palaeoecological data to understand human-environment interactions. *Frontiers in Ecology and Evolution* **4**, pp. 109, 10.3389/fevo.2016.00109.

Pickett, S.T.A., Ostfeld, R.S., Shachak, M. and Likens, G.E., editors. 1997, The Ecological Basis of Conservation; Heterogeneity, Ecosystems, and Biodiversity. Chapman and Hall, New York.

Pinto, I., Jack, C. and Hewitson, B., 2018, Process-based model evaluation and projections over southern Africa from coordinated regional climate downscaling experiment and coupled model intercomparison project phase 5 models. *International Journal of Climatology* **38**, pp. 4251–4261, 10.1002/joc.5666.

Pohl, B., Macron, C. and Monerie, P.-A., 2017, Fewer rainy days and more extreme rainfall by the end of the century in Southern Africa. *Scientific Reports* **7**, pp. 1–7, 10.1038/srep46466.

Rachmayani, R., Prange, M. and Schulz, M., 2015, North African vegetation–precipitation feedback in early and mid-Holocene climate simulations with CCSM3-DGVM. Climate of the Past **11**, 10.5194/cp-11-175-2015.

Richer, S. and Gearey, B., 2018, From Rackham to REVEALS: reflections on palaeoecological approaches to woodland and trees. *Environmental Archaeology* **23**, pp. 286–297, 10.1080/14614103.2017.1283765.

Rogers, K.H., Luton, R., Biggs, H., Biggs, R.O., Blignaut, S., Choles, A.G., Palmer, C.G. and Tangwe, P., 2013, Fostering Complexity Thinking in Action Research for Change in Social–Ecological Systems. *Ecology & Society* **18**, p. 31 http://dx.doi.org/10.5751/ES-05330-180231.

Seddon, A., Froyd, C., Leng, M., Milne, G. and Willis, K., 2011, Ecosystem Resilience and Threshold Response in the Galápagos Coastal Zone. *PLoS ONE* **6**, e22376, 10.1371/journal.pone.0022376.

Skead, C.J., Boshoff, A., Kerley, G.I.H. and Lloyd, P., 2007, Historical incidence of the larger land mammals in the broader Eastern Cape. Centre for African Conservation Ecology, Nelson Mandela Metropolitan.

Tierney, J.E., Pausata, F.S. and deMenocal, P.B., 2017, Rainfall regimes of the Green Sahara. *Science advances* **3**, e1601503, 10.1126/sciadv.1601503.

Vallè, F., Brunetti, M., Pini, R., Maggi, V. and Ravazzi, C., 2018, Testing pollen-climate models over the last 200 years in N-Italy using instrumental data. *Alpine and Mediterranean Quaternary* **31**, pp. 165–168.

van Asperen, E.N., Kirby, J.R. and Shaw, H.E., 2020, Relating dung fungal spore influx rates to animal density in a temperate environment: Implications for palaeoecological studies. *The Holocene* **30**, pp. 218–232, 10.1177/0959683619875804.

van Wilgen, B.W., Govender, N., Smit, I.P. and MacFadyen, S., 2014, The ongoing development of a pragmatic and adaptive fire management policy in a large African savanna protected area. *Journal of Environmental Management* **132**, pp. 358–368, 10.1016/j.jenvman.2013.11.003.

Veldman, J.W. 2016, Clarifying the confusion: old-growth savannahs and tropical ecosystem degradation. *Philosophical Transactions of the Royal Society B: Biological Sciences* **371**, pp. 20150306, 10.1098/rstb.2015.0306.

Veldman, J.W., Aleman, J.C., Alvarado, S.T., Anderson, T.M., Archibald, S., Bond, W.J., Boutton, T.W., Buchmann, N., Buisson, E. and Canadell, J.G., 2019, Comment on "The global tree restoration potential". *Science* **366**, 10.1126/science.aay7976, eaay7976.

Venter, Z.S., Hawkins, H.J. and Cramer, M.D., 2017, Implications of historical interactions between herbivory and fire for rangeland management in African savannas. *Ecosphere* **8**, pp. e01946, 10.1002/ecs2.1946.

Wigley, B.J., Bond, W.J. and Hoffman, M.T., 2009, Bush encroachment under three contrasting land-use practices in a mesic South African savanna. *African Journal of Ecology* **47**, pp. 62–70, 10.1111/j.1365-2028.2008.01051.x.

Williams, J.W. and Jackson, S.T., 2007, Novel climates, no-analog communities, and ecological surprises. *Frontiers in Ecology and the Environment* **5**:475–482, 10.1890/070037.

Willis, K.J., Gillson, L. and Brncic, T., 2004, How "Virgin" is Virgin Rainforest. *Science* **304**, pp.402–403, 10.1126/science.1093991.

Wolfe, B.B., Armitage, D., Wesche, S., Brock, B.E., Sokal, M.A., Clogg-Wright, K.P., Mongeon, C.L., Adam, M.E., Hall, R.I. and Edwards, T.W., 2007, From isotopes to TK interviews: towards interdisciplinary research in Fort Resolution and the Slave River Delta, Northwest Territories. Arctic, pp.75–87.

Regional Index

Subject Index

Palaeoecology of Africa

International Yearbook of Landscape Evolution and Palaeoenvironments

ISSN: 2372–5907

Volume 1-12 *Out of Print*

13. Palaeoecology of Africa and the Surrounding Islands
    Editors: J.A. Coetzee & E.M. van Zinderen Bakker
    1981, ISBN: 978-90-6191-203-3

14. Palaeoecology of Africa
    Editors: J.A. Coetzee & E.M. van Zinderen Bakker
    1982, ISBN: 978-90-6191-204-0

15. Palaeoecology of Africa
    Editors: J.A. Coetzee, E.M. van Zinderen Bakker, J.C. Vogel,
    E.A. Voigt & T.C. Partridge
    1982, ISBN: 978-90-6191-257-6

16. Palaeoecology of Africa
    Editors: J.A. Coetzee & E.M. van Zinderen Bakker
    1984, ISBN: 978-90-6191-510-2

17. Palaeoecology of Africa
    Editors: J.A. Coetzee & E.M. van Zinderen Bakker
    1986, ISBN: 978-90-6191-625-3

18. Palaeoecology of Africa
    Editor: K. Heine
    1987, ISBN: 978-90-6191-689-5

19. Palaeoecology of Africa – Out of Print
    Editors: K. Heine & J.A. Coetzee
    1988, ISBN: 978-90-6191-834-9

20. Palaeoecology of Africa
    Editor: K. Heine
    1989, ISBN: 978-90-6191-880-6

21. Palaeoecology of Africa – Out of Print
    Editors: K. Heine & R.R. Maud
    1990, ISBN: 978-90-6191-997-1

22. Palaeoecology of Africa
    Editors: K. Heine, A. Ballouche & J. Maley
    1991, ISBN: 978-90-5410-110-9

23. Palaeoecology of Africa and the Surrounding Islands
    Editor: K. Heine
    1993, ISBN: 978-90-5410-154-3

24. Palaeoecology of Africa
    Editor: K. Heine
    1996, ISBN: 978-90-5410-662-3

25. Palaeoecology of Africa – Out of Print
    Editors: K. Heine, H. Faure & A. Singhvi
    1999, ISBN: 978-90-5410-451-3

26. Palaeoecology of Africa and the Surrounding Islands
    Editors: K. Heine, L. Scott, A. Cadman & R. Verhoeven
    1999, ISBN: 978-90-5410-476-6

27. Palaeoecology of Africa and the Surrounding Islands: Proceedings
    of the 25th Inqua Conference, Durban, South Africa, 3-11 August 1999
    Editors: K. Heine & J. Runge
    2001, ISBN: 978-90-5809-350-9

28. Dynamics of Forest Ecosystems in Central Africa During the Holocene:
    Past – Present – Future
    Editor: J. Runge
    2007, ISBN: 978-0-415-42617-6

29. Holocene Palaeoenvironmental History of the Central Sahara
    Editors: R. Baumhauer & J. Runge
    2009, ISBN: 978-0-415-48256-1

30. African Palaeoenvironments and Geomorphic Landscape Evolution
    Editor: J. Runge
    2010, ISBN: 978-0-415-58789-1

31. Landscape Evolution, Neotectonics and Quaternary Environmental Change in
    Southern Cameroon
    Editor: J. Runge
    2012, ISBN: 978-0-415-67735-6

32. New Studies on Former and Recent Landscape Changes in Africa
    Editor: Jürgen Runge
    2014, ISBN: 978-1-138-00116-9

33. Changing Climates, Ecosystems and Environments within
    Arid Southern Africa and Adjoining Regions
    Editor: Jürgen Runge
    2016, ISBN: 978-1-138-02704-6

34. The African Neogene – Climate, Environments and People
    Editor: Jürgen Runge
    2018, ISBN: 978-1-138-06212-2

T - #0073 - 111024 - C438 - 246/174/20 [22] - CB - 9780367755089 - Gloss Lamination